THE LIBRARY
ST. MARY'S COLLEGE OF MARYLAND
ST. MARY'S CITY, MARYLAND 20686

T5-BYU-760

MARINE ORGANISMS
Genetics, Ecology, and Evolution

NATO CONFERENCE SERIES

I	Ecology
II	Systems Science
III	Human Factors
IV	Marine Sciences
V	Air–Sea Interactions
VI	Materials Science

IV MARINE SCIENCES

Volume 1 Marine Natural Products Chemistry
edited by D. J. Faulkner and W. H. Fenical

Volume 2 Marine Organisms: Genetics, Ecology, and Evolution
edited by Bruno Battaglia and John A. Beardmore

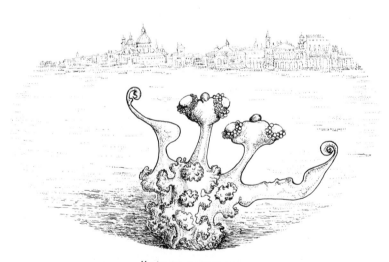

Neptunus serenissimus

MARINE ORGANISMS
Genetics, Ecology, and Evolution

Edited by
Bruno Battaglia
University of Padua
Padua, Italy

and
John A. Beardmore
University College of Swansea
Swansea, United Kingdom

Published in coordination with NATO Scientific Affairs Division

PLENUM PRESS · NEW YORK AND LONDON

Library of Congress Cataloging in Publication Data

Nato Advanced Study Research Institute on the Genetics, Evolution, and Ecology of Marine Organisms, Fondazione Giorgio Cini, 1977.
Marine organisms.

(NATO conference series: IV, Marine sciences; v. 2)
Includes index.
1. Marine fauna. I. Battaglia, Bruno, 1923-   II. Beardmore, John Alec. III. Title.
QL121.N28 1977   591.9'2   78-9715
ISBN 0-306-40020-0

Proceedings of a NATO Advanced Study Research Institute on the Genetics, Evolution, and Ecology of Marine Organisms held in the Fondazione Giorgio Cini, Isola San Giorgio Maggiore, Venice, Italy, March 24—April 4, 1977, sponsored by the NATO Special Program Panel on Marine Sciences

© 1978 Plenum Press, New York
A Division of Plenum Publishing Corporation
227 West 17th Street, New York, N.Y. 10011

All rights reserved

No part of this book may be reproduced, stored in a retrieval system, or transmitted, in any form or by any means, electronic, mechanical, photocopying, microfilming, recording, or otherwise, without written permission from the Publisher

Printed in the United States of America

PREFACE

The conference on which this book is based was conceived during the Fifth European Marine Biology Symposium held in Venice in 1970. Several free-ranging discussions on genetical and ecological aspects of marine biology took place at that time and from these emerged the idea that a meeting devoted solely to these topics would be a welcome and exciting newcomer on the scientific scene.

The study of the evolutionary and ecological genetics of marine organisms was until recently a largely unexplored field although Schmidt, over sixty years ago, was an early investigator of quantitative variation in Gadus and Zoarces while in the mid-twenties Ford and Huxley studied aspects of developmental genetics in the shrimp Gammarus chevreuxi. One of the main reasons for the lack of progress in this area undoubtedly lay in the difficulties associated with the laboratory culture and breeding of marine creatures and other difficulties were posed by lack of ecological knowledge. However, gradually, interest in the genetics of marine creatures awakened, stimulated in part no doubt by increasing recognition of the importance of marine ecosystems.

The advent of electrophoretic and allied techniques, so informative in providing descriptions of some properties of the gene-pools of man, mouse and fruit-flies provided a visible boost for work on the evolutionary genetics of marine species. Selander has claimed, in our view rightly, that this new technology has, during the last decade, revitalised evolutionary biology. It seems now likely that marine organisms will also contribute significantly to the 'Age of Expansion' envisaged by Ayala in which there will be not merely growth in technical and methodological knowledge but important conceptual advances as well.

In drawing up a proposal for a N.A.T.O. Advanced Research Institute in 1977, we felt that the time was ripe for an international meeting which would bring together a sample of those working on the genetics, ecology and evolution of marine organisms. The main thrust, we felt, should be genetical mainly because we believe that marine biology will benefit considerably from increased recognition of the contributions genetical knowledge has to make to that discipline. Additionally, we hoped for the more conventional goals of cross-fertilisation of ideas, information exchange and identification of areas in which international collaborative research might profitably be carried out.

The papers in this book range over a wide variety of topics and the individual approaches vary equally widely in the degree to which they are generalist or specialist contributions. We think that the headings given to the five sections of the book are self explanatory. However, it must be conceded that the choice of the appropriate section for some contributions is to some extent arbitrary. Nevertheless we hope that the reader will find that our taxonomy has some coherence and logic to commend it.

We received invaluable assistance in organising the Institute from Dr. G. A. Danieli. We would like also to acknowledge the support of the National Research Council of Italy (C.N.R.) in making available the O. V. 'Laguna' and the help received from the captain and crew of that vessel.

The generous financial support we received from the North Atlantic Treaty Organisation through the Marine Sciences programme of that body is gratefully acknowledged.

We are grateful to Mrs. Margaret Morris and Dr. J. Thorpe without whose patient help this volume could not have been realised and we are particularly indebted to Dr. R. D. Ward for help, given unstintingly and in manifold ways over many long hours, which has eased considerably the burdens of editorship.

                                              Bruno Battaglia

                                              John Beardmore

CONTENTS

SECTION 1: MEASUREMENT AND MAINTENANCE OF GENETIC VARIATION IN POPULATIONS OF MARINE ORGANISMS

The Maintenance of Polymorphisms by
       Natural Selection 3
  R. Milkman

Genetic Variation and Resource Stability
       in Marine Invertebrates 23
  F. J. Ayala and J. W. Valentine

Genetic Variability in Relation to the
       Environment in Some Marine Invertebrates . . 53
  B. Battaglia, P. M. Bisol and G. Fava

Protein Variation in the Plaice, Dab and
       Flounder, and Their Genetic Relationships. . 71
  R. D. Ward and R. A. Galleguillos

Genetic Variability in Deep-Sea Fishes
       of the Genus _Sebastolobus_ (Scorpaenidae) . . 95
  J. F. Siebenaller

Genetic Variation and Species Coexistence
       in _Littorina_ 123
  J. A. Beardmore and S. R. Morris

Electrophoretic Variability and Temperature
       Sensitivity of Phosphoglucose Isomerase
       and Phosphoglucomutase in Littorinids and
       Other Marine Molluscs. 141
  N. P. Wilkins, D. O'Regan and E. Moynihan

Genetic Divergence Between Populations of
       Thais lamellosa (Gmelin) 157
  C. A. Campbell

Geographic Variation in Protein Polymorphisms
    in the Eelpout, Zoarces viviparus (L) . . .   171
  F. B. Christiansen and V. Simonsen

Genetics of the Colonial Ascidian, Botryllus
    schlosseri   195
  A. Sabbadin

Biochemical Aspects of Genetic Variation at
    the Lap Locus in Mytilus edulis   211
  R. K. Koehn

Experimental Mortality Studies and Adaptation
    at the Lap Locus in Mytilus edulis   229
  J. Levinton and H. H. Lassen

SECTION 2: ECOLOGY, LIFE HISTORIES AND ADAPTIVE
            STRATEGIES

Genetic Consequences of Different Reproductive
    Strategies in Marine Invertebrates   257
  D. J. Crisp

Ecological, Physiological and Genetic
    Analysis of Acute Osmotic Stress   275
  R. W. Doyle

The Evolution of Competing Species   289
  T. Fenchel, J.-O. Frier and S. Kolding

On the Relationship Between Dispersal of
    Pelagic Veliger Larvae and the Evolution
    of Marine Prosobranch Gastropods   303
  R. S. Scheltema

Adaptive Strategies in the Sea   323
  J. W. Valentine and F. J. Ayala

Life Histories and Genetic Variation in
    Marine Invertebrates   347
  J. F. Grassle and J. P. Grassle

Is the Marine Latitudinal Diversity Gradient
    Merely Another Example of the Species
    Area Curve?   365
  T. J. M. Schopf, J. B. Fisher and C. A. F. Smith III

CONTENTS

SECTION 3: GENETIC VARIATION AND TAXONOMY

Genetic Variability in the European Eel
   Anguilla anguilla L 389
  E. Rodinó and A. Comparini

Taxonomy, Interspecific Variation and Genetic
   Distance in the Phylum Bryozoa 425
  J. P. Thorpe, J. A. Beardmore and J. S. Ryland

The Systematics and Evolution of Mytilus
   galloprovincialis Lmk. 447
  R. Seed

Genetic Aids to the Study of Closely Related
   taxa of the Genus Mytilus 469
  D. O. F. Skibinski, J. A. Beardmore and M. Ahmad

Chromosome Evolution in Some Marine
   Invertebrates 487
  D. Colombera and I. Lazzaretto-Colombera

SECTION 4: SEX DETERMINATION, BREEDING SYSTEMS AND ISOLATING MECHANISMS

Variability of the Modes of Sex Determination
   in Littoral Amphipods. 529
  H.-P. Bulnheim

Genetics of Sex Determination in Ophryotrocha
   (Annelida Polychaeta). 549
  G. Bacci

A New Ophryotrocha Species of the Labronica
   Group (Polychaeta, Dorvilleidae)
   Revealed in Crossbreeding Experiments . . 573
  B. Åkesson

Breeding Systems, Species Relationships and
   Evolutionary Trends in Some Marine
   Species of Euplotidae (Hypotrichida
   Cilitata) 591
  R. Nobili, P. Luporini and F. Dini

A Study of Reproductive Isolation within the
   Super-Species Tisbe clodiensis
   (Copepoda, Harpacticoida) 617
  B. M. Volkmann, B. Battaglia and V. Varotto

Genetics of Ethological Isolating Mechanisms
in the Species Complex _Jaera_ _albifrons_
(Crustacea, Isopoda) 637
  M. Solignac

SECTION 5: SOME APPLIED ASPECTS

Effects of Organophosphate Pesticides on
Fish Esterases: A Case for an
Isozyme Approach 667
  M. Krajnović-Ozretić and W. de Ligny

Quantitative Genetic Variation in Fish-
Its Significance for Salmonid Culture . 679
  M. Holm and G. Naevdal

The Extended Series of _Tf_ Alleles in
Atlantic Cod _Gadus_ _morhua_ L 699
  A. Jamieson and R. J. Turner

Contributors 731

AUTHOR INDEX 735

ORGANISM INDEX 745

SUBJECT INDEX 753

SECTION 1

MEASUREMENT AND MAINTENANCE OF GENETIC VARIATION IN POPULATIONS OF MARINE ORGANISMS

THE MAINTENANCE OF POLYMORPHISMS BY NATURAL SELECTION

R. Milkman

Department of Zoology, University of Iowa

Iowa City, Iowa 52242, U.S.A.

Darwinian evolution has two major aspects: adaptive change and ancestral relationship. The next decade may well be the most important for the study of evolution since Darwin's time, because we now have methods for exploring each of these aspects at a fundamental level. Marine organisms are already playing a central role in this work.

I should like briefly to support these assertions and then in more detail to describe some approaches to the nature and significance of genic polymorphisms and their relation to evolutionary change.

In a genetically isolated species there is only one source of genetic variation, namely, mutation; and there are only two processes, acting separately or in concert, that can account for genic polymorphism and for genetic change. The first process is random genetic drift: in the absence of selective differences among genotypes, the sampling process forces a departure from the status quo. The dynamics of this process have been described especially by Wright (1969) and by Kimura (1968; Crow and Kimura, 1970). The second process is of course selection.

The suggestion that in general two or more alleles coexisting in natural populations are adaptively indistinguishable (i.e. neutral with respect to one another) has been called the neutral hypothesis. This important hypothesis has been tested in two important macromolecular theatres: amino acid sequence in proteins and, more recently, nucleotide sequence in nucleic acids. The predictions of the neutral hypothesis are formulated in the terms of random genetic drift.

Decisive evidence (Salser and Isaacson, 1976; Grunstein et al., 1976) has been obtained that certain alternative bases in the third position of codons are indeed adaptively equivalent. Thus the neutral hypothesis is correct in at least one vastly important molecular domain. As such, it is best joined to the appropriate elements of drift theory and called the neutrality principle.

Neutrality is required for the establishment of common ancestry: for example, the worm shape is taxonomically useless because it is so highly adaptive. Convergence is seen in proteins as well; more important, it is an ever-present possibility as an explanation of similarities. For this reason, the recent nucleotide sequencing of Salser and Isaacson, and of Grunstein et al., has provided another fundamental theatre, one in which evolution's other major feature - ancestral relationship - can be studied. Salser and Isaacson (1976) showed that while in a sizeable segment rabbit and human β-hemoglobin messengers differ by one non-third base (corresponding to the one amino acid difference between the polypeptides), they differ by 6 third-bases. Even more striking is the work of Grunstein and Schedl (1976) and Grunstein et al. (1976) on histone IV messenger in two sea urchin genera, Lytechinus and Strongylocentrotus. While these messengers dictate identical proteins (and do not differ in bases at the first or second positions), they differ in some 30% of the third bases studied, 16 in all. This finding confirms that, as Crick's (1966) Wobble Hypothesis suggests, certain third bases are really equivalent. Until these observations were made, the possibility existed that differences in tRNA-binding efficiency between two acceptable third bases might lead to adaptively significant differences in protein synthesis rates, or in error-frequency. Moreover, if we accept the estimate that the sea urchins studied had a common ancestor 60 million years ago, then the results indicate roughly 3×10^{-9} nucleotide substitutions per third position per generation (=year). With the current relative ease of nucleotide sequencing (Sanger et al., 1977; Maxam and Gilbert, 1977), this work is likely to be the forerunner of a vast number of detailed studies on the concordance of similarities in adaptively neutral traits, the hallmark of common ancestry.

On the other hand, the evidence now appears decisive (Ohta, 1974; Kimura, 1976; Milkman, 1976b; Selander, 1976) that the neutral hypothesis in pure form (the "infinite allele" model) does not hold generally for protein variation. That is, we can now say that most allozymic differences are <u>potentially</u> adaptively significant. Ohta (1974) points out, however, that small selection coefficients will be ineffectual in small populations, rendering the corresponding genotypes effectively neutral. In order to silence random genetic drift, the required value of $\underline{s}$ is on the order of $\frac{1}{N_e}$, where $\underline{N_e}$ is effective population size. (As generally stated, the requirement is that $\underline{s} \gg \frac{1}{4 N_e}$.) In the present context population size fluctuations are relatively unimportant.

Accordingly, the following variables must be estimated with considerable assurance: effective size of the population undergoing selection; possible magnitude of selection coefficients; and the effective number of alleles at each locus.

Protein differences, which electrophoretic analysis has demonstrated abundantly in marine and other organisms, are as yet rarely understood in adaptive terms. Indeed, anecdotal examples could not constitute conclusive evidence for the adaptive significance of protein polymorphisms in general. Instead, the predictions of the neutrality principle have been refuted in electrophoretic studies on <u>E. coli</u> (Milkman, 1973, 1975, 1976b) and in the sequencing of cytochrome c from <u>Pseudomonas aeruginosa</u> (Ambler, 1974). In brief, far less variation has been found than the neutrality principle requires in these cases; these organisms are among the very few that cannot be disqualified on various grounds, notably that of having gone through a recent bottleneck; and it seems clear that proteins in a highly differentiated plant or animal would face adaptive requirements at least as stringent and specific as those of bacterial enzymes. Thus it appears that throughout the living world protein polymorphisms reveal unit adaptive differences or (where they differ by more than one amino acid) near-unit adaptive differences. It is now of great interest to determine whether these adaptive differences are in general great enough to be of selective importance, in essentially all common population sizes. There is good evidence for the adaptive significance of differences in nontranslated DNA as well (Wilson, 1976); in any event, protein variation provides one fundamental level at which to explore adaptation.

Marine organisms have an array of general and special properties that account for their recent and prospective importance for the analysis of the evolutionary process. In general, they differ from most temperate organisms in being available for study in a given place all year round. Their accessibility is relatively easy, and their numbers are often vast. It is probably also easier to describe their habitats and niches than those of most other forms. Finally, the extreme properties of certain marine environments, such as the physical constancy in ocean trenches, are of great use in testing specific hypotheses having to do with the nature of genetic variation and its maintenance.

The amount that can be learned about the genetic structure of marine species and the far greater amount waiting to be learned is evidenced by the numerous papers on <u>Mytilus</u> in this volume, as well as many others, including some recent work of R. K. Koehn and myself (Koehn <u>et al</u>., 1976; Milkman and Koehn, 1977). Genetic variation, temporal, spatial, and age-related, is seen in rich detail.

While all organisms are obviously at the mercy of the environment, environmental effects on marine organisms seem particularly direct. Consider for example the heat capacity of water versus that of air. Further, the vast array of sessile animals (including intertidal forms) provides opportunities to study adaptations that preserve complex, dynamic organisms, many of which can neither encyst nor run away. The following finding is therefore perhaps of special interest to the study of marine organisms.

I have recently been able to show (Milkman, 1978) the generality of a little-used relationship between the population geneticists' s (selection coefficient on a genotype) and the quantitative geneticists' i (standard selection differential). The two are related as follows: $s \simeq i\,g$, where g is the standard differential contribution of one genotype to the phenotype. Falconer (1960) discusses a quite similar relationship briefly, in a limited context.

A brief description of these variables may be useful. The selection coefficient on a genotype is the proportionate difference between its reproductive rate and that of a reference genotype.

The selection differential is the difference between the mean phenotypic value of a selected sample and that of the entire population. <u>Standard</u> selection differential, i, is expressed in units of the standard deviation (square root of the variance of the phenotypic value in question); since the population mean is then 0, the standard selection differential is equal to the standard sample mean.

Finally, g is the mean difference in phenotypic value, again in units of σ, between the individuals having the given genotype and those with the reference genotype. For example, if from a population of 1-year-old mussels we select a group with banded shells and they average 3 mm larger than the population mean, and if the standard deviation of size is 5 mm, then $i = 3/5 = 0.6$. Now if the AA individuals are largest on the average, and if aa individuals average 0.10 mm smaller ($g_{aa} = -0.10/5 = -0.02$), then $s_{aa} = 0.6 \times (-0.02)$. A selec- coefficient this large would of course silence random genetic drift in all but rather small populations, since the required value of s is on the order of $\frac{1}{N_e}$.

As I have described in detail elsewhere (Milkman, 1978), $s \simeq ig$ wherever fitness is a non-decreasing function of a ranked normally-distributed phenotypic value. This condition applies at the very least to a major portion of selection. In natural selection we can define a phenotype fitness potential as the weighted sum of values associated with all the properties of an organism relevant to its fitness. Unlike a breeder, who establishes his own weighting system, we cannot detail the components of fitness potential in advance, though we can determine from observations the contribution, g, of a genotype.

The distinction between fitness potential and fitness is important to the understanding of selection. Fitness potential comes before selection, and fitness comes after selection. Selection is a competitive process, and the properties important in the competition can be weighted and combined into one value: fitness potential. A breeder has to assign values to various properties in order to develop a quality index, which is an overall value on which he bases artificial selection, choosing individuals with the best scores. As a result of this selection, fitness values emerge for individuals and for genotypic classes; these are their relative reproductive rates, as permitted by the breeder. Animals with a certain index value might be selected on one farm but culled on another, either because the latter farm had more animals with higher values or because a smaller proportion of its individuals were selected to reproduce.

In nature the same process takes place. In general, no single property assures survival or elimination (disease resistance is an occasional exception). Those individuals prosper who have the best combinations of properties. Differential prosperity is an aspect of selection. Moreover, in competition for resources, some surviving individuals are left with more "scope for growth" (Bayne et al., 1976; Warren and Davis, 1967) than others. That is, there is more substance and energy remaining above and beyond that required for maintenance. Greater growth eventually means greater reproduction, and this amounts to greater fitness.

Thus selection in nature is unlike that practiced by many breeders in that there is no truncation of the population into two groups, those who die before reproducing and those who live to reproduce at a rate common to all. There are a number of reasons for this, each sufficient in itself. First and foremost, there is variation in reproductive rate. Next, the proportion surviving from locality is likely to vary even when the adaptive ranking of individuals follows the same rules in all localities, since the severity of competition is likely to vary. And finally, phenotypic properties are not likely to be uniform in adaptive value, that is, in contribution to fitness potential, in all localities. Thus no sharp cut off is likely to be seen in nature. This is why it was important to generalize the proof that $\underline{s} \simeq \underline{ig}$.

Clearly the effect of fitness potential on fitness is realized in the context of the fitness potential of competitors. The term <u>adaptive value</u> once held a meaning like that of fitness potential, before <u>adaptive value</u> was synonymized with <u>selective value</u> and <u>fitness</u> over the objections of Fisher (1958) and others.

In the present context, fitness potential is of interest as the phenotype upon which selection is based. In natural selection, fitness potential is assumed to be normally distributed; indeed, it can be defined that way, and appropriate units will follow. Thus standard selection differential, $\underline{i}$, can be calculated from a standard normal distribution table. This is easy for truncation models, and it is fairly easy for all selection models in which fitness is a non-decreasing function of fitness potential (Milkman, 1978). The latter category includes all competitive selection models, and, as noted previously, that means almost all selection.

In artificial selection, fitness potential is simply the criterion phenotype, subject to the qualification that it can be modified by uncontrolled differences in the reproductive rate of the selected individuals. Thus, if survival of a short exposure to high temperature is the subject of an experiment in artificial selection, then **heat resistance** (or this specific form of heat **resistance**) becomes the major component of fitness potential, since no qualities other than

MAINTENANCE OF POLYMORPHISMS BY NATURAL SELECTION

reproductive differences have an opportunity to intervene.

It now becomes possible to begin to relate genotype to fitness potential in nature by the following sort of experiment. Suppose a certain allele seems on the basis of its distribution, and perhaps seasonal changes in frequency, to be associated with resistance to acute high temperature. Samples from a genetically heterogeneous population may now be exposed to high temperatures with varying degrees of survival. This may be done by marking individuals at the time of death until none remain alive, or by exposing replicate samples to treatments of different intensities. Changes in genotype frequency would be expected to increase with the severity of the treatment if indeed the genotype had anything to do with heat resistance. Consider, for example, a hypothetical case where allele $\underline{A}$ is dominant — that is, the $\underline{AA}$ and $\underline{Aa}$ genotypes behave identically. The case is illustrated in Table 1.

We assume that heat-resistance, $\underline{x}$, is normally distributed,

TABLE 1

Hypothetical heat-treatment experiment

Treatment	Proportion surviving	i	$\underline{s}_{aa}$	$\underline{s}/\underline{i} = \underline{g}$
A	0.70	0.497	0.10	0.20
B	0.50	0.798	0.15	0.19
C	0.30	1.159	0.25	0.22
D	0.20	1.400	0.30	0.21
E	0.10	1.754	0.35	0.20

and we suppose that heat-treatments of five different intensities are tried. For each intensity, we supply values for survivorship and for the selection coefficient, $\underline{s}_{aa}$, against aa. In an actual experiment, survivorship would be observed directly, and $\underline{s}_{aa}$ would be calculated from an independent observation: the difference in genotype frequencies between the original population and the survivors. We call the survival threshold value of heat resistance x_T; thus, the proportion surviving is $1 - F(x_T)$. The values for $\underline{i}$ are derived

from the proportion surviving, according to a well-known
formula for the mean phenotypic value of a selected sample:

$$i \equiv x_S = \frac{f(x_T)}{1 - F(x_T)}$$

The frequency-density values, $f(x_T)$, can be found easily in
a standard normal distribution table. In the first row,
then,

$$\underline{i} = \frac{0.3877}{0.7000} = 0.497$$

It appears that $g_{aa} \approx 0.20$ in this hypothetical case.
Now if the $\underline{A}$- and $\underline{aa}$ genotypes occur in equal frequencies,
$P = Q = 0.5$, the variance related to the $\underline{A}$ locus is $PQg^2 = 0.5 \times 0.5 \times 0.20^2 = 0.01$ of the entire phenotypic variance.
If half the phenotypic variance were genetic in origin, then
the $\underline{A}$ locus could be one of $0.5/0.01 = 50$ independent loci
contributing equally to heat resistance. Thus artificial
selection of this sort is potentially a very sensitive test
of the possible adaptive significance of particular genotypes.

Observations in nature introduce the possibility that
a genotype has more than one important adaptive consequence,
and that they conflict. Nevertheless, the opportunity exists
to determine the importance of heat resistance in natural
situations. Suppose that 5% of a population dies in a hot
two-week period. Here $\underline{i} = 0.109$. If death were due solely
to heat, as in the artifical selection experiment, then we
would expect to find $s_{aa} = 0.109 \times 0.20 \approx 0.02$. Selection
coefficients this small require large samples to verify.

We can proceed further, however. It is reasonable to
expect that some survivors are in better shape than others,
and that subsequent deaths might also reflect the after-
effects of differential contributions to heat resistance by
the genotypes at the $\underline{A}$ locus. Accordingly, after another
month, a total of, say 0.15 of the population might have died.
Then $i = 0.274$ and s_{aa} might be as high as $0.274 \times 0.20 = 0.055$. On the other hand, a lower observed value of $\underline{s}$ would
indicate that other properties not affected in the same way
by the $\underline{A}$ genotypes were involved in selection. Note that
the ability to observe restricted episodes of this kind make
it possible to sort out opposing forces that are separated
in time. In summary, one can use this simple relationship
to search for the adaptive significance of particular geno-

types. The sensitivity of the technique means that one is not looking for a needle in a haystack; one of many equal factors can be detected. By the same token, the failure to achieve fairly dramatic results in an artificial selection experiment is good evidence against the importance of the relationship being tested.

An example of considerable relevance is provided by an experiment by Johnson and Powell (1974) on alcohol dehydrogenase allozymes in D. melanogaster. After a heat shock severe enough to kill 90% of the flies (30 min at 41°), the frequency of the slow (Adh^S) allele rose from 0.66 to 0.71. The value of i corresponding to 90% mortality is 1.754, and if we assume that Adh^S is dominant with respect to heat resistance,

$$q = -0.05 = \frac{(0.66 \times 0.34)(0.34s)}{1 - 0.34^2 s}$$

and $s = 0.61$. Thus $g = 0.61/1.754 = 0.35$. It appears that the Adh locus is quite important here. But how many other loci could have a similar effect on heat resistance? The contribution of the Adh locus to the phenotypic variance (i.e., variance for heat resistance) is $(0.34^2)(1 - 0.34^2)g^2 = 0.0125$. If half the variance were due to genetic causes, then $0.5/0.0125 = 40$ loci could have comparable effects. If Adh^S were recessive, s_{F-} would be 0.28 and g_{F-} would be 0.16. The locus variance would be $(0.66^2)(1 - 0.66^2)g^2 = 0.0063$. Again if half the phenotypic variance were genetic, $0.5/0.0063 = 79$ loci could participate equally and independently. Thus the dramatic change in frequency observed for the Adh alleles does not indicate a unique or even predominant role in heat resistance. Once again, if allelic differences are really important in relation to a particular environmental factor, artificial selection will produce dramatic results.

In describing Johnson and Powell's results, I combined the data from several parallel experiments involving different stocks. Some of these were even more dramatic, and others produced no evidence of selection. Whether the disparities between experiments were due to cryptic (i.e. non-electrophoretic) differences among the Adh allozymes present, or to other factors, is uncertain at this point.

Clearly, this approach holds considerable promise for the study of the genetic basis of adaptation, and so it may be worthwhile to detail its application to the study of natural and artificial selection processes.

A major question today is the number of genic polymorphisms that can be maintained by natural selection. Their maintenance requires the nullification of random genetic drift; this in turn requires selection coefficients greater than $1/4 N_e$, where N_e is effective population size. Now while we have already concluded that most genic polymorphisms that affect protein structure are potentially adaptive (in other words, that allozyme differences are potentially adaptively significant), it remains of interest to determine how large the arena of selection must be. That is, are selection coefficients great enough to explain polymorphisms on the basis of selection in local populations, or must we fall back on the vast numbers that make up the entire species? The answer to this question bears on the ease with which we can expect to find and characterize the environmental factors that impose selection.

How many polymorphisms can be maintained by selection in an isolated locality with an effective population of 1000? Since $\frac{1}{4N_e} = 0.00025$, we can require $\underline{s}$ favouring a certain rare genotype to be 0.0010, which is considerably greater than $\frac{1}{4N_e}$. If selection eliminates 90% of the potential reproduction of the population, 10% is left (subject of course to further non-selective depredations). Assuming a normal distribution of fitness potential, $\underline{i} \equiv \bar{x}_s$ is known to be $\frac{f(x_T)}{1 - F(x_T)}$, where $1 - F(x_T) = 0.10$. Thus $\underline{i} = 1.754$. In turn, $\underline{g} = \frac{0.0010}{1.754} = 0.0006$. We haven't yet specified the number of phenotypically distinct genotypes at the locus under consideration, nor their frequencies. If there were 3 genotypes of frequencies of 0.09, 0.42, and 0.49, with 0.42 being the frequency of the heterozygote and with no dominance, then this locus would contribute 0.1050 g^2 or 0.000000038 to the population's variance in fitness potential, which is set to total 1. If 90% of this phenotypic variance were due to non-genetic causes (environment, chance), then the remaining 0.1 could be divided evenly among 0.1/0.000000038, or 2.6 million loci. If N_e were 100 and selection removed only 50% of the reproductive potential, $\underline{s}$ could be set at 0.01, $\underline{i}$ would be 0.798, $\underline{g}$ would be 0.0125, the locus contribution to variance would be 0.0000164, and over 6000 loci could still be kept polymorphic by selection. This makes it clear that local factors need not be excluded as possible causes of genic polymorphism, since selection even in small populations can be effective enough to maintain many polymorphisms. Obviously migration could easily add to the number of polymorphisms; the size of the source populations and the role of selection in their genetic composition would then come into consideration.

It is interesting to note that truncation selection is the most efficient form of selection possible. This conclusion is clear from the fact that one must transfer reproductive activity from a higher-ranked class to a lower-ranked class to modify truncation selection into anything else. It is also interesting that a linear relationship between fitness and rank of fitness potential over the entire range (this leads to a cost of 0.50) yields a selection intensity, i, of 0.565, while truncation of half the population (cost = 0.50, also) results in i = 0.798. It turns out that a great variety of forms of rank-order selection, as Wills (1978) has called it, are roughly comparable in their efficiency (within a factor of 2) to truncation selection.

There are, of course, models of lower efficiency: if fitness rises linearly from 0 to 1 as fitness potential (rather than ranked fitness potential) goes from -3σ to $+3\sigma$, i = 0.332 and i/cost is 0.332/0.500 = 0.664. And if fitness rises from 0 to 1 as standard fitness potential goes from $-\infty$ to $+\infty$, the efficiency is near 0. But when fitness is made to vary linearly with ranked fitness potential, a reasonable model, the range of cost efficiency is surprisingly narrow. Figure 1 illustrates one example, in which fitness is 0 when $F(x) < a$; it rises linearly to 1 between $F(x) = a$ and $F(x) = b$; and it remains 1 thereafter. By varying a and b, one can in this simple way approximate a great variety of more complex but non-decreasing fitness functions. Thus it is of interest to compute the cost-effectiveness of selection for a number of specific cases of this model. Cost-effectiveness will be expressed as the ratio of standard selection differential, i, to cost of selection. Table 2 lists some values. The difference between b and a is first designated (0, of course, represents truncation), and then the cost chosen determines where a and b are placed. For example, if $b - a$ = 0.2, and a cost of 0.8 is chosen, we calculate as follows: Between b and a half the reproductive potential is lost. (In a simple survival model one can say half the individuals with fitness potential between a and b are killed.) This means a cost of 0.2/2 = 0.1. The rest of the cost must be found to the left of a. So a = 0.7, and b = 0.7 + 0.2 = 0.9. Now the standard selection differential must be estimated. This can be done by dividing the distribution into 20 equal parts and computing the mean fitness potential, $\bar{x}$, for each class. The estimate is simple: it is 20 times the difference between the $F(x)$ values for the lower and upper limits of the class. These

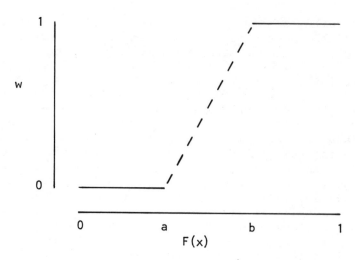

Figure 1. Fitness, w, in relation to ranked fitness potential, F(x). An example.

TABLE 2

Ratio of $\underline{i}$ to cost in various selection schemes in which $w = 0$ when $F(x) < a$, $w = \dfrac{F(x) - a}{b - a}$ when $a < F(x) < b$, and $w = 1$ when $F(x) > b$.

	b - a				
cost	0	0.2	0.5	0.8	1
0.1	1.95	1.81	--	--	--
0.2	1.75	1.71	--	--	--
0.25	1.69	1.66	1.50	--	--
0.3	1.66	1.63	1.50	--	--
0.4	1.61	1.59	1.49	1.29	--
0.5	1.60	1.58	1.49	1.31	1.12
0.8	1.75	1.71	--	--	--
0.9	1.95	1.81	--	--	--
0.95	2.17	--	--	--	--

MAINTENANCE OF POLYMORPHISMS BY NATURAL SELECTION 15

values can be found in a standard normal distribution table; the factor of 20 reflects the division of the difference by the class size, 0.05, to get the mean value. Now one simply averages the contributions from each class (weighted by the class fitness), and the result is $\underline{i}$. Finally, $\underline{i}$/cost is the efficiency given in Table 2. It is impressive that the highest value is less than twice the lowest.

The purpose of these calculations is to show that over a wide range of selection schemes the cost-efficiency doesn't vary a great deal, and so one need not shy away from problems where cost is involved. Furthermore, since $\underline{s} \propto \underline{ig}$, the $\underline{s}$/cost ratio can be found or estimated via $\underline{g}$, and this leads, with a little additional input to $\Delta\underline{q}$.

Heterogeneous environments are likely causes of polymorphisms, and Levene's (1953) model illustrates the results of selection in opposing directions within a panmictic population. The three-genotype situation described on p. 10 could be maintained in a two-phase environment in which the rare genotype (A_1A_1) is favoured in phase 1 and the other homozygote in phase 2. In this example, the homozygote favoured in a given phase is partially dominant - that is, the heterozygote's fitness is better than intermediate between the two homozygotes. If we state the relative fitnesses as follows:

	A_1A_1	A_1A_2	A_2A_2
phase 1	1.003	1.000	0.987
phase 2	0.983	1.000	1.007

then if the two phases hold equal numbers of individuals, $w_{A_1A_1} = 0.993$, $w_{A_1A_2} = 1.000$, and $w_{A_2A_2} = 0.997$. The equilibrium frequency, p, of A_1 will be $\dfrac{0.003}{0.007 + 0.003} = 0.30$, which is the value we chose to start with.

By considering a patchy environment, we have reduced the cost-efficiency of selection, but not drastically. The cost is 0.0071 in the two-phase environment, as opposed to 0.0021 in the corresponding case of simple heterosis.

A further implication of this relationship is the essential constancy of selection coefficients, even when genotype frequencies are changing considerably, provided that the phenotypic variance remains constant. And of course what we observe in species over periods of a few years is generally not change but lack of change. Furthermore, even if mean values change, selection coefficients will remain constant if the variance does not change.

To summarize the consequences of the relationship $s \propto ig$ and the implications of the values obtained, it should be possible to investigate the adaptive significance of particular genotypes in a fairly straightforward manner: by observing changes in natural populations and by performing artificial selection for properties suspected to have importance in nature.

Where migration can be excluded and mating is random, changes in genotype frequency are due almost solely to selection. In this context, mutation rates are now computed on a per site basis: 10^{-8} -10^{-9} for allozyme substitution. The mutation rate per locus, 10^{-5} -10^{-6}, clearly does not apply to genic polymorphisms except in the complicating factors of lethals and null alleles. With the increased resolution of allelic differences (Singh et al., 1976; Wright and MacIntyre, 1965; Milkman and Koehler, 1976; Coyne, 1976; Sampsell, 1977; Coyne and Felton, 1977), however, an additional agent of change comes into play, and this is intragenic recombination (Watt, 1972). Two common alleles may differ at two or more widely-spaced sites, and this could lead to the production of additional alleles. Even if these were somewhat inferior, their frequency could easily reach 1%. Using r for recombination in place of the usual u (mutation), if $\hat{q}$ were to be 0.01 and r 10^{-5}, s against recessive homozygotes would be 0.1, since $r = s\hat{q}^2$. The assumption that common alleles differ by only one amino acid is based on very little positive evidence, and it is contradicted by some observations (Petra et al., 1969). In any event, alleles at frequencies near 1% seem less propitious for investigation than do commoner alleles.

Returning to the analysis of selection in natural and in artificial populations, what can we do with the $s \propto ig$ model that we could not have done before? The model is helpful both in outlook and in method.

The following influences on our concept of selection can be noted:

1. The phenotype, and specifically the fitness-phenotype (fitness potential), emerges as a biologically realistic combination of factors. This can be distinguished from a series of games of genetic roulette in which the losers die and the winners emerge unscathed.

2. Effective selection is cheap enough to conduct simultaneously on many loci.

3. Selection coefficients can be realistically considered as constants in a broad range of circumstances.

4. We are reminded that reproduction simply cannot be ignored. As to methodology, the reminder about reproduction makes it highly desirable to choose organisms for study whose reproductive effort can be monitored easily and precisely. In addition, three specific lines of inquiry become obvious:

1. We can have confidence in artificial selection as a means of determining whether a genotype at a locus has a particular adaptive significance. If it has, the effect on genotype frequency will be substantial. If the effect on gene frequency is slight or dubious, we should look elsewhere.

2. We can determine the importance of a locus, in a given adaptive context, relative to all other factors and estimate the total number of loci involved to a significant extent.

3. We can determine the minimum size of a population in which selection can be effective in maintaining a given polymorphism.

What is contributed is support for these concepts and approaches, which are not novel in themselves.

Recently the improvement of electrophoretic techniques and the application of other criteria of allozyme variation (e.g. thermostability) have revealed new alleles at numerous loci. These findings have had an impact on several areas of investigation. First of all, the question has been raised as to whether we must rethink our current conclusions as to the mechanism of the maintenance of allelic variation. So far, however, the effective number of alleles (a good measure of allelic variation) has not increased to an extent that would be significant in this regard. The scores of rare hemoglobin variants known from clinical screening are not significant contributions to natural variation; similarly, the recently discovered allelic variants (though not quite so rare as the hemoglobin variants) are not frequent enough to alter greatly the level of heterozygosity in natural populations. Moreover it appears from studies at several loci (Coyne and Felton, 1977; Milkman, 1976a; Milkman and Koehler, 1976) that the most common allele remains the most common allele after additional variants have been found. Thus the suggestion of Ohta (1974), that genic polymorphism stems from the small magnitude of selection pressures against certain alleles, may apply to this distinct class of additional variants.

Nevertheless, the new distinctions do have promise in the study of species structure. For example, a heat-resistant Adh-fast allele, Adh^{Fr}, in Drosophila melanogaster is quite rare in most parts of the United States but is at a frequency of 7% in Ellsworth, Maine (Sampsell, 1977). Whereas this observation does not convey an immediate clue to its adaptive significance, it does provide new information on local differences. In this regard, it should be extremely interesting to apply the newly-increased resolving power available to the analysis of heterogeneous species in which time's arrow is obscured: species in which introgression and divergence are possible alternative explanations for striking local differences in correlated properties. The Mytilus edulis/Mytilus galloprovincialis question discussed elsewhere in this volume might well profit from the further resolution of allozyme differences. For example, several different nests of galloprovincialis among the edulis might differ from one another in previously unresolved allozymes.

A variety of means have recently been explored to resolve previously undetected allozymic differences. These include heat treatments (Bernstein et al., 1973; Milkman, 1976a; Sampsell, 1977; Singh et al., 1976; Thörig et al., 1975; Wright and MacIntyre, 1965), exposure to hydrogen-bond-breaking agents, immunological analysis, molecular hybridization studies (where multimeric enzymes are constituted from monomers, including a test monomer) (Cobbs, 1976), molecular sieving (Johnson, 1977), electrophoresis under new conditions (Coyne, 1976; Coyne and Felton, 1977), electrofocusing (Milkman and Koehler, 1976), and a variety of measurements of chemical properties (see Singh et al., 1976; because of the increasing interest in these approaches, current journals are likely to be the most efficient sources of information).
In all of these cases, it is necessary to establish that the basis of the difference lies in the structural gene of the protein in question, and genetic mapping is the best method available for this. Of course, electrophoresis and electrofocusing (a closely allied technique) require less additional study than the other techniques because of the physical separation of the allozymes.

Clearly, parallel but independent divergences from a common origin are much more likely to be distinguishable by these subtle means than are separate relics of a common introgressive ancestor.

In summary, it appears that new genetic techniques are likely to advance our understanding of the evolution and ecology of marine organisms, and that the application of these techniques to marine organisms is likely to provide information of major general interest.

Acknowledgements

This work was supported in part by NSF Grant DEB 76-01903. I thank Robin Spicher for typing.

REFERENCES

Ambler, R. P. 1974. The evolutionary stability of cytochrome C-551 in Pseudomonas aeruginosa and Pseudomonas flavescens biotype C. Biochem. J. 137, 3-14.

Bayne, B. L., Widdows, J. and Thompson, R. J. 1976. Physiological integration. In, Marine Mussels: Their Ecology and Physiology. Bayne, B. L. (ed). pp. 261-291. Cambridge University Press, Cambridge.

Bernstein, S. C., Throckmorton, H. L. and Hubby, J. L. 1973. Still more genetic variability in natural populations. Proc. Nat. Acad. Sci. U.S.A. 70, 3928-3931.

Cobbs, G. 1976. Polymorphism for dimerizing ability at the esterase-5 locus in Drosophila pseudoobscura. Genetics, 82, 53-62.

Coyne, J. A. 1976. Lack of genetic similarity between two sibling species of Drosophila as revealed by varied techniques. Genetics 84, 593-607.

Coyne, J. A. and Felton, A. A. 1977. Genic heterogeneity at two alcohol dehydrogenase loci in Drosophila melanogaster and Drosophila persimilis. Genetics 87, 285-304.

Crick, F. H. C. 1966. Codon-anticodon pairing: The Wobble Hypothesis. J. Mol. Biol. 19, 548-555.

Crow, J. F. and Kimura M. 1970. An Introduction to Population Genetics Theory. Harper and Row, New York.

Falconer, D. S. 1960. Quantitative Genetics. p. 206. Ronald, New York.

Fisher, R. A. 1958. Polymorphism and natural selection. J. Ecol. 46, 289-293.

Grunstein, M. and Schedl, P. 1976. Isolation and sequence analysis of Sea Urchin (Lytechinus pictus) histone H4 messenger RNA. J. Mol. Biol. 104, 323-349.

Grunstein, M., Schedl, P. and Kedes, L. 1976. Sequence analysis and evolution of Sea Urchin (Lytechinus pictus and Strongylocentrotus purpuratus) histone H4 messenger RNAs. J. Mol. Biol. 104, 351-369.

Johnson, F. M. and Powell, A. 1974. The alcohol dehydrogenases of Drosophila melanogaster: Frequency changes associated with heat and cold shock. Proc. Natl. Acad. Sci. U.S.A. 71, 1783-1784.

Johnson, G. 1977. Evaluation of the stepwise mutation model of electrophoretic mobility: comparison of the gel sieving behaviour of alleles at the esterase-5 locus of Drosophila pseudoobscura. Genetics 87, 139-157.

Kimura, M. 1968. Evolutionary rate at the molecular level. Nature 217, 624-626.

Kimura, M. 1976. Molecular evolution. In, Discussion forum. Trends in Biochemical Sciences 1:N152-N154.

Koehn, R. K., Milkman, R. and Mitton, J. B. 1976. Population genetics of marine pelecypods. IV. Selection, migration, and genetic differentiation in the blue mussel, Mytilus edulis. Evolution 30, 2-32.

Levene, H. 1953. Genetic equilibrium when more than one ecological niche is available. Am. Nat. 87, 131-133.

Maxam, A. and Gilbert, W. 1977. A new method for sequencing DNA. Proc. Natl. Acad. Sci. U.S.A. 74, 560-564.

Milkman, R. 1973. Electrophoretic variation in E. coli from natural sources. Science 182, 1024-1026.

Milkman, R. 1975. Allozyme variation in E. coli of diverse natural origins. In, Isozymes, Vol. IV: Genetics and Evolution. Markert, C.L. (ed). pp. 273-285. Academic Press, New York.

Milkman, R. 1976a. Further evidence of thermostability variation within electrophoretic mobility classes of enzymes. Biochem. Genet. 14, 383-388.

Milkman, R. 1976b. Molecular evolution. In, Discussion forum. Trends in Biochemical Sciences 1:N152-N154.

Milkman, R. 1978. Selection differentials and selection coefficients. Genetics 88, in press.

Milkman, R. and Koehler, R. 1976. Isoelectric focusing of MDH and 6-PGDH from Escherichia coli of diverse natural origins. Biochem. Genet. 14, 517-522.

Milkman, R. and Koehn, R. K. 1977. Temporal variation in the relationship between size, numbers, and an allele frequency in a population of Mytilus edulis. Evolution 31, 103-115.

Ohta, T. 1974. Mutational pressure as the main cause of molecular evolution and polymorphism. Nature 252, 351-354.

Petra, P. H., Bradshaw, R. A., Walsh, K. A. and Neurath, H. 1969. Identification of the amino acid replacements characterizing the allotypic forms of bovine carboxypeptidase A. Biochemistry 8, 2762-2768.

Salser, W. and Isaacson, J. S. 1976. Mutation rates in globin genes: the genetic load and Haldane's dilemma. Prog. in Nucl. Acid Res. and Mol. Biol. 19, 205-220; see p. 206.

Sampsell, B. 1977. Isolation and genetic characterization of alcohol dehydrogenase thermostability variants occurring in natural populations of Drosophila melanogaster. Biochem. Genet. 15, 971-988.

Sanger, F., Air, G. M., Barrell, B. G., Brown, N. L., Coulson, A. R., Fiddes, J. C., Hutchinson III, C. A., Slocombe, P. M. and Smith, M. 1977. Nucleotide sequence of bacteriophage ϕX174 DNA. Nature 265, 687-695.

Selander, R. 1976. Genic variation in natural populations. In, Molecular Evolution. Ayala, F. J. (ed). pp. 21-45. Sinauer Associates, Sunderland, Massachusetts.

Singh, R. S., Lewontin, R. C. and Felton, A. A. 1976. Genetic heterogeneity within electrophoretic "alleles" of xanthine dehydrogenase in Drosophila pseudoobscura. Genetics 84, 609-629.

Thörig, G. E. W., Schoone, A. A. and Scharloo, W. 1975. Variation between electrophoretically identical alleles at the alcohol dehydrogenase locus in *Drosophila melanogaster*. Biochem. Genet. 13, 721-730.

Warren, C. E. and Davis, G. E. 1967. Laboratory studies on the feeding, bioenergetics and growth of fish. In, *The Biological Basis of Freshwater Fish Production*. Gerking, S. D. (ed). Blackwell, Oxford, (Quoted in Bayne, Widdows, and Thompson, 1976).

Watt, W. B. 1972. Intragenic recombination as a source of population genetic variability. Am. Nat. 106, 737-753.

Wills, C. 1978. Rank-order selection is capable of maintaining all genetic polymorphisms. Genetics: (in press).

Wilson, A. C. 1976. Gene regulation in evolution. In, *Molecular Evolution*. Ayala, F. J. (ed). pp. 225-234. Sinauer Associates, Sunderland, Massachusetts.

Wright, S. 1969. *Evolution and the Genetics of Populations. II. The Theory of Gene Frequencies*. University of Chicago, Chicago.

Wright, T. R. F. and MacIntyre, R. J. 1965. Heat-stable and heat-labile esterase-6^F enzymes in *Drosophila melanogaster* produced by different Est-6^F alleles. J. Elisha Mitchell Scient. Soc. 81, 17-19.

GENETIC VARIATION AND RESOURCE STABILITY IN MARINE INVERTEBRATES

F. J. Ayala and J. W. Valentine

Depts. of Genetics and Geology

Univ. of California, Davis, California U.S.A.

Descriptive and Causal Hypotheses

Science may be defined as the systematic organization of knowledge about the universe on the basis of explanatory principles subject to the possibility of empirical falsification (Dobzhansky et al., 1977, p. 476). Science seeks to formulate explanations for natural phenomena by identifying the conditions that account for their occurrence. But an explanatory hypothesis is scientific only if it can be subject to empirical testing, i.e., to the possibility of empirical falsification. Scientific hypotheses allow predictions to be made about the empirical world; a hypothesis is tested by finding out whether or not precise predictions derived as logical consequences from the hypothesis agree with the state of affairs found in the empirical world.

Two broad stages can be recognized in the development of explanatory hypotheses in a branch of science. In the first stage, which might be called "descriptive", explanatory hypotheses are based on observed correlations, and formulate connections between states or values of different variables. If the state or value of a certain variable is known, the hypothesis predicts what state or value some other variable -- the dependent variable -- will have. A more advanced stage of scientific explanation is achieved when hypotheses are advanced that provide causal explanations for observed correlations. This second stage of scientific development may be called "causal" or "mechanistic" -- it strives to ascertain the underlying processes, mechanisms, or causal relations responsible for the correlations between empirical variables. Like descriptive explanations, causal hypotheses make it possible to predict the state or value of a dependent variable from knowledge of the state or value of some other variable; moreover, causal hypotheses make predictions about the processes responsible for the values of the variables observed. Thus, causal explanations have greater explanatory power, and more predicting capacity, than descriptive hypotheses.

The two stages of scientific advance do not necessarily occur as successive, temporally discrete steps; they often come hand in hand within a given branch of science, with some hypotheses being primarily descriptive, others formulated in causal terms. Moreover, it is sometimes the case that causal hypotheses are, at a certain stage of scientific development, tested only with respect to their descriptive component. That is, although a causal explanation proposes that certain processes underlie the relationships between two variables, it happens sometimes that the hypothesis is tested by examining whether the predicted correlations prevail, while the underlying processes are not directly subject to empirical testing. In such case, although causal hypotheses are formulated, the field of science remains at the descriptive state with respect to their empirical testing.

This state of affairs largely obtains at present in many domains of ecology, particularly with respect to what have been called by Levins (1968) "adaptive strategies". An adaptive strategy is a suite of population parameters that is adaptive to the temporal and spatial characteristics of an environmental regime. Some hypotheses concerning adaptive strategies are descriptive: they simply postulate what population parameters -- morphological, physiological, behavioral, genetic, and so on -- will occur in given environmental regimes. These hypotheses may be seen as inductive generalizations subject to further testing. Other hypotheses are formulated in causal terms: they propose the processes by which a population acquires certain states in response to given environmental characteristics. It is usually the case, however, that the underlying mechanisms of the population response are not subject to empirical testing, but only the predicted correlation between the values of the environmental parameters and the morphological, physiological, genetic, etc., characteristics of the population.

In this paper, we shall be concerned with <u>genetic</u> adaptive strategies. The genetic system of a population is largely responsible for its morphological, physiological, behavioral, and other adaptive characteristics. It is therefore reasonable to expect that the genetic constitution of populations will reflect the characteristics of the environmental regimes in which they live. In particular, a variety of hypotheses have been advanced that postulate higher or lower levels of genetic variation in response to levels of heterogeneity and other environmental characteristics (Carson, 1960; Grassle, 1967, 1972; Levins, 1968; Bretsky and Lorenz, 1969, 1970; Ayala et al., 1973, 1975a; Valentine 1976). These hypotheses often postulate specific relationships between environmental parameters and organisms that would account for the predicted levels of genetic variation. However, by and large, at the present state of ecological and genetic knowledge, the environment-genotype relationships cannot directly be investigated; consequently the hypotheses are tested only by ascertaining whether the predicted correlations between given environmental regimes and levels of genetic variation obtain.

Gel Electrophoresis

The possibility of testing hypotheses concerning adaptive genetic strategies was much enhanced during the last decade through the application of the techniques of gel electrophoresis and enzyme assay. These techniques make it possible to estimate the amount of genetic variation in gene pools. The procedure is schematically represented in Figure 1. Tissue samples from organisms are individually homogenized to release the enzymes and other proteins. The homogenate supernatants are placed in a gel made of starch, polyacrylamide, or some other jellylike substance. The gel with the tissue samples is then subjected for a given length of time to an electric current. Each protein in the gel migrates in a direction and at a rate that depend on net electric charge and molecular size. After removing the gel from the electric field, it is treated with a solution that contains a specific substrate for the enzyme to be assayed, and a coloring salt that reacts with the product of the reaction catalyzed by the enzyme. At the place in the gel to which the specific enzyme has migrated, a reaction will take place which can be symbolized as follows:

$$\text{Substrate} \xrightarrow{\text{Enzyme}} \text{Product} + \text{Salt} \rightarrow \text{Colored spot}$$

The usefulness of the method is due to the fact that the genotypes of the individuals in a sample are simply given by

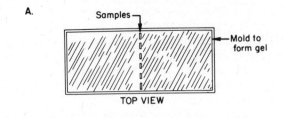
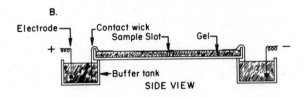
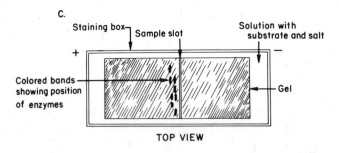

Figure 1. Techniques of gel electrophoresis and enzyme assay used to measure genetic variation in natural populations.
A. A tissue sample from each of the organisms to be surveyed is homogenized to release the proteins in the tissue. The homogenate supernatants are placed in a gel made of starch, agar, polyacrylamide, or some other jellylike substance.
B. The gel with the tissue samples is then subjected, usually for a few hours, to an electric current. Each protein in the samples thus will migrate in a direction at a rate which depends on the protein's net electrical charge and molecular size. C. After removing the gel from the electrical field, it is treated with an appropriate chemical solution containing a substrate specific for the enzyme to be assayed, and a salt. The enzyme catalyzes the reaction from the substrate to its product and this product then couples with the salt giving colored bands at the positions where the enzymes had migrated. Enzymes with different mobilities in the electric field have different structures and therefore are coded by different genes. The genotype at the gene locus coding for a given enzyme can thus be established for each individual from the number and position of the bands in the gels, as shown in Figures 2 and 3.

the patterns observed in the gels. Two illustrative gels are shown in Figures 2 and 3. The gel in Figure 2 contains the tissue homogenates of 22 Drosophila pseudoobscura flies assayed for the enzyme phosphoglucomutase; the gene locus coding for this enzyme may be represented as Pgm. The first and third individuals in the gel, starting from the left, have enzymes with different electrophoretic mobilities, and thus with different amino acid sequences. Let us represent the alleles coding for the enzymes in the first and third individuals as Pgm^{100} and Pgm^{104}, where the superscripts indicate that the enzyme coded by allele Pgm^{104} migrates 4 mm farther in the gel than the enzyme coded by Pgm^{100}. Since the first and third individual each exhibit only one band, we infer that they are homozygotes with genotypes $Pgm^{100/100}$ and $Pgm^{104/104}$ respectively. The second individual in Figure 2 exhibits two phosphoglucomutase bands, one of the bands has the same migration as that of the third individual, but the other migrates 8 mm farther; the allele coding for the latter enzyme may be represented as Pgm^{112}. We infer that this individual is heterozygous with the genotype $Pgm^{104/112}$. Proceeding similarly, the genotypes of the 22 individuals in Figure 2 going from left to right are as follows (for simplicity, only the superscripts identifying the alleles are given): 100/100, 104/112, 104/104, 100/112, 104/104, 104/112, 104/104, 100/112, 100/100, 104/112, 104/104, 100/104, 100/100, 104/104, 100/100, 104/104, 100/100, 100/104, 100/100, 100/104, 100/100, 100/112.

Figure 3 shows a gel with tissue samples from 21 Drosophila equinoxialis flies assayed for the enzyme malate dehydrogenase. The first two individuals exhibit only one band each. If we represent the gene locus coding for malate dehydrogenase as Mdh, their genotypes can be written as $Mdh^{104/104}$ and $Mdh^{94/94}$. The active form of malate dehydrogenase is, however, a dimer consisting of two units, and as a consequence heterozygous individuals exhibit three bands. Why this is so can be simply understood: if an individual has two types of units, say a and b, of a dimer enzyme, there are three possible associations of two units, namely aa, ab, and bb. In Figure 3, individuals 4, 5, 6, 8, 14, 18, 20, and 21 are heterozygotes, with genotype $Mdh^{94/104}$, individuals 2, 9, 11, and 17 have genotype $Mdh^{94/94}$, and individuals 1, 3, 7, 10, 12, 13, 15, 16, and 19 have genotype $Mdh^{104/104}$.

Tetramer and even higher order enzymes are known. The genetic interpretation of electrophoretic phenotypes is based on the same principles just illustrated. In general, the number of bands in a heterozygous individual is one more than the number of subunits making up the enzyme (assuming that the subunits are coded by only one gene locus); for example a

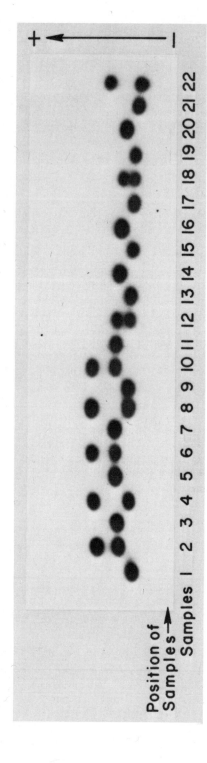

Figure 2. An electrophoretic gel stained for the enzyme phosphoglucomutase. The gel contains tissue samples from each of 22 individuals of a single species. Organisms with only one colored band are inferred to be homozygotes; those with two bands are inferred to be heterozygotes. Enzymes with different migration are different in amino acid sequence and thus are coded by different alleles. There are three different bands in the gel; which may be represented as 100 (the band closest to the origin), 104 (the intermediate band, which migrates about 4 mm farther than 100), and 112 (the farthest moving band, which migrates about 12 mm farther than 100). These numbers may also be used to represent the alleles coding for the three enzymes. If we represent the gene locus coding for phosphoglucomutase as Pgm, the genotypes of the first five individuals, starting from the left, are as follows:

Pgm 100/100, Pgm 104/112, Pgm 104/104, Pgm 100/112, and Pgm 104/104.

The genotypes of the other 17 organisms can be inferred from the gel patterns in the same fashion.

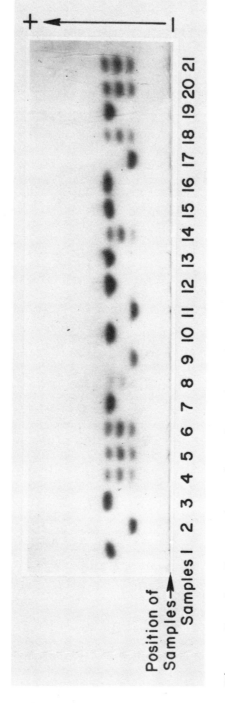

Figure 3. An electrophoretic gel stained for the enzyme malate dehydrogenase. The gel contains tissue samples from each of 21 organisms of a single species. Flies with only one colored band are, as in Figure 2, inferred to be homozygotes; but the heterozygotes exhibit three bands because malate dehydrogenase is a dimeric enzyme. If the gene locus coding for this enzyme is represented as Mdh, the genotype of the first and third individuals in inferred to Mdh$^{104/104}$, the genotype of the second individual is inferred to be Mdh$^{94/94}$, individuals fourth, fifth and sixth all have the heterozygous genotype Mdh$^{94/104}$, and so on. As in Figure 2, the number representing alleles refer to different amounts of migration of the enzymes coding for them.

heterozygote for a tetramer enzyme will exhibit five electrophoretic bands, namely, aaaa, aaab, aabb, abbb, bbbb.

Electrophoretic techniques permit detection of allelic variants in individuals genes. Variant as well as invariant gene loci can be identified, and therefore a random sample of genes with respect to variation is possible. Proteins and enzymes for which the appropriate assay techniques exist can be chosen for study without knowing a priori whether or not they are variable, or how variable they are. A moderate number of proteins studied in a moderately large number of individuals is sufficient to estimate the amount of variation over the whole genome in a population. However, not all allelic variants are detectable by gel electrophoresis. Generally, only those amino acid substitutions which alter the net charge of the protein will change its mobility in an electrophoretic gel. Considerations of the genetic code and of electrical properties of amino acids suggest that only about one third of all amino acid replacements are detectable by gel electrophoresis. Moreover a given amino acid may be coded by two or more codons, and therefore not all changes in the nucleotide sequence of the DNA result in different amino acid sequences in the corresponding protein. Consequently, the amount of allelic variation is underestimated.

Measures of Genetic Variation

The basic information obtained from electrophoretic surveys of protein variation consists of the genotypic or the allelic frequencies at each locus studied in a given population. For simplicity, only the allelic frequencies are usually given, because the number of different genotypes is usually greater than the number of different alleles; if there are n different alleles at a locus, the number of possible different genotypes is $n(n + 1)/2$.

A variety of measures can be used to **express** in a single statistic the amount of genetic variation in a population. For a random mating population, the most informative measure is the overall incidence of heterozygosity, which can be expressed in a variety of ways. The proportion of polymorphic loci in a population is another commonly used measure, but one which is less precise and informative. A third measure of genetic variation is the average number of alleles per locus; however, this measure is strongly dependent on the number of individuals sampled, since the more individuals are examined, the greater the probability of finding additional alleles.

GENETIC VARIATION AND RESOURCE STABILITY

In random mating populations, the <u>expected</u> frequency of heterozygotes, $\underline{H}$, at a locus can be directly calculated from the allelic frequencies. If there are $\underline{n}$ alleles with frequencies $\underline{f}_1$, $\underline{f}_2$, $\underline{f}_3$, ..., $\underline{f}_n$, the expected frequency of heterozygotes is

$$\underline{H} = 1 - (\underline{f}_1^2 + \underline{f}_2^2 + \underline{f}_3^2 + \ldots + \underline{f}_n^2)$$

Alternatively, one may measure heterozygosity using the <u>observed</u> frequency of heterozygotes, rather than the expected frequency. The observed frequency of heterozygotes is calculated by counting the number of heterozygous individuals, and dividing it by the total number of individuals in a sample. For the sample of flies in Figure 2, the observed frequency of heterozygotes is 8/22 = 0.364; for the sample of flies in Figure 3, it is 8/21 = 0.381. In large samples of individuals from random mating populations, the observed and the expected frequency of heterozygotes usually agree quite well, although differences may exist because of natural selection or other factors.

Once the heterozygosity, $\underline{H}$, has been calculated for each of the gene loci under study, the overall amount of variation in a population is estimated by the average frequency of heterozygotes per locus, $\overline{\underline{H}}$. This is simply obtained by averaging $\underline{H}$ over all loci sampled. $\overline{\underline{H}}$ may be expressed with its standard error, which reflects the amount of heterogeneity among the loci sampled.

The heterozygosity of a population can also be expressed as the average frequency of heterozygous loci per individual, $\overline{\underline{H}}_i$. This is estimated by averaging (over all individuals) the proportion of heterozygous loci observed in each individual. The values of $\overline{\underline{H}}$ and $\overline{\underline{H}}_i$ are the same but their variance and standard error will generally be different. The variance of $\overline{\underline{H}}_i$ reflects the heterogeneity among individuals for the set of loci studied. The variance of $\overline{\underline{H}}$ measures heterogeneity among loci. In a random mating population $\overline{\underline{H}}_i$ will be normally distributed, and its variance will be relatively speaking small. There is no <u>a priori</u> reason, however, why $\overline{\underline{H}}$ should be normally distributed, and generally it is not; the variance of $\overline{\underline{H}}$ is generally larger than the variance of $\overline{\underline{H}}_i$. In order to estimate the amount of genetic variation in a population <u>over the whole genome</u>, $\overline{\underline{H}}$ with its standard error is preferable since this statistic reflects the great amount of heterogeneity among loci, and its large standard error indicates that $\overline{\underline{H}}$ may vary in value substantially when different sets of loci are sampled.

Genetic variation can also be measured by the proportion of polymorphic loci, $\underline{P}$, in a population. This statistic is to a certain extent arbitrary and imprecise. It is arbitrary because it must first be decided when a locus will be considered polymorphic. Two criteria commonly used are (1) when the frequency of the most common allele in a population is no greater than 0.950; (2) when it is no greater than 0.990. Criterion (1) is more restrictive than criterion (2); every locus polymorphic by criterion (1) is also polymorphic according to (2), but not <u>vice versa</u>.

$\underline{P}$ is also imprecise, because for each locus it establishes whether or not it is polymorphic, but not how polymorphic it is. A locus with two alleles with frequencies 0.95 and 0.05, and a second locus with 10 alleles each with a frequency of 0.10, contribute equally to $\underline{P}$, although the latter locus has more genetic variation. If several populations of a species are studied, the average proportion of polymorphic loci per population $\underline{P}_p$, can be calculated as the average of $\underline{P}$ over all populations. Alternatively, the proportion of populations in which a locus is polymorphic may be calculated first, and the average, $\underline{P}_\ell$ estimates the average proportion of populations in which a locus is polymorphic. Generally, $\underline{P}_p$ has a smaller variance than $\underline{P}_\ell$

Genetic Polymorphisms in <u>Tridacna maxima</u>

Our interest in the problem of adaptive genetic strategies was sparked by a hypothesis advanced by Bretsky and Lorenz (1969, 1970) claiming that some extinction patterns observed in the fossil record of marine invertebrates are due to the occurrence of less variable gene pools in more stable environments. We selected the "giant clam," <u>Tridacna maxima</u> Roding, as an organism living in a stable environment that therefore, according to the hypothesis of Bretsky and Lorenz, should have a low level of genetic variation. We shall describe in some detail the results of such study as an illustration of the use of electrophoretic techniques to estimate genetic variation in natural populations.

According to Bretsky and Lorenz, selection in relatively stable environments should favor a rather narrow range of phenotypes, and thus alleles that cause variation from the favored phenotypic mode would tend to be eliminated. The hypothesis suggests that after a certain period of evolution in a stable environment gene pools should be largely depleted of alternate alleles owing to selection for the most favorable genotype mix. Now if a relatively stable environmental regime were to change towards instability, a different genetic strategy would be favored, one involving the evolution of multiple alternative genotypes to provide phenotypes adapted to a wider range of conditions.

It appears that some major extinctions registered in the fossil record are associated with trends towards reduced environmental stability, and involved the disappearance of arrays of shallow-marine lineages; specialized lineages have been particularly hard hit. We searched for a living ecological analog of these sorts of lineages. *Tridacna maxima* seemed a satisfactory analog (Ayala et al., 1973). *Tridacna* belongs to a highly specialized, even aberrant family of cardioid pelecypods that is large for a pelecypod (one species is the largest pelecypod known to be living) and probably has a high trophic stability in its present environmental regime, for it contains abundant zooxanthellae in its mantle tissues, which provide an energy supplement. *Tridacna* has been previously employed as an ecological analog of certain large extinct brachiopod lineages, owing to a number of inferred functional features that they share (Cowen, 1970). In short, *Tridacna* represents the sort of specialized organism in a rather constant environment that has so often been swept away during extinctions.

The species studied, *T. maxima*, lives at low latitudes throughout the Indo-Pacific, from East Africa to Pitcairn Island at $25°$ latitude S. and $130°$ longitude W. This broad geographic range is actually greater than the range of other living species of *Tridacna*, which might imply a certain adaptive flexibility, at the very least affecting dispersal. However, *T. maxima* is restricted to a single biotic province that happens to range widely because of the broad longitudinal spread of islands and shelves through the tropical Indo-Pacific region, without the occurrence of major dispersal barriers. Many of the species that disappeared during mass extinctions were similarly widespread, so that the analogy of *T. maxima* with those lineages is biogeographically plausible.

A total of 107 specimens of *T. maxima* were collected at Enewetak, Marshall Islands, and brought live to our laboratory where they were frozen until used for electrophoreses. A total of 39 gene loci coding for enzymes and non-enzymatic proteins were assayed in the animals, using a variety of tissues (demibranch, kidney, mantle, adductor muscle, and stomach). The results are summarized in Table 1, which gives for each gene locus the number of genes sampled, the

TABLE 1

Allelic variation at 30 gene loci in a natural population of the tropical pelecypod *Tridacna maxima*. N is the number of genes sampled at each locus.

Locus	N	Allelic frequencies					Frequency of heterozygotes		Is the locus polymorphic?[†]
							observed	expected	
Adk-1	150	100/1.00					0.000	0.000	no
Adk-2	184	98/0.05	100/0.90	102/0.04	104/0.005		0.174	0.182	yes
Adk-4	204	98/0.01	100/0.02	102/0.06	110/0.005		0.137	0.147	yes
Adk-5	198	100/0.99	104/0.005	112/0.005			0.020	0.020	no
Adk-6	116	96/0.04	100/0.81	104/0.08	106/0.06		0.310	0.331	yes
Est-2	174	95/0.05	98/0.07	100/0.69	102/0.01	105/0.18	0.414	0.438	yes
Est-4	116	96/0.02	98/0.08	100/0.88	102/0.03		0.241	0.220	yes

Table 1 (continued)

Locus	N								
Est-5	214	98/0.005	100/0.99	102/0.005	112/0.005		0.028	0.028	no
Est-6	134	98/0.01	100/0.86	103/0.03	105/0.09	110/0.01	0.254	0.254	yes
Gdh-2	176	98/0.01	100/0.92	103/0.07			0.136	0.147	yes
G3pdh	214	94/0.01	100/0.70	106/0.29	110/0.005		0.477	0.434	yes
Idh	144	98/0.10	100/0.82	102/0.07	110/0.01		0.333	0.313	yes
Ldh	210	98/0.02	100/0.96	102/0.02			0.086	0.085	no
Lap-1	210	97/0.01	100/0.78	102/0.20	106/0.02		0.257	0.359*	yes
Lap-3	202	98/0.01	100/0.61	103/0.26	106/0.11		0.327	0.547**	yes
Mdh-1	214	100/1.00					0.000	0.000	no
Mdh-2	206	96/0.01	100/0.71	102/0.01	104/0.20	106/0.02 108/0.04	0.456	0.454	yes

Table 1 (continued)

Locus								
Mdh-3	176	97 / 0.08	100 / 0.87	104 / 0.04	106 / 0.01	0.239	0.236	yes
Mdh-4	62	100 / 1.00	100			0.000	0.000	no
Me	204	98 / 0.03	100 / 0.97			0.059	0.057	no
Odh	110	97 / 0.11	100 / 0.84	103 / 0.05		0.327	0.286	yes
Pgm-1	214	98 / 0.09	100 / 0.62	103 / 0.22	105 / 0.07	0.439	0.551*	yes
Pgm-2	212	92 / 0.02	96 / 0.12	100 / 0.83	102 / 0.04	0.283	0.303	yes
Pt-1	208	100 / 1.00				0.000	0.000	no
Pt-2	212	98 / 0.20	100 / 0.67	102 / 0.11	104 / 0.01	0.415	0.497	yes
Pt-3	150	98 / 0.03	100 / 0.97			0.053	0.052	no
Pt-4	184	100 / 1.00				0.000	0.000	no

Table 1 (continued)

Locus	N					Obs. Het.	Exp. Het.	Poly.+
To	210	92/0.005	100/0.98	112/0.01	118/0.005	0.038	0.038	no
Tpi-1	214	94/0.04	100/0.92	106/0.05		0.168	0.158	yes
Tpi-2	64	96/0.03	100/0.72	104/0.23	112/0.02	0.375	0.427	yes
Average	176±8					0.202±0.029	0.220±0.034	0.633

+A locus is considered polymorphic when the frequency of the most common allele is ≤ 0.95.

*The difference between the observed and the expected number of heterozygous individuals is statistically significant, $P < 0.05$.

**Same, $P < 0.001$.

alleles found and their frequencies, the frequency of
heterozygous individuals, and whether or not the locus
is polymorphic. In diploid organisms, the number of genes
sampled is simply twice the number of individuals. The
frequency of heterozygotes is given as observed, and also
as expected assuming Hardy-Weinberg (<u>i.e.</u>, random mating)
equilibrium. A locus is considered polymorphic when the
frequency of the most common allele is no greater than 0.95.

No allelic variation was detected at five of the 30 gene
loci. At the other 25 loci the number of alleles per locus
varies from two to six, with an average over the 30 loci of
3.33 ± 0.24. At 19 loci (63.3 percent), the frequency of
the most common allele is no greater than 0.95. The
observed frequency of heterozygous individuals per locus
ranges from zero, at the five invariant loci, to 47.7 percent
at the <u>G3pdh</u> locus. The average percent frequency of
heterozygotes, $\bar{H}$, is 20.2 ± 2.9 or 22.0 ± 3.4, depending on
whether the observed or the expected frequencies are used.

Figure 4 shows the distribution of loci with respect
to observed heterozygote frequency. The distribution has a
mode at the lower end (0-5 percent heterozygotes), but is
rather uniform for the range from 5 to 50 percent. This type
of distribution, with the mode at the lower end, is typical
of virtually all organisms that have been studied electrophoretically. On the other hand, the distribution of heterozygous loci per individuals, shown in Figure 5, is typically

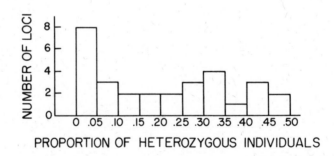

Figure 4. Distribution of 30 gene loci studied in <u>Tridacna
maxima</u> with respect to the proportion of heterozygous
individuals per locus.

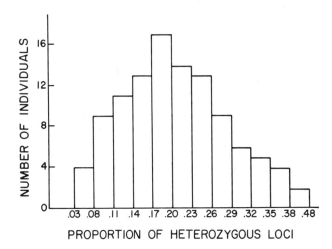

Figure 5. Distribution of 107 individuals of _Tridacna maxima_ according to the proportion of heterozygous loci per individual.

normal, with the mode around the mean. The mean of the observed frequency of heterozygous loci per individual, $\bar{H}_i$, is 20.2 ± 0.8 percent. As expected $\bar{H}_i$ has the same value but a smaller standard error than $\bar{H}$ (0.8 versus 2.9 percent).

A second study of genetic variation in _T. maxima_ has been carried out by Campbell et al. (1975) using a sample of 50 individuals from Heron Island, Great Barrier Reef, Australia. This population proved to be genetically very similar to the one from Enewetak. The value of $\bar{H}$ for the 28 gene loci studied is 21.6 ± 4.3 or 20.3 ± 3.7 percent, depending on whether the observed or the expected heterozygosities are used. Seven loci were studied in the Heron Island population that had not been examined in the Enewetak population. Combining the data from both studies, we have a total of 37 gene loci sampled, with $\bar{H}$ being 20.9 ± 2.7 percent and 21.6 ± 3.2 percent for the observed and expected heterozygosities, respectively.

Classification of Environmental Variation

Tridacna maxima, contrary to the prediction derived from the hypothesis of Bretsky and Lorenz (1969, 1970), exhibits a very high level of genetic variation although it lives in a fairly stable environment. Other authors (see references given above) have advanced hypotheses predicting a positive correlation between environmental heterogeneity and genetic variation, so that more variable gene pools would be expected in more heterogeneous environments. However, before proceeding any further with hypothesis testing, it is useful to examine the meanings of terms, such as "heterogeneity", "stability", "predictability", and so on, since these are predicated of the environment with meanings that are not always made explicit, and are not always consistent from author to author.

Table 2 classifies various terms used in the ecological literature concerning environmental variation. Some features of the table deserve attention. Environmental homogeneity may have considerably different consequences depending on whether they apply to variation in space or to variation in time, and thus they are distinguished in the table. Predictability is sometimes confused with stability, although a highly seasonal environment is unstable but may be highly predictable. Environmental homogeneity, stability, etc. may refer to some but not other components of the environment; in discussions of genetic variability, three have been specifically distinguished: physical, biotic, and trophic. The trophic component is, of course, a subclass of the biotic environment. There are aspects of the environment that have not been included in Table 2, such as complexity of the trophic web, age of the environment, and the like. The distinctions made should be sufficient for the present purposes.

Trophic Stability Versus Environmental Instability

The observation of high levels of genetic variability in T. maxima suggests that environmental instability does not correlate positively with amount of genetic variation. Among several possible environmental characteristics of the tropical coral reef environment of Tridacna that could be conjectured responsible for high levels of genetic variation, the following two seem worthy of attention: (1) stability of the physical environment, and (2) stability of trophic resources. In tropical coral reef environments seasonality is low with respect to both the physical and the trophic components of the environment. Moreover, as pointed out above, T. maxima contains abundant zooxanthellae in its mantle tissues, which provide an energy supplement contributing to high trophic stability.

TABLE 2

Types of environmental variation and terms used to refer to them

Environmental state	Environmental variable	Environmental component		
		Physical	Biotic	Trophic
Homogeneous	space	uniform	uniform	uniform
	time	stable or constant	stable or constant	stable or constant
Heterogeneous	space	patchy or diversified	patchy or diversified	patchy or diversified
	time	unstable or variable	unstable or variable	unstable or variable
Stable or seasonal	time	predictable	predictable	predictable
Erratic	time	unpredictable	unpredictable	unpredictable
Abundant	resources	habitat-rich	diversity-high	rich
Scarce	resources	habitat-poor	diversity-low	poor

Two organisms were studied to decide between the two proposed hypotheses: the temperate intertidal phoronid Phoronopsis viridis, and the Antarctic brachiopod Liothyrella notorcadensis. P. viridis, is known from widely scattered intertidal localities in California embayments. We examined electrophoretically 39 presumptive gene loci in a sample of about 100 individuals collected in Bodega Bay, California (Ayala et al., 1974). The distribution of L. notorcadensis is restricted to Antarctica; we surveyed 34 presumptive gene loci in some 100 individuals collected subtidally in Arthur Harbor, Anvers Island, at a depth of about 17 meters (Ayala et al., 1975a).

With respect to physical components, the marine environment of Antarctica is fairly stable; seasonal temperatures at -17 m. in Arthur Harbor vary only from about $-1.8^{o}C$ in winter to about $+1.0^{o}C$ in summer. This is far more stable than in mid-latitudes, such as the Californian coast, and rivals the tropical marine waters in stability. Thus if high levels of genetic variability correlate directly with the stability of the physical environment, we would expect higher levels of genetic variation in Liothyrella than in Phoronopsis.

On the other hand, the Antarctic environment is quite unstable with respect to the trophic resource supply. Primary productivity is highly seasonal and effectively disappears during the long winter's night, from about May to November; zooplankton is similarly seasonal disappearing from Arthur Harbor during the winter. Suspension feeders, such as articulate brachiopods, that rely upon living phytoplankton face a period of starvation during winter, or must switch to other food sources such as bacteria and suspended detritus. Thus, if levels of genetic variation are positively correlated with the stability of trophic resources, L. notorcadensis should have little genetic variation, i.e. less than the temperate P. viridis and of course much less than T. maxima.

The results of our electrophoretic surveys of L. notorcadensis and P. viridis are summarized in Table 3. Liothyrella exhibits very low levels of genetic variation, expected H = 3.9 + 1.9 percent; while Phoronopsis exhibits a moderate level, expected H = 9.4 + 2.5 percent. Both organisms have considerably less genetic variation than T. maxima, with an expected H of 22.0 + 3.4 percent. The amount of genetic variation found in P. viridis is comparable to that observed in a previously studied temperate invertebrate, the horse-shoe crab Limulus polyphemus (H = 6.7 percent; Selander et al., 1970). The results are consistent

with the hypothesis proposing a direct correlation between
levels of genetic variation and stability of trophic resources, but contradict any hypothesis predicting a direct
correlation between genetic variation and stability of the
physical environment. The results are also inconsistent
with other hypotheses, such as any one proposing a positive
correlation between the predictability of the environment and
the amount of genetic variation; it would seem that the
environmental regime of Liothyrella is no less predictable
than that of Tridacna.

Trophic Stability Versus Resource Abundance

However, other environmental variables besides the
stability of trophic resources can be identified that
correlate well with the levels of genetic variation observed
in Tridacna, Phoronopsis, and Liothyrella. A possible one
might be the abundance of trophic resources. In order to
decide between the two hypotheses--abundance versus stability
of trophic resources--an environment can be chosen where the
trophic supply is stable but scarce. The deep sea meets this
requirement. The deep sea is among the most temporally stable
environments on earth, both in terms of physical parameters
and of trophic resource supply (Sanders, 1968; Hessler and
Jumars, 1974); seasonality appears to be nearly or completely absent. On the other hand, trophic resources are
low; primary productivity is absent, and deep-sea animals
depend on the organic remnants reaching them through the
water column.

We have studied electrophoretically six species of
organisms recovered by otter trawl in the San Diego Trough
at a depth of 1,244 meters: the articulate brachiopod
Frieleia halli (Valentine and Ayala, 1975), the ophiuran
Ophiomusium lymani (Ayala and Valentine, 1974), and four
asteroids, Nearchaster aciculosus, Pteraster jordani,
Diplopteraster multipes, and Myxoderma sacculatum ectenes
(Ayala et al., 1975b). The results are summarized in
Table 4. These organisms are genetically very polymorphic
($\bar{H} \simeq 17$ percent), which is consistent with the hypothesis
proposing that levels of genetic variation correlate with the
stability of the resource supply. The results are inconsistent with the hypothesis proposing a direct correlation
between abundance of trophic resources and amount of geneticc
variation. Other authors also have found high levels of
genetic polymorphism in populations of deep-sea organisms
(Schopf and Gooch, 1971, 1972; Gooch and Schopf, 1972;
Doyle, 1971, 1972).

TABLE 3

Genetic variation in two marine invertebrates. Phoronopsis viridis is a temperate zone intertidal organism, while Liothyrella notorcadensis lives subtidally in Antarctica. N is the average number of genes sampled per locus. A locus is considered polymorphic when the frequency of the most common allele is (1) ≤ 0.95, or (2) ≤ 0.99.

Organism	Loci sampled	N	Average heterozygosity observed	expected	Proportion of polymorphic loci (1)	(2)	Alleles per locus
P. viridis	39	197+9	0.073+0.007	0.094+0.025	0.282	0.487	2.23+0.21
L. notorcandensis	34	157+8	0.038+0.018	0.039+0.019	0.117	0.285	1.56+0.20

Trophic Stability and Genetic Variation in Krill.

One difficulty with the studies so far reviewed of genetic variation in marine invertebrates is that they involve only a limited number of species belonging to widely different taxonomic groups. If phylogenetic relationships have a substantial effect on levels of genetic variation, they might be confounding the environmental correlations sought here. There is, however, some evidence indicating that the correlations observed are not primarily the result of phylogeny. Two brachiopod species have been studied, L. notorcadensis from Antarctica has little genetic variation while F. halli from the deep sea is very polymorphic. The deep-sea asteroids are very polymorphic, while two species (Asterias forbesi and A. vulgaris), from the northwestern Atlantic where seasonality in primary productivity is relatively high, have been shown by Schopf and Murphy (1973) to have very low levels of genetic variation.

The effects of any possible phylogenetic component affecting levels of genetic variation will, of course, tend to cancel out as more and more species from the same taxonomic group but different environmental regimes are studied. Nevertheless, the contribution of the environmental regime to levels of genetic variation may be best evaluated by studying closely related species--ideally, species of the same genus--living in different environmental regimes. In order to meet this requirement, we selected for study three species of krill of the genus Euphausia (Valentine and Ayala, 1976). The species are closely related, belonging to the same subgenus, but one is circumantarctic, another is temperate, and the third is tropical. All three species are pelagic.

Euphausia superba was collected about 19 km west of Arthur Harbor, Antarctica. It is a species with high fecundity and occurs in enormous numbers. It lives under the antarctic ice and ranges northward to the Antarctic Convergence; primary productivity is highly seasonal in these waters. E. mucronata was collected off northern Chile in transition water of the Peru Current at approximately $21°$ S, $71°$ W; the species is distributed along the western coast of South America from the southern tropical to the temperate region. Somewhat irregular upwelling occurs in the region, but seasonality is considerably less pronounced than in circumantarctic waters. E. distinguenda is restricted to tropical waters in the eastern Pacific, north and south of the equator; it was collected at two sites ($7°$ 32' N, $92°$ 16' W and $10°$ N, $93°$ 45' W) in Pacific Equatorial water over the Guatemalan Basin. In this region upwelling

TABLE 4

Genetic variation in deep-sea organisms from the San Diego Trough. Four asteroid species are included: Nearchaster aciculosus, Pteraster jordani, Diplopteraster multipes, and Myxoderma sacculatum ectenes. N is the average number of genes sampled per locus. A locus is considered polymorphic when the frequency of the most common allele is (1) ≤ 0.95, or (2) ≤ 0.99.

Organism	Loci sampled	N	Average heterozygosity		Proportion of polymorphic loci		Alleles per locus
			observed	expected	(1)	(2)	
Frieleia halli	16	100+13	0.169+0.043	0.169+0.046	0.750	0.563	2.50+0.29
Ophiomusium lymani	15	476+20	0.166+0.054	0.170+0.054	0.733	0.533	3.67+0.53
Asteroids	24	54+1	0.176+0.035	0.164+0.032	0.833	0.583	3.21+0.39

associated with the dynamics of the equatorial current systems raises productivity above the tropical average, but productivity is nevertheless more stable than in the waters where the other two species were sampled.

The levels of genetic variation found in the three *Euphausia* species are summarized in Table 5. The predicted trend prevails: levels of genetic variation correlate well with the trophic stability of the environmental regime. The expected heterozygosities are 5.7 ± 1.9 percent for *E. superba*, 14.1 ± 2.5 percent for *E. mucronata*, and 21.3 ± 3.4 percent for *E. distinguenda*. It is worth noting that the marine invertebrates reviewed earlier are benthic species: the present result corroborates for the pelagic realm the hypothesis proposing a direct correlation between the stability of trophic resource supply and levels of genetic variation.

It would seem appropriate to conclude that the evidence presented here is consistent with the hypothesis that the stability of trophic resources plays a significant role in determining levels of genetic variation in populations of marine invertebrates. This hypothesis, however, should at present be considered only as a working hypothesis which needs further testing. Additional evidence, obtained from studies made by various investigators, is reviewed by Valentine and Ayala in this volume. Needless to say, the implication of the hypothesis is not that trophic resource stability is the only parameter determining genetic variation, but rather that it plays a significant role; doubtless, levels of genetic variation are affected by a variety of factors (Selander and Kaufman, 1973; Somero and Soulé, 1974).

One final comment. The empirical tests reviewed here were exclusively concerned with the descriptive component of the alternative hypotheses considered; that is, with the question whether or not correlations exist between states of environmental parameters and levels of genetic variation. The possible mechanisms underlying the correlation between trophic stability and genetic variation have been discussed elsewhere (e.g. Ayala et al., 1975a; Valentine, 1976) and will be examined in the article by Valentine and Ayala in this volume. As shown there, the proposed causal process implies that organisms in trophically stable environments perceive the environment as "patchy" or "coarse grained", i.e., that temporal homogeneity in time (= stability) with respect to trophic resources makes possible for organisms to perceive the environment as spatially heterogeneous. This provides additional ways of testing the hypothesis, since other

TABLE 5

Genetic variation in three species of krill of the genus Euphausia. $\overline{N}$ is the average number of genes sampled per locus. A locus is considered polymorphic when the frequency of the most common allele is (1) ≤ 0.95, or (2) ≤ 0.99.

Species	Loci sampled	$\overline{N}$	Average heterozygosity observed	Average heterozygosity expected	Proportion of polymorphic Loci (1)	Proportion of polymorphic Loci (2)	Alleles per locus
E. superba	36	127	0.058±0.018	0.057±0.019	0.139	0.361	1.81±0.14
E. mucronata	28	50	0.153±0.028	0.141±0.025	0.571	0.679	2.54±0.30
E. distinguenda	30	110	0.211±0.033	0.213±0.034	0.700	0.828	3.20±0.28

factors besides stability of trophic resources affect the
grain or patchiness of the environment as perceived by
organisms.

Acknowledgement

The research reported here has been supported in part
by ERDA Contract PA 200-14 Mod# 4, and NSF grant DEB 74-
0892612.

REFERENCES

Ayala, F.J. and J.W. Valentine. 1974. Genetic variability in
a cosmopolitan deep-water ophiuran, Ophiomusium lymani.
Marine Biol. 27, 51-57.

Ayala, F.J., Hedgecock, D, Zumwalt, G.S. and Valentine, J.W.
1973. Genetic variation in Tridacna maxima an ecological
analog of some unsuccessful evolutionary lineages.
Evolution 27, 177-191.

Ayala, F.J., Valentine, J.W., Barr, L.G. and Zumwalt, G.S. 1974.
Genetic variability in a temperate intertidal phoronid,
Phoronopsis viridis.   Biochem. Genet. 18, 413-427.

Ayala, F.J., Valentine, J.W., DeLaca, T, and Zumwalt, G.S.
1975a. Genetic variability of the Antarctic brachiopod
Liothyrella notorcadensis and its bearing on mass
extinction hypotheses. J. Paleontology 49, 1-9.

Ayala, F.J., Valentine, J.W., Hedgecock, D. and Barr, L.G.
1975b. Deep-sea asteroids: high genetic variability in
a stable environment. Evolution 29, 203-212.

Bretsky, P.W. and Lorenz, D.M.   1969. Adaptive response to
environmental stability: a unifying concept in
paleoecology. Proc. N. Amer. Paleo. Convention, Pt.
E, 522-550.

Bretsky, P.W. and Lorenz, D.M. 1970. An essay on genetic-
adaptive strategies and mass extinctions. Geol. Soc.
Amer. Bull. 81, 2449-2456.

Campbell, C.A., Ayala, F.J. and Valentine, J.W. 1975. High
genetic variability in a population of Tridacna maxima
from the Great Barrier Reef. Marine Biol. 33, 341-345.

Carson, H.L. 1960. Genetic conditions which promote or retard the formation of species. Cold Spring Harbor Symp. Quant. Biol. 24, 87-105.

Cowen, R.C. 1970. Analogies between the recent bivalve Tridacna and the fossil brachiopods Lyttoniacea and Richthofeniacea. Palaeogeogr., Palaeoclimatol., Palaeocol. 8, 329-344.

Dobzhansky, Th., Ayala, F.J., Stebbins, G.L., Valentine, J.W. 1977. Evolution. W.H. Freeman & Co., San Francisco. Chapter 16.

Doyle, R.W. 1971. Genetic differentiation of ophiuroid populations on the lower continental slope. Ecol. Soc. Amer. Bull. 52(2), 45.

Doyle, R.W. 1972. Genetic variation in Ophiomusium lymani (Echinodermata) populations in the deep sea. Deep-sea Res. 19, 661-664.

Gooch, J.L. and Schopf, T.J.M. 1972. Genetic variability in the deep-sea; relation to environmental variability. Evolution 26, 545-552.

Grassle, J.F. 1967. Influence of environmental variation on species diversity in benthic communities on the continental shelf and slope. Ph.D. Thesis, Duke Univ., Durham, North Carolina, 193 pp.

Grassle, J.F. 1972. Species diversity, genetic variability and environmental uncertainty. Fifth European Marine Biological Symposium, Piccin Editore, Padua, 19-26.

Hessler, R.R. and Jumars, P.A. 1974. Abyssal community analysis from replicate box cores in the north central Pacific. Deep-sea Res. 21, 185-209.

Levins, R. 1968. Evolution in Changing Environments. Princeton Univ. Press, Princeton, New Jersey.

Sanders, H.L. 1968. Marine benthic diversity: a comparative study. Amer. Nat. 102, 243-282.

Schopf, T.J.M. and Gooch, J.L. 1971. A natural experiment using deep-sea invertebrates to test the hypothesis that genetic homozygosity is proportional to environmental stability. Biol. Bull. 141-401.

Schopf, T.J.M. and Gooch, J.L. 1972. A natural experiment to test the hypothesis that loss of genetic variabiltiy was responsible for mass extinctions of the fossil record. J. Geol. 80, 481-483.

Schopf, T.J.M. and Murphy, L.S. 1973. Protein polymorphism of the hybridizing seastars Asterias forbesi and Asterias vulgaris and implications for their evolution. Biol. Bull. 145, 589-597.

Selander, R.K. and Kaufman, D.W. 1973. Genic variability and strategies of adaptation in animals. Proc. Nat. Acad. Sci. USA 70, 1875-1877.

Selander, R.K., Yan, S.Y., Lewontin, R.C. and Johnson, W.E. 1970. Genetic variation in the horseshoe crab. (Limulus polyphemus), a phylogenetic "relic". Evolution 24, 402-414.

Somero, G.N. and Soule, M. 1974. Genetic variation in marine fishes as a test of the niche-variation hypothesis. Nature 249, 670-672.

Valentine, J.W. 1976. Genetic strategies of adaptation. In Molecular Evolution, F.J. Ayala, ed. Sinauer Associates, Sunderland, Massachusetts. pp. 78-94.

Valentine, J. W. and Ayala, F.J. 1975. Genetic variation in Frieleia halli, a deep-sea brachiopod. Deep-sea Res. 22, 37-44.

Valentine, J.W. and Ayala, F.J. 1976. Genetic variability in krill. Proc. Nat. Acad. Sci. USA 73, 658-660.

GENETIC VARIABILITY IN RELATION TO THE ENVIRONMENT IN SOME

MARINE INVERTEBRATES

B. Battaglia, P. M. Bisol and G. Fava

Ist. di Biologia del Mare, Venezia and

Ist. di Biologia Animale, Università di Padova, Italy.

Introduction

One of the central topics of modern evolutionary biology concerns the relationships between the degree of genetic polymorphism (or, sensu latu, genetic variability) and the diversification of environmental factors in time as well as in space.

The analysis and resolution of the above relationships will help to elucidate, in general, the mechanisms involved in the choice of adaptive strategies and, more specifically, the patterns of genetic adaptation upon which the evolutionary process is largely grounded.

Marine environments provide excellent opportunities for the study of interrelations between genetics and ecology. In fact, in some habitats (e.g. littoral areas, estuaries, lagoons, tide pools) the relations between the genetic structure of populations and certain environmental parameters can be expressed at a microgeographical level. Moreover, as the control of environmental parameters is less complex in a liquid medium, a marine environment offers some advantages for the experimental analysis of these relations. This is especially true in the study of organisms with a short life cycle and suitable for laboratory culture.

Studies in marine organisms, which, to a certain extent, are connected with the main problem of genetic variability in relation to the environment, began a considerable time ago.

We shall here briefly mention some and comment on some of the more relevant results.

Sexton and Clark (1936) showed that in the amphipod <u>Gammarus chevreuxi</u> recessive mutants are quite abundant, especially in populations which had migrated from stable environments into environments subject to frequent ecological changes. On the basis of these results, Bacci (1954) put forward the hypothesis that a high frequency of recessive mutations would be a general feature of brackish water populations. As we shall see later, these observations and predictions - probably because they were not sufficiently supported by an adequate genetic analysis (which was not feasible at that time) - have not been confirmed by further investigations.

Battaglia (1954) compared geographic populations of the harpacticoid copepod <u>Porcellidium fimbriatum</u> inhabiting marine and brackish water environments. This species is characterized by a striking colour polymorphism. The main result of this research was to show that populations from Roscoff and Naples, (both typical marine environments), possessed a larger number of morphs than the population from the lagoon of Venice, a brackish water habitat showing remarkable fluctuations in physical characters and with lower biotic diversity.
In <u>Porcellidium</u>, the lower number of morphs in the lagoon population was accompanied by a much smaller variability in certain biometric traits (Battaglia, 1959). Similar results were obtained by comparing marine and brackish water populations of the isopod <u>Sphaeroma serratum</u> and the copepod <u>Tisbe reticulata</u>. The latter species proved to be better than the others for experimental study of the genetic basis of the polymorphism. It was shown that the polymorphism in <u>Tisbe reticulata</u> is genetically controlled, has an adaptive role, and has a balanced nature (Battaglia, 1958; 1959). Already in this early work, the hypothesis was advanced that a lower degree of genetic polymorphism would be a common feature of populations inhabiting brackish water basins. This was ascribed to the smaller number of niches offered by the lagoon, but also to the exacting nature of this marginal environment, where the chances of survival depend primarily upon the ability to tolerate the inconstancy of the medium. In other early work (Battaglia, 1961; 1965) the role of several possible strategies adopted by marine species to confront unpredictable environmental changes was discussed. In particular, adaptive devices based upon genetic flexibility, phenotypic plasticity, stored genetic variability, and the possible joint action of various mechanisms, were considered. A prediction was that in the harpacticoid copepod <u>Tigriopus</u>, inhabiting rock-pools,

the adaptation to the extraordinarily challenging conditions of such a peculiar environment, where the extreme fluctuations are the norm rather than accidental events, must have been achieved by means of individual plasticity. We shall see if, and to what extent, these early hypotheses are supported.

The growing application of electrophoretic techniques to population genetics has justified a renewed interest in the problem of relationships between genetic and environmental variability.

The perspectives disclosed by this new and more rigorous approach are very promising, although the dispute between the 'neutralist' and the 'selectionist' theories of enzyme polymorphisms has not yet been resolved.

In the last few years, several theories have been formulated to account for certain observed or suspected relations between genetic and environmental variability and some predictions have been made (see Ayala and Valentine; Valentine and Ayala; Grassle, this volume).

Most of the observational and experimental work leading to the formulation of the various theories and hypotheses, has been carried out on species belonging to different phyla, and whose life histories and ecological requirements were largely unknown. Moreover, the environments more often considered were the 'stable' ones, such as the deep-sea, whereas the observations on more marginal, physically variable marine environments are quite scanty. However, exceptions to both or either of the above limitations are, for example, the recent studies of Somero and Soulé (1974) on fishes, of Levinton (1975) on Macoma, of Valentine and Ayala (1976) on Euphausia, of Wilkins (1977) on limpets.

The theories which have emerged from a number of investigations carried out with many species of organisms inhabiting a variety of environments, are often controversial.

Scope and Methods of the Research

The main purpose of the present research is to fill certain environmental gaps, by including extremely variable environments such as, for instance, rock-pools. This has made possible the comparison between a wide range of habitats: marine, brackish-water, rock-pools, showing a progressive trend from maximum physical stability and biotic diversity (coral reefs, Gulf of Eilat) to maximum physical instability and biotic homogeneity (Mediterranean rock-pools). Moreover, all the species utilized belong to the same class (Crustacea), have well known life histories, and can easily be reared and bred in the laboratory. For one organism (Tisbe holothuriae) a comparison has also been made between geographic populations of the same species occupying different habitats.

The study of genetic variability was carried out mainly by means of gel electrophoresis. The staining techniques are those of Brewer (1970), Shaw and Prasad (1970), and Ayala et al. (1972), with modifications of pH and quantities of substrates (Bisol, 1976).

The possibility of crossing these species in the laboratory enables us to demonstrate the genetic nature of the various electromorphs. For instance, in the amphipod Gammarus insensibilis (Battaglia and Bisol, 1973; Bisol et al., 1977a) the genetic control of an esterase system at the level of a single locus with three alleles has been successfully demonstrated.

However, the application of the method to individual specimens of the harpacticoid copepods Tisbe and Tigriopus (whose size is about 1mm) proved to be much more laborious. Nevertheless a suitable technique has been worked out for these copepods (Fig.1), and the results of breeding experiments (Bisol et al., 1976) confirm its accuracy.

A total of 21 enzymes were analysed electrophoretically; the number of loci scored varied, according to the different species or populations, between 15 and 24 (Table 1). For each enzyme, 60 to 70 individuals from each population sampled were analyzed. Because of the small size of these organisms, particularly the Copepods, it was impossible to analyse single individuals for more than, at most, a few enzyme systems. This, in addition to the fact that in some cases homogeneous sub-samples have been pooled together, accounts for the comparatively high total numbers of individuals per species which had to be used (Table 2).

Furthermore, the small size of these organisms imposes a limitation to the method, since this allows an estimate of the average heterozygosity over loci, but not, as would be desirable, the distribution of heterozygosity among individuals.

In a few cases the analyses were carried out on specimens collected from nature. Otherwise laboratory populations were used, which were started with numbers of individuals

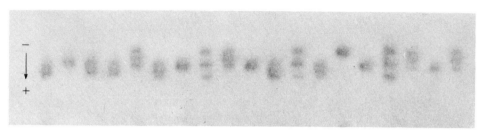

Figure 1. Electrophoretic variants for aminopeptidase in Tisbe holothuriae.

large enough to represent the natural gene pool. An exception was perhaps T. biminiensis from Eilat which, due to the scantiness of its individuals in natural populations, was collected in comparatively small numbers (ca. 40).

Previous experiments carried out with a marine and a brackish water population of the copepod Tisbe holothuriae indicated a significantly different degree of inbreeding depression in the two populations compared (Battaglia, 1970). Since inbreeding depression is believed to give, at least in certain cases, a glimpse of the genetic load of the organisms considered (Wallace, 1970), a study has been made of genetic load in populations of Tisbe holothuriae, expressed as the number of lethal equivalents (b) present in different populations (for the method see Morton, Crow and Muller, 1956). The limitations of this approach consist mainly in the fact that the values of b are given by the sum of the effects of n more or less detrimental genes, but do not permit an estimate of n. The approach, nevertheless, seems a reasonable one to apply to populations of the same species and differences in the values of b may thus be taken to reflect differences in the values of n.

Insofar as the genetic load may provide, though indirectly, an estimate of hidden genetic variation, this method, has also been applied to several populations of Tisbe in addition to the electrophoretic analysis, (Fava et al., 1976; Lazzaretto-Colombera et al., 1976).

TABLE 1

Enzymes assayed and number of loci scored in electrophoretic assay of six species of Crustaceans. (T. bim = Tisbe biminiensis; T. clod = Tisbe clodiensis; T. hol = Tisbe holothuriae; G. ins = Gammarus insensibilis; T. brev = Tigriopus brevicornis; T. ful = Tigriopus fulvus.)

Enzymes	T.bim	T. clod	T. holo	G. ins	T. brev	T. ful
Acid phosphatase	1	1	–	1	1	1
Alcohol dehydrogenase	–	1	1	–	–	1
Alkaline phosphatase	–	2	2	2	2	2
Aminopeptidase	3	2	2	2	3	1
Esterase	5	2	5	3	3	5
Glucose-6-phosphate dehydrogenase	–	–	–	1	1	–
Glutamateoxalacetictransaminase	1	–	–	2	–	–
Glyceraldehyde-3-phosphatede	–	–	–	1	1	–
Hexokinase	–	–	–	1	–	–
Isocitrate dehydrogenase	–	–	–	1	1	–
Lactate dehydrogenase	–	1	3	3	2	1
Leucinoaminopeptidase	1	1	3	–	1	1
Malate dehydrogenase	3	3	2	2	1	2
Malic enzyme	1	1	1	–	1	1
Nothing dehydrogenase	–	–	–	1	1	1
Peroxidase	–	–	–	1	–	–
Phosphoglucomutase	1	1	1	1	1	1
Phospho.hexoseisomerase	1	1	1	1	1	1
Sorbitol dehydrogenase	–	–	1	–	1	1
Tetrazolium oxidase	–	–	–	2	3	1
Xanthine dehydrogenase	1	–	–	1	1	–

TABLE 2

Summary of electrophoretic variation in six species of Crustaceans. (T.bim = Tisbe biminiensis; T. clod = Tisbe clodiensis; T. hol = Tisbe holothuriae; G. ins = Gammarus insensibilis; T. brev = Tigriopus brevicornis; T. ful = Tigriopus fulvus.)

Parameter \ Species	T. bim (Eilat)	T. clod (Banyuls)	T. hol (Banyuls)	T. hol (Sigean)	G. ins (Venice)	T. brev (Tavvallich)	T. ful (Leghorn)
Number of enzymes	10	10	10	10	15	15	13
Number of loci	19	15	19	19	24	23	19
Number of individuals	592	590	672	710	282	905	1893
Mean genes sampled per locus	117.37 ±8.96	121.60 ±6.82	132.22 ±10.09	136.16 ±7.57	147.33 ±11.23	114.52 ±6.24	339.58 ±26.77
Mean alleles per locus	1.789 ±0.211	2.000 ±0.276	2.158 ±0.268	2.000 ±0.268	1.417 ±0.163	1.435 ±0.164	1.158 ±0.115
Polymorphic loci (%) *	52.632	60.00	57.90	52.63	20.83	26.02	10.52
Polymorphic loci (%) **	42.105	40.33	57.90	49.99	16.67	21.74	10.52
H ***	0.165 ±0.048	0.181 ±0.063	0.241 ±0.058	0.240 ±0.061	0.062 ±0.031	0.071 ±0.032	0.054 ±0.036

* A locus is considered polymorphic when the frequency of the most common allele is ≤ 0.99
** A locus is considered polymorphic when the frequency of the most common allele is ≤ 0.95
*** H (gene diversity) calculated according to Nei (1975)

Ecology and Life Histories of the Species Studied

The work has used the following species of the harpacticoid copepods <u>Tisbe</u> and <u>Tigriopus</u>: <u>Tisbe biminiensis</u>, <u>Tisbe holothuriae</u>, <u>Tisbe clodiensis</u>, <u>Tigriopus fulvus</u>, <u>Tigriopus brevicornis</u>; and the amphipod <u>Gammarus insensibilis</u>.

We shall briefly describe their life histories, studied under laboratory conditions, according to their marine, brackish water or rock-pool origin.

1. <u>Tisbe biminiensis</u>. This species was collected from the coral reefs of Eilat, Red Sea, an environment extremely stable in time as to physical and chemical factors such as temperature and salinity, and at the same time extremely diversified in space, a common feature of all coral reefs. In the laboratory, the minimum generation time is ca. 14 days, the life span <u>ca</u>. 40 days, and the length of the reproductive period ca. 20 days. The species is restricted to the marine habitat, where its populations usually have low densities.

2. <u>Tisbe holothuriae</u>. This has a wide, almost cosmopolitan geographic distribution, and great ecological versality. All the populations which have been studied so far are perfectly interfertile and in many cases crosses produce heterotic offspring (Battaglia and Volkmann-Rocco, 1973). The species is practically ubiquitous, being found in typically marine habitats as well as in greatly variable brackish water environments. Its minimum generation time is the shortest among all the species considered (<u>ca</u>. 12 days). Females produce a large number of offspring per egg-sac. Longevity and duration of the reproductive period are almost the same as in <u>Tisbe biminiensis</u>. Two populations have been compared electrophoretically: one from Banyuls-sur-Mer, a typically Mediterranean marine habitat, and the other from the brackish water lagoon of Sigean. The latter environment, about 40 miles north of Banyuls, is very shallow and therefore, exhibits wide fluctuations of salinity and temperature due to fresh and sea water inflow, rainfall and evaporation. Biotic diversity is rather low, and the species is often present at high density. A few more populations have been compared only as to their concealed genetic variation. These populations were collected in the following localities: Split, Yugoslavia (typical marine habitat); Alberoni (a station, in the Lagoon of Venice, topographically and ecologically very close to the adjacent sea); Malamocco (an inner station in the Lagoon of Venice, characterized by a lower biotic diversity compared with Alberoni); Grado (a brackish water lagoon similar in many respects to Sigean).

3. **Tisbe clodiensis**. This species also has a wide geographic distribution and can be found both in marine and brackish water habitats. However, gene flow appears to be remarkably low compared with **Tisbe holothuriae**, due to the frequent occurrence of partial or total inter-deme reproductive isolation (Volkmann, Battaglia and Varotto, this volume). The life cycle parameters are similar to those of **T. holothuriae**. Only the marine population from Banyuls-sur-Mer has been examined in this study.

4. **Gammarus insensibilis**. This amphipod lives exclusively in brackish water habitats. It is a viviparous organism, with a minimum generation time of ca. 35 days, a long life span (ca. 120 days), and a reproductive period of ca. 100 days. The population we have examined comes from the Lagoon of Venice, a typical brackish water basin with considerable variation in physical and chemical parameters.

5. **Tigriopus brevicornis and T. fulvus**. The copepod **Tigriopus** inhabits the most demanding and severe habitat which can be found in the marine environment: that of the rock- or tide-pools. In **T. fulvus** (a Mediterranean species), the reproductive stage is reached in about 20 days, a period slightly longer than that of **Tisbe**. The ability to reproduce is maintained for at least three months, and the mean longevity is the greatest (ca. 210 days) among all the species investigated in the present work. Its physiological adaptability is extraordinary. The species can live in conditions ranging from almost fresh-water up to salinity oversaturation; this made possible through a sort of dormancy from which the animals recover when conditions become less severe (Issel, 1914). Similar life-cycle properties and physiological adaptabilities have recently been described for an Atlantic species, **T. brevicornis** (Bisol et al., 1977b). Two species of **Tigriopus** were studied: **T. fulvus**, from Leghorn, Italy and **T. brevicornis**, from Tavvallich, Scotland, where the environmental fluctuations, although still quite high, are much less marked than in the Mediterranean rock-pools.

Results and Discussion

The electrophoretic data are summarized in Table 2. The genetics of some systems has been verified by breeding; similar tests are being applied to each of the other systems.

The degree of heterozygosity is expressed, according to Nei (1975), in the form of 'gene diversity'. This index is based on expected values and therefore does not consider the effects of factors, such as inbreeding, which might disturb equilibria. For the purposes of the present study, these factors may probably legitimately be ignored for the moment.

In most cases, the observed distributions of genotypes fit Hardy-Weinberg expectations. In the few cases where significant deviations were found, they almost always take the form of heterozygote deficiency. The causal aspects of these deviations will be discussed elsewhere.

For most of the Tisbe species studied, genetic variability has possibly been underestimated. This applies particularly to T. biminiensis, for which we cannot ignore the effects of sampling errors. However, this underestimation should not affect the results of our comparison, as the species which exhibit the lower values of genetic variation (Gammarus, Tigriopus fulvus) are those (1) whose natural populations have larger sizes, and (2) where electrophoretic analysis was applied directly to specimens collected from the natural environment.

The only legitimate generalization to be made from the data in Table 2, is that genetic variation in the copepods and the amphipod tested decreases with the temporal constancy of the environment. This trend appears more clearly in Table 3, where the data are pooled according to the marine, brackish and rock-pools nature of the environments of the various populations sampled. Such an order of habitats is characterized by a progressive increase in temporal fluctuation of physical ecological parameters and a parallel decrease in spatial heterogeneity or diversification of living communities.

In other words, the more exacting the environment, the greater the chances of a strategy based upon individual flexibility rather than on genetic plasticity. When this hypothesis was first formulated (Battaglia, 1959), it was not possible to decide whether the increase of phenotypical flexibility was to be ascribed to genetic homeostasis (sensu Lerner) or to fixation of 'generalist' alleles. For Tigriopus, we can now state that the second alternative is likely to be the correct one.

It cannot be denied that Tigriopus populations may often go through severe bottlenecks. In order to measure the extent to which the founder effect or the genetic drift might account for the low genetic variability observed in T. fulvus, we have compared samples from several rock-pools of various sizes, situated at different levels above the sea, or at the same level but at various distances from each other. In

TABLE 3

Mean genetic variation in species in three different environments.

Environment and Species Parameter	Marine Tisbe clodiensis Tisbe holothuriae* Tisbe biminiensis	Brackish-water Tisbe holothuriae** Gammarus insensibilis	Rock-pools Tigriopus brevicornis Tigriopus fulvus
Mean number of alleles per locus	1.982	1.788	1.297
Mean percentage of loci polymorphic (0.99)	56.84	42.03	18.27
Mean percentage of loci polymorphic (0.95)	46.78	38.89	16.13
H	0.196	0.151	0.063

* Banyuls-sur-Mer
** Sigean

each of the pools, the population size was of the order of many thousands of individuals. Since no appreciable differences in genetic structure were found between these samples these data have been pooled. Further confirmation of the negligible role of drift as a factor responsible for the low genetic variability detected in <u>Tigriopus</u>, emerges from the following observation. The same rock-pools in the Leghorn area where the first samples of a <u>T. fulvus</u> were collected, in early autumn, were sampled more than a year later and in a different season (late winter). The results indicate that the genetic structure of the population remained practically unaltered. On the same occasion samples were collected from other pools, situated a few miles away from the first ones. In this case, moreover, where individuals collected from the pools were electrophoresed directly, substantially the same picture was obtained. More detailed information on these experiments will be provided elsewhere.

A similar difference of genetic variation between marine and brackish water habitats, can be inferred at the intraspecific level also by the data on genetic load. The comparison involves six populations of the copepod <u>Tisbe holothuriae</u>, samples in various localities. The first two populations compared were those from Banyuls-sur-Mer and from Sigean. The values of $\underline{b}$ were, 1.897 and 1.067 respectively. Further comparisons concerned two populations from the Lagoon of Venice (Alberoni and Malamocco), one population from Split, and one from Grado. For these four populations the values of $\underline{b}$ were, 1.779, 1.244, 2.265 and 1.065 respectively. As the station 'Alberoni' is somewhat intermediate between a typically marine and a typically brackish water habitat, and summarising the data in a figure in which $\underline{b}$ is plotted against the type of environments (Fig.2), it is apparent that the degree of concealed genetic variation is lowest for populations inhabiting ecologically and geographically marginal habitats, and increases as the habitat becomes marine, attaining its maximum values in typical marine conditions. This trend agrees well with the isozyme data obtained with the populations from Banyuls and Sigean and is indirect evidence that parts of the genome which may be different and are assayed in different ways respond in substantially the same way.

Electrophoretic data from the Split, Alberoni, Malamocco, and Grado populations are not yet available.

In comparing the populations of <u>Tisbe holothuriae</u> from Banyuls and Sigean, we found that they differed not only in the degree of genetic variation but also in the relative frequencies of some alleles. The most significant differences

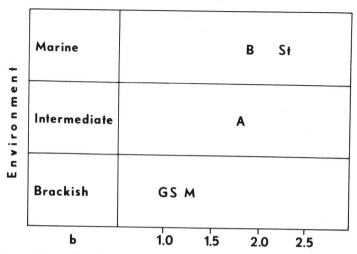

Figure 2. Correlation between "concealed genetic variability" and environmental stability in Tisbe holothuriae. The letters indicate the various localities of sampling. B = Banyuls-sur-Mer; St = Split; A = Alberoni (Lagoon of Venice); M = Malamocco (Lagoon of Venice); S = Sigean; G = Grado.

are reported in Table 4. Studies are now being carried out to investigate the possible selective meaning of these differences.

TABLE 4

Local distribution of some allelic frequencies in Tisbe holothuriae.

Locus	Alleles	Banyuls	Sigean
Aph-1	b	0.897	0.671
Adh-1	b	0.730	0.431
Me-1	a	0.300	0.728

It is likely that the trend towards lower genetic variability in those marine habitats characterized by wide ecological fluctuations, reflects a general condition. In fact,

this tendency has emerged from research carried out by means
of different techniques and methods on a number of species
and populations. The species studied are phylogenetically
close to each other, representative of the respective habitats,
well known as to their ecology and life histories, and easy
to rear in the laboratory. The latter fact has helped considerably in the interpretation of results. In particular,
the knowledge of the biological features of the species studied
had made it possible to ascertain the ways in which each species
perceives the environmental grain, which gives a clue to the
understanding of how natural selection has built up certain
adaptive devices.

In fact, life histories, including parameters such as
geographic distribution and larval biology, may be largely
responsible for the choice of adaptive strategies. Let us
briefly mention, in this respect, the case of the copepods
Tisbe holothuriae and Tigriopus fulvus, the organisms exhibiting the highest and the lowest genetic variability,
respectively. Of all the species considered in the present
work, T. holothuriae has the shortest generation time, the
widest geographic and ecological distribution, and the least
barrier to gene flow between its populations.

As shown by the studies on concealed genetic variation,
by the results of research on physiological racial differentiation (Battaglia, 1970) and those concerning the diversification of gene frequencies in different habitats, the adaptive
strategy of Tisbe holothuriae must be based upon a multiplicity
of mechanisms, among which individual flexibility and genotypic
plasticity are likely to play the major role. This strategy
is probably the optimal compromise for the species to conquer
a great variety of ecological niches, spatial as well as temporal, physical as well as biotic. If a cost has to be imposed for a strategy of this kind, it is perhaps the one deriving from a higher genetic load of ecological origin.
In Tigriopus, however, the prevailing strategy consists in
the fixation of alleles which confer the highest possible
individual flexibility in environments which are extremely
challenging in their short term fluctuations of physical
characters. The price which the species has to pay for such
a strategy might be its low tolerance of biotic diversity,
that is, in a lower competitive ability. As a matter of
fact, the species can hardly be found in habitats other than
rock-pools, and laboratory experiments show that among all
the harpacticoid copepods tested for this character Tigriopus
is the poorest competitor (unpublished data).

The finding of low genetic variability in Tigriopus agrees
with the view, shared by several authors, according to which

in challenging habitats little genetic variation may be permitted, natural selection acting in favour of a few genotypes underlying flexible phenotypes.

In conclusion, any attempt to correlate genetic variation with the degree of environmental diversification in time and space, should be supported by an adequate set of information on the properties of the environment and, above all, on the most significant biological characteristics of the species considered. These characteristics largely affect the way in which the species perceives the environmental grain, which is intimately related in the choice of the adaptive strategy. In the absence of information of this sort, wide-ranging generalizations are liable to be incorrect or inappropriate. The controversy surrounding a number of recent hypotheses and theories on this matter, arises in part from an insufficiency of such relevant knowledge.

Summary

This paper considers the problem of relationships between genetic variability and heterogeneity of ecological variables in marine environments.

Species in three types of habitat are compared using, as measures of genetic variation, electrophoretic assay and estimates of genetic load. The environments studied were marine, brackish water and rock-pools representing a series ranging from conditions of relative physicochemical stability and biotic diversity to conditions of relative physicochemical instability and biotic homogeneity.

The species used are all crustaceans (Amphipods and Copepods) and can be readily bred and maintained in laboratory conditions. The comparisons include geographical populations of the same species, species of the same genus and different genera.

The results indicate that, using either measure, genetic variation is highest in the populations of marine habitats, intermediate in brackish water populations and lowest in rock-pools. This suggests that, for these species, temporal environmental inconstancy tends to lead to relatively lower levels of genetic diversity.

Acknowledgements

This work was supported by the National Research Council of Italy (C.N.R.). For help in obtaining material and in providing facilities we wish to thank the Directors and staff of the following laboratories:- Laboratoire 'Arago', Banyuls-sur-Mer; Dunstaffnage Marine Research Laboratory, Oban; H. Steinitz Marine Laboratory, Eilat; Centro Interuniversitario di Biologia Marina, Livorno.

REFERENCES

Ayala, F. J., Powell, J. R., Tracey, M. L. Murão, C. A. and Pérez-Salas, S. 1972. Enzyme variability in the Drosophila willistoni group. IV. Genic variation in natural populations of Drosophila willistoni. Genetics 70, 113-139.

Bacci, G. 1954. Alcuni rilievi sulle faune di acque salmastre. Pubbl. Staz. Zool. Napoli 25, 3-19.

Battaglia, B. 1954. Note sulla variabilità geografica di alcuni copepodi bentonici marine. Confronto biometrico di popolazioni diverse e osservazioni sul policromatismo. Atti IX Congr. Int. Genetica, Caryologia, suppl. Vol. 6, 770-774.

Battaglia, B. 1958. Balanced polymorphism in Tisbe reticulata, a marine Copepod. Evolution 12, 358-364.

Battaglia, B. 1959. Il polimorfismo adattativo e i fattori della selezione nel Copepode Tisbe reticulata Bocquet. Archo Oceanogr. Limnol. 11, 19-69.

Battaglia, B. 1961. Problemi di adattamento di popolazioni nell'ambiente marino. Boll. Zool. 28, 125-173.

Battaglia, B. 1965. Advances and problems of ecological genetics in marine animals. Proc. XI Intern. Congr. of Genetics, Pergamon Press, 451-463.

Battaglia, B. 1970. Cultivation of marine copepods for genetics and evolutionary research. Helgolander wiss. Meeresunters, 20, 385-392.

Battaglia, B. and Bisol, P. M. 1973. Polimorfismi enzimatici in Gammarus insensibilis della laguna veneta. Atti Ist. Ven. SS. LL. AA. 101, 441-448.

Battaglia, B. and Volkmann-Rocco, B. 1973. Geographic and reproductive isolation in the marine harpacticoid copepod *Tisbe*. Mar. Biol. 19, 156-160.

Bisol, P. M. 1976. Polimorfismi enzimatici ed affinità tassonomiche in *Tisbe* (Copepoda, Harpacticoida). Atti Accad. Naz. Lincei 60, 864-870.

Bisol, P. M., Varotto, V. and Battaglia, B. 1976. Controllo genetico della fosfoesosoisomerasi (*Phi*) in *Tisbe clodiensis* (Copepoda, Harpacticoida). Atti Accad, Naz. Lincei 60, 499-504.

Bisol, P. M., Battaglia, B. and Bovo, G. 1977a. Polimorfismi enzimatici in *Gammarus insensibilis.* II. Controlle genetico e caratterizzazione delle esterasi. Atti Accad. Naz. Lincei 62, 261-266.

Bisol, P. M., Renier, M., Tombolan, E. and Varotto, V. 1977b. Ciclo biologico di *Tigriopus brevicornis* a diverse temperature e salinità. Atti IX Congr. Naz. S.I.B.M., Lacco Ameno 19-22 Maggio 1977, in press.

Brewer, G. J. 1970. *Introduction to Isoenzyme Techniques.* Academic Press, New York, 186 pp.

Fava, G., Lazzaretto-Colombera, I. and Carvelli, M. 1976. Carico genetico in *Tisbe* (Copepoda, Harpacticoida). II *Tisbe holothuriae* e *Tisbe bulbisetosa* della laguna di Venezia. Atti Accad. Naz. Lincei 60, 699-708.

Issel, R. 1914. Vita latente per concentrazione dell'acqua (anabiosi osmotica) e biologia delle pozze di scogliera. Pubbl. Staz. Zool. Napoli 22, 191-254.

Lazzaretto-Colombera, I., Fava, G. and Gradenigo-Denes, M. 1976. Carico genetico in *Tisbe* (Copepoda, Harpacticoida) I. *Tisbe holothuriae* di due popolazioni del Mediterraneo occidentale. Acc. Naz. Lincei 60, 691-698.

Levinton, J. S. 1975. Levels of genetic polymorphism at two enzyme encoding loci in eight species of the genus *Macoma* (Mollusca: Bivalvia). Mar. Biol. 33, 41-48.

Morton, N. E., Crow, J. F. and Muller, H. J. 1956. An estimate of the mutational damage in man from data on consanguineous marriages. Proc. Nat. Acad. Sci., USA. 42, 855-863.

Nei, M. 1975. Molecular Population Genetics and Evolution. North-Holland/American Elsevier, Amsterdam 288 pp.

Sexton, E. W. and Clark, A. R. 1936. A summary of the work on the Amphipod Gammarus chevreuxi Sexton carried out at the Plymouth Laboratory. J. Mar. Biol. Assoc. U.K. 21, 357-414.

Somero, G. N. and Soulé, M. 1974. Genetic variation in marine fishes as test of the niche-variation hypothesis. Nature 249, 670-672.

Shaw, C. R. and Prasad, R. 1970. Starch gel electrophoresis of enzymes. A compilation of recipes. Biochem. Genet. 4, 297-320.

Valentine, J. W. and Ayala, F. J. 1976. Genetic variability in krill. Proc. Nat. Acad. Sci. U.S.A. 73, 658-660.

Wallace, B. 1970. Genetic Load. Prentice-Hall, Inc. Englewood Cliffs. N. J. 116 pp.

Wilkins, N. P. 1977. Genetic variability in littoral gastropods: Phosphoglucose isomerase and phosphoglucomutase in Patella vulgata and P. aspersa. Mar. Biol. 40, 151-156.

PROTEIN VARIATION IN THE PLAICE, DAB AND FLOUNDER, AND THEIR

GENETIC RELATIONSHIPS

R. D. Ward and R. A. Galleguillos

Dept. of Genetics, University College of Swansea

Swansea, United Kingdom.

Introduction

Over the last ten years or so, the use of electrophoretic techniques in screening variation at the protein level has shown that natural populations of most plant and animal species are genetically highly variable. Whether or not this variation is of adaptive significance to the species remains a highly controversial topic, and an unequivocal resolution of this question seems no nearer now than ten years ago. The opposing arguments are discussed by Lewontin (1974), but for a more recent statement of the neutralist position, Ohta (1974) may be consulted. The discovery of this variation has led to other interesting questions, such as whether different functional or structural classes of enzymes show different average heterozygosities, and whether evolutionary trees derived from a comparison of enzyme patterns correspond well with those derived from morphological and anatomical comparisons.

We have been investigating some of these questions through a study of protein variation in three species of marine flatfish; the dab, the plaice and the flounder. These species are abundant in British waters. Originally they (together with all other then recognised flatfish) were grouped into the single genus Pleuronectes by Linnaeus, and a few books still use this terminology, but most European taxonomists today split them into three genera and designate the species Limanda limanda, Pleuronectes platessa and Platichthys flesus respectively. This is the convention used throughout this

article. Russian taxonomists put the plaice into the genus
Platichthys and the flounder into the genus Pleuronectes.
The three species belong to the family Pleuronectidae and to
the order Pleuronectiformes (Heterosomata). Thus the single
genus of Linnaeus has been elevated to an order.

In the present paper, we relate enzyme variation in these
three fish species to theories predicting correlations of
heterozygosity with function or structure, particularly quaternary structure (the number of polypeptide chains or subunits
making up the active enzyme molecule). Secondly, bearing in
mind the different ecological niches occupied by the species,
we examine whether overall correlations exist between enzyme
variation and niche similarity. Thirdly, we compare the
evolutionary relationships derived from protein comparisons
with those from morphological and anatomical comparisons.
Much of the work dealing specifically with protein variation
in the plaice, Pleuronectes platessa, has already been published (Beardmore and Ward, 1977; Ward and Beardmore, 1977).

Materials and Methods

Plaice and dab were trawled from Carmarthen Bay (off the
south coast of Wales), and flounder were trawled from the
Loughor estuary (Figure 1). Details of the preparative procedures and electrophoretic techniques are given in Ward and
Beardmore (1977). The designation of enzymes as glucose-
metabolizing or non-glucose metabolizing, and as monomeric,
dimeric or tetrameric, is also given in that paper. The
allele coding for the most common allozyme of a particular
enzyme in the plaice is designated 100, and other alleles are
designated relative to that allele. For example, the enzyme
specified by allele Odh^{75} migrates approximately 75% as far
as that specified by Odh^{100}.

The formal genetics of most of the protein systems have
not been proven, although breeding data does exist for a few
loci (Purdom, Thompson and Dando, 1976; Ward and Beardmore,
1977). Zymogram patterns are similar to those described
for other fish species, genotype distributions at all loci
accord with Hardy-Weinberg expectations, and there is no reason to doubt the genetic interpretations given to the zymogram
patterns.

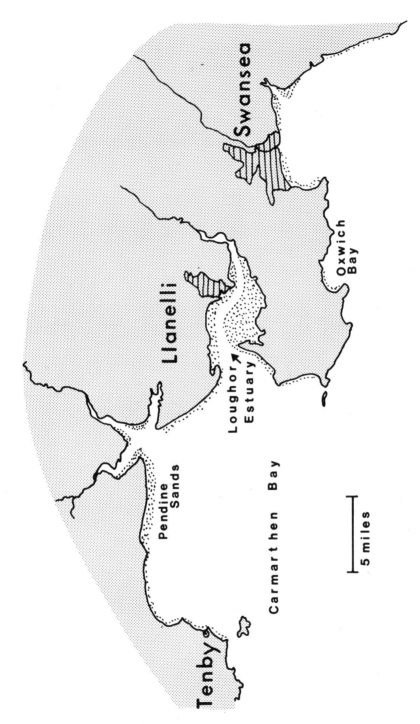

Figure 1. Map showing locations of Carmarthen Bay and the Loughor Estuary.

Results

(1) <u>Amount of protein variation</u>. Tables 1 and 2 give the basic data on the amount of genetic heterogeneity at the 31 protein loci that have been studied in common in the three species. In addition, two further loci have been studied in the dab (alcohol dehydrogenase and alkaline phosphatase - both monomorphic) and fifteen further loci in the plaice (see Ward and Beardmore, 1977).

Table 3 summarizes the extent of protein variation in these species. It can be seen that in the plaice an individual fish is expected to be heterozygous at approximately 10% of its loci, and that, using the definition of polymorphism which depends upon the frequency of the most common allele not exceeding 0.99, approximately 50% of gene loci are polymorphic in the Bristol Channel population. However, there are two major reasons why such estimates should be viewed with some caution. Firstly, they assume that the loci screened are representative of the genome as a whole. This is unlikely to be true. So far only a few score of protein loci can be screened out of the tens of thousands that higher organisms are thought to possess, and in this small sample loci coding for the membrane-bound proteins and the structural proteins are underrepresented. Even so, it is apparent that average heterozygosities vary from locus to locus. Secondly, electrophoresis only detects a fraction of all possible mutations, a fraction commonly held to be about 0.30. Substitutions not altering the electrostatic charge on the molecule are not detected. This consideration implies that we may be seriously underestimating the total amount of genetic variation, and indeed recent studies using heat treatment and other techniques have been picking up enzyme variants not detected by standard electrophoretic methods (Bernstein, Throckmorton and Hubby, 1973; Singh, Lewontin and Felton, 1976; Trippa, Loverre and Catamo, 1976). The dab and flounder appear to be a little less variable than the plaice, but confirmation of this requires that many more loci be screened.

(2) <u>Enzyme function and heterozygosity</u>. Soon after electrophoresis became widely used as a method for screening genetic variation, it became clear that certain enzymes were on average more variable than others. On the basis of their work with flies of the genus <u>Drosophila</u>, Gillespie and Kojima (1968; Kojima, Gillespie and Tobari, 1970) proposed that enzyme heterozygosity was related to function. It was suggested that those enzymes involved with glucose metabolism (catalysing steps in, or adjacent to, the glycolytic pathway and the tricarboxylic acid cycle) were on average less variable than

PROTEIN VARIATION AND GENETIC RELATIONSHIPS

TABLE 1
Sample sizes and average heterozygosities per locus

Locus	Plaice N	Plaice Het	Dab N	Dab Het	Flounder N	Flounder Het
Odh	573	0.056	74	0.000	63	0.111
αGpdh-1	2275	0.234	150	0.066	20	0.500
αGpdh-2	294	0.466	90	0.000	47	0.000
Sdh	320	0.431	89	0.022	60	0.000
Ldh-1	654	0.000	128	0.000	49	0.000
Ldh-2	418	0.024	32	0.000	49	0.082
Ldh-3	401	0.000	22	0.000	49	0.000
Ldh-4	58	0.000	22	0.000	49	0.000
Mdh-1	2202	0.006	143	0.000	40	0.000
Mdh-2	2275	0.216	145	0.098	40	0.000
Idh-1	949	0.573	103	0.000	40	0.025
Idh-2	742	0.026	66	0.000	49	0.041
6Pgdh	2114	0.047	130	0.000	35	0.000
G3pdh-2	353	0.000	15	0.000	33	0.000
Sod	1058	0.010	93	0.022	60	0.000
Ck-1	55	0.000	15	0.000	47	0.021
Ck-2	55	0.000	15	0.000	47	0.000
Ak	70	0.000	47	0.000	47	0.000
Pgm-1	2273	0.483	123	0.463	40	0.425
Est-2	1039	0.024	55	0.655	34	0.353
Est-4	953	0.058	43	0.000	36	0.389
Est-D	241	0.012	118	0.000	21	0.000
Lap	58	0.052	42	0.000	20	0.000
Ap-1	412	0.000	20	0.000	36	0.000
Ap-2	237	0.000	48	0.000	36	0.000
Ada	2272	0.418	117	0.538	40	0.300
Pgi-1	2270	0.058	130	0.000	40	0.000
Pgi-2	2275	0.110	59	0.508	40	0.425
Prot-3	67	0.060	40	0.000	40	0.000
Prot-4	52	0.000	40	0.000	40	0.000
Prot-5	40	0.000	40	0.000	40	0.000

Legend to Table 1

Abbreviations

N - number of individuals sampled per locus. Het-observed heterozygosity. Odh - octanol dehydrogenase. αGpdh - αglycerophosphate dehydrogenase. Sdh - sorbitol dehydrogenase. Ldh - lactate dehydrogenase. Mdh - malate dehydrogenase. Idh - isocitrate dehydrogenase. 6Pgdh - 6 - phosphogluconate dehydrogenase. G3pdh - glyceraldehyde-3-phosphate dehydrogenase. Sod - superoxide dismutase (tetrazolium oxidase). Ck - creatine kinase. Ak - adenylate kinase. Pgm - phosphoglucomutase. Est - esterase. Lap - Leucine aminopeptidase. Ap - aminopeptidase. Ada - adenosine deaminase. Pgi - phosphoglucose isomerase. Prot - non-enzymic protein.

TABLE 2

Allele frequencies in the three species

Locus	Allele	Plaice	Dab	Flounder
Odh	119	0.020	-	0.064
	100	0.980	-	0.929
	75	-	1.000	0.007
αGpdh-1	135	0.060	-	-
	100	0.870	-	-
	73	0.070	-	-
	45	-	-	0.390
	40	-	0.966	0.580
	20	-	0.034	0.030
αGpdh-2	115	0.360	-	-
	100	0.605	1.000	1.000
	79	0.035	-	-
Sdh	137	0.342	0.011	-
	100	0.655	-	-
	99	-	0.989	1.000
Ldh-1	165	-	1.000	-
	100	1.000	-	1.000
Ldh-2	120	-	-	0.021
	100	0.990	-	0.959
	75	0.110	1.000	0.020
Ldh-3	100	1.000	1.000	-
	94	-	-	1.000
Ldh-4	100	1.000	1.000	-
	97	-	-	1.000
Mdh-1	100	1.000	1.000	1.000
Mdh-2	100	0.875	-	-
	95	-	0.013	-
	77	0.125	-	1.000
	75	-	0.952	-
	50	-	0.035	-
Idh-1	118	0.105	-	0.020
	100	0.510	-	0.980
	90	-	1.000	-
	81	0.375	-	-
Idh-2	100	0.985	-	0.987
	88	0.015	1.000	0.013

Table 2 (continued)

Locus	Allele	Plaice	Dab	Flounder
6Pgdh	121	-	1.000	-
	117	0.020	-	-
	100	0.975	-	-
	86	0.005	-	-
	76	-	-	1.000
G3pdh-2	100	1.000	1.000	1.000
Sod	220	-	0.011	-
	140	-	0.989	-
	100	1.000	-	1.000
Ck-1	100	1.000	-	-
	96	-	1.000	-
	95	-	-	0.989
	80	-	-	0.011
Ck-2	100	1.000	1.000	1.000
Ak	140	-	1.000	-
	100	1.000	-	1.000
Pgm-1	228	-	0.024	-
	186	-	-	0.025
	160	-	0.695	-
	140	0.375	-	0.750
	120	-	0.281	-
	100	0.625	-	0.225
Est-2	121	-	0.050	-
	115	-	0.100	-
	103	-	0.450	-
	102	-	0.290	0.235
	100	1.000	0.090	0.765
	97	-	0.020	-
Est-4	103	0.005	-	-
	100	0.970	-	0.778
	97	0.025	1.000	0.083
	92	-	-	0.139
Est-D	120	-	1.000	-
	100	1.000	-	1.000
Lap	100	0.974	1.000	1.000
	95	0.026	-	-
Ap-1	100	1.000	1.000	1.000
Ap-2	100	1.000	1.000	-
	93	-	-	1.000
Ada	118	-	0.017	-
	112	-	0.542	-
	108	0.110	-	-
	104	-	0.342	-
	100	0.745	-	0.787
	98	-	0.090	-

Table 2 (continued)

Locus	Allele	Plaice	Dab	Flounder
	91	0.140	–	0.213
	84	–	0.009	–
	80	0.005	–	–
Pgi-1	100	0.975	–	1.000
	13	0.025	1.000	–
Pgi-2	110	–	–	0.738
	104	0.040	0.246	–
	100	0.945	0.746	0.262
	90	0.015	0.008	–
Prot-3	100	0.970	–	–
	94	0.030	–	–
	90	–	–	1.000
	75	–	1.000	–
Prot-4	100	1.000	–	1.000
	92	–	1.000	–
Prot-5	100	1.000	1.000	–
	86	–	–	1.000

As can be seen from Table 1, sample sizes for many loci in P. platessa are large, and consequently more rare alleles have been detected in this species. In order to remove this possible source of bias, allele frequencies for this species are here re-estimated from a random sample of approximately 100 individuals.

TABLE 3

Levels of protein variation

	No. Loci	Mean Heterozygosity per locus ± SE	Proportion of Loci Polymorphic	
			A	B
plaice	46	0.102 ± 0.026 (0.108 ± 0.031)	0.48 (0.52)	0.26 (0.26)
dab	33	0.072 ± 0.031 (0.077 ± 0.033)	0.24 (0.26)	0.12 (0.13)
flounder	31	0.086 ± 0.029	0.33	0.21

Figures in parenthesis for plaice and dab are for those loci scored in common with the flounder. Proportion of loci polymorphic: A and B give figures for those loci where the most common allele is at a frequency > 0.99 and 0.95 respectively.

those involved in other reactions. The latter class of enzyme contained many that were relatively non-specific with respect to substrate (e.g. esterases and phosphatases), and furthermore many of the substrates of these enzymes originated from the environment. It was concluded that substrate heterogeneity was reflected in enzyme heterogeneity. Subsequent work in a variety of taxa has supported this relationship (see reviews by Powell, 1975, and Selander, 1976), although it appears that in vertebrates taken as a whole the non-glucose enzymes are only marginally more variable than the glucose enzymes (Ward, 1977). Table 4 shows that of the three species considered here, two show no apparent differences in the relative heterozygosities of the two enzyme groups, whilst in P. platessa, the glucose enzymes are substantially more variable than the non-glucose enzymes.

(3) Enzyme structure and heterozygosity. It has been suggested that the hybrid molecules (heteromultimers) formed by individuals heterozygous for multimeric enzymes (those composed of more than one subunit) may provide a basis for heterosis: the hybrid molecule is then assumed to possess unique properties which make the carrier adaptively superior to the homozygotes, and the balance of selective forces would then maintain a stable polymorphism at this locus. The hypothesis predicts that as a result multimers should show more polymorphism than monomers. Zouros (1976) tested this prediction but found that multimers were in fact less variable than monomers; the hypothesis is not supported by the evidence.

Such results are predicted by a different theory. Amino acid sequences of certain regions of enzyme molecules are highly conserved from species to species. These regions include the active site, sites for binding regulatory molecules, and sites of critical importance in maintaining tertiary and quaternary structure: amino acid substitutions in these regions are likely to be deleterious and will therefore be selected against. These ideas have been considered in greater detail by Zuckerkandl (1976a and b). It is not easy to measure the degree to which different enzymes are subjected to different structural constraints, but one component that can be quantified in an approximate manner is determined by the need to maintain a particular quaternary structure. The more subunits make up the active enzyme, the greater the degree of structural constraint. Furthermore, multimeric enzymes will have a higher volume/surface ratio and therefore a higher proportion of hydrophobic amino acids than monomeric enzymes. Variation in such amino acids is likely to go undetected by electrophoresis. Thus monomeric enzymes are expected to be more variable than dimeric forms, which in turn

TABLE 4

The Relationship between Enzyme Heterozygosity, Function and Quaternary Structure.

Enzyme Class		Plaice	Dab	Flounder
Glucose metabolizing	n	20	15	15
	het	0.169	0.065	0.095
Non-Glucose metabolizing	n	18	15	13
	het	0.062	0.088	0.094
Monomer	n	10	6	6
	het	0.164	0.286	0.246
Dimer	n	19	16	14
	het	0.132	0.035	0.079
Tetramer	n	8	6	6
	het	0.057	0.004	0.013

The quaternary structures of the enzymes coded for by the two aminopeptidase (AP) loci are unknown. The function of an enzyme in the plaice termed "nothing dehydrogenase" is unknown.

should be more variable than tetrameric enzymes of similar subunit chain length. These expectations were confirmed by an analysis of available data on protein variation in both invertebrates and vertebrates by Ward (1977) and for man by Harris, Hopkinson and Edwards (1977). It can be seen from Table 4 that data from the three flatfish species are also in accord with this theory.

(4) *Enzyme variation and ecological variation.* If enzyme variation is adaptively significant, species occupying similar environments may be more similar to each other in their genetic variation than they are to species occupying different habitats. That is associations may exist between variability at specific loci and environmental variables. Such ideas are clearly related to, but are different from, those proposing correlations between heterozygosity *per se* and the environment.

The ecological niches of the plaice and dab are similar. The two species coexist over large areas and are trawled together in large numbers from Carmarthen Bay. Furthermore they have very similar diets, and in many places the dab is a serious food competitor of the commercially more valuable plaice. On the other hand, the flounder is to be found in large numbers in estuaries or close inshore, and in our study area specimens are only very rarely caught when trawling for dab and plaice.

The model predicts a higher correlation coefficient of heterozygosity over all 28 enzyme loci for the plaice-dab pair than for the plaice-flounder or dab-flounder pairs. Parametric tests are not appropriate in testing the hypothesis, the heterozygosity values being far from normally distributed, and we have chosen to use Kendall's coefficient of rank correlation. The correlation coefficient, r, has a range -1 (total inverse correlation) through 0 (no correlation) to +1 (total correlation). Results of such an analysis are set out in Table 5. The precision of the test is reduced by the many zero heterozygosity values leading to an abundance of ties, but it is clear that the hypothesis is not supported. The highest correlation is between the dab and the flounder, which gives a value significantly greater ($0.05 > P > 0.02$) than either the plaice-dab or plaice-flounder pairs. The significance of these results is not clear. It may be that the high correlation between dab and flounder is to some extent a sampling artefact, or it may be that there is some important aspect in their biology common to each but different from the plaice. The coefficient of rank correlation is significantly greater than 0 for each species pair. This probably results from the general observation that some enzymes are more likely to be polymorphic than other (sections 2 and 3 above), or possibly from the fact that all three species are flatfishes, closely related, and share broadly similar life histories.

(5) Genetic relationships between the three species. The allele frequency data given in Table 2 can be used to estimate the degree of divergence of the three species. The rationale is simple: the more closely are a pair of species related, the more similar will be their allelic distributions. The usefulness of such approaches in tackling systematic problems is now widely recognised, and has been reviewed by Avise (1974). Phylogenetic trees derived from protein analysis are generally similar to those derived from examination of morphological and anatomical characteristics. There are several formulae that can be used to estimate genetic similarities over a number of loci, but that described by Nei (1972) has gained most widespread acceptance in recent years.

TABLE 5

Kendall's coefficients of rank correlation, r, for the locus heterozygosities of the three paired species' comparisons.

	plaice dab	plaice flounder	dab flounder
r	0.326	0.346	0.756
P	0.015	0.010	$\approx 2 \times 10^{-8}$

P is the probability of fit to the null hypothesis that there is no significant correlation. The same 28 enzyme loci were used in each species pair comparison.

It is calculated as follows:-

The normalized genetic identity of genes between two populations at the j locus is defined as

$$I_j = (\Sigma x_i y_i)/\sqrt{\Sigma x_i^2 \, \Sigma y_i^2}$$

where x_i and y_i are the frequencies of the i allele in populations X and Y respectively. The normalized identity of genes when all loci are considered is defined as

$$I = Jxy/\sqrt{JxJy}$$

where Jx, Jy and Jxy are the arithmetic means over all loci of Σx_i^2, Σy_i^2 and $\Sigma x_i y_i$ respectively. I values range from 0 (no alleles shared between the two populations at any locus) to 1 (all alleles at all loci present in equal frequencies in the two populations). The genetic distance between X and Y is then defined as

$$D = -\log_e I$$

D estimates the accumulated number of codon substitutions per locus since the time of divergence of the two populations. An approximate measure of the standard error of D is given by Nei (1971) as $\sqrt{(1 - I)} / In$, where n is the number of proteins examined. This is the measure used here. A more precise but more complex formula is given by Nei and Roychoudhury (1974a), but the two estimates in fact give similar values except at high or very low values of $\underline{D}$.

Estimates of $\underline{I}$ and $\underline{D}$ for the three species of flatfish are given in Table 6.

If certain assumptions are made, $\underline{D}$ is a measure of divergence time, although the value of the constant required to derive the time scale is a matter of some controversy. Nei and Roychoudhury (1974b; Nei, 1975) hold that a $\underline{D}$ value of 1 is equivalent to approximately 5 million years of divergence, whereas another group of workers believe it to approximate to about 18 million years (Yang, Soulé and Gorman, 1974; Gorman and Kim, 1976; Gorman, Kim and Rubinoff, 1976). A phylogenetic tree based upon the genetic distance data is given in Figure 2, together with the different estimates of evolutionary time.

Discussion

Pleuronectes platessa, *Platichthys flesus* and *Limanda limanda* have average heterozygosities per locus of 0.102 ± 0.026, 0.072 ± 0.031 and 0.086 ± 0.029 respectively, and each of these values is increased if the non-enzymic proteins are excluded from the analysis. These figures are all higher than the average of a range of fish species (0.058 ± 0.006, from the review of Powell, 1975), and are also higher than that of vertebrate species as a whole (0.050 ± 0.004, Powell, 1975). Why this should be so is not clear: it may reflect either sampling error (the particular loci scored not being representative of the gene pool as a whole) or, perhaps more probably, some important stochastic or deterministic aspect of the biology of the species.

The high heterozygosity values we have recorded do not seem to accord with the predictions of Ayala and Valentine (Ayala et al., 1975; Valentine, 1976). They have proposed that in tropical waters, where trophic resource stability is greatest (i.e. the least seasonal), microhabitat heterogeneity is most efficiently exploited by having a variety of genotypes: genetic variability is favoured. On the other hand, where trophic resources are highly seasonal (i.e. in circumpolar waters), one or two "generalist" alleles, able to cope with the temporal fluctuations of the environment, are favoured at each locus: genetic variability is therefore low. Most

PROTEIN VARIATION AND GENETIC RELATIONSHIPS

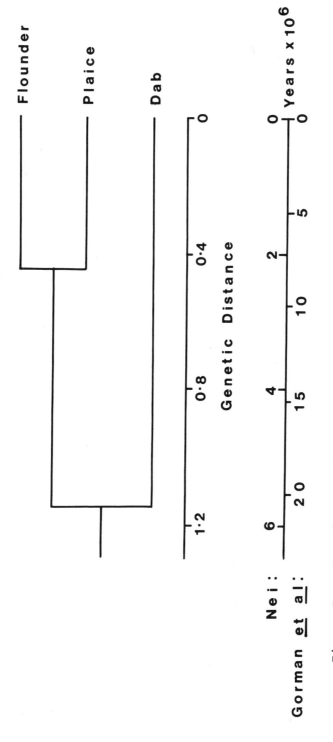

Figure 2. Genetic relationships between the plaice, flounder and dab. Two different estimates of divergence time are given.

TABLE 6

Genetic Identity (I) and Genetic Distance (D) (Nei, 1972), based upon data from 31 protein loci.

	Plaice Flounder	Plaice Dab	Flounder Dab
I	0.637	0.365	0.276
D	0.451 ± 0.136	1.008 ± 0.237	1.287 ± 0.291

The standard errors of D are from Nei (1971).

of the evidence in favour of this hypothesis comes from studies on marine invertebrates, but it should equally apply to vertebrate species. It predicts that given the trophic resource pattern of British waters, species from these waters should have heterozygosity levels approximating to the average of their taxon. This prediction does not agree with our observations, even though it is undoubtedly true that these fish perceive their trophic resource supply to be highly seasonal. Feeding almost ceases during the winter. Of course, other factors undoubtedly influence mean heterozygosity. Population size is an obvious example. Small populations, reflecting the consequences of genetic drift or founder effects, will in general be less variable at the protein level than larger populations. It seems that flatfishes as a group may show higher levels of genetic variation than other fish species. Johnson and Utter (1976) assayed a different range of flatfish caught in the Puget Sound, Washington, U.S.A., and found that they had higher average heterozygosities (0.050, 8 species) than other fish species caught at the same time (0.009, 7 species). The mean heterozygosities of the flatfish species Platichthys stellatus and Kareius bicoloratus have been estimated by Fujio (1977) to be 0.097 and 0.089 respectively.

The Gillespie-Kojima hypothesis (Gillespie and Kojima, 1968; Kojima et al., 1970), which predicts that glucose-metabolizing enzymes should be less variable than the non-glucose metabolizing enzymes, is not supported by our data. The general validity of this model has also been questioned by others, for example Frydenberg and Simonsen (1973) and Band (1975).

Attention has recently focused upon relationships between

enzyme structure and heterozygosity, and, in particular, upon possible correlations between subunit number and heterozygosity (Zouros, 1976; Ward, 1977; Harris et al., 1977). The finding that monomeric enzymes are on average more variable than dimeric enzymes which are themselves more variable than tetrameric forms is further substantiated by the data given here. It should be emphasized that such results do not in any way discriminate between the selectionist and neutralist models of protein variation. Amino acid variation in the less critical, less highly conserved, regions of enzyme molecules may or may not be of adaptive significance. Harris et al. (1977), for man, further found that those multimeric enzymes not forming interlocus hybrid molecules were considerably more variable than those that did form such hybrids. This is not the case for the three species discussed here. Of the 69 loci coding for multimeric enzymes, the 29 not forming interlocus hybrids have a mean heterozygosity of 0.072, whereas the 40 forming hybrids have a value of 0.070. Clearly more work needs to be done in this area. It is also apparent that hypotheses relating function to heterozygosity should be reconsidered in the light of these findings. For example, the "non-regulatory" class of Johnson (1974), which is the least variable of his three functional classes, contains a high proportion of tetrameric enzymes.

The hypothesis that the plaice and dab, by virtue of occupying similar niches, should show more similar patterns of enzyme variation than either does with the flounder, a more estuarine species, is not supported by a test using Kendall's coefficient of rank correlation. Nonetheless, it may be worth applying this type of test in other situations. Providing that homologous loci are being compared, it should be possible to examine the effect of any particular environmental variable on genetic variation in any group of species. Since the test depend upon the rank values of the locus heterozygosity estimates, differences in average heterozygosity between species should not affect its outcome.

The evolutionary relationships between the dab, flounder and plaice have been described by Norman (1934). The dab represents one of the earlier forms of the sub-family Pleuronectinae; the lower pharyngeals are narrow and separate, and each is armed with two sets of conical teeth. In the plaice, the lower pharyngeals are firmly united but the round molar-like teeth retain traces of the primitive biserial arrangement of Limanda. The shape of the lower pharyngeals of the flounder is similar to that of the plaice, but the teeth (molariform) are now scattered in a mosaic fashion. From these anatomical comparisons, the proposed evolutionary trend

is Limanda → Pleuronectes → Platichthys, and this agrees well
with the genetic distance data of Table 6 and the dendrogram
of Figure 2.

Two estimates of evolutionary time are given alongside
Figure 2. There are no a priori grounds for preferring one
to another in this particular case, since no independent estimates of divergence times are available. In the few cases
where geological evidence does provide independent estimates,
the relationship 1.0D = 18×10^6 years seems to be more appropriate (Yang et al., 1974; Gorman et al., 1976). This gives
a divergence time of about 8 million years between the plaice
and flounder, yet reproductive isolation between these species
is still not complete and hybridisation occurs in some areas.
Speciation in marine animals is frequently thought to be a
relatively slow process, due to the comparative lack of geographic barriers between populations, but nevertheless an
estimate of about 2 million years based upon the D/time relationship of Nei (1975) may be preferred. Indeed, it is
probably not realistic to assume a constant relationship
between D and time over all species: some situations may promote more rapid divergence than others. There are also other
problems to be considered. Unless large numbers of loci are
compared, the standard error of D is large. Furthermore,
when D is greater than 1, divergence time may be seriously
underestimated (Nei, 1975).

Electrophoretic comparisons of gene products necessarily
over-estimate similarities between species and under-estimate
differences, since only a proportion of amino acid substitutions can be detected. For example, the Ldh-1 locus of
plaice and flounder produces enzymes that are electrophoretically indistinguishable in the buffer scheme we use, yet
they appear to differ at some point in their amino acid sequence: the flounder enzyme is highly stable at $-20^{\circ}C$ or
$-75^{\circ}C$, whereas the plaice enzyme forms a series of faster
isozymes upon storage. But since such considerations are
already taken into account when relating D to time, they
should not bias the time estimates.

Inspection of the genetic distance data indicates that
the plaice and flounder are more closely related to one another than either is to the dab. This is in accord with the
known facts about the species' ability to hybridize: plaice
and flounder hybridize quite freely in the Danish Belt Sea
and western parts of the Baltic, and artificial crosses give
fertile progeny (see Sick, Frydenberg and Nielsen, 1963),
whereas there seem to be no substantiated cases of plaice x
dab or flounder x dab hybrids from the sea. We have observed

no plaice x flounder hybrids in the Bristol Channel, presumably because here the species spawn in different areas, or perhaps at different times. Nor did any of the individuals we assayed have the genetic constitution expected of plaice x dab or flounder x dab hybrids. The plaice-flounder genetic distance of 0.451 is quite typical of congeneric species, whereas values for the plaice-dab and flounder-dab comparisons, 1.008 and 1.287 respectively, may be more typical of species from different but related genera. However, intrageneric estimates of D vary widely between taxa. This is to be expected since speciation processes may, in principle, be accompanied by much or little genetic change; a topic discussed further in Nei, (1975). Riley and Thacker (1969) conclude their article on the experimental production of dab x flounder hybrids by saying that perhaps a return to Linnaeus' classification of these three species into the single genus Pleuronectes should be considered carefully. We believe that our data support the regrouping of the plaice and the flounder into a single genus, Pleuronectes, but consider that the retention of the dab in a separate genus, Limanda, is justified.

Summary

The plaice, Pleuronectes platessa L., flounder, Platichthys flesus (L.) and dab, Limanda limanda (L.), show average heterozygosities for electrophorectically detectable genetic variation of 10.2% (46 loci), 8.6% (31 loci) and 7.2% (33 loci) respectively. Enzymes involved with glucose metabolism are no less polymorphic than other enzymes. For each species, the average heterozygosity of the monomeric enzymes is higher than that of the dimeric enzymes, which in turn is higher than that of the tetrameric enzymes. A hypothesis predicting that the two species occupying the most similar habitats should be more similar genetically than each should be to the third species is not supported. Genetic distance estimates between the species are calculated and agree with the postulated evolutionary relationships based on anatomical and morphological characteristics. The occurrence of hybrids in certain seas and the close genetic similarity between Pleuronectes platessa and Platichthys flesus suggests that these two species should be regrouped into the genus Pleuronectes.

Acknowledgement

R.A.G. was supported by a grant from the World University Service.

REFERENCES

Avise, J. C. 1974. Systematic value of electrophoretic data. Syst. Zool. 23, 465-481.

Ayala, F. J., Valentine, J. W., Hedgecock, D. and Barr. L. G. 1975. Deep-sea asteroids: high genetic variability in a stable environment. Evolution 29, 203-212.

Band, H. T. 1975. A survey of isozyme polymorphism in a Drosophila melanogaster natural population. Genetics 80, 761-771.

Beardmore, J. A. and Ward, R. D. 1977. Polymorphism, selection, and multi-locus heterozygosity in the plaice, Pleuronectes platessa L. In, Measuring Selection in Natural Populations. Christiansen, F. B. and Fenchel, T. M. (eds). Springer-Verlag, Berlin.

Bernstein, S. C., Throckmorton, L. H. and Hubby, J. L. 1973. Still more genetic variability in natural populations. Proc. Natl. Acad. Sci. U.S.A. 70, 3928-3931.

Frydenberg, O. and Simonsen, V. 1973. Genetics of Zoarces populations. V. Amount of protein polymorphism and degree of genic heterozygosity. Hereditas 75, 221-233.

Fujio, Y. 1977. Natural hybridization between Platichthys stellatus and Kareius bicoloratus. Jap. J. Genet. 52, 117-124.

Gillespie, J. H. and Kojima, K. 1968. The degree of polymorphism in enzymes involved in energy production compared to that in non-specific enzymes in two Drosophila ananassae populations. Proc. Natl. Acad. Sci. U.S.A. 61, 582-585.

Gorman, G. C. and Kim, Y. J. 1976. Anolis lizards of the eastern Caribbean: a case study in evolution. II. Genetic relationships and genetic variation of the bimaculatus group. Syst. Zool. 25, 62-77.

Gorman, G. C., Kim, Y. J. and Rubinoff, R. 1976. Genetic relationships of three species of Bathygobius from the Atlantic and Pacific sides of Panama. Copeia 1976, 361-364.

Harris, H., Hopkinson, D. A. and Edwards, Y. H. 1977.

Polymorphism and the subunit structure of enzymes: A contribution to the neutralist-selectionist controversy. Proc. Natl. Acad. Sci. U.S.A. 74, 698-701.

Johnson, A. G. and Utter, F. M. 1976. Electrophoretic variation in intertidal and subtidal organisms in Puget Sound, Washington Anim. Blood Grps. biochem. Genet. 7, 3-14.

Johnson, G. B. 1974. Enzyme polymorphism and metabolism. Science 184, 28-37.

Kojima, K., Gillespie, J. and Tobari, Y. N. 1970. A profile of Drosophila species' enzymes assayed by electrophoresis. 1. Number of alleles, heterozygosities, and linkage disequilibrium in glocose-metabolizing systems and some other enzymes. Biochem. Genet. 4, 627-637.

Lewontin, R. C. 1974. The Genetic Basis of Evolutionary Change. Columbia University Press, New York.

Nei, M. 1971. Interspecific gene differences and evolutionary time estimated from electrophoretic data on protein identity. Amer. Natur. 105, 385-398.

Nei, M. 1972. Genetic distance between populations. Amer. Natur. 106, 283-292.

Nei, M. 1975. Molecular Population Genetics and Evolution. North-Holland Publishing Company, Amsterdam.

Nei, M. and Roychoudhury, A. K. 1974a. Sampling variances of heterozygosity and genetic distance. Genetics 76, 379-390.

Nei, M. and Roychoudhury, A. K. 1974b. Genic variation within and between the three major races of man, Caucasoids, Negroids, and Mongoloids. Amer. J. Hum. Genet. 26, 421-443.

Norman, J. R. 1934. A Systematic Monograph of the Flatfishes (Heterosomata). Vol. 1. Psettodidae, Bothidae, Pleuronectidae. Printed by order of the trustees of the British Museum, London.

Ohta, T. 1974. Mutational pressure as the main cause of molecular evolution and polymorphism. Nature 252, 351-354.

Powell, J. R. 1975. Protein variation in natural populations of animals. Evol. Biol. 8, 79-119.

Purdom, C. E., Thompson, D. and Dando, R. P. 1976. Genetic analysis of enzyme polymorphisms in plaice (Pleuronectes platessa). Heredity 37, 193-206.

Riley, J. D. and Thacker, G. T. 1969. New intergeneric cross within the Pleuronectidae, dab and flounder. Nature 221, 484-486.

Selander, R. K. 1976. Genic variation in natural populations. In, Molecular Evolution. Ayala, F. J. (ed). Sinauer Associates. Inc., Sunderland, Massachusetts.

Sick, K., Frydenberg, O. and Nielsen, J. T. 1963. Haemoglobin patterns of plaice, flounder and their natural and artificial hybrids. Nature 198, 411-412.

Singh, R. S., Lewontin, R. C. and Felton, A. A. 1976. Genetic heterogeneity within electrophoretic "alleles" of xanthine dehydrogenase in Drosophila pseudoobscura. Genetics 84, 609-629.

Trippa, G., Loverre, A. and Catamo, A. 1976. Thermostability studies for investigating non-electrophoretic polymorphic alleles in Drosophila melangaster. Nature 260, 42-44.

Valentine, J. W. 1976. Genetic strategies of adaptation. In, Molecular Evolution. Ayala, F. J. (ed). Sinauer Associates, Inc., Sunderland, Massachusetts.

Ward, R. D. 1977. Relationship between enzyme heterozygosity and quaternary structure. Biochem. Genet. 15, 123-135.

Ward, R. D. and Beardmore, J. A. 1977. Protein variation in the plaice, Pleuronectes platessa L. Genet. Res. 30, 45-62.

Yang, S. Y., Soulé, M. and Gorman, G. C. 1974. Anolis lizards of the eastern Caribean: a case study in evolution. 1. Genetic relationships, phylogeny, and colonization sequence of the roquet group. Syst. Zool. 23, 387-399.

Zouros, E. 1976. Hybrid molecules and the superiority of the heterozygote. Nature 262, 227-229.

Zuckerkandl, E. 1976a. Evolutionary processes and evolutionary noise at the molecular level. I. Functional density in proteins. J. Mol. Evol. 7, 167-183.

Zuckerkandl, E. 1976b. Evolutionary processes and evolutionary noise at the molecular level. II. A selectionist model for random fixations in proteins. J. Mol. Evol. 7, 269-311.

GENETIC VARIABILITY IN DEEP-SEA FISHES OF THE GENUS SEBASTOLOBUS (SCORPAENIDAE)

J. F. Siebenaller

Scripps Institution of Oceanography

La Jolla, California, U.S.A.

Introduction

The relationship of physical environment variation with genetic variability has been viewed as a key to the determination of the adaptive mechanisms by which protein polymorphisms are maintained (Schopf and Gooch, 1971; Nevo and Bar, 1976). Treatments have dealt with the potential importance of spatial heterogeneity (e.g. Levins, 1968; Bryant, 1974; Gillespie, 1974) or temporal variation (e.g. Haldane and Jayakar, 1963; Kojima, 1970; Levinton, 1973, 1975) in maintaining polymorphisms. Other models dealing specifically with the maintenance of protein polymorphisms also stress the variability of the external environment (Gillespie and Langley, 1974; G. B. Johnson, 1974, 1976). Variation in temperature and/or salinities has been found to be correlated with allelic frequency changes (e.g. M. S. Johnson, 1972; Merrit, 1972; Mitton and Koehn, 1975; Koehn et al., 1976). However, such correlations may not identify factors acting directly on the locus which is examined (e.g. Levinton and Fundiller, 1975; G.B. Johnson, 1976).

Attempts to correlate the physical and chemical parameters of the marine environment, such as temperature and hydrostatic pressure, with genetic variation are appealing, since these parameters are known to influence protein structure and function: higher orders of protein structure and stability, rates of enzymic catalysis and enzyme-ligand interactions are sensitive to pressure and temperature changes (Somero and Hochachka, 1976a,b). Because these factors are "felt" by the enzymes of marine ectotherms, these parameters may act as an important selective force on that portion of the genome which can be assayed with electrophoretic techniques.

Consideration of the potential importance of physical factors in maintaining protein polymorphisms has led to predictions that heterogeneous physical environments will maintain larger pools of genetic diversity than more homogeneous environments (Bretsky and Lorenz, 1970; Grassle, 1972; Levinton, 1973; Nevo, 1976). One means of examining the role of protein polymorphisms in adaptation to varying physical environments has been to compare levels of heterozygosity of species from regimes with differing degrees of environmental variability (e.g. Somero and Soulé, 1974; Valentine, 1976).

The deep sea is an environment of highly constant and uniform physical and chemical parameters, placing it at one extreme of a continuum of environmental heterogeneity. Temperature and salinity fluctuations are negligible, and these parameters are uniform over broad expanses. Light does not penetrate to these depths and there are no seasons. The environment has persisted over geological time (Rokop, 1974 and references therein). The hypothesized relationship of physical environment variation with genetic variability and the environmental constancy of the deep sea have led to predictions that populations in this environment would maintain little genetic variability (Manwell and Baker, 1970; Grassle, 1972; Grassle and Sanders, 1973). The application of techniques of gel electrophoresis (as well as other techniques (Schopf, 1976)) has revealed that deep-sea invertebrate populations are far from genetically depauperate. Deep-sea representatives of a number of invertebrate phyla and classes maintain levels of genetic variability which are average or above average for marine invertebrates (Gooch and Schopf, 1972; Ayala and Valentine, 1974; Valentine and Ayala, 1975; Ayala et al., 1975; Murphy et al., 1976).

A second means of assessing the role of environmental heterogeneity in maintaining protein polymorphisms is to examine the patterning of genetic variability with respect to environmental parameters of potential selective importance (e.g. M. S. Johnson, 1972; Merrit, 1972). In the deep sea hydrostatic pressure and, to a lesser extent, temperature are depth related; every 10 m of depth change represents a change of approximately one atmosphere of hydrostatic pressure. Bathyal species may have bathymetric distributions resulting in 100-200 atm of hydrostatic pressure difference between populations (Ekman, 1953).

Few studies of the geographic and depth patterning of genetic variability in deep-sea species have been undertaken. Doyle (1972) reported heterogeneity of genic and genotypic frequencies at a diallelic esterase locus in the cosmopolitan ophiuran Ophiomusium lymani sampled over a 1000 m depth range off the North Carolina (USA) coast. Murphy et al. (1976) detected no population differentiation over 600 m in samples of O. lymani from the Hudson Canyon (USA) area.

No information has been available on the patterns of genetic variability in deep-sea fishes. Bathyal species have far-ranging geographic distributions and broad depth ranges, although the life histories of many of these deep-sea species are poorly known (e.g. Iwamoto, 1975). The life histories of two deep-living species of the genus Sebastolobus have been described (Moser, 1974). These scorpaenid congeners have very similar life histories and geographic distributions, but differ in their bathymetric distributions. The protein differentiation of these species may reflect the extent to which their divergence in depth distributions have exposed that portion of the genomes encoding proteins to divergent selective pressures. What is the patterning of genetic variability in these species, geographically and bathymetrically, in light of their life histories and the distribution of temperature and pressure through their depth ranges? Are there functional differences in the enzymes of these species which may adapt them to their different environments?

Materials and Methods

Two species of the genus Sebastolobus are found in the eastern North Pacific Ocean, S. altivelis and S. alascanus. Sebastolobus alascanus is reported from northern Baja California to the Bering Sea and the Commander Islands off the Asian mainland. Sebastolobus altivelis has been taken from the southern tip of Baja California to the Aleutian Islands (summarized in Moser, 1974). The bathymetric ranges of these species overlap to some degree and the two may occasionally be taken in the same trawls (Hubbs, 1926 and personal observation). The distribution of abundances with depth are such that throughout their latitudinal range S. altivelis always occurs deeper than S. alascanus. Sebastolobus altivelis is common between 550 and 1300 m (known depth range: 200-1550 m). The 352 specimens used in this study were taken at 10 stations in the Southern California Continental Borderland ranging in depth from 463 to 1415 m (Table 1). Sebastolobus alascanus generally occurs from 180 to 440 m (known depth extremes: 18-1524 m). Sixty-three

TABLE 1

Station data for Sebastolobus altivelis and S. alascanus

Station	Date	Depth (m)	Latitude (N)	Longitude (W)	General area	No. of specimens altivelis	alascanus
AFTERTHOT II							
A15	V-6-75	1332	32°32.0' 32°32.7'	118°53.1' 118°53.6'	East Cortez Basin	29	--
A24	V-8-75	1069	32°52.1' 32°52.3'	117°42.1' 117°41.9'	San Diego Trough	13	--
A26	V-9-75	444	33°02.7' 33°02.5'	117°24.5' 117°24.5'	Off Solana Beach, California	--	5
A27	V-9-75	759	32°59.6' 33°00.3'	117°34.1' 117°34.5'	Off Solana Beach, California	73	2
ALLOZYME I							
A28	VIII-6-75	1156	32°57.2' 32°57.5'	118°24.0' 118°24.0'	Santa Catalina Basin	41	--
A29	VIII-6-75	1123	33°04.0' 33°04.3'	118°16.5' 118°17.0'	Santa Catalina Basin	44	--

GENETIC VARIABILITY IN DEEP-SEA FISHES

Table 1 (continued)

A30	VIII-7-75	1415	32°48.8' 119°31.9' 32°49.1' 119°32.25'	Tanner Basin	29	--
A31	VIII-7-75	1341	32°47.4' 119°32.8' 32°48.3' 119°32.9'	Tanner Basin	39	--
A33	VIII-8-75	1202	32°34.7' 117°32.8' 32°35.7' 117°32.4'	San Diego Trough	8	1
AGASSIZ CRUISE						
E1	III-8-76	364– 392	33°28.8' 118°28.3' 33°27.8' 118°26.5'	North of Santa Catalina Island	--	21
E3	III-9-76	541– 560	33°34.0' 118°41.5' 33°33.1' 118°37.5'	North of Santa Catalina Island	47	4
AGASSIZ CRUISE						
B1	V-22-75	229	33°33.7' 118°02.6'	Off Newport Beach, California	--	30
VANTUNA CRUISE						
BLB	XI-19-76	463	33°30.1' 117°47.9' 33°30.5' 117°47.6'	Off Laguna Beach, California	29	--
Total					352	63

specimens from six stations (229-1202 m) were examined (Table 1).
The two species were taken in the same trawls at three stations
(541-1202 m, Table 1).

The life histories of these two deep-water species are
documented on the basis of mid-water trawl collections of the
pelagic juvenile stages and benthic otter trawl samples of the
adults (Moser, 1974). These life histories are similar in
many respects. The adults of both species are benthic.
Both spawn in a 4-5-mo period peaking in April. The planktonic larvae of both species transform into pelagic juveniles
from July to December. Sebastolobus alascanus spends approximately 14-15 mo in the water column from spawning to
settlement; S. altivelis, 20 mo. These two species are morphologically similar. The key characters used in separating
them are the size and shape of the spinous dorsal fin (S.
altivelis: third dorsal spine longest; S. alascanus: fourth
or fifth dorsal spine longest), and the coloration of the
lining of the gill cavity (S. altivelis: dark grey or black;
S. alascanus: pale) (Miller and Lea, 1972). Other meristic
characters separate the species as well (Gilbert, 1915).

Samples of benthic adults of Sebastolobus from the
Southern California Continental Borderland were taken by otter
trawl on five cruises of the R/V Agassiz and the R/V Vantuna
(station data in Table 1). The Southern California Continental Borderland differs from the typical continental shelf in
consisting of a number of bathyal depth fault basins with
sediment-covered bottoms (Emery, 1960). Electrophoretic and
staining procedures are essentially those followed by Somero
and Soulé, (1974). Electrostarch (Electrostarch Co., Madison,
Wisconsin, USA) (12.2% w:v) lots 371 and 307 were used. No
differences in banding patterns of the proteins of the scored
loci were observed in the two lots of starch. Twenty individuals were assayed in each gel; S. alascanus individuals were
always run with S. altivelis individuals for comparison.
Individuals heterozygous or homozygous for rarer alleles at
the PGM locus were rerun to confirm the scoring. The systems
assayed in each tissue, the number of loci scored and the
electrophoretic buffers are given in Table 2.

Results and Discussion

Genetic variability in deep-sea fishes. Of the 20
presumptive gene loci surveyed in both species of Sebastolobus,
10% (2) of the loci were polymorphic at the 0.05 level (the
most common allele less frequent than 0.95). in S. altivelis.
By a less restrictive criterion of polymorphism, i.e. that

TABLE 2

Loci assayed and electrophoretic buffers for Sebastolobus. Buffers are described in Somero and Soulé, 1974.

Enzyme system	Buffer system	No. of loci scored	Locus	Tissue
Acid Phosphatase (ACPH)	Tris-citrate II pH 8.0	1	Acph	Liver
Esterase (EST)	Tris-citrate II pH 8.0	1	Est-1	Liver, Muscle, Eye, Heart
General Protein (GP)	Poulik pH 8.6/8.1	3	Gp-1, Gp-2, Gp-3	Muscle
Glutamate Oxalate Transaminase (GOT)	Tris-maleate pH 7.4	2	Got-1, Got-2	Muscle
α-glycerophosphate dehydrogenase (αGPD)	Tris-maleate pH 7.4	1	αGpd	Muscle
Lactate dehydrogenase (LDH)	Poulik pH 8.6/8.1	3	Ldh-1	Muscle, Heart, Eye
			Ldh-2	Eye
			Ldh-3	Liver
Malate dehydrogenase (MDH)	Tris-maleate pH 7.4	2	Mdh-1	Muscle
			Mdh-2	Muscle, Heart
Malic enzyme (ME)	Tris-citrate II pH 8.0	1	Me	Liver

Table 2 (continued)

Octanol dehydrogenase (ODH)	Poulik pH 8.6/8.1	1	_Odh_	Muscle
Phosphoglucomutase (PGM)	Poulik pH 8.6/8.1	1	_Pgm_	Muscle
Phosphoglucose isomerase (PGI)	Poulik pH 8.6/8.1	2	_Pgi-1_, _Pgi-2_	Muscle
Tetrazolium oxidase (TO)	Poulik pH 8.6/8.1	1	_To_	Liver (Muscle, Eye)
Xanthine dehydrogenase (XDH)	Tris-maleate pH 7.4	1	_Xdh_	Liver

the most common allele be less frequent than 0.99, 30% (6) of the loci were polymorphic. Sebastolobus alascanus was polymorphic at 15% (3) of the 20 loci at the 0.05 level and 20% (4) loci at the 0.01 level (Table 3). The differences in levels of polymorphism at the 0.01 level may simply reflect the much larger sample size of S. altivelis.

The average proportion of loci at which an individual is heterozygous is 0.0466 $\pm$ 0.0449 in S. altivelis and 0.0489 $\pm$ 0.0484 in S. alascanus. Under Hardy-Weinberg equilibrium, pooling all samples, the proportion of loci expected to be heterozygous is 0.0447 in S. altivelis and 0.0536 in S. alascanus. Levels of genetic variability in shallow-living members of the family Scorpaenidae range from 0.018 to 0.038 (Johnson et al., 1973). However, this study of three Sebastes species was heavily weighted with general protein loci (40% of the 24, 24, 23 loci examined). General proteins tend to be monomorphic (Selander, 1976), and this bias toward these loci may account for the low levels of variability observed.

The two Sebastolobus species, deep-living representatives of a typically shallow-water family, may not reflect levels of genetic variability of species from truly deep-sea families. The macrourid (rattail) fishes are a family rich in deep-sea species and in many areas comprise the dominant component of the bathyal fish fauna (Marshall, 1965; Pearcy and Ambler, 1974). The rattails are one component of the deep-sea fauna which is highly motile and adapted to scavenge large organic falls, as seen from its rapid attraction to bait. Coryphaenoides acrolepis is benthopelagic, although it may occasionally be taken in midwater, and has a generalized diet (Pearcy and Ambler, 1974). Smith and Hessler (1974) studied its respiration in situ at 1230 m and reported oxygen utilization values only 4% of those of a phylogenetically related species living at similar temperatures in shallow water.

A survey of 27 specimens of C. acrolepis taken from a depth range of 997-1766 m revealed average individual heterozygosities of 0.033 over 25 loci; four of the loci had variants. (This contrasts with a preliminary report of higher heterozygosity for this species based on six loci (Somero and Soulé, 1974).) No general conclusions can be drawn from the heterozygosities of these three deep-living species regarding the levels of genetic variability typical of deep-sea fishes. Because population size may explain a large proportion of the variation in heterozygosities among species, comparisons of species must be made with caution (Soulé, 1976). If there are intrinsic differences in levels

TABLE 3a

Monomorphic loci in <u>Sebastolobus</u>; number = number of individuals scored

Locus	S. altivelis	S. alascanus
	Loci monomorphic and identical	
Acph	243	56
αGpd	300	58
Est-1	259	33
Gp-1	343	58
Gp-3	342	58
Got-2	324	58
Mdh-2	343	51
Me-2	227	43
Odh	290	37
Xdh	281	37
	Loci monomorphic and fixed for different alleles	
Got-1	342	52
Gp-2	335	58

of genetic variability between families of fish, comparisons of heterozygosities characterizing species in various environments will be further complicated.   (For example, although the heterozygosities of the Sebastolobus species are below average for fish in general, they are higher than previous reports of genetic variability in other species of the family.)   However, it is clear that deep-sea fish species are not genetically depauperate.

Patterning of genetic variability in Sebastolobus altivelis. The distribution of genetic variability among populations of S. altivelis seems determined by the wide-ranging dispersal of this species, rather than by strong short-term selection by depth-related factors acting on the benthic adults.   Allelic and zygotic frequencies are homogeneous over all of the localities (Table 4) (with the exception of the Pgm locus at Sta. A15) (Homogeneity χ^2 of Workman and Niswander, 1970).   The populations studied cover the entire typical depth range of the species and encompass 70% of the reported depths where any specimens have been taken. There is, however, no apparent patterning of genetic variability with depth.   Samples were pooled by depth (E3 and BLB; A27; A24, A28, A29, A33, A15, A30, A31) and analyzed for homogeneity.   Allelic frequencies were homogeneous with depth (Pgm, log likelihood ratio, G_6 = 9,803, $0.5 < p < 0.1$ (Sokal and Rohlf, 1969); Pgi-1 G_6 = 4.348, $0.9 < p < 0.5$).   Inspection reveals no trend of allelic or zygotic frequencies with depth.

Specimens from all of the "A" stations (taken within a period of 3 mo) were pooled to examine for trends in allelic or genotypic frequencies related to size classes.   The length frequency data, possibly because of sampling biases, did not permit the partitioning of the data into discrete size classes.   Individuals were pooled into four size classes of unequal length (245.5-175.5 mm, 175.5-130.5 mm, 130.5-95.5 mm, 95.5-50.5 mm), but of approximately equal numbers as a conservative procedure.   The Pgm allelic frequencies are heterogeneous over these classes (G_6 = 17.80, $p < 0.01$; Table 5).   The Pgm-1 genotypic classes are heterogeneous with length (G_6 = 14.639, $p < 0.025$; Table 5), but allelic frequencies are not (G_3 = 2.776, $0.5 < p < 0.1$).   In both instances the heterogeneity stems mainly from the largest size class (245.5-175.5 mm), which, for the Pgm locus, contains an excess of the 1.01 allele and for Pgi-1 an excess of individuals with the 1.02/1.02 genotype.

There is no directionality to this heterogeneity with

TABLE 3b

Allelic frequencies for the polymorphic loci of Sebastolobus

Species	n	0.85	0.91	0.92	0.96	0.97	0.98	1.00	1.01	1.02	1.03	1.04	1.06	1.09
altivelis	350			0	0.003			0.041	0.845	0.104	0.007			
alascanus	62		0.008	0				0.032	0.936	0.024	0			
Pgi-1														
altivelis	339							0.659	0.341					
alascanus	52							0.442	0.558					
Pgi-2														
altivelis	338					0.009	0.004	0.981			0.006			
alascanus	50					0.020	0.010	0.970			0			
Ldh-1														
altivelis	352							0.969		0.031				
alascanus	63							0.992		0.008				
Ldh-2														
altivelis	316	0.001						0.986		0.013				
alascanus	57	0						1.000		0				
Ldh-3														
altivelis	331							0.983				0	0	0.017
alascanus	53							0				0.755	0.245	0

Table 3b (continued)

			Mdh-1	
altivelis	338		0.003	0.997
alascanus	57		1.000	0
			To	
altivelis	340	0.002		0.998
alascanus	56	0		1.000

TABLE 4

Geographic distribution of allelic frequencies of 3 polymorphic loci in Sebastolobus altivelis.

Station	n	\multicolumn{5}{c}{Allele}				
		0.96	0.98	1.00	1.01	1.02
\multicolumn{7}{c}{Pgm}						
A15	28	0	0	0.9821	0	0.0179
A24 & A33	21	0	0.0476	0.7381	0.1905	0.0238
A27	72	0.0069	0.0347	0.8125	0.1458	0
A28 & A29	85	0.0059	0.0588	0.8294	0.1000	0.0059
A30 & A31	68	0	0.0294	0.8676	0.0882	0.0147
E3	47	0	0.0532	0.8511	0.0957	0
BLB	29	0	0.0517	0.8448	0.1034	0
\multicolumn{7}{c}{Pgi-1}						
A15	29			0.7568		0.2414
A24 & A33	21			0.6190		0.3810
A27	73			0.6575		0.3425
A28 & A29	81			0.6790		0.3210
A30 & A31	63			0.5962		0.4048
E3	43			0.6163		0.3837
BLB	29			0.7414		0.2586
\multicolumn{7}{c}{Ldh-1}						
A15	29			1.0000		0
A24 & A33	21			1.0000		0
A27	73			0.9589		0.0411
A28 & A29	85			0.9118		0.0883
A30 & A31	68			1.0000		0
E3	47			0.9894		0.0106
BLB	58			1.0000		0

TABLE 5

Genotypic and genic variation with length in Sebastolobus altivelis. Number in parentheses is number expected for an homogeneous distribution with length.

Size class (mm)	Pgi-1 Genotype			Pgm Allele		
	1.02/1.02	1.02/1.00	1.00/1.00	1.00	1.01	Others
175.5–245.5	13 (7.69)	17 (24.39)	28 (25.92)	81 (90.16)	21 (10.15)	4 (5.69)
130.5–175.5	7 (9.15)	39 (29.01)	23 (30.84)	121 (119.08)	13 (13.41)	6 (7.51)
95.5–130.5	9 (9.15)	24 (29.01)	36 (30.84)	126 (119.08)	7 (13.41)	7 (7.51)
50.5– 95.5	6 (9.02)	31 (28.59)	31 (30.39)	116 (115.68)	9 (13.03)	11 (7.30)

G = 14.638, 6df; $p < 0.025$ G = 17.880, 6 df; $p < 0.01$

increasing size when all the size classes are considered; there is no indication of strong directional selection acting on the benthic adults. Two mutually compatible hypotheses may explain the heterogeneity among these size classes: (1) There is differentiation on much a broader geographic scale than has been examined. Temporal changes in current patterns and intensities could result in differential contributions of these populations to the study area. There is some support for this hypothesis. One hundred specimens of S. alascanus from the Queen Charlotte Sound, B. C., Canada, were found to be monomorphic at the Pgm locus (Johnson et al., 1972). Four alleles were found in a smaller sample of S. alascanus taken from the Southern California Continental Borderland (see Table 3). (2) Variation in selective patterns during the pelagic dispersal stages could also account for such a pattern.

Despite the hydrostatic pressure differences (88 atm) between populations, strong short-term selection adapting S. altivelis populations to different depths is not apparent. This contrasts with the strong short-term selection which may be acting in other widely dispersing organisms (Williams et al., 1973; Williams, 1975; Koehn et al., 1973) over environmental gradients. Sebastolobus altivelis seems to adapt to the environmental conditions of its depth range by means of a eurytolerant protein strategy, in which a single form of an enzyme functions adequately over the entire depth range. Contrasting with this would be a polymorphic strategy, in which enzyme variants, each well suited to a narrow portion of the total environmental range would be used to adapt populations (see Somero, 1975a for a discussion of these types of strategies).

Differentiation of the Sebastolobus species and the role of hydrostatic pressure. The two Sebastolobus species share electrophoretically detected alleles at 17 of the 20 loci assayed (Table 3). The genetic distance between the species, calculated with Nei's (1972) genetic distance statistic, is $0.2260 = D$ $(= -\ln I)$; $I = 0.7977$. The average genetic distance between the three Sebastes species (based on 21, 21, and 20 of the loci reported by Johnson et al., (1973), with unpublished data on comparative electrophoretic mobilities generously provided by Dr. F. M. Utter) is $\bar{D} = 0.194 \pm 0.047$, $\bar{I} = 0.824 \pm 0.0331$. The two Sebastolobus species have apparently become adapted to much more divergent physical regimes than those of the Sebastes species (Miller and Lea, 1972) without substantially more electrophoretically detectable protein differentiation.

Because electrophoretic mobility classes may conceal variation, estimates of genetic distance are minimal estimates (cf. Avise, 1976). Recent studies have demonstrated considerable genetic heterogeneity within electrophoretic mobility classes (Coyne, 1976; Singh et al., 1976). The degree to which such electrophoretically cryptic variation will affect our view of patterns of genetic variability is unclear. A comparison of the functional characteristics of the muscle lactate dehydrogenases (E.C. 1.1.1.27, lactate: NAD^+ oxidoreductase) encoded by the Ldh-1 1.00 allele reveals differences between the species which can be related to their adaptations to different depths. Zymogram patterns of the muscle lactate dehydrogenases show a single band which is electrophoretically indistinguishable between the two species under a variety of conditions (pH 7.0–8.1, starch concentrations of 12.2%–17% w:v). Although electrophoretically identical, the two enzymes have different pressure sensitivities, differentially suiting them to high hydrostatic pressures.

The physical basis of pressure effects on chemical processes stems from volume changes: processes proceeding with a positive volume change will be inhibited by increased pressure; processes with negative volume changes will be pressure-enhanced; processes with no net volume change will be pressure-insensitive. At a constant pressure and temperature, this may be expressed as follows:

$$\Delta H = \Delta E + P\Delta V$$
$$\Delta G^O = \Delta E + P\Delta V - T\Delta S$$
$$\Delta G^{\ddagger} = \Delta E^{\ddagger} + P\Delta V^{\ddagger} - T\Delta S^{\ddagger}$$

These equations relate the enthalpy change (ΔH) to the change in internal energy (ΔE) and to the pressure-volume work ($P\Delta V$), and the effect of these terms on the standard free energy change (ΔG^O) and the free energy change of the transition from the ground state to the activated enzyme-substrate complex ($\Delta G^{\ddagger}$). T is temperature; ΔS is change in entropy. (The nature of these volume effects are discussed in detail in F. H. Johnson and Eyring, 1970; Low and Somero, 1975a, b, c; and Somero and Hochachka, 1976a.) Equilibria will be affected by volume changes in going from products to reactants; volume changes associated with the transition to the activated complex, consisting of volume changes associated with the void volume of the enzyme, as well as the degree of exposure of amino acid side chains and peptide linkages (Low and Somero, 1975a,b), will influence the susceptibility of catalytic rates to pressure increases.

To maintain proper regulatory and catalytic functions, enzymes must maintain their proper affinity for substrate and cofactors in the face of temperature and pressure changes which can potentially disrupt the weak bonding upon which such affinities depend (Hochachka and Somero, 1973). Apparent K_m's (the half-saturation concentration of substrate and a measure of enzyme affinity for substrate) are known to vary with temperature, yet when organisms from different thermal environments are compared, each at its in situ temperature, the K_m's are found to be similar (Somero, 1975b).

The apparent K_m's for pyruvate of the LDH's of the two Sebastolobus species were determined at three temperatures, using enzyme purified by affinity chromatography (Spielmann et al., 1973; O'Carra et al., 1974). At $5°$, $10°$, and $15°C$ (1 atm), using several buffers, the K_m's for pyruvate of the two species are highly similar (Table 6). Over this range of temperatures, the two enzymes do not differ in their affinity for substrate. However, application of hydrostatic pressure typical of depths at which the deeper-living S. altivelis is common causes a divergence in the affinities of the two enzymes for substrate (Table 6). The affinity for pyruvate of the LDH of S. altivelis is not affected by hydrostatic pressure typical of the depth at which it is found, but the LDH of the shallower-living S. alascanus has a lessened affinity for pyruvate at this pressure compared to its affinity at 1 atm.

A comparison of the activation volumes ($\Delta V^{\ddagger}$) of the LDH reactions of the two species shows that of S. alascanus to have a larger positive volume change over a range of salt concentrations (50-150 mM) than does that of S. altivelis (Fig.1). Because $V^{\ddagger}$ is salt-sensitive, it is necessary to measure this parameter over a range of salt concentrations to approximate intracellular conditions (Low and Somero, 1975a, b, c). The enzyme of S. altivelis is less pressure sensitive, i.e. catalytic rates and affinity for substrate are less sensitive to increased pressures than is the enzyme of the shallower-living S. alascanus, and is thus better adapted to function at greater depths. The two enzymes have diverged in function, adapting them to the pressure regimes which they experience, without diverging electrophoretically.

The $\Delta V^{\ddagger}$ (100 mM KCl) of the shallow-living halibut (Hippoglossus stenolepis) is 12.7 cc/mole (Somero and Low, 1977). The tuna (Thunnus thynnus) has a ΔV (100 mM KCl) of 10.7 cc/mole (Somero, personal communication). The range of $\Delta V^{\ddagger}$ (100 mM KCl) for four mesopelagic vertically migrating myctophid fishes is 0.3 - 7.4 cc/mole (Somero,

TABLE 6

Apparent K_m for pyruvate (mM) of LDH in <u>Sebastolobus altivelis</u> and <u>S. alascanus</u>. Assay: Buffer, .15 mM NADH, varying concentrations of pyruvate. The high pressure optical cell used is described by Low and Somero (1975a, b). Reactions were initiated by addition of enzyme and followed by monitoring absorbance at 340 nm. K_m was calculated with a weighted linear regression according to the technique of Wilkinson (1961). For each species, the first set of numbers indicates K_m (mM pyruvate) $\pm$ SE; the second set indicates 95% confidence limits for K_m.

Temperature (°C)	pH	Pressure (atm)	S. altivelis	S. alascanus
		66.7 mM Potassium phosphate buffer		
5	7.4	1	0.1662+0.0066	0.1992+0.0098
			0.1822-0.1501	0.2244-0.1740
10	7.4	1	0.2537+0.0079	0.2461+0.0059
			0.2739-0.2335	0.2606-0.2316
15	7.4	1	0.3571+0.0160	0.3282+0.0087
			0.3949-0.3193	0.3303-0.3260
		80 mM Imidazole buffer*		
5.4	7.27	1	0.2379+0.0018	0.2359+0.0011
			0.2425-0.2333	0.2387-0.2331
10	7.17	1	0.3111+0.0044	0.3210+0.0026
			0.3232-0.2990	0.3271-0.3149
15	7.05	1	0.3658+0.0072	0.3606+0.0098
			0.3842-0.3474	0.3847-0.3365

* Imidazole data provided by P.H. Yancey.

Table 6 (continued)

			80 mM Tris/HCl	100 mM KCl
5	7.5	1	0.3947±0.0114	0.3986±0.0109
			0.4216-0.3678	0.4253-0.3719
5.6	7.5	68	0.4064±0.0156	0.5154±0.0197
			0.4446-0.3682	0.5637-0.4671
10	7.5	68	0.3953±0.0166	0.5745±0.0487
			0.4539-0.3547	0.6940-0.4557

personal communication). To the extent that adaptational adjustments to depth of activation volume and affinity for substrate characterize other enzyme systems in bathyal species, these data may shed light on factors involved in the observed depth zonation of deep-sea species. Deep-sea species have been observed to have characteristic depth ranges throughout their geographical distributions. Because there are no obvious physical barriers imposing depth restrictions, it has been unclear to what degree these patterns result from ecological interactions or physiological limitations in coping with temperature and hydrostatic pressures (Sanders and Hessler, 1969). If adaptations to specific pressures and temperatures are necessary at a number of enzyme systems, then then physical environmental influences may be an important determinant of a species' characteristic depth distribution.

Summary

Levels of genetic variability in invertebrate species of a number of phyla and classes from the physically stable and seasonless deep-sea environment have been found to be high, contradicting hypotheses predicting that species from such environments would be genetically depauperate. Utilizing techniques of starch gel electrophoresis, levels of genetic variability were examined in three deep-sea teleost species from two families: Sebastolobus altivelis (352 specimens, 20 gene loci), S. alascanus (63 specimens, 20 gene loci) (Scorpaenidae), and Coryphaenoids acrolepis (27 specimens, 25 gene loci) (Macrouridae). Average individual heterozygosities in these three species are 4.5%, 4.9% and 3.3%,

GENETIC VARIABILITY IN DEEP-SEA FISHES

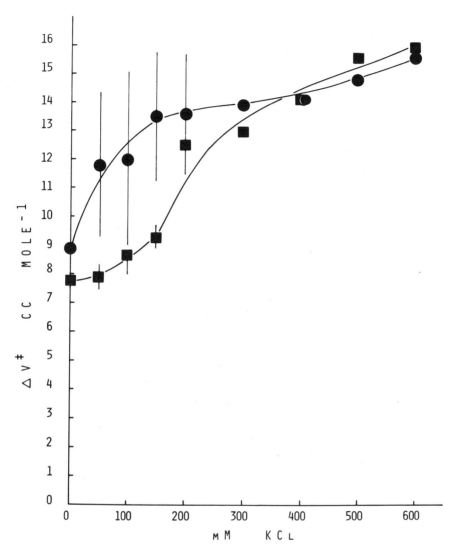

Figure 1. The effects of different concentrations of KCl on the activation volumes $(\Delta V)^{\ddagger}$ of the Sebastolobus altivelis (squares) and the Sebastolobus alascanus (circles) lactate dehydrogenase reactions. Assay: 80 mM Tris/HCl pH 7.5 at the assay temperature of 10°C, 0.15 mM NADH, and varying concentrations of pyruvate and neutral salts. The high pressure optical cell is described by Low and Somero, 1975a, b. Assays were done with crude homogenate preparations. The points plotted are means of 2-10 determinations. 95% confidence limits are shown.

which are below the average heterozygosities generally found for fish.

The morphologically similar congeners, S. altivelis and S. alascanus, have similar life histories. The adults of both species are benthic, but the eggs and larval stages are dispersed through the plankton. The adults of these two species have different bathymetric distributions, although they co-occur geographically. Sebastolobus altivelis is typically found between 550 and 1,300 m; S. alascanus is common between 180 and 440 m. Hydrostatic pressure and temperature, depth-related parameters of the physical environment, are known to exert an important influence on protein structure and function. Allelic frequencies are uniform over ten samples of populations of S. altivelis taken from depths of 463 to 1,415 m in the Southern California Continental Borderland. No trend of electrophoretically detectable variation with depth was found. This species apparently adapts to a wide range of hydrostatic pressures by maintaining eurytolerant protein forms, rather than polymorphic variants suited to a narrow portion of the total range of hydrostatic pressures which the species encounters.

The genetic distance between the two Sebastolobus species, based on electrophoretic data is 0.2260 ($I = 0.7977$). However, a comparison of the functional characteristics of the muscle lactate dehydrogenases of these two species, which by electrophoretic criteria are homologous and identical, suggests that these species are more distinct genetically than indicated by the electrophoretic data. The muscle LDH forms differ in their sensitivity to hydrostatic pressure, an important environmental parameter. Apparent K_m's for pyruvate are similar in the two species at $15°$, $10°$ and $5°C$ (1 atm). However, the K_m's of the two species diverge when examined at hydrostatic pressures characteristic of the depth range of S. altivelis (68 atm). At physiological salt concentrations, the LDH of S. altivelis has a smaller positive volume change associated with the activation event of catalysis. The enzyme of S. altivelis is less pressure inhibited than that of the shallow-living S. alascanus and thus is better adapted to function at depth.

Acknowledgements

I thank R. R. Hessler and G. N. Somero for helpful advice and discussion; G. N. Somero, F. M. Utter and P. H. Yancey for providing unpublished data; M. Neubauer for competent laboratory assistance and the many people who helped in obtaining specimens. This research was supported by grants DES 74-21506 and PCM-09498 from the National Science Foundation.

REFERENCES

Avise, J.C. 1976. Genetic differentiation during speciation, pp. 106-122. In, Molecular Evolution. Ayala, F. J. (ed). Sinauer Associates, Sunderland, Massachusetts.

Ayala, F.J., and Valentine, J.W. 1974. Genetic variability in a cosmopolitan deep-water ophiuran Ophiomusium lymani. Mar. Biol. 27, 51-57.

Ayala, F.J., Valentine, J.W., Hedgecock D, and Barr, L.G. 1975. Deep-sea asteroids: High genetic variability in a stable environment. Evolution Lancaster, Pa. 29, 203-212.

Bretsky, P.W., and Lorenz, D.M. 1970. An essay on genetic-adaptive strategies and mass extinctions. Bull. Geol. Soc. Am. 81, 2449-2456.

Bryant, E.H. 1974. On the adaptive significance of enzyme polymorphisms in relation to environmental variability. Am. Nat. 108, 1-19.

Coyne, J.A. 1976. Lack of genic similarity between two sibling species of Drosophila as revealed by varied techniques. Genetics 84, 593-607.

Doyle, R.W. 1972. Genetic variation in Ophiomusium lymani (Echinodermata) populations in the deep sea. Deep-sea Res. 19, 661-664.

Ekman, S. 1953. Zoogeography of the sea. Sidgwick and Jackson, London 417p.

Emery, K.O. 1960. The sea off southern California. John Wiley & Sons, Inc., New York. 366p.

Gilbert, C.H. 1915. Fishes collected by the United States Fisheries Steamer Albatross in southern California in 1904. Proc. U.S. Natl. Mus. 48, 305-380.

Gillespie, J. 1974. Polymorphism in patchy environments. Am. Nat. 108, 145-151.

Gillespie, J. and Langley, C.H. 1974. A general model to account for enzyme variation in natural populations. Genetics 76, 837-848.

Gooch, J.L., and Schopf, T.J.M. 1972. Genetic variability in the deep sea: Relation to environmental variability. Evolution 26, 545-552.

Grassle, J.F. 1972. Species diversity, genetic variability and environmental uncertainty, pp. 19-26.   In, Fifth Europ. Mar. Biol. Symp. Battaglia, B. (ed). Piccin Editore, Padua.

Grassle, J.F., and Sanders, H.L. 1973. Life histories and the role of disturbance. Deep-Sea Res. 20, 643-659.

Haldane, J.B.S., and Jayakar, S.D. 1963. Polymorphism due to selection of varying direction. J. Genet. 58, 237-242.

Hochachka, P.W., and Somero, G.N. 1973. Strategies of biochemical adaptation. W.B. Saunders Co., Philadelphia, Pennsylvania. 358p.

Hubbs, C.L. 1926. The supposed intergradation of two species of Sebastolobus (a genus of scorpaenoid fishes) of western America. Am. Mus. Novit. 216, 1-9.

Iwamoto, T. 1975. The abyssal fish Antimora rostrata (Günther). Comp. Biochem. Physiol. 52B, 7-11.

Johnson, A.G., Utter, F.M. and Hodgins, H.O. 1972. Electrophoretic investigation of the family Scorpaenidae. U.S. Nat. Mar. Fish. Ser. Fish. Bull. 70, 403-413.

Johnson, A.G., Utter, F.M. and Hodgins, H.O. 1973. Estimate of genetic polymorphism and heterozygosity in three species of rockfish (genus Sebastes). Comp. Biochem. Physiol. 44B, 397-406.

Johnson, F.H., and Eyring, H. 1970. The kinetic basis of pressure effects in biology and chemistry, pp. 1-44. In, High Pressure Effects on Cellular Processes. Zimmerman, (A. M. (ed). Academic Press, New York.

Johnson, G.B. 1974. Enzyme polymorphism and metabolism. Science 184, 28-37.

Johnson, G.B. 1976. Genetic polymorphism and enzyme function, pp. 46-59. In, Molecular Evolution. Ayala, F. J. (ed). Sinauer Associates, Sunderland, Massachusetts.

Johnson, M.S. 1971. Adaptive lactate dehydrogenase variation in the crested blenny, Anoplarchus, Heredity 27, 205-226.

Koehn, R.K. 1969. Esterase heterogeneity: Dynamics of a polymorphism. Science 163, 943-944.

Koehn, R.K., Milkman, R. and Mitton, J.B. 1976. Population genetics of marine pelecypods. IV. Selection, migration and genetic differentiation in the blue mussel Mytilus edulis. Evolution 30, 2-32.

Koehn, R.K., Turano, F.J. and Mitton, J.B. 1973. Population genetics of marine pelecypods. II. Genetic differences in microhabitats of Modiolus demissus. Evolution 27, 100-105.

Kojima, K. 1971. Is there a constant fitness for a given genotype? No! Evolution 25, 281-285.

Levene, H. 1949. On a matching problem arising in genetics. Ann. Math. Stat. 20, 91-94.

Levins, R. 1968. Evolution in changing environments. Princeton Univ. Press, Princeton, New Jersey, 120 p.

Levinton, J.S. 1973. Genetic variation in a gradient of environmental variability: Marine bivalvia (Mollusca) Science 180, 75-76.

Levinton, J.S. 1975. Levels of genetic polymorphism at two enzyme encoding loci in eight species of the genus Macoma (Mollusca:Bivalvia). Mar. Biol. 33, 41-47.

Levinton, J.S. and Fundiller, D. L. 1975. An ecological and physiological approach to the study of biochemical polymorphisms. pp. 165-178. In, Proc. 9th Europ. Mar. Biol. Symp. Barnes, H. (ed). Aberdeen Univ. Press.

Low, P. S. and Somero, G.N. 1975a. Activation volumes in enzymic catalysis: Their sources and modification by low-molecular-weight solutes. Proc. Natl. Acad. Sci. USA 72, 3014-3018.

Low, P.S. and Somero, G.N. 1975b. Protein hydration changes during catalysis: A new mechanism of enzymic rate-enhancement and ion activation/inhibition of catalysis. Proc. Natl. Acad. Sci. USA 72, 3305-3309.

Low, P.S. and Somero, G.N. 1975c. Pressure effects on enzyme structure and function *in vitro* and under simulated *in vivo* conditions. Comp. Biochem. Physiol. 52B, 67-74.

Manwell, C. and Baker, C.M.A. 1970. *Molecular biology and the origin of species*. Univ. Washington Press. Seattle. 394 p.

Marshall, N.B. 1965. Systematic and biological studies of the macrourid fishes (Anacanthini - Teleostii). Deep-Sea Res. 12, 299-322.

Merrit, R. 1972. Geographic distribution and enzymatic properties of lactate dehydrogenase allozymes in the fathead minnow. *Pimephales promelas*. Am. Nat. 106, 173-184.

Miller, D.J. and Lea, R.N. 1972. Guide to the coastal marine fishes of California. Calif. Dep. Fish Game, Fish Bull. 157, 235 p.

Mitton, J. B. and Koehn, R.K. 1975. Genetic organization and adaptive response of allozymes to ecological variables in *Fundulus heteroclitus*. Genetics 79, 97-111.

Moser, H.G. 1974. Development and distribution of juveniles of *Sebastolobus* (Pisces: Family Scorpaenidae). U.S. Nat. Mar. Fish. Ser. Fish. Bull. 72, 865-884.

Murphy, L. S., Rowe, G.T. and Haedrich, R. L. 1976. Genetic variability in deep-sea echinoderms. Deep-Sea Res. 23, 339-348.

Nei, M. 1972. Genetic distance between populations. Am. Nat. 106, 283-292.

Nevo, E. 1976. Adaptive strategies of genetic systems in constant and varying environments. pp. 141-158. In, *Population Genetics and Ecology*. Karlin, S. and Nevo, E. (eds). Academic Press, New York.

Nevo, E. and Bar, Z. 1976. Natural selection of genetic polymorphisms along climatic gradients. In, *Population Genetics and Ecology*. Karlin, S and Nevo, E (eds), Academic Press, New York, London.

O'Carra, P., Barry, S. and Corcoran, E. 1974. Affinity chromatographic differentiation of lactate dehydrogenase isoenzymes on the basis of differential abortive complex formation. FEBS Letters 43, 163-168.

Pearcy, W.G. and Ambler, J.W. 1974. Food habits of deep-sea macrourid fishes off the Oregon coast. Deep-Sea Res. 21, 745-759.

Rokop, F.J. 1974. Reproductive patterns in the deep-sea benthos. Science 186, 743-745.

Sanders, H. L. and Hessler, R.R. 1969. Ecology of the deep-sea benthos. Science 163, 1419-1424.

Schopf, T.J.M. 1976. Environmental versus genetic causes of morphologic variability in bryozoan colonies from the deep-sea. Paleobiology 2, 156-165.

Schopf, T.J.M. and Gooch, J. 1971. Gene frequencies in a marine ectoproct: A cline in natural populations related to sea temperature. Evolution 25, 286-289.

Selander, R.K. 1976. Genic variation in natural populations. pp.21-45. In, Molecular Evolution. Ayala, F. J.(ed). Sinauer Associates, Sunderland, Massachusetts.

Singh, R.S., Lewontin, R.C. and Felton, A.A. 1976. Genetic heterogeneity within electrophoretic "alleles" of xanthine dehydrogenase in Drosophila pseudoobscura. Genetics 84, 609-629.

Smith, K.L., Jr., and Hessler, R.R. 1974. Respiration of benthopelagic fishes: In situ measurements at 1230 meters. Science 184, 72-73.

Sokal, R.R. and Rohlf, F.J. 1969. Biometry. W. H. Freeman and Company, San Francisco. 776 p.

Somero, G.N. 1975a. The roles of isozymes in adaptation to varying temperatures. pp. 221-234. In, Isozymes II. Physiological Function. Markert C. L. (ed). Academic Press, New York.

Somero, G. N. 1975b. Temperature as a selective factor in protein evolution: The adaptational strategy of "compromise". J. Exp. Zool. 194, 175-188.

Somero, G.N. and Hochachka, P.W. 1976a. Biochemical adaptations to pressure. pp. 480-510. In, Adaptation to environment: Essays on the physiology of marine animals. Newell, R.C. (ed). Butterworths, London.

Somero, G. N. and Hochachka, P.W. 1976b. Biochemical adaptations to temperature. pp. 125-190. In, Adaptation to environment: Essays on the physiology of marine animals. Newell, R.C. (ed). Butterworths, London.

Somero, G. N. and Low, P. S. 1977. Enzyme hydration may explain catalytic efficiency differences among lactate dehydrogenase homologues. Nature 66, 276-278.

Somero, G. N. and Soulé, M. 1974. Genetic variation in marine teleosts: A test of the niche-variation hypothesis. Nature 249, 670-672.

Soulé, M. 1976. Allozyme variation: Its determinants in space and time. pp. 60-77. In, Molecular Evolution. Ayala, F. J. (ed). Sinauer Associates, Sunderland, Massachusetts.

Spielmann, H., Erickson, R.P. and Epstein, C. J. 1973. The separation of lactate dehydrogenase X from other lactate dehydrogenase isozymes of mouse testes by affinity chromatography. FEBS Letters 35, 19-23.

Valentine, J.W. 1976. Genetic strategies of adaptation. pp. 78-94. In, Molecular Evolution. Ayala, F. J. (ed). Sinauer Associates, Sunderland, Massachusetts.

Valentine, J.W. and Ayala, F. J. 1975. Genetic variation in Frieleia halli, a deep-sea brachiopod. Deep-Sea Res. 22, 37-44.

Wilkinson, G. N. 1961. Statistical estimation in enzyme kinetics. Biochem. J. 80, 324-334.

Williams, G. C. 1975. Sex and Evolution. Monographs in population biology 8. Princeton Univ. Press, Princeton, New Jersey. 200 p.

Williams, G. C., Koehn, R. K. and Mitton, J. B. 1973. Genetic differentiation without isolation in the American eel Anguilla rostrata. Evolution 27, 192-204.

Workman, P.L. and Niswander, J. D. 1970. Population studies on southwestern Indian tribes. II. Local genetic differentiation in the Pagopago. Am. J. Hum. Genet. 22, 24-49.

GENETIC VARIATION AND SPECIES COEXISTENCE IN LITTORINA

J. A. Beardmore and S. R. Morris

Dept. of Genetics, University College of Swansea

Swansea, United Kingdom.

Introduction

A satisfactory explanation of the high levels of genetic polymorphism characterising most populations of outbreeding populations is still lacking (Dobzhansky et al., 1977) and the central importance which this problem occupies in evolutionary genetics is the justification for the scale of contemporary research designed to seek such an explanation.

Attempts to account for variation in levels of genetic variation have frequently concentrated on the concept of the ecological niche (see Soulé (1976) for a good summary). Thus for example, positive correlations between morphological variation and niche width have been observed in a number of species of birds by Van Valen (1965). Nevo (1976) argued that the low levels of genetic variation in two species of Pelobates are a consequence of narrowness of niche and McDonald and Ayala (1974) in laboratory experiments produced evidence tending to support this view. Ultimately we would hope to establish an inventory of causal relationships between specific components of genetic variation and specific variables of the environment but progress towards this goal is extremely slow. At present a broader approach seems likely to be more informative. Elsewhere in this volume Ayala and Valentine point out the difficulties that lie in the way of those trying to work out general principles underlying the relationships between characteristics of the environments inhabited by a given species and the characteristics of the gene pool. As they say, hypotheses are testable only by looking to see whether predictable correlations exist between the environmental regimes of populations and the levels of genetic variation within populations.

The presence of competing species is one of the more important biotic variables of the environment of a population. Character displacement is one way in which closely related species may differ in regions where they are sympatric. Such differences are often relatable to isolating mechanisms as in the cases of mating calls in Amphibia described by Blair (1955) and Littlejohn (1965). In some cases, however, it is evident that large differences in parameters of characters not involved in isolating mechanisms are associated with sympatry of two or more related species (Schoener, 1965, and Fenchel et al., this volume). Fenchel and Christiansen (1977) have an interesting discussion of the significance of character displacement and conclude that variance tends to be reduced by competition with other species, a finding borne out by the results of a study by Johnson (1973) on Hawaiian species of Drosophila.

The work reported here is an attempt to explore the relative effects of coexistence of closely related species in terms of genetic variation. Winkles of the genus Littorina which exist in strictly sympatric and allopatric communities and in intermediate situations in populations relatively close together on the coast of S. Wales provide very suitable material for such a study. The degree of competition for food (or other resources) which exists in sympatric populations is not known but it is probable that there is some overlap in this respect (Newell, 1958).

Materials and Methods

Samples from seven localities on the coast of South Wales (Fig.1) were studied. The study focused upon Littorina littorea and Littorina rudis, one or both of which were the only Littorina species found to be present at six sites, although at one site L. nigrolineata is also present.

Winkles were collected between Spring 1976 and Spring 1977. At each site samples were taken within a measured area of either one or two square metres. For each collection the tide level of the sampling site was estimated from Admiralty Tide Tables. At Musselwick West, Oxwich, Watwick South, and Southerndown three tide levels were sampled, each sample being sited along a line roughly orthogonal to the contours of the shore at that point. At Wentloog, Saundersfoot

GENETIC VARIATION AND SPECIES COEXISTENCE

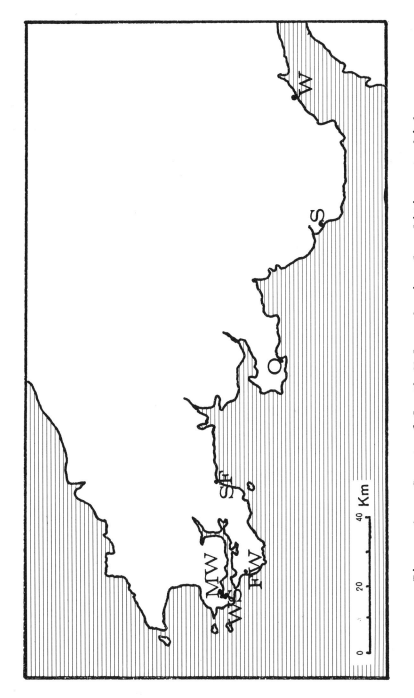

Figure 1. Coast of South Wales showing localities at which samples of _Littorina_ were taken.

and Freshwater West only one tide level was sampled.

Phenotypes for ten enzymatic proteins were determined using horizontal starch gel electrophoresis. The buffers and stains used for electrophoresis will be described elsewhere (Morris and Beardmore, in prep.). Descriptions of the enzymes used in the present study are given in Table 1, though those found to be monomorphic in all populations are not included here. An enzyme is defined as being polymorphic if the frequency of the most common allele is less than 0.95 in ore or more of the populations analysed.

Measures of genetic variation were based on heterozygosity estimates calculated as weighted means over all polymorphic loci, so that in the few cases of unequal sample sizes, no single locus would contribute disproportionately to the estimate. Estimates were made of $\bar{H}$, the observed level of heterozygosity

TABLE 1

Loci found to be polymorphic in Littorina of the S. Wales Coast

Locus	L. rudis	L. littorea	L. nigrolineata
Pgi	Polymorphic	Polymorphic	Monomorphic
Est-2	Polymorphic	Polymorphic	Polymorphic
Lap-1	Polymorphic	Polymorphic	Polymorphic
Lap-2	Monomorphic	Polymorphic	Monomorphic
Sdh	Polymorphic	Polymorphic	Polymorphic
6-Pgdh	Polymorphic	Polymorphic	Polymorphic
Ldh	Polymorphic	Monomorphic	Polymorphic
Mdh-1	Polymorphic	Monomorphic	Monomorphic
Pgm	Polymorphic	Polymorphic	Polymorphic
Ap	Polymorphic	Polymorphic	Polymorphic

and of $\hat{H}$, the level of heterozygosity expected from calculated gene frequencies.

A simple morphometric index (I) was calculated for each shell, being the quotient of the maximum mouth diameter by the maximum shell height. (No measurements were made on material from Saundersfoot.) From these figures the variance σ_I^2 was calculated for each sample.

Exposure values were assigned to each shore utilising the biological scale of Ballantine (1961). Where it was seen that the shore had an intermediate value on this scale, for the purposes of analysis it has been assigned a value 0.5 above the lower of the two values between which it falls on the scale. The scalar significance of this is ill-defined as Ballantine does not consider his scale to be necessarily linear.

Overlap of species on the shore was estimated using two methods. The first is a crude method utilising the measurement of the linear range of a species on the shore. The estimate of the overlap is given as the percentage of this range which is shared with congeneric species. The second is an application of the measure of niche overlap proposed by Pielou (1972) each tide level sampled being treated as a separate habitat.

Then the niche overlap is given by

Niche overlap = $v_j h_j(a)$

where v_j = $\dfrac{x_{.j}}{N}$ the proportion of total occurrences that were in the jth habitat

$h_j(a)$ = the species diversity within the jth habitat.

This is calculated by

Niche overlap = $\sum \dfrac{x_{.j}}{N} \left(\dfrac{1}{x_{.j}} \log x_{.j} - \dfrac{1}{x_{.j}} \sum_i \log x_{ij} \right)$

This gives a total overlap value for all species of <u>Littorina</u> on a given shore. To estimate the proportion of the overlap which is contributed by a given species the total overlap is multiplied by u_i, the proportion of total occurrences which are of the ith species.

Results

Estimates of $\bar{H}$, $\hat{H}$ and σ_I^2 for populations of <u>Littorina</u> from seven localities are given in Table 2. Estimates of Ballantine exposure values, Pielou Niche overlaps (total and proportional) and linear overlap are given in Table 3.

Correlations between Exposure, Pielou value, Linear overlap and $\bar{H}$ and $\hat{H}$ were calculated using Kendall's rank correlation. The results of these comparisons are given in Table 4.

The two measures of overlap show a significant positive correlation with each other for both species. This tends to indicate that, broadly, both are estimates of the same ecological variation. Correlations with σ_I^2 are all non-significant. Correlations of H with overlap estimates are all positive and are, on the whole, statistically significant though the correlation between $\bar{H}$ and species overlap for <u>L. rudis</u> is marginally outside the 0.05 boundary. There is a highly significant correlation between $\bar{H}$ and $\hat{H}$ for <u>L. littorea</u> but for <u>L. rudis</u> the correlation has an attached probability of 0.068.

Further analysis of the data used analyses of variance. For this purpose linear overlap was arcsin transformed, all other parameters were found to be distributed in a manner not significantly different from normality (Shapiro-Wilk W test). The results show the analyses for overall mean heterozygosities weighted for unequal sample size at each locality. This was considered to be the most realistic estimate of overall population heterozygosity. In those cases where the animals were sampled only at one tide level this was because they either did not occur, or were too rare for effective sampling, at the other tide levels. However, analyses were also carried out treating each sampling point separately and this procedure gives results broadly similar to those described here.

Table 5 shows the regression analyses for the two measures of genetic variation ($\bar{H}$ and $\hat{H}$) on the two measures of overlap in the two species. The data are also represented graphically in Figure 2. The most striking feature of the table is the consistent and highly significant nature of the joint regressions which is coupled with a general lack of heterogeneity between regressions. There are also striking differences in mean heterozygosities in some analyses.

Table 6 shows the differences in estimates of heterozygosity (both observed and estimated) in samples taken at

TABLE 2

Mean observed and mean expected heterozygosities for each species at seven localities (n.d. = not determined)

Locality	Level	Species	No. Genes scored	$\bar{H}$	$\hat{H}$	σ_I^2
SF	MTL	L. littorea	390	0.222	0.209	n.d.
FW	MHW	L. rudis	390	0.233	0.302	0.956
W	MHW	L. rudis	720	0.265	0.334	1.767
O	MLW	L. littorea	200	0.075	0.075	0.837
	MTL	L. littorea	289	0.146	0.191	0.993
		L. rudis	284	0.139	0.215	1.650
	MHW	L. rudis	200	0.205	0.219	0.800
S	MLW	L. littorea	200	0.055	0.050	n.d.
	MTL	L. littorea	220	0.095	0.102	0.535
		L. rudis	210	0.176	0.191	1.160
	MHW	L. littorea	212	0.106	0.125	1.482
		L. rudis	302	0.224	0.210	1.236
MW	MLW	L. littorea	380	0.142	0.160	1.383
	MTL	L. nigrolineata	400	0.090	0.096	0.980
		L. rudis	424	0.179	0.220	0.780
	MHW	L. nigrolineata	120	0.096	0.112	n.d.
		L. rudis	280	0.209	0.234	0.686
WS	MLW	L. littorea	350	0.094	0.110	n.d.
	MTL	L. littorea	260	0.124	0.121	n.d.
		L. nigrolineata	460	0.054	0.067	n.d.
	MHW	L. rudis	310	0.258	0.273	0.740

TABLE 3

Exposure values, Pielou total niche overlap and proportional niche overlap for _L. littorea_, _L. rudis_ and _L. nigrolineata_.

Locality	Ballantine Exposure	Pielou Tot. Niche O'lap	_L. littorea_ Prop. Niche Lin. O'lap	_L. littorea_ Niche Lin. O'lap	_L. rudis_ Prop. Niche Lin. O'lap	_L. rudis_ Niche Lin. O'lap	_L. nigrolineata_ Prop. Niche Lin. O'lap	_L. nigrolineata_ Niche Lin. O'lap
Saundersfoot (S)	5.5	0	0	0	–	–	–	–
Freshwater W. (FW)	3.5	0.024	–	–	0.02	5	–	–
Wentloog (W)	6	0	–	–	0	0	–	–
Oxwich (O)	5.5	0.17	0.08	40	0.09	50	–	–
Southerndown (S)	4.5	0.26	0.101	70	0.15	100	–	–
Musselwick W. (MW)	7	0.24	0.021	5	0.12	80	0.096	100
Watwick S. (WS)	5	0.16	0.05	50	0.04	20	0.062	60

TABLE 4

Sign and significance of correlations (Kendall's Rank Correlation).
(n.d. = not determined, ** $P = <0.01$, * $P = <0.05$, o $P = >0.05$, oo $P = >0.2$).

Correlations for *L. littorea*

	$\bar{H}$	$\hat{H}$	Linear Overlap	Pielou value	Exposure	σ^2_I
$\bar{H}$		+ ***	− **	− *	− *	+ oo
$\hat{H}$	+ o		− **	− *	− o	+ oo
Linear Overlap	+ o	− **		+ *	+ o	n.d.
Pielou value	+ o	− **	+ **		+ o	n.d.
Exposure	+ oo	− oo	− oo	− oo		n.d.
σ^2_I	− oo	− oo	n.d.	n.d.	n.d.	

Correlations for *L. rudis*

TABLE 5

Analyses of variance of regressions of measures of genetic variation on measures of habitat overlap in two species of Littorina.

Comparisons	Source	ssq	d.f.	F	P
$\bar{H}$ & $\hat{H}$/Pielou L. rudis	Joint regression Bet. regressions Bet. means Residual	0.0155 0.0012 0.0046 0.0044	1 1 1 8	28.18 2.18 8.29 -	<<0.001 >0.2 <0.025 -
	Total	0.0257	11	-	-
$\bar{H}$ & $\hat{H}$/Linear o'lap L. rudis	Joint regression Bet. regressions Bet. means Residual	0.01384 0.00132 0.00456 0.00592	1 1 1 8	18.70 1.78 6.16 -	<0.001 >0.2 <0.05 -
	Total	0.02564	11	-	-
$\bar{H}$ & $\hat{H}$/Pielou L. littorea	Joint regression Bet. regressions Bet. means Residual	0.0152 0.0002 0.0002 0.0034	1 1 1 6	26.82 0.35 0.40 -	<<0.001 >0.5 >0.5 -
	Total	0.0190	9	-	-
$\bar{H}$ & $\hat{H}$/Linear o'lap L. littorea	Joint regression Bet. regressions Bet. means Residual	0.0169 0.0000 0.0002 0.0021	1 1 1 6	48.28 - 0.657 -	<<0.001 - >0.25 -
	Total	0.0192	9	-	-
$\bar{H}$ & $\hat{H}$ in L. rudis & L. littorea/ Pielou	Joint regression Bet. regressions Bet. means Residual	0.0293 0.0028 0.0596 0.0078	1 3 3 14	52.59 5.03 106.97 -	<<0.001 <0.025 <<0.001 -
	Total	0.0995	21	-	-
$\bar{H}$ & $\hat{H}$ in L. **rudis** & L. littorea/ Linear o'lap	Joint regression Bet. regressions Bet. means Residual	0.0284 0.0037 0.0596 0.0080	1 3 3 14	49.58 2.13 104.04 -	<<0.001 >0.10 <<0.001 -
	Total	0.0997	21	-	-

TABLE 6

Differences in heterozygosity estimates of samples of *Littorina* from different tide levels

Species	Tide levels compared	H	Oxwich	Locality Southerndown	Musselwick West	Watwick South
L. littorea	MTL – MLW	$\hat{H}$	+0.116	+0.023		+0.011
		$\bar{H}$	+0.071	+0.011		+0.030
L. littorea	MHW – MTL	$\hat{H}$		+0.052		
		$\bar{H}$		+0.040		
L. rudis	MHW – MTL	$\hat{H}$	+0.004	+0.004	+0.014	
		$\bar{H}$	+0.066	+0.048	+0.037	
L. nigrolineata	MHW – MTL	$\hat{H}$			+0.016	
		$\bar{H}$			+0.004	

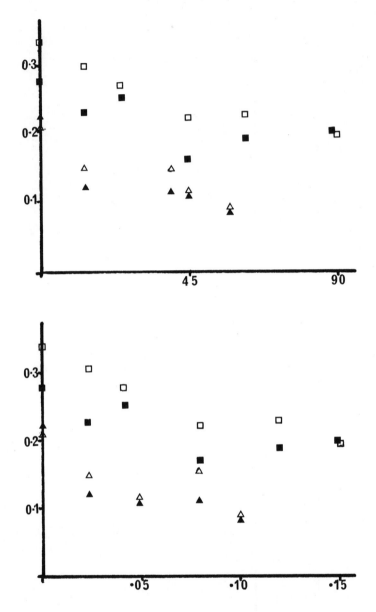

Figure 2. Heterozygosity and species overlap in <u>Littorina</u>
■ = <u>L. rudis</u>   □ = <u>L. littorea</u>
upper: X – $\bar{H}$ (solid) $\hat{H}$ (open)
Y – arcsin linear overlap

lower: X – $\bar{H}$ (solid) $\hat{H}$ (open)
Y – Pielou proportional niche overlap.

different tide levels. In each of the eight comparisons the sample from the higher tide level has a higher mean observed heterozygosity and a higher mean estimated heterozygosity than that from the lower tide level.

Discussion

The analysis shows that there is considerable variation in the total heterozygosity at the ten loci assayed, both within and between populations on the shores sampled. On the whole L. littorea is less heterozygous than L. rudis as Wilkins et al. (this volume) also observe.

No correlation exists between variation in heterozygosity and degree of exposure. However, significant correlations are seen between heterozygosity and overlap of species. The relationship appears to be consistent in both L. rudis and L. littorea. Both species feed in a similar manner as does also L. nigrolineata. A character displacement of ingested food particles as described for overlapping Hydrobia spp by Fenchel et al. elsewhere in this volume, is a possible method by which the available niche width may be partitioned by the species, but no evidence bearing on this is available for Littorina. That some reduction in the niche width available to a species is caused as a result of the overlap seems probable. This reduction has been roughly quantified by the two independent measures of overlap and if this is accepted we must conclude that reduction in niche width leads to a general reduction in heterozygosity over the ten loci examined. As the pattern of overlap is not related in any discernible manner to the geographical distribution of the sample sites, this effect is not related to a spatial cline and is, therefore, unlikely to be a relict effect of historical population movements as suggested for clinal variation by Lewontin (1974). We propose therefore that the relationship observed results from a mode of normalising selection in which, in populations containing both species, peripheral genotypes are selected against in the face of increasing competition from congeners equally or more suited to the microgeographical niches occupied by such peripheral genotypes.

The relationship between heterozygosity, whether calculated or observed, and coexistence is least evident for $\bar{H}$ in L. rudis. This may be the result of a consistent significant deficit of heterozygotes observed in L. rudis. Experiments by G. E. Newell (1958) on L. littorea have shown that the behaviour of this animal involves daily foraging from, and return to, roughly the same "home" position. This seems likely to be broadly applicable to other members of the genus. This parochial behaviour must therefore limit random mating in the population and increase the probability of inbreeding. The effects of such inbreeding are, however, countered in **L. littorea** by the pelagic larval form which will mix genotypes. In **L. rudis** the young are liveborn (Lebour, 1937) in the vicinity of the parents, thus inbreeding and depression of observed levels of heterozygosity must occur. A more extensive discussion of this phenomenon in these and other British members of the genus will appear elsewhere (Morris and Beardmore, in prep.). The effect of inbreeding in depressing $\bar{H}$ leads to a difference between $\bar{H}$ and $\hat{H}$. However, the difference between $\hat{H}$ and $\bar{H}$ diminishes as the overlap increases. This, if it is a real effect, may be explained in terms of the increased competition and the increased mixing of species in the ecospace available which might tend to increase the movement necessary to find a suitable mate, thus tending to reduce inbreeding.

The apparent clinal distribution of heterozygosity within each population in relation to vertical distribution on the shore is believed to be related to size/age heterogeneities such as those reported by Moore (1939) and Vermeij (1976) and not to differential intensity of competition over the vertical range of a population. These clinal patterns are almost certainly the result of selective processes but these processes cannot explain the relationship between heterozygosity and coexistence which we have detected. **L. rudis** tends to occupy habitat higher up the shore than **L. littorea**. Thus the parallel clinal pattern of heterozygosity observed in the two species will in one tend to reinforce the coexistence effect and in the other case to counteract it.

Despite the effective lack of correlation of heterozygosity with exposure, it is nevertheless possible that exposure is a factor in determining levels of genetic variation. That exposure has an influence on patterns of zonation in this and other genera has been noted on a number of occasions (Delf, 1942; Lewis, 1953; Ballantine, 1961 and Moyse and Nelson-Smith, 1961). The difficulty lies in the lack of availability of a very general exposure scale related to overall shore characteristics. Effective exposure for different species is very sensitive to topographical variations of rocky shores, thus for instance at Southerndown, the presence of extensive cracks and runnels in an almost horizontal bedrock must surely reduce the effective exposure for **Littorina** species considerably. This problem remains irresolvable until a more effective scale is developed.

While little weight can be put on comparisons of genetic variation with variance of only one morphological character, the lack of any association between morphological variance and genetic variation agrees with the findings of Ayala et al. (1975) who however made only interspecific comparisons. In some situations, moreover, it is to be expected that there would be negative correlations between genetic variation and variation in morphological characters. Such negative correlations are frequently found for fitness characters (Beardmore, 1970).

The most interesting relationship to emerge from this study is a general positive association between genetic variation and the extent of presence of a closely related species. The effect which we describe is, in one sense, contrary to that described by Battaglia et al. elsewhere in this volume in which genetic variation is seen to be greater in populations from habitats of greater biotic diversity. However, as they point out, other factors such as environmental instability or marginality are also important. Ayala and Valentine, in their contribution to this book, propose that stability of trophic resources is directly relatable to levels of genetic variation. It is difficult to say whether the presence of a competing species is likely, on average, to increase or decrease the stability of trophic resources and this poses problems in relating our data directly to this hypothesis. However, Valentine (1976), considering benthic marine invertebrates, estimated relative trophic resource stability by species richness and it might therefore be argued that the average effect of presence of competing species is to increase trophic stability. Johnson (1971) observed a strong negative correlation between the frequency of an Ldh allele in Anoplarchus purpurescens and the frequency of a related species in an area of sympatry. The effect on heterozygosity parallels that found by us but it would be unwise to read too much into observations of this kind on a single locus. It seems probable that, as more investigations are carried out, further correlations of levels of genetic variation with variables of the environment will emerge. As Battaglia et al. (op cit) comment, adequate information on the most significant biological characteristics of the species being considered is a prerequisite for the establishment of valid generalist theories. At present, as Soulé (1976) points out, all the niche-related theories of heterozygosity suffer from significant defects and it seems to us likely that it will turn out that the magnitude of the individual fractions of total genetic variation explicable by variation in different ecological factors will vary considerably from species to species and life form to life form.

Summary

Species of <u>Littorina</u> are favourable material for the study of relationships between species coexistence and genetic variation. Ten loci polymorphic in one or both of the species <u>L. littorea</u> and <u>L. rudis</u> have been assayed in samples from seven sites on the coast of South Wales. <u>L. rudis</u> is generally more heterozygous than <u>L. littorea</u>. Observed levels of heterozygosity ($\bar{H}$) are consistently lower than expected values ($\hat{H}$) in <u>L. rudis</u> but not in <u>L. littorea</u>. Levels of heterozygosity in both species (and in <u>L. nigrolineata</u>) are greater higher on the shore than lower down and this is possibly related to heterogeneity of age distribution on the shore.

Significant regressions exist of measures of heterozygosity ($\bar{H}$ $\hat{H}$) on the extent of species overlap but no significant correlations are found between H and exposure index or variance of one morphometric character. We conclude that in <u>Littorina</u> it is likely that the coexistence of other closely related and competing species decreases effective niche width which, through a process of normalising selection, leads to a reduction of genetic variation which is relatable to the extent of coexistence.

Acknowledgement

The award of a research studentship from the Natural Environment Research Council to S.R.M. is gratefully acknowledged.

REFERENCES

Ayala, F. J., Valentine, J. W., De Laca, T. E. and Zumwalt, G. S. 1975. Genetic variability of the Antarctic brachiopod Liothyrella notorcadensis and its bearing on mass extinction hypotheses. J. Paleontol. 49, 1-9.

Ballantine, W. J. 1961. A biologically defined exposure scale for the comparative description of rocky shores. Field Studies 1, 1-19.

Beardmore, J. A. 1970. Ecological factors and the variability of gene pools in Drosophila. 299-313. In, Essays in Evolution and Genetics in Honour of Th. Dobzhansky. Evolutionary Biology Suppl.

Blair, W. F. 1955. Mating call and stage of speciation in the Microhyla olivacea - M. carolinensis complex. Evolution 9, 469-480.

Delf, E. M. 1942. The significance of the exposure factor in relation to zonation. Proc. Linn. Soc. Lond. 154, 234-237.

Dobzhansky, Th., Ayala, F. J., Stebbins, G. L. and Valentine, J. W. 1977. Evolution. San Francisco, W. H. Freeman.

Fenchel, T. M. and Christiansen, F. B. 1977. Selection and interspecific competition. 477-498. In, Measuring Selection in Natural Populations. Christiansen, F.B. and Fenchel, T. M. (eds). Berlin, Springer-Verlag.

Johnson, G. B. 1973. Relationship of enzyme polymorphism to species diversity. Nature 242, 193-194.

Johnson, M. S. 1971. Adaptive lactate dehydrogenase variation in the crested blenny Anoplarchus. Heredity 27, 205-226.

Lebour, M. V. 1937. The eggs and larvae of the British prosobranchs with special reference to those living in the plankton. J. Mar. Biol. Ass. U.K. 22, 105-166.

Lewis, J. R. 1953. The ecology of rocky shores around Anglesey. Proc. Zool. Soc. Lond. 123, 481-549.

Lewontin, R. C. 1974. The Genetic Basis of Evolutionary Change. New York, Columbia University Press.

Littlejohn, M. J. 1965. Premating isolation in the *Hyla ewingi* complex (Anura, Hylidae). Evolution 19, 234-243.

McDonald, J. F. and Ayala, F. J. 1974. Genetic response to environmental heterogeneity. Nature 250, 572-574.

Moore, H. B. 1939. The biology of *Littorina littorea*. Part II. Zonation in relation to other gastropods on stony and muddy shores. J. Mar. Biol. Ass. U.K. 24, 227-237.

Moyse, J. and Nelson-Smith, A. 1963. The zonation of animals and plants on rocky shores around Dale, Pembrokeshire. Field Studies 1, 1-31.

Nevo, E. 1976. Genetic variation in constant environments. Experientia 32, 858-859.

Newell, G. E. 1958. The behaviour of *Littorina littorea* (L.) under natural conditions and its relation to position on the shore. J. Mar. Biol. Ass. U.K. 42, 229-237.

Pielou, E. C. 1972. Niche width and niche overlap: A method for measuring them. Ecology 53, 688-692.

Schoener, Th. W. 1965. The evolution of bill size differences among sympatric congeneric species of birds. Evolution 19, 189-213.

Soulé, M. 1976. Allozyme variation: its determinants in time and space. 60-77. In, *Molecular Evolution*. Ayala, F. J. (ed). Sinauer Associates, Inc., Sunderland, Massachusetts.

Valentine, J. W. 1976. Genetic strategies of adaptation. 78-94. In, *Molecular Evolution*. Ayala, F. J. (ed). Sinauer Associates, Inc., Sunderland, Massachusetts.

Van Valen, L. 1965. Morphological variation and width of ecological niche. Am. Nat. 99, 377-390.

Vermeij, G. J. 1972. Intra-specific shore level size gradients in intertidal Mollusca. Ecology 53, 693-700.

ELECTROPHORETIC VARIABILITY AND TEMPERATURE SENSITIVITY OF PHOSPHOGLUCOSE ISOMERASE AND PHOSPHOGLUCOMUTASE IN LITTORINIDS AND OTHER MARINE MOLLUSCS

N. P. Wilkins, D. O'Regan and E. Moynihan

Dept. of Zoology, University College

Galway, Ireland

Introduction

Natural populations of invertebrate species exhibit high levels of genetic variability. Marine invertebrates in particular constitute the most highly genetically variable group of organisms known. What is not known is the way in which this high level of variability helps organisms to adapt to their environment. One approach favoured by many population geneticists to the investigation of this important problem is to estimate levels of variability at numerous loci in relatively small numbers of individuals of a few species, and to relate the average variability of a species (generally expressed as mean heterozygosity over all loci investigated) to selected features of the environment. Where genetic variability is found to be correlated with environmental conditions, organisms are said to exhibit a "genetic strategy of adaptation". Valentine (1976) has recently discussed and reviewed genetic strategies of adaptation in marine organisms.

Some of the assumptions of this approach, which stresses heterozygosity per se, are that the species studied are typical of the environment or region sampled; that the differing sets of enzymes investigated in the different species are representative of the genome, are comparable between species, and are environmentally meaningful; and that the "genetic strategies" of different taxa are similar, at least as a first approximation. All of these assumptions may not be fully justified: electrophoretic analysis of enzyme polymorphisms may detect less than 0.1% of the total variability in the genome of eucaryotes (Powell, 1975), and the extent of variability actually observed is influenced by the relative proportions of different classes of enzymes analysed. For this reason other approaches to the problem of the adaptive significance of genetic variability are necessary to complement that on broad adaptive strategies.

One such complementary approach involves a more detailed study of specific variability at one or a few loci in larger numbers of individuals. These may be from a single species ranging over large geographic areas (e.g. Koehn, 1975) or from a number of species which occur together in one locality but which exhibit different ecological preferences or differing physiological, ethological or reproductive adaptations. In this case the rationale is that since species living sympatrically are subjected to the same macroecological conditions, any genetic differences observed between them within a single locality may reflect differences in their microecological preferences or requirements. For example, overall air and water temperature fluctuations are similar for all intertidal organisms in a given locality, and differences in the extent of their genetic variability at certain loci may correlate with the relative effectiveness with which their behavioural or other known adaptations (e.g. zonal distribution, downshore migration, infaunal habit etc.) dampen the extremes of temperature they actually experience.

The major advantage of this approach is that it may be possible to identify those environmental factors which influence variability at discrete loci and ultimately to understand the mechanism of their action. A further advantage of this approach is an entirely practical one: the identification of variants at discrete loci which affect the survival or success of organisms under known environmental conditions may be used to hasten the process of genetic improvement of maricultured stocks, especially under managed conditions.

In this paper evidence is presented of electrophoretic variability at the Pgi and Pgm loci in sympatric populations of winkles (Littorina littorea (L), L. saxatilis (Olivi) and L. littoralis (L)) from the coasts of Ireland, together with data on the thermostability of PGI in these and other intertidal and shallow water species of molluscs. The observations are discussed with reference to the ecology of these species and their significance for other studies is considered.

Materials and Methods

A total of almost 1800 individuals comprising 615 L. littorea, 771 L. littoralis and 402 L. saxatilis were collected during the periods September 1973 to July 1974 and December 1975 to July 1976. The 1973/4 samples were collected from four localities, viz. Galway, Spiddal, Carna & Killary, situated in or near Galway Bay on the west coast of Ireland. In 1975/6, further samples were collected from three of these localities, and two new localities, one near Cork on the south coast and one near Dublin on the east coast, were sampled. Samples of other species of molluscs mentioned here were collected from these, or nearby sites, at various times in 1977.

Aliquots of muscle tissue were extracted, subjected to electrophoresis and stained as previously described (Wilkins, 1977).

Certain individuals were maintained in aquaria for periods of 9 to 16 days at high ($16^{\circ}C$) or low ($0.5^{\circ}C$) water and air temperatures. These are referred to as warm acclimated and cold acclimated individuals respectively.

For heat inactivation experiments, equal volumes of the crude extract of single individuals of two, or more, species were mixed together to form a single mixed sample. This mixed sample was divided into a number of aliquots. Each aliquot was heated in a water bath for 15 minutes at $45^{\circ}C$, or $50^{\circ}C$, or $55^{\circ}C$, or $60^{\circ}C$. One aliquot was maintained untreated as a control. Certain aliquots were subjected to a single, or to six, freeze-thaw cycles ($-22^{\circ}C$ to $+25^{\circ}C$). All aliquots were analyzed by electrophoresis in the usual fashion after these treatments, the control aliquot being analysed side by side with them.

Note on taxonomy of Littorinids. The taxonomy of littorinids is currently undergoing re-examination and the numbers, characters and status of putative species and sub-species are confused. Littorina littorea (L) is treated by all authorities as a single, good species. Littorina littoralis (L) has been split by Sacchi and Rastelli (1967) into two species L. obtusata & L. mariae. L. mariae occurs much lower on the shore than L. obtusata. The material studied here contains few, if any, L. mariae. The well known "varieties" of L. saxatilis, which is extremely polymorphic for shell characteristics, are considered by Heller (1975) to comprise distinct sympatric species viz. L. rudis, L. patula, L. nigrolineata and L. neglecta, at least on the coast of Wales. L. neglecta is too small to have been included in the L. saxatilis collections discussed here. Recent morphological studies at the collection sites sampled here indicate that both L. saxatilis rudis and L. saxatilis nigrolineata may have been sampled as L. saxatilis. Even if this is so, it does not materially alter the conclusions of this paper since: (a) all samples analysed were in Hardy Weinberg equilibrium and (b) small samples analysed during this survey and subsequent extensive sampling which is still in progress, proves that the common PGI alleles, and all the PGM alleles, are present in "pure" samples of L. saxatilis rudis and L. saxatilis nigrolineata, and in the same rank order of frequency. For the heat inactivation experiments certain individuals were identified according to Heller (1975). These are referred to in Heller's nomenclature where appropriate, and no electrophoretic or thermostability differences were observed between them and undifferentiated L. saxatilis.

Results

Electrophoretic variability. Two alleles at a single locus encoded PGI in all populations of L. littorea. The allozymic fractions observed in this species migrated more anodally than those of the other two species. Three alleles (F^1, F, S) encoded PGI in all populations of L. littoralis. Four Pgi alleles were observed in L. saxatilis; two of these (F,S) were very common in all populations, and two rare alleles (F^1, S^1) were observed to occur in some, but not all, populations. The allozymic products of the F^1, F and S alleles in L. littoralis and L. saxatilis were electrophoretically identical. The S^1 allele of L. saxatilis encoded the slowest fraction observed.

PGM was encoded by three alleles at a single locus in the five populations of L. littorea which were tested, by two alleles in the four populations of L. littoralis and by three alleles in the single population of L. saxatilis analysed at this locus. The PGM allozymes of L. littorea migrated much less anodally than those of L. littoralis or L. saxatilis; the products of the two common alleles in both of the latter two species were electrophoretically identical.

Table 1 summarises the data on PGI in all the samples. All the Pgi alleles were expressed in each population of L. littorea and L. littoralis in both the 1974 and 1976

PHOSPHOGLUCOSE ISOMERASE AND PHOSPHOGLUCOMUTASE

TABLE 1

Variability at the Pgi locus in three species of winkles from Irish coastal sites. N_t: number of individuals analysed. N_a: actual number of alleles observed. p = frequency of most common allele. H_o = observed heterozgosity.

Species, Locality and year		N_t	N_a	p	H_o
Littorina littorea					
A. Galway	1974	20	2	0.925	0.150
	1976	56	2	0.938	0.089
B. Spiddal	1974	69	2	0.949	0.073
	1976	146	2	0.932	0.123
C. Carna	1974	71	2	0.965	0.070
	1976	59	2	0.975	0.051
D. Killary	1974	46	2	0.967	0.065
E. Cork	1976	58	2	0.931	0.140
F. Dublin	1976	90	2	0.944	0.089
Littorina littoralis					
A. Galway	1974	45	3	0.667	0.378
	1976	69	3	0.732	0.507
B. Spiddal	1974	47	3	0.798	0.362
	1976	171	3	0.757	0.374
C. Carna	1974	71	3	0.756	0.409
	1976	77	3	0.721	0.364
D. Killary	1974	39	3	0.718	0.410
E. Cork	1976	97	3	0.690	0.467
F. Dublin	1976	160	3	0.728	0.475
Littorina saxatilis					
A. Galway	1974	68	2	0.846	0.250
B. Spiddal	1974	48	2	0.885	0.188
C. Carna	1974	39	3*	0.680	0.385
	1976	63	3*	0.818	0.302
D. Killary	1974	39	2	0.872	0.224
E. Cork	1976	42	3	0.833	0.262
F. Dublin	1976	103	3	0.626	0.534

*The third, rare, allele at this locality in 1974 was S^1 and and in 1976, F^1. There are, therefore 4 distinct Pgi alleles in this species.

material, and within each species allele frequencies were very similar among the different year classes at each locality. In L. saxatilis, only the two common alleles were observed in all populations in both year classes. The rare S^1 allele was observed at locality C in 1974 and locality F in 1976, at frequencies of 0.01 and 0.005 respectively. The rare F^1 allele was observed at localities C & E in 1976 at frequencies of 0.008 and 0.05 respectively. The observed distributions of phenotypes were in Hardy-Weinberg equilibrium in all the samples.

The extent of genetic variability at this locus can be assessed from the observed number of alleles, the frequency value of the commonest allele, and the observed heterozygosity in each population. L. littorea, in which heterozygosity at this locus never exceeds 0.15, is clearly less variable than the other two species.

At the Pgm locus also (Table 2), L. littorea is much less polymorphic than the other two species. Although three Pgm alleles are expressed in L. littorea, heterozygosity never exceeds 0.125 and the frequency of the commonest allele is never less than 0.93. In contrast, heterozygosity in L. littoralis exceeds 0.34 in all populations, and the frequency of the commonest allele never exceeds 0.78.

Tables 1 and 2 indicate a further difference between L. littorea and the other two species. Judged by the criteria of geographic and chronological variability in allele frequencies, and the occurrence of rare alleles in different geographic isolates, there is little genetic differentiation evident among the different populations of L. littorea, but there is among the populations of L. littoralis, and among those of L. saxatilis. This is confirmed by calculating Hedrick's (1971) index of genotypic identity (I) for all interlocality comparisons for both loci in each species. Values of I for the Pgi locus are highest and most uniform in L. littorea and smaller and less uniform in L. littoralis and L. saxatilis. At the Pgm locus also, I values are higher and more uniform in L. littorea than in L. littoralis. At the three localities (A,B,C) which were sampled both in 1974 and in 1976, inter-year comparisons indicate higher and more uniform values of I in L. littorea (0.996, 0.997, 1.000 resp.) than for L. littoralis (0.939, 0.992, 0.994 resp.) or L. saxatilis (0.945 locality C only).

Heat sensitivity. In all three species, PGI was observed to be remarkably stable after incubation at high temperatures in vitro. Enzyme activity was reduced, but still

TABLE 2

Variability at the Pgm locus in three species of winkles from Irish coastal sites in 1976. N_t, N_a, p, H_o as in Table 1

Species and Locality	N_t	N_a	p	H_o
Littorina littorea				
A. Galway	56	3	0.955	0.089
B. Spiddal	42	3	0.940	0.119
C. Carna	60	3	0.942	0.117
E. Cork	57	3	0.956	0.088
F. Dublin	40	3	0.938	0.125
Littorina littoralis				
A. Galway	72	2	0.625	0.583
C. Carna 7	77	2	0.636	0.468
E. Cork	96	2	0.568	0.552
F. Dublin	127	2	0.772	0.347
Littorina saxatilis				
F. Dublin	47	3	0.777	0.383

appreciable, after incubation at 55°C for 15m. (Fig.1). Of fourteen other molluscan species analysed, only one showed similar resistance to heat denaturation (see below). Complete loss of activity occurred in all species after incubation for

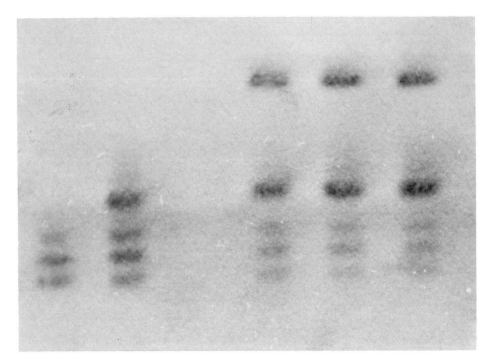

Figure 1. Effect of incubation at elevated temperature on PGI activity in L. littorea, L. littoralis and P. vulgata. Slots 1 and 2: mixed sample of P. vulgata and L. littoralis, control (slot 2) and incubated at 55°C for 15 min. (slot 1). Slots 3 - 6: mixed sample of L. littorea, P. vulgata and L. littoralis, incubated for 15 min. at 60°C (slot 3), 50°C (slot 4) 45°C (slot 5) and control (slot 6). Note how temperatures up to 55°C have no effect on the Littorina species, but activity is reduced in Patella vulgata at 50°C and absent entirely at 55°C.
E = L. littorea; V = P. vulgata; L = L. littoralis, FS heterozygote.

15m. at 60°C. Heat sensitivity experiments were completed on 152 individual winkles comprising 75 L. littorea, 34 L. littoralis and 43 L. saxatilis. Of the L. saxatilis individuals, 8 were "pure" L. saxatilis rudis and 6 were "pure" L. saxatilis nigrolineata. The L. littoralis sample included 2 L. mariae. Homozygous and heterozygous individuals were included for each species. In general, all individuals responded similarly to the heat treatment i.e. enzyme activity was reduced, but still easily detected after 15m. at 55°C. Freezing of crude extracts, on the other hand, had no noticeable effect on any of the species or individuals tested. Of the individuals tested, 58 (28 L. littorea, 15 L. littoralis and 15 L. saxatilis) had been warm acclimated and 47 (28 L. littorea, 6 littoralis and 13 L. saxatilis) had been cold acclimated. No obvious difference was detected in their temperature sensitivity when heated to 55°C and subsequently stained either at 39°C or 1°C. Hence the thermostability characteristics of PGI were independent of acclimation state. There was some indication that in certain individuals in L. saxatilis the FF isozyme may be less heat stable than the FS and SS isozymes of the same individual. However, since enzyme concentrations were standardised only in an imprecise way, and densitometric measurements of staining intensity have not yet been undertaken, the possible occurrence of different thermosensitive PGI alleles in this species remains uncertain.

These experiments were carried out at temperatures of 45°, 50°, 55° & 60°C. In the absence of very precise standardisation of enzyme concentrations it is unlikely that much finer temperature sub-divisions would yield more exact heat denaturation end points. Never the less, the technique, which is simple and in which direct species comparisons are possible, amply demonstrates the relatively high thermostability of PGI in the littorinids and permits a ranking of different species according to the temperature sensitivity of their PGI allozymes. The results obtained with seventeen species of molluscs (9 gastropods and 8 bivalves) are presented in Table 3. No individual differences in sensitivity were observed among the 32 Patella vulgata individuals analysed, nor did acclimation state (10 warm acclimated and 10 cold acclimated) have any noticeable effect. No effect of acclimation state was noted on smaller numbers of individuals of Mytilus edulis, Tapes decussatus and Nucella lapillus.

The species analysed can be classified according to their characteristic distribution in the intertidal region. Some species extend their distribution into the upper reaches of the shore; others occur principally on the lower shore extending, at most, up to mid tide level and still others occur predominantly in the sublittoral region and in permanently moist, low shore habitats. These groups are classified U, L, and S respectively in Table 3. In general, temperature stability was greater in upper shore species than in lower shore and sublittoral species: only one of the six upper shore species (Mytilus edulis) lacked PGI activity after 15m. at 50°C, whereas five of the eleven (45%) lower shore and sublittoral species had lost all PGI activity after this treatment. Estimates are available of heterozygosity at the Pgi locus in most of these species and these are also presented in Table 3. Mean heterozygosity at this locus was highest among those species which were least temperature resistant (no activity after 15m. at 50°C) and lowest among the most resistant species (those unaffected by treatment at 50°C). This negative correlation between mean heterozygosity and heat resistance was evident not only when all species were considered, but also within the low intertidal and sublittoral species considered together; among the upper intertidal species, heterozygosity was greatest in Mytilus edulis, the only heat sensitive species of this group.

In L. littorea, the thermostability of the PGM allozymes was similar to that of PGI, and was greater than that of PGM in L. saxatilis and L. littoralis. The latter two species lost all PGM activity after incubation for 15m. at 55°C, whereas some PGM activity was still evident in L. littorea, even when heated at 60°C for 15 minutes. Crassostrea gigas, Ostrea edulis and Venus verrucosa still exhibited PGM activity after heating to 55°C, but no activity was evident in Dosinia exoleta or Glycymeris glycymeris. PGM in Gibbula cineraria was less heat stable than in G. umbilicalis.

Discussion

Compared with the other two species, L. littorea is considerably less heterozygous at the loci examined, and its allele frequencies are more uniform from locality to locality and from year to year. Snyder & Gooch (1973) who studied the species L. saxatilis alone, observed three common and one rare Pgi allele, extensive heterozygosity and significant interlocality differentiation in that species. The greater degree of interlocality differentiation observed in both L. saxatilis and L. littoralis reflects the limited dispersal ability of these species, both of which lack a planktonic larva. Low dispersal ability retards gene flow and often results additionally in smaller effective population size: both these factors act together to enhance genetic differentiation of geographic isolates.

TABLE 3

Heat-sensitivity of phosphoglucose isomerase in intertidal and shallow water marine molluscs. U, L, S indicate upper shore, lower shore and sublittoral distribution as discussed in text. Het: observed heterozygosity at the PGI locus. V indicates that variability occurs at this locus, but heterozygosity values are not available. Ref: source of heterozygosity data, as follows: 1 Wilkins (1977); 2 Wilkins (1976); 3 Levinton (1973); 4 Wilkins (1975). Where no reference number is indicated, the data are presented here for the first time. +++ enzyme activity undiminished; ++ enzyme activity reduced; - no enzyme activity observed.

Species	Distribution	Activity after heating			Het.	Ref.
		45°C	50°C	55°C		
GASTROPODA						
Littorina littorea	U	+++	+++	++	0.088	
Littorina littoralis	U	+++	+++	++	0.412	
Littorina saxatilis	U	+++	+++	++	0.336	
Gibbula umbilicalis	U	+++	++	-	0.372	
Gibbula cineraria	L	+++	++	-	0.667	
Calliostoma zizyphinum	L	+++	++	-	V	
Patella vulgata	U	+++	++	-	0.005	1
Patella aspera	S	++	-	-	0.275	1
Nucella lapillus	L	++	-	-	V	
BIVALVIA						
Crassostrea gigas	S	+++	+++	++	0.266	2
Ostrea edulis	S	+++	++	-	0.046	2
Mya arenaria	L	+++	++	-	0.412	3
Tapes decussatus	L	+++	++	-	0.609	2
Venus verrucosa	S	++	-	-	0.783	4
Dosinia exoleta	S	++	-	-	0.682	
Glycymeris glycymeris	S	++	-	-	0.368	
Mytilus edulis	U	++	-	-	0.744	2

The distribution of these species on eastern North Atlantic shores is similar, Thorson (1965). All three occur sympatrically from Murmansk in the North to Gibraltar in the South. Whilst considerable overlap can occur on some shores, the various species have characteristic distributions and habitats in the intertidal region at any locality. L. littoralis is always found on, and among, fucoid algae in the mid shore region. L. saxatilis occurs higher on the shore, generally in crevices and among weeds. In contrast, L. littorea is widely distributed from the sublittoral region to MHWN and occurs on a wide variety of physical substrata e.g. on rocks, among small stones, on wooden structures and even on sand and mud. It is the only one of these species which has a pelagic larva. On these grounds it can be inferred that the ecological niche of L. littorea is broader than that of the other two species. Since heterozygosity at both the Pgi and Pgm loci is less in L. littorea, genetic variability appears to be negatively correlated with niche breadth in these related winkles. This result is in good agreement with that on limpets from the same area: genetic variability at these two loci is much less in Patella vulgata than in its ecologically more restricted congener P. aspera (Wilkins, 1977). Furthermore other studies on Winkles lend support to this view (Beardmore & Morris this volume).

The reduced electrophoretic variability of the PGI of P. vulgata is associated with a greater degree of heat resistance: P.aspera is electrophoretically more variable and also less resistant to high temperature. No such association between electrophoretic variability and heat sensitivity is noted among the Littorina species, all of which are highly resistant to heat denaturation. In general, heat stability of PGI tends to be greater in upper intertidal species, than in low intertidal - sublittoral species, although there are notable exceptions e.g. Mytilus edulis and Crassostrea gigas. The C. gigas analysed was from a hatchery produced population and it would be interesting to know whether its unusually high heat stability is a result of hatchery breeding or is characteristic of wild populations as well. Mean heterozygosity at the PGI locus is less in the upper intertidal species than in the low intertidal-sublittoral, although the range of heterozygosities is great within both groups. Sufficient tests have not been carried out to investigate similar trends at the PGM locus, but it does appear that PGM in L. littorea is more heat resistant than in the other Littorina species.

The way in which different heat stability at $55^{\circ}C$ reflects differences in enzyme function at physiological or environmental temperatures is not known. If high stability at 55°C reflects a greater ability of the enzyme to function efficiently at temperatures close to, or slightly above, "normal" environmental temperatures, then the observed trends suggest that high intertidal species have fewer, more adaptive phenotypes (i.e. phenotypes capable of functioning over a wider temperature range) than do low intertidal-sublittoral species. This observation resembles the molecular model proposed by Valentine (1976) to account for differing levels of genetic variability in variable environments. This report, and that on the limpets (Wilkins, 1977) are the first to present experimental evidence in support of this hypothesis as applied to marine invertebrates.

It can be concluded from these studies that closely related species living sympatrically do not necessarily exhibit similar degrees of heterozygosity at homologous loci. Different species, and even different loci in a single species, show differential susceptibility to the influence of specific environmental factors like temperature. The analysis of these differences may contribute to an understanding not only of different levels of electrophoretic heterogeneity between species, but also of the factors maintaining the observed distributions and habitat preferences of littoral and shallow sublittoral organisms.

Summary

Phosphoglucose isomerase is encoded by two alleles in Littorina littorea, by three alleles in L. littoralis and by four alleles in L. saxatilis. Phosophoglucomutase is encoded by three alleles in L. littorea, by two alleles in L. littoralis and by three alleles in L. saxatilis. At both loci, L. littorea is much less heterozygous than the other two species, in samples from six sites around Ireland where these species occur sympatrically. PGM is very heat stable in winkles, and L. littorea which is the least heterozygous species at this locus is also the most heat resistant. In all three species, PGI is very heat stable, enzyme activity remaining appreciable after incubation at 55°C for 15 mins. Heat resistance is not influenced by acclimation state. When other intertidal and shallow water molluscs are compared with these winkles it appears that heat resistance of PGI is greater in upper shore species than in lower shore and sublittoral species. Mean heterozygosity, also, is less in heat stable species than in heat resistant.

Acknowledgements

We thank Miss B. Sherlock for invaluable technical assistance. This work was supported by the National Science Council of Ireland.

REFERENCES

Hedrick, P.W. 1971. A new approach to measuring genetic similarity. Evolution, Lancaster Pa. 25, 276-280.

Heller, J. 1975. The taxonomy of some British Littorina species, with notes on their reproduction (Mollusca: Prosobranchia). Zool. J. Linn Soc 56, 131-151.

Koehn, R.K. 1975. Migration and population structure in the pelagically dispersing marine invertebrate Mytilus edulis. In, Isozymes, IV. Genetics & Evolution. Markert, C.L. (ed). Academic Press, New York.

Levinton, J.S. 1973. Genetic variation in a gradient of environmental variability: marine Bivalvia (Mollusca). Science N.Y. 180, 75-76.

Powell, J.R. 1975. Isozymes and non-Darwinian evolution: a re-evaluation. In, Isozymes, IV. Genetics & Evolution. Markert, C.L. (ed). Academic Press, New York.

Sacchi, C. and Rastelli, M. 1967. Littorina mariae nov. sp. Les differences morphologiques et ecologiques entre "nains" et "normaux" chez L'espece L.Obtusa (L.) (Gastro. Prosobr.) et leur signification adaptative et evolutive. Atti Societa Italiana di Scie. 105, 351-370.

Snyder, T.P. and Gooch, J.L. 1973. Genetic differentiation in Littorina saxatilis (Gastropoda). Mar. Biol. 22, 177-182.

Thorson, G. 1965. The distribution of benthic marine mollusca along the N.E. Atlantic shelf from Gibraltar to Murmansk. Proc. First Europ. Malac. Congr. 1962, 5-25.

Valentine, J.W. 1976. Genetic strategies of adaptation. In, Molecular evolution. Ayala, F.J. (ed). Sinauer Associates, Inc., Sunderland, Massachusetts.

Wilkins, N.P. 1965. Phosphoglucose isomerase in marine molluscs. In, Isozymes, IV. Genetics & Evolution. Markert, C.L. (ed). Academic Press, New York.

Wilkins, N.P. 1976. Genic variability in marine Bivalvia: implications and applications in molluscan mariculture. In, *Proceedings of the 10th European Symposium on Marine Biology. Vol 1. Mariculture.* Persoone, G. & Jaspers, E. (eds). Universa Press, Wetteren, Belgium.

Wilkins, N.P. 1977. Genetic variability in littoral Gastropods: Phosphoglucose isomerase and phosphoglucomutase in *Patella vulgata* and *P. aspera*. Mar. Biol. 40, 151-155.

GENETIC DIVERGENCE BETWEEN POPULATIONS OF THAIS LAMELLOSA (GMELIN)

C. A. Campbell

*Department of Genetics, University of California

Davis, California, U.S.A.

Introduction

The marine gastropod Thais lamellosa has received considerable attention from taxonomists and ecologists because of it's tremendous morphological variation. To the taxonomist, such a vast array of intergrading forms presents a challenge to binominal systematics. At times, the species has been subdivided into a number of taxonomic units. Dall (1915) responded to this great variation by designating six varieties, while Kincaid (1957) considered the species to be composed of numerous races. It is now generally accepted as a single, highly polymorphic species. Ecological studies have sought to assess the extent to which diverse morphological forms reflect differing environmental conditions. This question is still unresolved. Of basic interest to both taxonomy and ecology is the degree to which the morphological differentiation is underlain by genetic differentiation. Are the morphologically distinct populations genetically divergent to the point that they should be regarded as separate subspecies or even species? Does this morphological diversity represent adaption to local environments or merely the random divergence of small, isolated populations?

The development of the techniques of gel electrophoresis during the last decade has provided a tool with which to estimate genetic distance between various taxa on the basis of soluble proteins specified by stuctural gene loci. Data have now accumulated from a variety of organisms related over a range of taxonomic levels. It is therefore possible to determine whether populations of <u>Thais lamellosa</u> are unusually divergent genetically, or whether morphological variability is maintained despite genetic similarities equivalent to those found within monotypic species.

Materials and Methods

<u>Samples</u>. <u>Thais lamellosa</u> has a wide geographic range, occurring between San Francisco, California, and the Bering Straits, Alaska. This range extends over $20°$ latitude, compared with an average of $9°$ for other marine molluscs from this region (Valentine, 1967). Samples of 12 populations were collected during the spring and summer months of 1973, 1974, and 1975 from localities shown in Figure 1. The populations were selected both to represent the latitudinal range of the species and also to sample a range of environments within limited areas. The sample localities are as follows:

1. San Francisco Bay, California. $37°49'$N $122°25'$W. On rock reef approximately 500 feet west of the corner of Beach and Polk Streets.

2. Cape Arago, Oregon. $43°21'$N $124°19'$W. Rock reef in North Cove, approximately 1000 feet due north of northern tip of Cape Arago.

3. Tatoosh Island, Washington. $48°24'$N $124°08'$W. Rocky Island lying approximately one mile northwest of Cape Flattery.

4. Guss Island, Washington. $48°35'$N $123°09'$W. Small rocky islet in the middle of Garrison Bay on the west side of San Juan Island.

5. Turn Rock, Washington. $48°32'$N $122°58'$W. One half mile east of Turn Island in the San Juan Channel.

6. Iceberg Point, Washington. $48°25'$N $122°53'$W. Promontory on the southwest corner of Lopez Island, in the Straits of Juan de Fuca.

GENETIC DIVERGENCE BETWEEN POPULATIONS

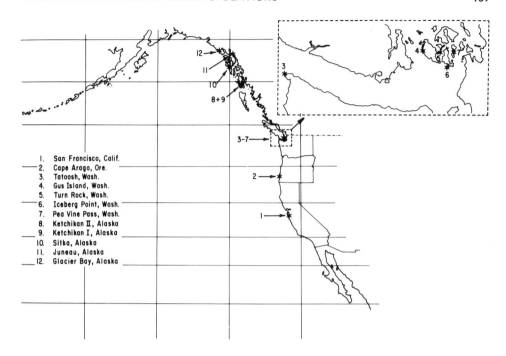

1. San Francisco, Calif.
2. Cape Arago, Ore.
3. Tatoosh, Wash.
4. Gus Island, Wash.
5. Turn Rock, Wash.
6. Iceberg Point, Wash.
7. Pea Vine Pass, Wash.
8. Ketchikan II, Alaska
9. Ketchikan I, Alaska
10. Sitka, Alaska
11. Juneau, Alaska
12. Glacier Bay, Alaska

Figure 1. Localities of twelve populations of Thais lamellosa sampled.

7. Peavine Pass, Washington. 48°35'N 122°48'W. Dredged between 26 and 28 fathoms in Rosario Straits east of Peavine Pass between Obstruction Island and Blakely Island.

8. Ketchikan II, Alaska. 55°21'N 131°44'W. Rocky gravel beach below airport on Gravina Island, just west of Ketchikan.

9. Ketchikan I, Alaska. Approximately one mile north of previous locality on rocky point near Lewis Point.

10. Sitka, Alaska. 58°07'N 135°22'W. Rocks on north side of small bay 2½ miles northwest of Sitka.

11. Juneau, Alaska. 58°26'N 134°45'W. Steep rocky cliffs north of Auke Bay.

12. Glacier Bay, Alaska. 57°20'N 136°45'W. On eastern side of the western arm of Torch Bay, Alaska, 3/4 mile southeast of mouth.

Electrophoretic Techniques. Snails were transported to the University of California, Davis, either live or frozen on dry ice. They were maintained at -70°C. until used for electrophoresis. Freezing did not seem to affect the mobility or functioning of the enzymes assayed. The spire of each shell was cut through immediately above the body whorl using a copper-bladed rock saw. The columnar muscle was detached and the animal removed from the shell and sexed. Experience indicated that the digestive glands and gonads were best suited for enzyme assay. In large individuals, only these two tissues were used, but in smaller individuals the whole body was employed. The tissue or whole body samples were placed in a test tube and homogenized in equal volumes of distilled water and toluene while the tube was immersed in an ice bath. The homogenate was centrifuged at 17,000 RPM for twenty minutes. The aqueous layer was removed and stored at -70°C. until used.

Fourteen zones of activity could be reliably scored on gels stained for the following enzymes: galactose-6-phosphate dehydrogenase (locus Gal6pdh), glucose-6-phosphate dehydrogenase (Glu6pdh), hexokinase (Hk), isocitrate dehydrogenase (Idh), lactate dehydrogenase (Ldh), leucine amino peptidase (Lap), malic enzyme (Me), octanol dehydrogenase (Odh), and xanthine dehydrogenase (Xdh). The enzymes were assayed using slight modifications of the techniques of Ayala et al., (1973); details of the methods will appear elsewhere (Campbell, in preparation).

Results

The observed allelic frequencies at each locus in the twelve populations are given in Table 1. Genetic similarities were estimated from this data using Nei's (1972) statistic of genetic identity, I. The probability that two alleles from different populations are the same is

$$I_j = \frac{\Sigma x_i y_i}{(\Sigma x_i^2 \Sigma y_i^2)^{1/2}}$$

where x_i and y_i represent the frequencies of the ith allele in populations x and y respectively. The mean genetic identity over several loci is

$$I = \frac{J_{xy}}{(J_x J_y)^{1/2}}$$

where J_x, J_y, and J_{xy} are the arithmetic means over all loci $\Sigma\, x_i^2$, $\Sigma\, y_i^2$, and $\Sigma\, x_i^2\, \Sigma\, y_i^2$ respectively

The pair-wise genetic identities between the twelve populations are given in Table 2. The values range from 0.749 to 0.997 with a mean genetic identity of 0.926 ± 0.062. Data on genetic similarities are available for a number of organisms related at various taxonomic levels (for reviews see Ayala, 1975 and Avise, 1976). An extensive study of the willistoni group of Drosophila (Ayala et al., 1974) reveals the following levels of genetic divergence calculated with Nei's identity statistic for taxa of increasing evolutionary divergence:

conspecific populations I = 0.970 ± 0.006

subspecies I = 0.795 ± 0.013

semi-species I = 0.798 ± 0.026

sibling species I = 0.517 ± 0.024

non-sibling species I = 0.352 ± 0.023

Studies of most other organisms have shown remarkable agreement with these figures. However, certain groups, notably mammals, have been found to exhibit considerably less genetic divergence (King and Wilson, 1975). This reduced evolutionary divergence has been interpreted as indicating that evolution has proceeded by differentiation of regulatory rather than structural genes in these animals.

Most of the pairwise comparisons of populations in the present study fall within the normal range for conspecific populations. However, lower values occur between certain pairs, notably those involving the Glacier Bay (number 12) and Tatoosh (number 3) populations, which are more typical of groups related at the subspecific or semi-specific level.

Three loci are responsible for most of the differentiation between the populations. At the Me-3 locus, seven bands of activity are present. Populations can be divided into two groups on the basis of whether they have alleles from the set of the three slower migrating bands or from the set of the four faster migrating ones. No population was found which expressed a mixture of alleles from the two sets. Four bands of activity appear at the Me-2 locus. In all populations except Glacier Bay the band designated Me-2$_{100}$ occurs at the highest frequency, although in four of the populations, few individuals express the locus at all. At

TABLE 1

Allele frequencies in samples from twelve populations of *Thais lamellosa*, numbered as in Figure 1.

Locus	Allele	Populations				
		1	2	3	4	5
Gal6pdh	102	0	0.02	0	0	0
	100	0.86	0.55	0.86	0.74	0.92
	98	0.10	0.31	0.02	0.09	0.08
	95	0.03	0.10	0.13	0.18	0
	94	0	0.02	0	0	0
Glu6pdh	102	0	0	0.10	0	0
	100	1.00	1.00	0.99	1.00	0.97
	98	0	0	0	0	0.03
Hk-2	102	0	0	0	0	0
	100	1.00	1.00	0.97	1.00	1.00
	98	0	0	0.03	0	0
Hk-3	102	0	0	0	0	0
	100	1.00	1.00	1.00*	1.00*	1.00*
	90	0	0	0	0	0
Idh-1	102	0	0	0.02	0	0
	100	1.00	1.00	0.98	1.00	1.00
	98	0	0	0	0	0
Idh-2	102	0	0	0	0	0
	100	1.00	1.00	1.00	1.00	1.00
	98	0	0	0	0	0
Lap-1	102	0	0	0	0	0
	100	1.00	1.00	1.00	1.00	1.00
	98	0	0	0	0	0
Lap-2	102	0	0.02	0.01	0	0.01
	100	1.00	0.95	0.96	1.00	0.99
	98	0	0.03	0.03	0	0
Me-1	102	0	0	0.02	0.03	0.04
	100	0.97	0.91	0.81	0.88	0.84
	98	0.03	0.09	0.16	0.09	0.12
	96	0	0	0.01	0	0
Me-2	105	0	0.19	1.00	0	0
	100	0.48	0.81	0	1.00	1.00
	99	0.02	0	0	1.00	1.00
	98	0.50	0	0	0	0
Me-3	102	0.17	0	0	0	0
	100	0.83	0	0	0	0.97
	98	0	0	0	0	0.03
	91	0	0	0.03	0	0
	90	0	0.64	0.85	0.71	0
	88	0	0.36	0.03	0.29	0
	86	0	0	0.09	0	0
Ldh	FIXED					
Odh	FIXED					
Xdh	FIXED					

*Where only a few individuals expressed the locus, it was assumed to be fixed for the allele most common to the species if at least one individual expressed that allele.

GENETIC DIVERGENCE BETWEEN POPULATIONS

6	7	8	9	10	11	12
0	0	0	0	0	0.01	0.02
0.92	0.77	0.83	0.60	0.50	0.45	0.67
0.03	0	0.02	0.40	0.48	0.18	0.27
0.06	0.23	0.15	0	0.02	0.33	0.04
0	0	0	0	0	0.02	0
0	0	0.01	0	0	0	0.02
1.00	0.98	0.89	0.96	1.00	0.95	0.95
0	0.02	0.10	0.04	0	0.05	0.04
0.01	0	0.01	0	0	0	0
0.99	1.00	0.98	1.00	1.00	0.99	0.99
0	0	0.01	0	0	0.01	0.01
0.04	0	0	0.11	0	0.05	0
0.96	1.00*	0	0.89	1.00	0.85	1.00
0	0	1.00	0	0	0.10	0
0	0	0	0	0	0	0
1.00	1.00	1.00	1.00	1.00	1.00	0.96
0	0	0	0	0	0	0.04
0	0	0	0.07	0	0	0
1.00	1.00	1.00	0.82	1.00	1.00	1.00
0	0	0	0.11	0	0	0
0	0	0	0	0	0	0.98
1.00	1.00	1.00	1.00	1.00	1.00	0
0	0	0	0	0	0	0.02
0.01	0	0	0.15	0.01	0	0.01
0.99	1.00	0.97	0.97	0.99	0.99	0.85
0	0	0.03	0.15	0	0.01	0.14
0.03	0	0.05	0	0	0	0.01
0.86	0.99	0.89	0.96	0 89	0.78	0.89
0.11	0.01	0.06	0.04	0.11	0.19	0.09
0	0	0	0	0	0.03	0
0	0	0	0	0.18	0.17	0
0.63	1.00*	1.00*	1.00*	0.68	0.75	0.75
0.08	0	0	0	0	0	0.01
0.29	0	0	0	0.14	0.08	0.24
0	0	0	0	0	0	0.01
0	0	0	0	0	0	0.94
0	0	0	0	0	0	0.05
0.03	0.06	0	0.08	0.02	0.02	0
0.58	0.94	1.00	0.61	0.95	0.98	0
0.38	0	0	0.27	0.02	0	0
0.02	0	0	0.04	0	0	0

TABLE 2

Nei's identity for pair-wise comparisons of twelve populations of <u>Thais lamellosa</u>, numbered as in Figure 1.

	1	2	3	4	5	6	7	8	9	10	11
2	0.913										
3	0.861	0.922									
4	0.912	0.992	0.903								
5	0.973	0.920	0.834	0.928							
6	0.938	0.982	0.920	0.990	0.925						
7	0.901	0.980	0.907	0.993	0.915	0.974					
8	0.897	0.976	0.907	0.991	0.912	0.973	0.912				
9	0.908	0.991	0.885	0.989	0.923	0.973	0.978	0.976			
10	0.900	0.986	0.929	0.976	0.894	0.967	0.977	0.976	0.979		
11	0.889	0.980	0.927	0.981	0.891	0.964	0.985	0.984	0.970	0.988	
12	0.892	0.829	0.749	0.828	0.898	0.831	0.816	0.810	0.827	0.814	0.800

the Gal6pdh locus, there are five alleles. The same allele is the most common in every population, but the frequency ranges from 0.92 to 0.45.

Discussion

The twelve populations of Thais lamellosa studied show a wide range of genetic differentiation based on allele frequencies at fourteen structural gene loci. Two populations in particular, Glacier Bay and Tatoosh, exhibit a great degree of differentiation. Using Nei's statistic of genetic identity, the average value of pairwise comparisons between Glacier Bay and the remaining eleven populations is 0.827 ± 0.040; the average of all pairwise comparisons with the Tatoosh population is 0.885 ± 0.054. In addition, five other pairwise comparisons involving three different populations have identity values under 0.900.

There is no evidence of latitudinal clines in allele frequency at any locus. Determining an association between certain alleles and environmental factors is, of course, difficult, but certainly none seems likely here. Me-2 provides the clearest separation of populations into two groups. Three populations are distinguished by having two or more of the three slower migrating alleles: San Francisco, Glacier Bay and Peavine Pass. These populations occur at the two ends and in the middle of the latitudinal range. There are no associations between this set of alleles and any features of shell form. Individuals from San Francisco Bay have medium-sized, heavy shells with little or no sculpture, those from Peavine Pass have large, highly sculptured, thick shells, and those from Glacier Bay have small, highly sculptured, thin shells. All three populations were living in relatively quiet water localities with little wave action. However, Guss Island and the Ketchikan populations are also from quiet water environments. It is not possible, of course, to rule out the possibility that selection is operating at the Me-2 locus, but the relevant environmental factor is not evident.

As no locus shows a definite correlation between allele frequency and either geographic or environmental factors, the possibility of random genetic differentiation between populations must be considered. In fact, the population structure of the species is such as to favour drift. Thais lamellosa does not have a pelagic larval stage. Eggs are laid in attached egg cases and the young emerge as crawl-aways. Adults are unable to cross soft substrates and are therefore confined to continuous rocky surfaces. Furthermore, adults congregate to form breeding aggregations and often return to the same breeding aggregation year after year (Spight, 1972), thereby reducing the effective population size and the amount of migration between demes. New populations may be established by one or a few adults or eggs being washed to a previously uninhabited rocky shore. Thus the existence of small, isolated populations and the founder effect may facilitate the action of random genetic drift.

The low mobility of Thais lamellosa suggests that these populations have evolved largely independently, and genetic differences should accumulate over time due both to selective and non-selective forces. Limited migration does occur between populations. An evolutionary tree model is not adequate for depicting relationships between conspecific populations because it does not permit hybridization or mixing of populations after an initial separation (Morton et al., 1968). A minimum length spanning network, which represents relationships between groups at a single time, is a more appropriate way to depict such populational relations. First proposed by Edwards and Cavalli-Sforza (1963) as a method of approximating minimum evolution based on present day gene frequencies, this method uses pairwise measures of genetic distances to construct a minimum length spanning network. The network is a representation of a single instant of time (Thompson, 1975).

The programme used, called MINITREE, was developed by E. A. Thompson and is a modification of earlier programmes by A. W. F. Edwards and A. Cornfield. The programme seeks the minimum spanning network between n populations in n-1 dimensional space.

The minimum length spanning network between the twelve populations of Thais lamellosa is projected in two dimensions in Figure 2. The populations are numbered as in Figure 1, with number 1 being the southernmost and number 12 being the northernmost population. Rotation is free around any of the internal nodes; thus it is not possible to tell the distance between two points separated by one or more nodes except by tracing along the lines connecting them.

As can be seen from Figure 2, the similarity of populations does not correlate with their geographic positions. Nor is it possible to attribute genetic similarity to any environmental parameter, such as wave exposure, substrate, food type, or temperature.

If random genetic drift were the primary factor determining allele frequency, one would expect smaller and more

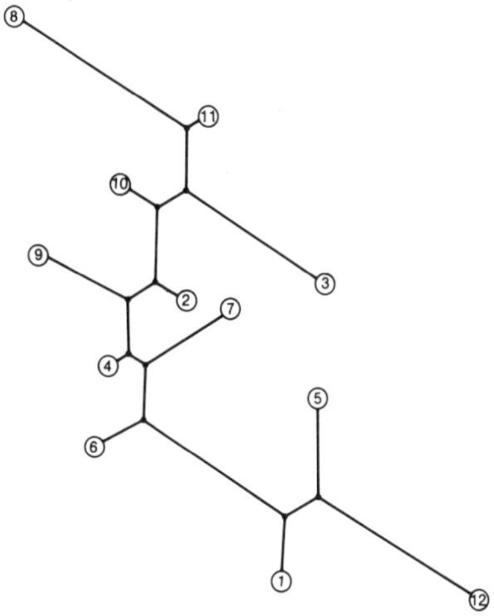

Figure 2. Minimum length spanning network between the twelve populations of Thais lamellosa. 1 = San Francisco Bay, California; 2 = Cape Arago. Oregon; 3 = Tatoosh Island, Washington; 4 = Guss Island, Washington; 5 = Turn Rock, Washington; 6 = Iceberg Point, Washington; 7 = Peavine Pass, Washington; 8 = Ketchikan II, Alaska; 9 = Ketchikan I, Alaska; 10 = Juneau, Alaska; 11 = Sitka, Alaska; 12 = Glacier Bay, Alaska.

isolated populations to be more differentiated than larger ones. However, there is no reliable indication of a correlation between genetic similarity and either estimated population size or isolation. Guss Island (number 4) and Tatoosh (number 3) are the two apparently smallest and most isolated populations. Guss Island is a small rocky outcrop in a bay, surrounded by sandy and muddy beaches, inhospitable localities for Thais lamellosa. Tatoosh Island is located approximately one mile off the Northwest tip of the Olympic peninsula, and only some scores of snails were found inhabiting the whole island. The Guss Island population clusters closely with the other populations ($I = 0.952 \pm 0.053$). Tatoosh, on the other hand, is one of the more divergent populations ($I = 0.856 \pm 0.055$). However, Glacier Bay (number 12) is also highly divergent, but is quite large, and other large populations occur nearby.

It is difficult to estimate population numbers reliably, however, and breeding aggregates may further distort the effective population size. Isolation may be determined more by capricious physical factors, such as storms, than by geographic distance. Furthermore, one cannot reconstruct the past history of a population to discover whether it has recently passed through a bottleneck.

The present study does not provide consistent evidence of a correlation between population size and divergence. Nevertheless, genetic drift cannot be discounted as one of the factors determining the genetic identity in Thais lamellosa.

Summary

The techniques of gel electrophoresis were employed to study genetic divergence between twelve populations of Thais lamellosa. The degree of differentiation, as estimated with Nei's (1972) statistic of genetic identity, varies greatly between populations. The majority of pairwise comparisons of the population samples yield identities of $I > 0.900$. However, certain populations gave consistently lower comparisons with values as low as $I = 0.749$. The degree of divergence does not follow a geographic pattern, nor were any evident environmental factors correlated with genetic similarity. Thus, there is no reason to assign the populations subspecific, racial or ecotypic status.

The range and the apparently random pattern of divergence suggest that non-selective mechanisms may be the primary influence determining the allelic frequencies in the populations. Certain rather unusual aspects of the life history pattern of this snail reinforce the maintenance of small, reproductively isolated populations. Thus, random genetic drift may influence the genetic structure of this species more than it does in other, panmictic organisms.

Acknowledgements

Dr. Elizabeth A. Thompson of the Statistical Laboratory and King's College, Cambridge University, made available the computer programmes used in this study and gave generously of her time and advice. Dr. F.J. Ayala provided continuing support and advice for the project. Dr. J.W. Valentine offered useful criticism of the manuscript. This paper was made possible by travel grants from the Cambridge Philosophical Society and Kings' College, Cambridge.

REFERENCES

Avise, J.C. 1976. Genetic differentiation during speciation. In, *Molecular Evolution*. Ayala, F. J. (ed). Sinauer Associates, Inc., Sunderland Massachusetts.

Ayala, F. J. 1975. Genetic differentiation during the speciation process. Evol. Biol. 8, 1-78.

Ayala, F. J., Hedgecock, D., Zumwalt, G. S. and Valentine, J. W. 1973. Genetic variation in *Tridacna maxima*, an ecological analog of some unsuccessful evolutionary lineages. Evolution 27, 177-191.

Ayala, F. J., Tracey, M. L., Barr, L. G., McDonald, J. F. and Perez-Salas, S. 1974. Genetic variation in natural populations of five *Drosophila* species and the hypothesis of the selective neutrality of protein polymorphisms. Genetics 77, 343-384.

Dall, W. H. 1915. Notes on the species of the molluscan subgenus *Nucella* inhabitating the northwest coast of America and adjacent regions. Proc. U.S. Nat. Mus. 49, (2124) 557-572.

Edwards, A. W. F. and Cavalli-Sforza, L. L. 1963. The reconstruction of evolution. Heredity 18, 553.

Kincaid, T. 1957. Local races and clines in the marine gastropod, *Thais lamellosa* Gmelin. A population study. Calliostoma Press, Seattle, Washington.

King, M.C. and Wilson, A. C. 1975. Evolution at two levels. Molecular similarities and biological differences between humans and chimpanzees. Science 188, 107-116.

Morton, N.E., Yasuda, N., Miki, C. and Yee, S. 1968. Bioassay of population structure and isolation by distance. Am. J. Hum. Genet. 20, 411-419.

Nei, M. 1972. Genetic distance between populations. Am. Nat. 106, 282-292.

Spight, T.M. 1972. Patterns of change in adjacent populations of an intertidal snail, *Thais lamellosa.* Ph.D. thesis, University of Washington.

Thompson, E.A. 1973. The method of minimum evolution. Ann. Hum. Genet. 37, 69-80.

Thompson, E. A. 1975. *Human Evolutionary Trees*. Cambridge University Press.

Valentine, J.W. 1967. The influence of climatic fluctuations on species diversity within the Tethyan Provincial System. In, *Systematics Association Publication No.7*, Aspects of Tethyan Biogeography. Adams, C.G. and Ager, D. V. (eds).

GEOGRAPHIC VARIATION IN PROTEIN POLYMORPHISMS IN THE EELPOUT, ZOARCES VIVIPARUS (L.)

F. B. Christiansen and V. Simonsen

Dept. of Ecology and Genetics, University of Aarhus

Dk 8000 Aarhus C, Denmark

Introduction

Marine organisms are in many respects well suited for population genetic investigations. Many are abundant and form large dense populations which may be sampled extensively without disturbing their natural population densities. However, as a population genetic investigation must be founded on knowledge of the formal genetic laws that govern the studied variation, difficulties arise as few marine organisms are well suited for extensive genetic analyses in the laboratory. This difficulty may be overcome in organisms with direct development and brood protection, given that the variation of interest may be recognized in the embryos. The protecting parent and its offspring may then be considered as a breeding experiment performed to disclose the mechanism of inheritance of the variation expressed in the population.

The teleostean fish the eelpout Zoarces viviparus (L.) (Percomorphi, Blennioidei, Zoarcidae), has this fortunate property of brood protection as it is viviparous. The eelpout (Fig.1) occurs in shallow water along the shores of Northern Europe from the White Sea to the English Channel and from the Baltic Sea to the east coast of Great Britain. It matures at an age of one and a half years at a size of about 25 cm. It mates in the late summer and the female delivers the young, about 5 cm in size, in the mid-winter. The brood size varies from about 50 in two year old females up to hundreds in older females.

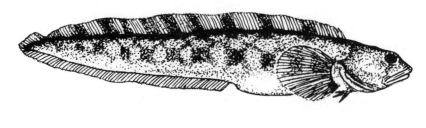

Figure 1 An adult and a newborn eelpout, Zoarces viviparus.

For the biochemically defined loci EstIII, HbI, PgmI and PgmII, which will be the major concern of this paper, the genotypes of the feti may be recognized from November, through January. Each of these four loci shows variation which is due mainly to the segregation of two co-dominant alleles. This assertion has been confirmed by extensive mother-offspring studies (Simonsen and Frydenberg, 1972; Hjorth, 1971, 1974, 1975). Furthermore, the analyses show that EstIII and HbI, EstIII and Pgm loci, and PgmI and and PgmII segregate independently.

We intend to review the data pertaining to the geographic variation in gene frequencies at the four loci mentioned. However, it seems appropriate first to discuss an additional characteristic of organisms in which samples including incomplete family data may be obtained, namely the possibility of a detailed analysis of selection (Christiansen and Frydenberg, 1973, 1976; Cooper, 1966, 1968).

As an eelpout litter may stem from more than one mating, analysis of segregation in the litters provides incomplete information concerning the fertilizing males. Nevertheless, rather detailed information about the population may be obtained by collecting samples including mother-offspring combinations (Table 1). A sample of adults is taken during the time when the females are carrying the young.

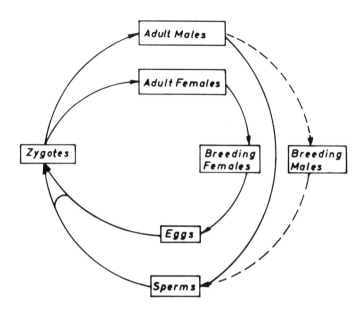

Figure 2   Life cycle stages represented in a simple mother-offspring sample.

TABLE 1

Structure of a mother-offspring sample

Adult genotypes	11	12	22	Sum	Adult ♀♀ NPF	Total	Adult ♂♂	All Adults
11	C_{11}	C_{12}		F_1	S_1	$A_{♀1}$	$A_{♂1}$	A_{O1}
12	C_{21}	C_{22}	C_{23}	F_2	S_2	$A_{♀2}$	$A_{♂2}$	A_{O2}
22		C_{32}	C_{33}	F_3	S_3	$A_{♀3}$	$A_{♂3}$	A_{O3}
Sum	C_{O1}	C_{O2}	C_{O3}	F_O	S_O	$A_♀$	$A_♂$	A_{OO}

The genotypes of the adults are determined and they are separated according to sex. Furthermore, the females are separated into pregnant and non-pregnant females (NPF) and for each pregnant female the genotype of <u>one</u> randomly chosen fetus is scored so that each pregnant female is given a mother-offspring type.

Three stages in the life cycle of the animal are represented in a population sample including mother-offspring combinations; <u>viz</u>., the population of zygotes, the population of adults and the populations of adults involved in reproduction (Fig.2). The adult population is evidently represented by the unsorted adult sample, and the zygote population is represented by the random sample of feti which may be estimated from the mother-offspring table if knowledge about the zygote production (brood size) of each female is available. The population of female breeders is represented simply by the pregnant females and the population of breeding males is represented by the fertilizing male gametes which may be identified in the mother-offspring combinations.

This partitioning of the life cycle as indicated in Figure 2 may be used for a detailed investigation of selection, as had been done for the <u>EstIII</u> polymorphism in the eelpout (Christiansen and Frydenberg, 1973; Christiansen, Frydenberg and Simonsen, 1973). Comparisons between zygotes and adults provide information about the viability of the genotypes during juvenile life; comparisons between adults and breeders provide information on the sexual performance of the genotypes and the analysis of the transition from breeders to zygotes provides information on fecundity selection and gametic selection.

However, the representation of the life cycle in Figure 2 applies only to an organism with discrete non-overlapping generations and the eelpout may breed for several consecutive years after maturity is reached. Therefore, an evaluation of selection working during adult life requires the sample of adults and breeders to be structured according to age so that the natural units of the population, the cohorts, may be followed (Christiansen and Frydenberg 1976). Such investigations have been made for the <u>EstIII</u> polymorphism (Christiansen, Frydenberg and Simonsen 1977a), and we will refer to the results later.

Geographic Variation in Polymorphism

<u>Large scale geographic variation</u>. In the years 1969-

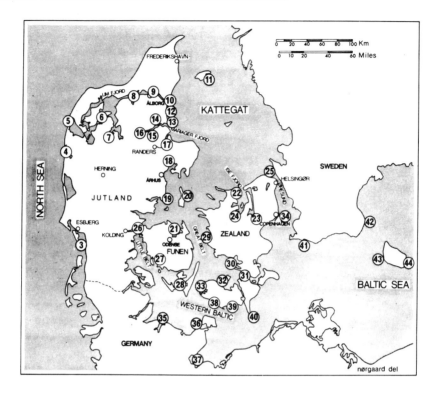

Figure 3. Location of sampling stations in the investigations 1969-1971. (Frydenberg et al., 1973).

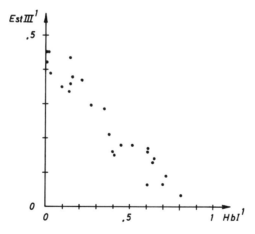

Figure 4. Relationship between gene frequencies of $EstIII^1$ and HbI^1 at locations in the region covered by location nos. 18-40 (Fig.3). (Data from Frydenberg et al., 1973; Hjorth and Simonsen, 1975, and Christiansen et al., 1977b).

1971, the geographic variation in the four loci EstIII, HbI, PgmI and PgmII was investigated in the waters surrounding Denmark. The area of investigation is depicted in Figure 3. Within this area the four loci studied fall into two groups according to the geographic variation of the polymorphisms. The two loci EstIII and HbI show large scale geographic variation in allele frequencies (Frydenberg et al., 1973; Hjort and Simonsen, 1975), whereas the two Pgm loci are characterized by fairly constant allele frequencies (Christiansen et al., 1976). The esterase variation is characterized by high frequencies of the allele $EstIII^1$ (0.35-0.45) in the populations in the Limfjord and Kattegat and by low frequencies (about 0.05) in the Baltic and in Holland. These areas of high and low frequencies are connected by clines, one along the North Sea coast and one through the Danish Belt Sea. The variation in hemoglobins in the inner Danish waters is very similar to that of the esterases. The HbI^1 allele is virtually absent in the North Sea, the Limfjord and Kattegat; whereas it rises values above 0.90 in the populations in the Baltic. These two areas are connected by a cline through the Danish Belt Sea which is of a shape similar to that of EstIII (Fig.4).

In contrast to the differentiation observed at the EstIII and HbI loci, the allele frequencies at the PgmI and PgmII loci show only minor variations. This dichotomy in the pattern of geographic variation was analyzed by Christiansen and Frydenberg (1974) by attempting to extract information about the phenomena responsible for the different patterns. The analysis is necessarily performed as a test of the null hypothesis that all four loci are, and have been, selectively neutral, so that the patterns have emerged from the interaction of genetic drift and migration. The parallelity of the EstIII and HbI clines are consistent with the hypothesis that two populations, a Kattegat population and a Baltic population, are diffusing into each other. The null hypothesis of neutrality would accordingly suffice to explain the present geographic patterns of allele frequencies if it explains the frequencies in these two populations as having originated by genetic drift in two semi-isolated populations (Table 2). Christiansen and Frydenberg (1974) argued that either the divergence observed at the EstIII and HbI loci or the concordance of the Pgm loci may be the result of drift of neutral polymorphisms through a hypothesized time span in two virtually isolated populations; however, it was clear that genetic drift is unlikely to account for both phenomena. Therefore, selective forces have to be introduced to explain one of the two types of geographic patterns.

However, three qualifications of this conclusion should be

TABLE 2

The hypothetical Kattegat and Baltic populations
(Christiansen and Frydenberg, 1974).

Allele	EstIII[1]	HbI[1]	PgmI[1]	PgmII[1]
Kattegat	0.40	0.00	0.60	0.15
Baltic	0.05	0.95	0.60	0.15

stressed. First, it provides no information as to which group of loci has been subjected to selection. Secondly, selection is the agent in at least one of the patterns, but this selection does not necessarily act in contemporary populations. Finally, the selective forces need not be functions of adaptive differences among genotypes at the locus considered. If an allele at one of our marker loci at some point in time has been associated with an allele at a closely linked locus subject to selection, then the marker allele may "hitch-hike" and inherit some of the selection working on the neighboring locus (Maynard Smith and Haigh, 1974; Thomson, 1977).

Microgeographic variation. In addition to the large scale geographic variations in gene frequencies, the data from 1969-1971 show some interesting microgeographic patterns.

Let us first reconsider the geographic constancy of the gene frequencies of the Pgm loci. As an example, the gene frequencies at the PgmII locus are shown in Figure 5, upper graph. Although the gene frequency variation over the investigated area is trivial compared to the variation at the EstIII and HbI loci, it is clear that neighboring populations are on the average more similar than expected from the observed variation in gene frequencies over the entire area studied. As a guide, we have encircled populations belonging to the same general area. From the left, the Limfjord, Kattegat, the Belt Sea, the Western Baltic and finally the Baltic are shown (Fig.3). The statistical analysis of this variation is shown in Figure 5, lower graph. This graph shows the standardized deviation of the gene frequency p_i observed at the i'th location from the mean gene frequency p_o of all samples:

$$d_i = (p_i - p_o) / \left[p_o (1-p_o)/2N_i \right]^{1/2}$$

where N_i is the sample size at the i'th location. Under the assumption that the gene frequency is the same in all locations, this deviation measure is approximately normally distributed with zero mean and unit variance. This approximation is expected to be very good provided that the number of locations is so large that we may consider p_o as a constant when compared to the probable fluctuation of p_i. Each deviation measure, d_i, may be considered as a test for the hypothesis that the gene frequency at this location is equal to the mean gene frequency. Assuming this hypothesis to

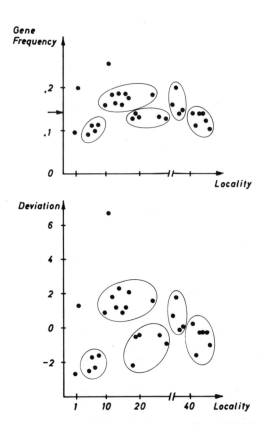

Figure 5. Untransformed (upper) and transformed (lower) gene frequencies of the PgmII[1] allele. The abscissa shows the location number as defined in Figure 3. (After Christiansen et al., 1976).

be true, the variation of the d_i's is predicted to form the normal distribution approximation, e.g., about 95 percent of the d_i's should fall in the interval -2 to 2. A glance at at Figure 5 reveals that the spread of the d_i's is too large, so that there is significant heterogeneity among gene frequencies. This heterogeneity can be fully explained by the observed differences between the above mentioned regions in that the variation of the gene frequencies within these regions is as expected from a hypothesis of regional homogeneity (Christiansen et al., 1976).

In addition to this regional variation, a more thorough analysis of the Pgm variation discloses some local differentiations (Christiansen et al., 1976). One example is apparent from Figure 5, viz. location 11 which is the island Laesø (p_i is six standard deviations from p_o).

A similar analysis of the loci EstIII and HbI is difficult since local or regional differentiations in these cases must be inferred as deviations from the common variation shown in Figure 4. (The reason for this is our ignorance about the populations structure of the studied area such that we are unable to predict the gene frequencies in the populations from their geographic location.)

However, outside the region of proper clinal variation and under special geographic circumstances, we may still be able to disclose local differentiations. Here one striking feature emerges from the EstIII data, namely that the frequency of the EstIII^1 allele seems to increase inside fjords. This is especially pronounced in a cline through Mariager Fjord (locations 13, 14, 15 and 16), where location 15 is actually the population with the highest frequency observed throughout the 1969-1971 investigations.

Analysis of three fjords. To further study the microgeographic variation, the two large Kattegat fjord systems, Mariager Fjord (Fig.6) and Isefjord/Roskilde Fjord (Fig.7), were investigated in 1974. In addition, the German fjord, Slien, just north of location 35 (Figs. 3 and 8), was investigated to further illuminate the patterns observed in the Kattegat fjord systems. The information concerning the gene frequencies in these fjords is summarized in Tables 3, 4 and 5 with the sample sizes in the different years, depicted in Table 6.

For the Pgm loci, this investigation was mainly a confirmation of the previously described patterns (Fig. 9). The regional constancy of the Northern Kattegat population was confirmed by new data but an additional atypical population was sampled in this area, viz., Roskilde Fjord location 1. This population shows a strongly deviant gene frequency in PgmI and heterogeneity between the sexes in PgmII with the males showing a deviant gene frequency. In addition, Slien defines a new region or atypical population as the PgmI gene frequencies were different from the Belt Sea population and the Western Baltic population.

The HbI locus is almost monomorphic in the two Kattegat fjord systems, but the HbI^1 frequency nevertheless shows significant variation in Isefjord and Roskilde Fjord, having a higher frequency at the mouth than in the inner part of the fjords. In Slien, where the HbI alleles are at polymorphic frequencies, there is again no sign of geographic heterogeneity.

The data from 1974 again show the pattern of higher $EstIII^1$ frequencies in the inner part of the two Kattegat fjord systems (Fig. 10) but no similar trend appeared in Slien.

Discussion

The investigation of eelpout populations shows that the geographic variation of polymorphisms is a powerful tool for disclosing the structure of population subdivision in a species. This method has been successfully employed in other marine fish species (Frydenberg et al., 1965; Sick, 1965a, b; Johnson 1971; Jamieson, this volume). We are in the fortunate position of being able to compare the genetic variation of the eelpout with its morphological variation. Johannes Schmidt (1917a, b, 1918; Smith 1921, 1922) investigated the geographic variation of four heritable metric characters. The pattern of variation in these characters is very similar to the pattern of variation at the Pgm loci, viz. there is great similarity within regions, but with differences between regions and with local irregular populations. However, one tendency was apparent in the variation of the metric characters, viz. clines of identical directions in the larger fjords, particularly in the three mentioned above. Thus the differentiation between the mouth and the inner part of the Kattegat fjords observed in EstIII is paralleled by a similar morphological differentiation. This parallel variation may or may not be accidental in that the two types of clines do not depend on the same causal factor. However, if there is a common cause of the two phenomena we expect the parallel variation to recur in

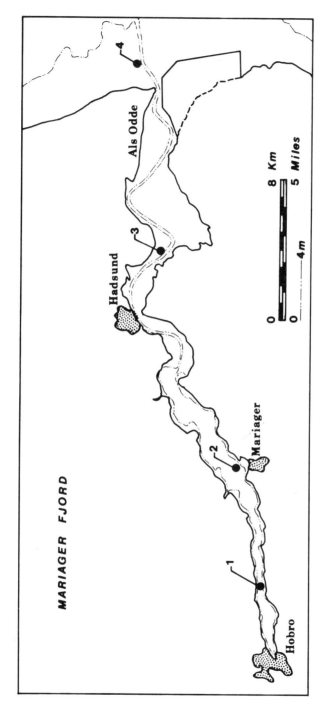

Figure 6. Sampling locations in Mariager Fjord. All the locations coincide with those sampled in 1969 and 1971 (From Christiansen et al., 1977b).

TABLE 3

Gene frequencies in Mariager Fjord based on samples 1969, 1971, 1974. Sample sizes shown in parentheses.

Data from Frydenberg et al., 1973, Hjorth and Simonsen, 1975; Christiansen et al., 1976, 1977b.

Location	$EstIII^1$	HbI^1	$PgmI^1$	$PgmII^1$
1	0.516 (224)	0.016 (222)	0.613 (221)	0.167 (225)
2	0.532 (391)	0.013 (200)^a	0.602 (382)	0.161 (383)
3	0.420 (194)	0.005 (103)	0.585 (188)	0.186 (188)
4	0.447 (850)^c	0.007 (681)	0.601 (390)^b	0.182 (393)^b
Σ	0.473 (1659)	0.009 (1206)	0.601 (1181)	0.173 (1187)

a: the 1969 sample omitted.
b: the Pgm loci are not typed in the 1971 sample.
c: an extra sample from 1976 included.

TABLE 4

Gene frequencies in Isefjord/Roskilde Fjord based on samples 1969, 1971 and 1974. Sample sizes are shown in parentheses.

Data from Frydenberg et al., 1973; Hjorth and Simonsen, 1975; Christiansen et al., 1976, 1977b.

Location	$EstIII^1$	HbI^1	$PgmI^1$	$PgmII^1$
1	0.452 (505)	0.009 (510)	0.731 (199)^a	0.140 (197)^{a,b}
2	0.422 (198)	0.010 (198)	0.642 (194)	0.180 (189)
3	0.390 (450)	0.034 (451)	0.598 (198)^a	0.178 (197)^a
4	0.449 (188)	0.016 (125)	0.617 (107)	0.183 (104)
Σ	0.427 (1341)	.019 (1284)	0.651 (698)	0.168 (687)

a; the Pgm loci are not typed in the 1971 sample.
b: here there is a significant difference between the gene frequencies of males and females.

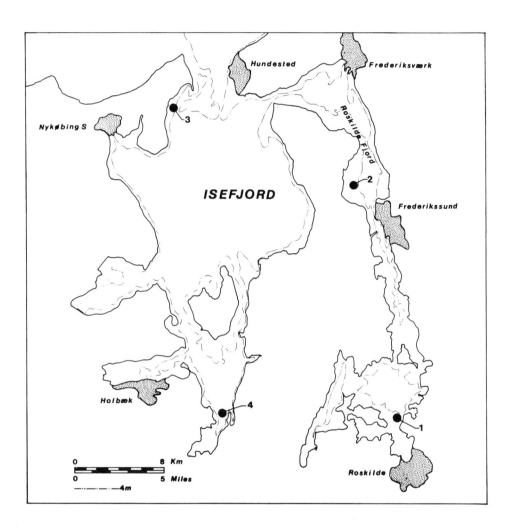

Figure 7. Sampling locations in Isefjord and Roskile Fjord. Locations 1, 3 and 4 coincide with those sampled in 1969 and 1971 (From Christiansen et al., 1977b).

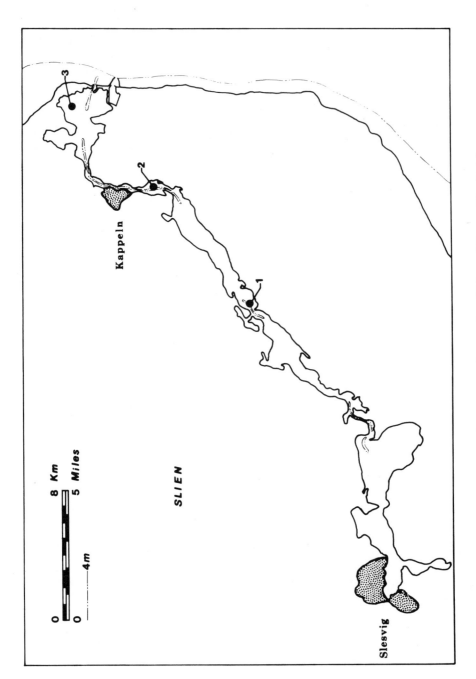

Figure 8. Sampling locations in Slien (From Christiansen et al., 1977b).

TABLE 5

Gene frequencies in Slien based on samples 1974.
Sample sizes are shown in parentheses.
Data from Christiansen et al., 1977b.

Location	EstIII^1	HbI^1	PgmI^1	PgmII^1
1	0.152 (161)	0.413 (160)	0.691 (160)	0.127 (161)
2	0.181 (166)	0.452 (166)	0.770 (165)	0.152 (165)
3	0.160 (210)	0.398 (210)	0.715 (209)	0.124 (210)
Σ	0.164 (537)	0.419 (536)	0.725 (534)	0.133 (536)

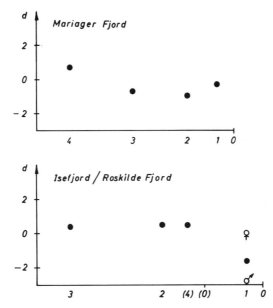

Figure 9. Transformed gene frequencies of PgmII^1 observed in the Kattegat fjord systems in 1974. The d statistics are a comparison between the observed gene frequency and the mean gene frequency of the Kattegat region. (Data from Christiansen et al., 1976, 1977b).

TABLE 6

Samples in the fjords.
Location number on Figure 3 shown in parentheses.

Location	Sample size in 1969	1971	1974	Σ
Mariager Fjord				
1 (16)	51		176	227
2 (15)	193		200	393
3 (14)	198			198
4 (13)	210	326	200	866[a]
Isefjord/Roskilde Fjord				
1 (23)		311	200	511
2			200	200
3 (22)		251	200	451
4 (24)	197			197
Slien				
1			161	161
2			167	167
3			210	210

a: including a sample from 1976.

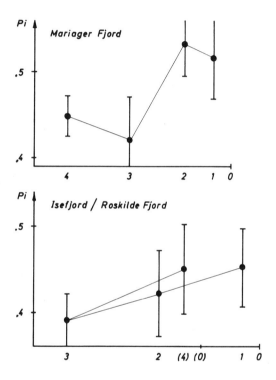

Figure 10. Variation in the EstIII[1] frequency through the Kattegat fjord systems. (Data from Frydenberg et al., 1973 and Christiansen et al., 1977b).

Slien, but no EstIII cline appeared there. This prediction
is founded on the assumption that the eelpout populations in
the fjords are comparable over the time span of fifty years,
an assumption which is not appropriate for Slien. Here
Schmidt·s samples are from two locations, one corresponding
to our position 3, the other from just outside Slesvig
(Fig. 8). However, at present eelpouts are sparse in the
inner parts of Slien, and we only obtained a sample from
location 1 with great difficulty.

Above and beyond the descriptive approach taken in
these investigations, we might try to interpret the observed
patterns in terms of adaptive differences between the genotypes
at the loci under consideration. Here it is most appealing
to interpret the clines as representing variation which may
be correlated with something else. However, for the clines
through the Danish Belts, this "something else" could be anything else, as any relevant environmental variation will vary
between the North Sea and the Baltic. Thus, a correlation
will not present any evidence concerning the interaction
between genetics and the environment and, furthermore, any
such correlation will be questioned by the North Sea cline in
EstIII. Nevertheless, in EstIII, the clines in the Kattegat
fjords may guide us towards a relevant environmental variable
as the clines in the fjords may be viewed as continuations of
the cline from the Baltic through the Belts. This immediately
rules out salinity as the relevant variable, but it suggests
temperature as the pattern of seasonal temperature variation
changes monotonically along the clines (Fig.11).

The population at location 18 (Fig. 3) has been subject
to an attempt to measure selection forces working on the
EstIII polymorphism. During the years 1969-1974, mother-
offspring samples were collected and analysed according to
the methodology described in the introduction (Christiansen
et al., 1973, 1974, 1977a). The result of this investigation
is the demonstration of selection against the heterozygote
in the time interval between birth and maturity. This is a
destabilizing selective force, so it does not explain the
polymorphism at the EstIII locus. However, a population
in a cline need not have the characteristics predicted by
the selective forces working at the location (Karlin and
Richter-Dyn, 1976), so the observation of inferior heterozygotes is not necessarily inconsistent with the observation
of polymorphism (Christiansen, 1977). Nevertheless, the
observations of underdominant selection at the EstIII locus
may serve as a warning against a too naive interpretation of
the geographic variation at this locus.

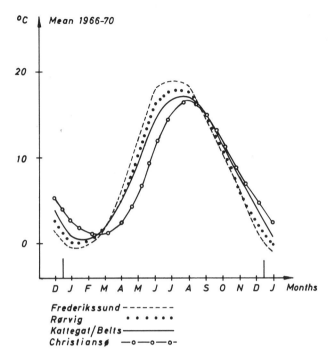

Figure 11. Monthly mean temperature at coast stations. ---: location 2 in Roskilde Fjord. ••••: location 1 in Isefjord. _____: mean of coast stations in Kattegat and the Belts. -•-•-: the Baltic off Bornholm. The means are calculated for the years 1966-1970 from the tables of "Oceanographic observations from Danish light-vessels and coastal stations, Det Danske Meteorologiske Institut, Charlottenlund".

Adaptive differences between the genotypes at the EstIII and HbI loci as a cause of the clines between Kattegat and the Baltic are acceptable. The two populations are virtually isolated from one another, so even rather small selective forces may through time lead to major genetic differentiations. It is an intriguing question whether the inner parts of the fjords may contain populations which to a similar extent are isolated from the rest of the species, or whether we have to postulate large selective forces to account for the observed differentiation. Here it is worth recalling that for the Kattegat fjords the EstIII differentiation of the inner parts of the fjords is paralleled by a differentiation in polygenic characters. Thus we must envisage the inner populations as genetically highly differentiated from the outer populations and accept, therefore, that the populations are rather isolated from one another even given the small geographic distance between them.

If parts of the genetic differentiation of the fjord populations are translated into phenotypic adaptations to the local environment, then the effect of immigration may be greatly reduced as descendants of immigrants to the fjords may have a lower fitness than the local breed (see Levinton and Lassen, this volume). The observed differentiation may thus be an effect of what may be termed accelerating differentiation. Imagine that the population in the inner part of the fjord is founded as a copy of the population outside the fjord, and assume that the fjord population is receiving immigrants at a non-trivial rate. We further assume that the fjord habitat is quite different from the habitat in the sea outside so that selection for adaptive changes is induced at a score of loci. Initially the limited isolation only allows for differentiation at loci where strong selection is induced. But this differentiation increases the fitness of the fjord population, and consequently the fitness of the immigrants are decreased which decreases the effective amount of migration. This increase in isolation of the inner fjord population allows for further differentiations at the already differentiated loci, but in addition it allows for minor adaptive adjustments at other loci. This again strengthens the isolation which in turn allows for more genetic differentiation.

This effect of accelerating differentiation is made possible primarily from the build up of linkage disequilibrium in the transition zone between two differentiated populations (Feldman and Christiansen 1975). This linkage disequilibrium will have the effect of amplifying the selective forces at single loci as correlations will exist between the occurrence of inner fjord alleles at different loci. This amplification of the selection for a specific allele in the fjord population will diminish the effective inflow of the alternative allele from outside the fjord (Slatkin 1975). These phenomena may be interesting to consider in a more general evolutionary context, because in the long run this kind of differentiation may conceivably induce other isolating mechanisms (Maynard Smith 1966; Balkau and Feldman 1973; Karlin and McGregor 1974: Christiansen and Feldman 1975).

Summary

The geographic variation of four protein polymorphisms, EstIII, HbI, PgmI and PgmII, has been studied extensively

over a part of the distribution range of the eelpout with
special emphasis on the inner Danish waters and the Baltic
Sea. The four loci fall into two contrasting groups, one
consisting of EstIII and HbI where the geographic variation
is dominated by a cline from Kattegat through the Danish Belt
Sea to the Baltic Sea and the other consisting of PgmI and
PgmII which is characterized by no appreciable variation in
allelic frequencies. The clines in EstIII and HbI are
remarkably parallel, justifying the conclusion that the two
aforementioned groups of loci express two different phenomena.

Apart from the major trends in the geographic variation,
detailed analyses show that a significant amount of local
differentiation occurs with loci of either group. To
illuminate these observations further, three large fjords
in the area have been studied more extensively.

Acknowledgements

We are indebted to Tom Fenchel for permission to use
Figure 1. Figures 2 and 4-11 were prepared by Arno Jensen.
Dr. Bengt Olle Bentson made valuable comments on the
manuscript.

REFERENCES

Balkau, B.J. and Feldman, M. W. 1973. Selection for migration modification. Genetics 74, 171-174.

Christiansen, F.B. 1977. Population genetics of Zoarces: A review. In, Measuring Selection in Natural Populations, Christiansen, F.B. and Fenchel, T.M. (eds). Lecture Notes in Biomathematics V19, 21-47, Springer Verlag, Berlin.

Christiansen, F.B. and Feldman, M. 1975. Subdivided populations: A review of the one-and two-locus deterministic theory. Theoret. Popul. Biol. 7, 13-38.

Christiansen, F. B. and Frydenberg, O. 1973. Selection component analysis of natural polymorphisms using population samples including mother-offspring combinations. Theoret. Popul. Biol. 4, 425-445.

Christiansen, F.B. and Frydenberg, O. 1974. Geographical Patterns of Four Polymorphisms in Zoarces viviparus as evidence of selection. Genetics 77, 765-770.

Christiansen, F. B. and Frydenberg, O. 1976. Selection component analysis of natural polymorphisms using mother-offspring samples of successive cohorts. In, Population Genetics and Ecology, Karlin, V. and Nevo, N. (eds). pp. 277-301. Academic Press, Inc., New York, San Francisco, London.

Christiansen, F. B., Frydenberg, O., Gyldenholm, A.O. and Simonsen, V. 1974. Genetics of Zoarces populations. VI. Further evidence, based on age group samples, of a heterozygote deficit in the EstIII polymorphisms. Hereditas 77, 225-236.

Christiansen, F.B., Frydenberg, O., Hjorth, J.P. and Simonsen, V. 1976. Genetics of Zoarces populations. IX. Geographic variation at the three phosphoglucomutase loci. Hereditas 83, 245-256.

Christiansen, F.B., Frydenberg, O. and Simonsen, V. 1973. Genetics of Zoarces populations. IV. Selection component analysis of an esterase polymorphism using population samples including mother-offspring combinations. Hereditas 73, 291-304.

Christiansen, F.B., Frydenberg, O. and Simonsen, V. 1977a. Genetics of Zoarces populations. X. Selection component analysis of the EstIII polymorphism using samples of successive cohorts. Hereditas, in press.

Christiansen, F.B., Frydenberg, O. and Simonsen, V. 1977b. Genetics of Zoarces populations. XI. Variations at four polymorphic loci in fjords. Hereditas, in preparation.

Cooper, D.W. 1966. A note on the examination of genotypic ratios in domestic animals using incomplete family data. Animal Production 8, 511-513.

Cooper, D.W. 1968. The use of incomplete family data in the study of selection and population structure in marsupials and domestic animals. Genetics 69, 147-156.

Feldman, M. and Christiansen, F.B. 1975. The effect of population subdivision on two loci without selection. Genet. Res., Camb. 24, 151-162.

Frydenberg, O., Gyldenholm, A.O., Hjorth, J.P. and Simonsen, V. 1973. Genetics of Zoarces populations. III. Geographic variations in the esterase polymorphism EstIII. Hereditas 73, 233-238.

Frydenberg, O., Møller, D., Naevdal, G. and Sick, K. 1965. Hemoglobin polymorphism in Norwegian cod populations. Hereditas 53, 257-271.

Hjorth, J.P. 1971. Genetics of Zoarces populations. I. Three loci determining the phosphoglucomutase isoenzymes in brain tissue. Hereditas 69, 233-242.

Hjorth, J.P. 1974. Genetics of Zoarces populations. VII. Fetal and adult hemoglobins and a polymorphism common to both. Hereditas 78, 69-72.

Hjorth, J.P. 1975. Molecular and genetic structure of multiple hemoglobins in the eelpout, Zoarces viviparus L. Biochem. Genet. 13, 379-391.

Hjorth, J.P. and Simonsen, V. 1975. Genetics of Zoarces populations. VIII. Geographic variations common to the polymorphic loci HbI and EstIII. Hereditas 81, 173-184.

Johnson, M.S. 1971. Adaptive lactate dehydrogenase variation in the crested blenny, Anoplarchus. Heredity 27, 205-224.

Karlin, S. and McGregor, J. 1974. Towards a theory of the evolution of modifier genes. Theoret. Popul. Biol. 5, 59-103.

Karlin, S. and Richter-Dyn, N. 1976. Some theoretical analyses of migration selection interaction in a cline: A generalized two range environment. pp. 659-706. In, Population Genetics and Ecology. Karlin, V. and Nevo, E. (eds). Academic Press, New York, San Francisco, London.

Maynard Smith, J. 1966. Sympatric Speciation. Amer. Natur. 100, 637-650.

Maynard Smith, J. and Haigh, J. 1974. The hitchhiking effect of a favourable gene. Genet. Res. 23, 23-35.

Schmidt, J. 1917a. Zoarces viviparus L. and local races of the same. C.R. Trav. Lab. Carlsberg 13(3), 277-397.

Schmidt, J. 1917b. Constancy investigations continued. C.R. Trav. Lab. Carlsberg 14(1), 1-19.

Schmidt, J. 1918. Racial Studies in Fishes I. Statistical Investigations with Zoarces viviparus L. J. Genet. 7, 105-118.

Sick, K. 1965a. Haemoglobin polymorphism of cod in the Baltic and the Danish Belt Sea. Hereditas 54, 19-48.

Sick, K. 1965b. Haemoglobin polymorphism of cod in the North Sea and the North Atlantic Ocean. Hereditas 54, 49-69.

Simonsen, V. and Frydenberg, O. 1972. Genetics of Zoarces populations II. Three loci determining esterase isoenzymes in eye and brain tissue. Hereditas 70, 235-242.

Slatkin, M. 1975. Gene flow and selection in a two locus system. Genetics 81, 787-802.

Smith, K. 1921. Statistical investigations on inheritance in Zoarces viviparus L. C.R. Trav. Lab. Carlsberg 14(11), 1-64.

Smith, K. 1922. Continued investigations on inheritance in Zoarces viviparus L. C. R. Trav. Lab. Carlsberg 14(19), 1-42.

Thomson, G. 1977. The effect of a selected locus on linked neutral loci. Genetics 85, 753-788.

GENETICS OF THE COLONIAL ASCIDIAN, BOTRYLLUS SCHLOSSERI

A. Sabbadin

Istituto di Biologia Animale

Università di Padova, Italy

Introduction

The colonial ascidian, Botryllus schlosseri, is suitable material for genetic studies in many respects. Each colony is a clone which can consist of thousands of zooids grouped into hundreds of systems. Replicas may be obtained by isolating and transplanting the systems; on the other hand, these systems can be processed sequentially for many characters, while the parental clone remains as a bank for the genetic markers that are gradually determined. The colonies can be readily cultured for indefinite periods in the laboratory, and are available for controlled breeding throughout the year (Milkman, 1967; Sabbadin, 1960; 1972).

Colonial structure and breeding habit. Many ascidians are solitary forms. It is difficult to say whether this habit or coloniality arose first (Millar, 1966). What can be established is that there has been an evolution in the colonial habit, the starting point being the simple contiguity of the colonial zooids, the end point being integration of the individuals into a morpho-physiological unit. The latter condition is that which has been achieved by the species of the genus Botryllus.

A colony of B. schlosseri consists of three coexisting blastogenic generations of zooids, the adults and two successive generations of buds, the stages of which are so correlated that the disappearance of the older generation coincides with the maturation of the next one and the initiation

of a new generation (Berrill, 1941). The zooids of each
generation behave as a single individual, because their
origin is confined to the same point in time in the life
cycle of the parents which are kept synchronized by a common
circulation. If the systems of zooids are separated, a de-
synchronization of their stages eventually appears; if then
the vascular connections are re-established, synchrony of
phase of the stages will follow.

At 18°C the blastogenic generations within the colony
succeed each other at weekly intervals. Ovaries and testes
form in the zooids of each generation. The eggs develop to
full growth in the buds, and are fertilized soon after the
opening of the siphons which allows the entry of sperm.
But, as shown by Milkman (1967), sperm are discharged by
the new adults somewhat later, so that protogyny is effective
in preventing self-fertilization in this species, in which
the zooids of any given generation are at the same sexual
stage. One colony crossed to another colony at a different
sexual stage acts as female and then as male (Sabbadin, 1971).
The embryonic development takes place inside the parental
body; free-swimming larvae are released slightly before the
parental zooids regress.

There is no sexual self-incompatibility, but self-
fertilization, if induced by interbreeding pieces of the same
colony at different sexual stages, results in a remarkable
inbreeding depression. This depression is expressed not
as a reduction of the percentage of fertilized eggs, but as
a reduction of the percentage of normal and metamorphosing
larvae, and in reduced survival and growth of the newly
founded colonies (Sabbadin, 1971).

<u>Genetics of colour polymorphism</u>. Natural populations
of <u>B. schlosseri</u> display a variety of colour morphs that in
the past were sometimes considered to represent different
species. The colour morphs result from a combination of
characters of two types: (a) the nephrocytes, silvery-
white in colour when seen by reflected light, are either dispersed or arranged in an intersiphonal band, double or single,
and/or in a peristomatic ring; (b) other types of pigmented
hemocytes - orange, bluish, reddish cells - may be either present or absent. On this basis, 48 colour morphs have been
identified (Sabbadin and Graziani, 1967).

Genetic analysis has shown that these characters are
controlled by five Mendelian loci, each with a dominant and
a recessive allele. The dominant allele is responsible for
the intersiphonal double band, the peristomatic ring, the

presence of orange and reddish pigments and the absence of blue pigment (Table 1).

Linkage tests have revealed the independence of locus A from B, B1, R, and of locus B1 from B and R (Sabbadin and Graziani, 1967). That is to say that at least 3 out of the 16 chromosomes, which form the haploid complement of the species (Colombera, 1969) probably carry pigmentation genes. More recently a close linkage has been demonstrated between the B and R loci (Sabbadin, 1977). Recombination occurs in both male and female gametogenesis, to the same extent. The total recombination between these two loci amounts to about 2% and these two genes thus represent the nucleus of the first linkage group discovered in this species.

Genetics of enzyme polymorphisms. Our familiarity with the population of Botryllus of the Venetian Lagoon, which for many years furnished the principal stock for our cultures, has long raised a question concerning its structure and dynamics. This problem could not be appropriately attacked using the colour polymorphism, as the heterozygous genotypes cannot be directly typed. We turned therefore to the study of enzyme polymorphisms.

Two enzymes were investigated by means of polyacrylamide gel electrophoresis: the supernatant malate dehydrogenase (MDH) and tetrazolium oxidase (TO). Both enzymes proved to be polymorphic in the Lagoon. The electrophoretic patterns suggest that each system is under the control of 4 codominant alleles at one locus ($\underline{Mdh-1}^a$, $\underline{-1}^b$, $\underline{-1}^c$, $\underline{-1}^d$; $\underline{To-1}^a$, $\underline{-1}^b$, $\underline{-1}^c$, $\underline{-1}^d$), with 10 single-banded and three-banded phenotypes involved. The hybrid band of the heterozygotes points to a dimeric structure for both types of molecules. We have successfully produced in vitro hybridization of the monomeric units encoded by $\underline{To-1}^b$ and $\underline{To-1}^c$.

At both loci frequency of the extreme alleles $\underline{-1}^a$ and $\underline{-1}^d$ hardly reaches 1% and this explains why some phenotypes have not yet been found in the wild. The phenotype To-1 A has however been obtained in the laboratory. The hypothesis of multi-allelic control of the two enzymes has been tested by breeding experiments, and the data presented in Table 2 fit the hypothesis.

The breeding experiments offered an opportunity of testing the linkage relationships between the two enzyme loci and between each of them and some of the loci which govern the polychromatism (Table 3). Table 3 shows clearly that in

TABLE 1

Genetic analysis of the colour polymorphism of **Botryllus schlosseri**. (Summarizing data from Sabbadin, 1959; Sabbadin & Graziani, 1967).

Character	Parental genotypes			Offspring phenotypes	
				A	a
orange pigment	a/a	x	a/a		344
	A/A	x	a/a	1,420	
	A/a	x	a/a	318	305
	A/a	x	A/a	246	75
				B1	b1
blue pigment	b1/b1	x	b1/b1		1,498
	B1/B1	x	b1/b1	796	
	B1/b1	x	b1/b1	1,059	1,031
	B1/b1	x	B1/b1	426	139
				R	r
reddish pigment	r/r	x	r/r		937
	R/R	x	r/r	76	
	R/r	x	r/r	515	504
	R/r	x	R/r	50	14
				B	b
double band	b/b	x	b/b		1,384
	B/B	x	b/b	928	
	B/b	x	b/b	880	853
	B/b	x	B/b	142	41
				F	f
perist. ring	f/f	x	f/f		87
	F/F	x	f/f	36	
	F/f	x	f/f	53	50
	F/f	x	F/f	32	13

TABLE 2

Genetic analysis of enzyme polymorphisms in
Botryllus schlosseri

Enzyme	Parental phenotypes	No. of crosses	Offspring phenotypes					
			B	BC	C	BD	CD	D
MDH	C x C	11			159			
	B x C	1		32				
	BC x BC	4	28	64	24			
	C x BC	6		83	75			
	BC x CD	1		4	5	–	3	
			A	AB	B	AC	BC	C
TO	B x B	6			140			
	B x C	2					19	
	BC x BC	3			14		27	15
	B x BC	9			117		122	
	C x BC	1					6	9
	AB x AB	1	2	4	–			
	AB x BC	1		3	1	1	3	

TABLE 3

Results of crosses testing linkage relationships between enzyme and pigmentation loci

Loci	Parental genotypes	Offspring phenotypes				Expected ratio
		BC,B	C,BC	BC,BC	C,B	
Mdh-1/To-1	b/c,b/c x c/c,b/b	5	10	3	6	1:1:1:1
		BC,A	C,A	BC,a	C,a	
Mdh-1/A	b/c,A/a x c/c,a/a	2	21	16	2	1:1:1:1
		2	16	13	0	
		BC,A	B,A	BC,a	B,a	
To-1/A	b/c,A/a x b/b,A/a	9	20	6	6	3:3:1:1
		13	9	1	4	
		BC,B	C,B	BC,b	C,b	
Mdh-1/B	b/c,B/b x c/c,b/b	14	7	11	7	1:1:1:1
		9	15	9	8	
		BC,B	B,B	BC,b	B,b	
To-1/B	b/c,B/b x b/b,b/b	7	4	12	9	1:1:1:1
	b/c,B/b x b/b,B/b	11	21	4	5	3:3:1:1
		BC,R	C,R	BC,r	C,r	
Mdh-1/R	b/c,R/r x c/c,r/r	9	8	9	15	1:1:1:1
		BC,R	B,R	BC,r	B,r	
To-1/R	b/c,R/r x b/b,r/r	12	9	7	4	1:1:1:1

most cases there is no evidence of linkage. Thus Mdh-1 and To-1 are independent, while To-1 appears to be independent of the pigmentation locus A, to which, however, Mdh-1 is linked (χ^2 for pooled data, 3df, = 54.1, p = < 0.001). This is the second linkage group to be demonstrated in the species, with a recombination value of about 7%. Mdh-1 is independent of the two linked genes B and R; the same holds true for To-1.

Factors affecting the genetic structure of Botryllus populations. There are some features in the colonial mode of life of B. schlosseri that should affect the genetic structure of populations, gene pool and flow. (a) There is a discontinuity in distribution of the animals. In the Venetian Lagoon the muddy bottom is not suitable for Botryllus, which is chiefly found on the pilings marking the channels and on local beds of Zostera. (b) The duration of the free-swimming larval stage is of the order of hours. Within the Lagoon the limited larval dispersal is conditioned by the tidal water mass which enters the harbour-channels and is preferentially distributed along the internal channels with minor overflowing into the adjacent shallower waters (Fig.1). (c) A preferential settlement of larvae in the proximity of the parental colonies is inferred from our lab-

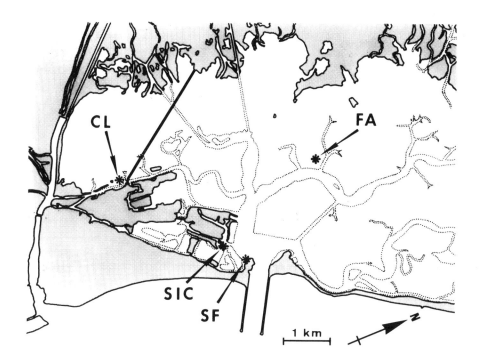

Figure 1. Chart of a section of the southern basin of the Venetian Lagoon near the city of Chioggia. Localities from which samples of Botryllus were taken. (The harbour-channel leads into the Adriatic Sea).

oratory observations. (d) Crowding of larvae is well documented in this species, as well as in many other ascidians. Crowding involves competition for space that in some species is known to entail great mortality. In Botryllus the competition ought to be greatly enhanced by colonial growth.

A mechanism of self/nonself-recognition might be considered the basic strategy adopted by Botryllus to oppose both the formation of highly inbred isolates and the impoverishment of their gene pool, which would result from the above mentioned phenomena. Colonies that come into contact by means of the vascular ampullae branching out from the marginal vessel in the tunic either fuse, or repel each other. It has been shown (Oka and Watanabe, 1957; Oka, 1970) that in B. primigenus this phenomenon is governed by a series of alleles at one locus, recalling the sexual self-incompatibility S gene of many flowering plants. Unpublished data show that a similar mechanism is effective in B. schlosseri. Fusion occurs when the apposed colonies share at least one allele. Each colony fuses with both the parental colonies, and the offspring of two incompatible colonies ($F^1F^2 \times F^3F^4$) segregate into four groups (F^1F^3, F^1F^4, F^2F^3, F^2F^4) each of them recognizing as self two other groups.

Fusion consists of the anastomosis of the vascular systems of the partners. The common circulation results in a synchronization of their sexual stages and this, in turn, thanks to protogyny, prevents fused colonies from intercrossing. The closer the relationship, the higher the probability of sharing common genes and thus fusing. Outbreeding is favoured.

A quantitative evaluation of the fusion process should now be possible by scoring the biochemical phenotypes for marker genes. A double phenotype related to one locus was occasionally met with in our electrophoretic analysis of colonies collected in the Lagoon, though only a small piece of each colony was used and the material had previously been scored for the pigmentation characters in order to avoid misinterpreting fused colonies for a single colony.

The composite colony resulting from the fusion of two or more colonies will preserve the genomes of the components, even if one or more of these components perishes. Actually, germ cells or stem cells of the germinal line are exchanged between the fused colonies, cycled through the common circulation, then transmitted to, and matured, by the zooids of successive blastogenic generations. This has been proved experimentally (Sabbadin, 1972). Colonies of the opposite genotypes AAbb and aaBB after remaining fused for some time were separated and each of them was crossed with the double recessive genotype aabb. For more than ten blastogenic generations we continued to find the phenotype Ab, in addition to the phenotype aB, in the offspring from the cross aaBB x aabb, and the phenotype aB, in addition to the phenotype Ab, in the offspring from the cross AAbb x aabb. This was true of both male and female lines.

Preservation and recycling of germ cells ought to be effective in counteracting the effect of a reduction in population size.

Population genetics. The results of our first approach to the population genetics of Botryllus in the Venetian Lagoon are presented.

Samples were taken in the southern basin of the Lagoon (Fig.1) from several sites and at different times, both from the natural habitat, pilings or beds of Zostera, and from artificial panels set out at critical periods in order to provide a check on the sexual generations subsequently sampled (Table 4). In the Venetian Lagoon sexual reproduction in Botryllus extends from April to November, that is the period during which the water temperature remains above 10°C (Sabbadin, 1955). Brunetti (1974) has shown that four sexual generations occur in one year: one in spring, two during the summer, and one in autumn. The latter is the overwintering generation which will reproduce in the next spring. Overlapping of generations is rather limited, because sexual maturity entails a metabolic crisis which usually results in the colonies dying out prior to the reproduction of the next generation.

The samples were examined for MDH and TO. The mean allele and heterozygote frequencies derived from the pooled samples at each locality are given in Table 5. To test the goodness-of-fit to the expected genotype distribution, on the assumption of Hardy-Weinberg equilibrium, the two rare alleles of each locus were pooled with their adjacent allele (Li and Horvitz, 1953). Two of the 26 χ^2 were found to be significant at the 5% level of probability, both of them referring to To-1 and paralleled by insignificant χ^2 values in the same samples examined for Mdh-1. On the whole the observed and the expected frequencies of heterozygotes are in close agreement and are indicative of random mating. The distribution of the mean frequencies at the different localities gives a rough idea of the similarities between the subpopulations.

TABLE 4

Distribution of samples of Botryllus
in time and space

Locality	Substrate	Date	Dominant generation	Sample size
SF	pilings	15 May	no. 4	87
	panels (3 May)	16 June	1	197
	" (16 June)	3 Aug.	2	45
	" (31 Aug.)	16 Dec.	4	71
				400
SIC	panels (20 May)	21 June	1	200
	" (" ")	31 Aug.	3	89
	" (31 Aug.)	20 Oct.	4	95
	pilings	8 Nov.	4	112
				496
CL	panels (20 May)	2 Aug.	2	199
	pilings	12 Oct.	4	101
	panels (2 Aug.)	30 Nov.	4	45
				345
FA	Zostera	20 May	4	110
	"	30 June	1	123
				233
				1474

TABLE 5

Mean allele and heterozygote frequencies at the To-1 and Mdh-1 loci from the pooled samples at each locality

Locality	No. of samples	Mean No. of genes	Mean allele frequencies (%)				Mean heterozygote frequencies (%)	
			a	b	c	d	observed	expected
				To-1				
SF	4	200	0.5	76.7	21.5	1.3	38.6	36.3
SIC	4	248	0.2	71.8	26.1	1.9	38.6	39.4
CL	3	172.5	0.6	79.0	20.1	0.3	33.2	33.4
FA	2	233	0.6	90.4	9.0	0.0	16.5	17.4
Total	13						33.9+6.5	33.7+2.3
				Mdh-1				
SF	4	197	0.6	13.8	84.8	0.7	24.7	25.8
SIC	4	239	1.6	13.0	84.6	0.7	24.7	26.3
CL	3	225	0.4	15.7	82.6	1.2	28.3	29.1
FA	2	232	0.2	5.5	94.0	0.2	11.9	11.2
Total	13						23.5+1.8	24.5+2.2
Grand Total	26						28.8+1.7	29.1+2.1

Both temporal and spatial variation patterns have been examined by testing the homogeneity of the different allelic frequency distributions (Table 6). No significant temporal variation occurs at the To-1 locus, but the pooled samples

TABLE 6

Homogeneity χ^2 values testing temporal and spatial variation of allelic frequencies at the To-1 and Mdh-1 loci. (*significant at the 0.05 level of probability; ** significant at the 0.01 level).

		SF	SIC	CL	FA	
		\multicolumn{4}{c	}{Within Localities}	
To-1		4.15	2.64	1.35	0.004	
Mdh-1		6.09	12.08**	0.99	2.28	
		\multicolumn{4}{c	}{Between Localities}	
		SF	SIC	CL	FA	
	SF	--	1.15	4.85*	43.09**	
	SIC	0.05	--	10.93**	57.09**	To-1
Mdh-1	CL	0.49	0.91	--	22.08**	
	FA	20.35**	12.36**	24.74**	--	

from the locality SIC gave a highly significant χ^2 value for the Mdh-1 data. Spatial variation is tested by means of pair-wise comparisons between localities; in Table 6 the χ^2 values for To-1 are in the upper right, those for Mdh-1 in the lower left. Two subpopulations (SF and SIC) are homogeneous at both loci and both differ slightly from the subpopulation CL. The remaining subpopulation FA is well separated from these three.

These differences and similarities between subpopulations are roughly correlated with distance (SF and SIC are only about 800 m apart), they may be related to the nature of the substrate (the subpopulation FA is on Zostera) and, I would hazard a guess, with the position of the stations relative to the channels which are likely to influence larval dispersal.

The conclusion that the population studied is subdivided into genetically different subpopulations is no more than tentative.

Summary

The colonial structure and breeding habit of B. schlosseri are considered with reference to its potential for genetic studies.

Earlier work on the genetic basis of colour polymorphism is discussed: 48 colour morphs have been identified; they are controlled by five Mendelian loci, two of them closely linked.

Two enzymes, MDH and TO, are polymorphic in the population of the Venetian Lagoon. Each of them is controlled by one locus with four codominant alleles; preliminary results of genetic analysis are presented.

The main factors which could affect the genetic structure of Botryllus populations are discussed, with an emphasis on the role of a self-nonself recognition mechanism and germ cell exchange between fused colonies.

An investigation of the population of the Venetian Lagoon has shown that this population is subdivided into distinct subpopulations significantly different in gene frequencies, though very close spatially.

Acknowledgments

The experimental work has been supported by C.N.R. grants through the Istituto di Biologia del Mare, Venice. I am indebted to W. J. Canzonier for reading through the English manuscript, to the staff of the Stazione Idrobiologica di Chioggia for facilities in sample collections, and to A. Tontodonati, M. Favaretto and B. Zenere for technical assistance.

REFERENCES

Berrill, N.J. 1941. The development of the bud in Botryllus. Biol. Bull. 80, 169-184.

Brunetti, R. 1974. Observations on the life cycle of Botryllus schlosseri (Pallas) in the Venetian Lagoon. Boll. Zool. 41, 225-251.

Colombera D. 1969. The karyology of the colonial ascidian Botryllus schlosseri (Pallas). Caryologia 22, 339-350.

Li, C. C. and Horvitz, D. G. 1953. Some methods of estimating the inbreeding coefficient. Amer. J. Hum Genet. 5, 107-117.

Milkman, R. 1967. Genetic and developmental studies on Botryllus schlosseri. Biol. Bull. 132, 229-243.

Millar, R. H. 1966. Evolution in Ascidians. In, Some contemporary studies in Marine Science, 519-534. Barnes, H. (ed). Allen & Unwin Ltd., London.

Oka, H. 1970. Colony specificity in compound ascidians. In, Profiles of Japanese Science and Scientists, 1970, 195-206, Kodansha, Tokyo.

Oka, H. and Watanabe, H. 1957. Colony specificity in compound ascidians as tested by fusion experiments. Proc. Jap. Acad. 33, 657-659.

Sabbadin, A. 1955. Il ciclo biologico di Botryllus schlosseri (Pallas) nella Laguna di Venezia. Arch. Ocean. Limnol. 10, 217-230.

Sabbadin, A. 1959. Analisi genetica del policromatismo di Botryllus schlosseri. Boll. Zool. 26, 221-243.

Sabbadin, A. 1960. Ulteriori notizie sull'allevamento e la biologia dei Botrilli in condizioni di laboratorio. Arch. Ocean. Limnol. 12, 97-107.

Sabbadin, A. 1971. Self - and cross-fertilization in the compound ascidian Botryllus schlosseri. Devl. Biol. 24, 379-391.

Sabbadin, A. 1972. Results and perspectives in the study of a colonial ascidian, Botryllus schlosseri. In, 5th Europ. Mar. Biol. Symp. 327-334. Battaglia, B. (ed).

Piccin, Padova.

Sabbadin, A. 1977. Linkage between two loci controlling colour polymorphism in the colonial ascidian, *Botryllus schlosseri*. Experientia 33, 876.

Sabbadin, A. and Graziani, G. 1967. New data on the inheritance of pigments and pigmentation patterns in the colonial ascidian, *Botryllus schlosseri* (Pallas). Riv. Biol. 60, 559-598.

BIOCHEMICAL ASPECTS OF GENETIC VARIATION AT THE LAP LOCUS IN

MYTILUS EDULIS

R. K. Koehn

Dept. of Ecology and Evolution, State Univ. of New York

Stony Brook, New York, U.S.A.

Introduction

Genetic variation at the leucine aminopeptidase (Lap) locus is phylogenetically widespread in marine bivalves (Koehn and Mitton, 1972; Levinton and Koehn, 1976). Five allelic forms can be observed in populations of Mytilus edulis, though only three are common and have been designated Lap^{98}, Lap^{96}, and Lap^{94}, in order of decreasing electrophoretic mobility of their products (Koehn et al., 1976).

Frequencies of these alleles vary among populations of Mytilus edulis on the east coast of North America (Koehn et al., 1976). In all non-estuarine samples from Virginia to Cape Cod, the frequency of Lap^{94} is generally never lower than 0.48 nor higher than 0.60, with an average value of about 0.56. North of Cape Cod, the frequency of this allele abruptly decreases to approximately 0.10 and intermediate frequencies can be observed in the Cape Cod Canal which connects these two major geographic regions. A Lap^{94} frequency of 0.10 occurs throughout the Gulf of Maine.

Although frequencies of Lap alleles are homogeneous throughout large geographic areas, there are significant differences in relative allele frequencies between coastal and estuarine populations. This difference is particularly dramatic among samples between Atlantic marine populations and Long Island Sound, a region of reduced salinity and other estuarine characteristics. Koehn et al. (1976) described the gentle clinal decrease in the frequency of Lap^{94} along the Connecticut shore from approximately 0.50 at the eastern entrance of the Sound to a minimum of 0.11 at the western end of the Sound. By contrast, samples from the north shore of Long Island were homogeneous at this low frequency. More recently Lassen and Turano (unpublished) have demonstrated that the differentiation between populations in the eastern Sound is actually more dramatic than originally described. The spatial change in frequency between Atlantic ($\approx$0.55) and Long Island Sound ($\approx$0.10) populations occurs in parallel on the Connecticut and eastern Long Island shores, over a distance of less than 20 miles.

This differentiation, especially the small geographic distance over which it occurs, similar differentiation of other estuarine populations connected to oceanic waters south of Cape Cod (Koehn et al., 1976), apparent differential viability among Lap phenotypes (Milkman and Koehn, 1977), and the concomitant spatial variation of frequencies at this locus in two different bivalve species (Koehn and Mitton, 1972), all suggest that these allele frequency variations are likely to be a consequence of natural selection. The extended larval dispersal in this species (3 to 7 weeks) would be expected to homogenize the genetic composition of populations in the same general geographic region but is apparently not sufficiently strong to counteract the differentiating forces of natural selection. The differences in Lap^{94} frequency occur between areas of normal oceanic and low estuarine salinity. However, the diminution of Lap^{94} frequency north of Cape Cod in normal oceanic waters, suggests that the selective regime is actually more complex and must involve at least temperature and salinity, or more likely, interactions of the two.

The variations at the Lap locus described above are not uniquely different from the numerous examples of systematic spatial variation in allozyme frequencies described in other organisms. Clines of various magnitudes have been commonly observed, and these are sometimes correlated with environmental variables (cf., Frydenberg et al., 1965; Koehn and Rasmussen, 1967; O'Gower and Nicol, 1968; Selander et al., 1969; Merritt, 1972: Williams et al., 1973; Hjorth and Simonsen, 1975; and others). Many of these studies have been used to infer the role of natural selection in the maintenance of spatial variation of allozyme frequencies. However, the real question that all these studies approach, but do not directly answer, is whether or not natural selection acts upon fitness differences that exist among the different allozyme genotypes that have been surveyed. In order to know this, significant functional differences must be described among the various allozymes. Ultimately, these differences must be demonstrable as differential survival or reproduction among the genotypes.

Evidence for the role of natural selection in the maintenance of enzyme polymorphism has emerged from studies which attempt to couple descriptive and/or experimental population genetics with direct evidence of molecular differences among allozymes (Koehn, 1969; Merritt, 1972; Vigue and Johnson, 1973; Thorig et al., 1975; Johnson, 1976). The investigations of leucine aminopeptidase variation in Mytilus edulis described below, were carried out with this general purpose in mind.

This paper describes phenotype-dependent leucine aminopeptidase enzyme activity, both in natural populations and as a response to the presence of certain divalent metal cations. These results jointly demonstrate that functional differences between the LAP allozymes exist. These differences may constitute the phenotypic diversity necessary for natural selection to act, even though at present, the detailed enzymological characterization of these allozymes has not been completed. Secondly, different levels of leucine aminopeptidase enzyme activity are examined in populations living in two different ecological situations and differing in Lap^{94} frequency. Lastly, temporal variation in overall enzyme activity is described, which further complicates interpretations of this system, but is important in ultimately deducing how the described differences in molecular function among individual phenotypes are physiologically manifested as an adaptation to environmental differences.

Methods and Materials

"Leucine aminopeptidase" has been used to describe a great diversity of enzymes, in a variety of organisms, whose only functional relatedness is the ability to hydrolyze the amino-bond between leucine and β-naphthylamine in the synthetic substrate L-leucyl-β-naphthylamide. These enzymes, in the absence of detailed knowledge of their catalytic characteristics and/or in vivo function, would be more properly described as "amino-acyl naphthylamidases". Although the true biochemical properties of many "naphthylamidases" have not been identified (Barrett and Dingle, 1971), some true proteinases have naphthylamidase activity (Otto, 1971).

Characterizations to date of Mytilus "leucine aminopeptidase" enzyme suggest that it is a neutral and aromatic aminopeptidase, with a thiol active site (unpublished). The enzyme is capable of cleaving a single neutral or aromatic residue from the N-terminus of oligopeptides and naphthylamide derivatives. Until the biochemical characterization of this enzyme has been completed, the name "leucine aminopeptidase" is retained.

Enzyme assays were performed on extracts of crude homogenates of hepatopancreas tissue from individual mussels that were prepared as follows. A small piece of hepatopancreas tissue was removed from freshly collected specimens, blotted dry, weighed, and diluted with assay buffer (see below) in a ratio of 1:3 (w/v). Dissected tissue specimens were held on ice through the entire assay procedure. Diluted tissue samples were sonicated at 0°C. for 5-10 seconds. Homogenates were centrifuged for 10 minutes at 0°C. and 48,000g. Following centrifugation, a constant volume of supernatant was removed and diluted further in a constant volume of buffer. Further dilution is necessary because of the high enzyme activity of tissue preparations, but the dilution factor varied as a function of the overall enzyme activity, in order to optimize the ultimate assay reaction.

The reaction mixture consisted of 1.0 ml of 0.20 M tris-HCl, pH 7.0 buffer and 0.25 ml 0.005 M L-leucyl-β-naphthylamide substrate in 0.20 M tris-HCl, pH 7.0 buffer and 20% glycerol. Substrate was predissolved in a few drops of dimethyl formamide. This reaction solution was chilled to 0°C., before the addition of 10 µl of diluted tissue extract. The reaction was initiated by placing the reaction tubes into a 37°C. water bath. The final concentration of substrate was 0.001 M (excess). After the reaction had proceeded for 15 minutes, it was stopped with a solution of 0.5% Fast Garnet GBC diazonium salt and 2% lauryl sulfate (SDS). The last compound solublizes the diazonium-naphthylamine complex, which develops as a red colour. Complete formation of this complex occurs within five minutes and is stable for several hours. The E_{510} of the complex was measured on a Gilford Model 2400 Automatic Spectrophotometer. When discussing the leucine aminopeptidase activity of single individuals, or their averages in population samples, activity is expressed as "units per ml", where one unit equals an E_{510} of 0.001, or the liberation from L-leucyl-β-naphthylamine of 6 µmoles of β-naphthylamine per minute at 37°C. and pH 7.0. All assays were done in duplicate. All chemicals were from Sigma Chemical (St. Louis, Mo.).

For each individual tissue extract, total protein was determined by the Folin method with a standard BSA curve. Specific activity is expressed as enzyme units per milligram total protein. Although determinations of specific activity were made, results presented here are not in these terms. In marine bivalves, specific activity of an enzyme is not a very meaningful parameter in crude preparations, since cellular protein levels differ between animals living in different salinities and/or sampled at different times of year. As a consequence, specific activity can be confounded by two fairly independent variables: enzyme activity, a function of both the catalytic properties of the enzyme and its concentration, and the total protein baseline with which this activity is compared. When samples are prepared by a standard method, activity can be expressed as units of activity per extract volume. Hence the measure "units per ml" is used exclusively in the presentation of these results.

Results

Levels of leucine aminopeptidase activity have been examined in two natural populations of Mytilus edulis. The genetic composition, vis a vis the Lap locus, of these two populations is markedly different (Table 1). This reflects the different allele frequencies at this locus that are characteristically observed in populations sampled from differing salinities (Koehn et al., 1976). Populations at Stony Brook are well within Long Island Sound, where salinity varies between 24 and 28 o/oo. The low frequency of Lap^{94} at this locality (Table 1) is representative of all samples taken throughout the inner reaches of Long Island Sound during 1971-1976. The second study population is at Shinnecock, on the south shore of the eastern portion of Long Island. It is a normal ocean salinity habitat (approximately 33 o/oo). This population, like all others in oceanic environments south of Cape Cod (Koehn et al., 1976), exhibits a high frequency of Lap^{94}, approximately 0.40 greater than populations living in Long Island Sound.

Since the spatial distribution of Lap alleles is correlated with salinity, the possible effect of salinity on the expression of LAP enzyme activity was investigated by characterizing enzyme activity within and between the Stony Brook and Shinnecock populations. So far, a total of 18 samples, nine from each population, have been collected (during 1975-1976). All comparisons of the average enzyme activity between Stony Brook and Shinnecock population samples were significantly different. One of these comparisons is given in Table 2. The average enzyme activity of individuals from Stony Brook (low salinity) was 249,800 units per ml. compared to 355,011 units per ml for individuals sampled from Shinnecock (high salinity). Since there are significant among-phenotype differences in average activity, with phenotypes of Lap^{94} exhibiting greatest activity (see below), it is formally possible that the difference in

TABLE 1

Representative samples illustrating the difference in genetic composition between populations of *Mytilus edulis* inside (Stony Brook) and outside (Shinnecock) Long Island Sound. See Koehn et al., 1976, for details

	N	Lap-98/98	-98/96	-96/96	-98/84	-96/94	-94/94	Lap^{98}	Lap^{96}	Lap^{94}
Stony Brook	76	22	28	14	6	2	4	0.513	0.382	0.105
Shinnecock	80	6	7	4	27	8	28	0.287	0.144	0.569

TABLE 2

Phenotype and salinity-dependent variations in average leucine aminopeptidase activity (units/ml) for six common phenotypes in two populations of M. edulis. Samples taken in December, 1975.

	Lap-98/98	-98/96	-96/96	-98/94	-96/94	-94/94	Total	F [§]
Stony Brook								
Mean	208,466	247,943	271,304	278,125	329,500	333,042	249,800	5.79**
Variance ($\times 10^{-9}$)	4.207	1.793	5.338	6.476	---	2.129	17.876 [ø]	
N	22	28	14	6	2	4	76	
Shinnecock								
Mean	266,903	356,405	320,833	360,467	378,042	366,594	355,011	5.62**
Variance ($\times 10^{-9}$)	4.519	2.299	5.123	1.025	3.579	2.865	12.073 [ø]	
N	6	7	4	27	8	28	80	
t Statistic	1.942	5.912	1.200	5.876	---	1.188	13.140*	
degrees of freedom	26	33	16	31	---	30	10	
P	>.05	<.001	>.20	<.001	---	>.20		

[§] One-way Analysis of Variance; * P < .005   ** P < .001

[ø] Among-group variance

overall activity between the two populations is due to the very different proportions of Lap^{94} phenotypes in each. However, this is not the case, as can be seen from a comparison of individual phenotypes sampled from the two populations. (Table 2). In addition to the grand means, two out of five comparisons of individual phenotypes from the two sites are significantly heterogeneous (Lap98/96 and Lap 98/94; Table 2). More importantly, individuals with higher overall enzyme activity, characteristics of high salinity populations, can be acclimated in approximately 11 days to the low activity observed in low salinity populations (Koehn et al., unpublished). The acclimation of general enzyme activity levels together with maintenance of phenotypic differences demonstrates that one is not a simple function of the other.

During the course of the above investigations on salinity-dependent variations in enzyme activity, it became apparent that there were substantial changes in overall enzyme activity with sampling date. Samples have been taken at irregular intervals since the fall of 1975. In the Shinnecock population (Fig.2), leucine aminopeptidase activity was lowest in mid-winter and highest in mid-summer. Between February and June there was an approximately nine-fold increase in overall enzyme activity. The same seasonal pattern of variation was observed in the Stony Brook population, although the differences in activity between them was maintained. Maximum activity was observed on July 1, 1976. In addition to population (i.e., salinity correlated) and seasonal differences in overall levels of leucine aminopeptidase enzyme activity, there are statistically significant differences in enzyme activity of individual phenotypes, in both populations (Table 2). Phenotype averages varied from about 200,000 to 330,000 units per ml in the Stony Brook sample and from 266,000 to 380,000 units per ml in the Shinnecock sample. Individuals with the Lap^{94} allele exhibited the highest average enzyme activity. Thus the temporal variation illustrated in Figure 2 could also be represented as a family of six parallel lines, each representing one of the Lap phenotypes.

Only a single sample from each of the two localities is presented in Table 2, but a total of 18 separate samples (9 at each) have been taken from the two localities during 1975 and 1976. In seven samples there was statistically demonstrable heterogeneity of the average enzyme activity among phenotypes. Since each phenotype is unequally represented in samples of a given size (see Table 2), phenotypic differences were not always demonstrable. The power of a particular test will vary with total sample size. In all samples, the average activity of phenotypes involving the Lap^{94} allele

BIOCHEMICAL ASPECTS OF GENETIC VARIATION

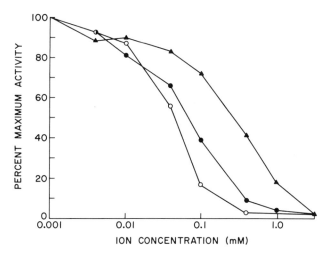

Figure 1. Inhibition of leucine aminopeptidase activity in $\underline{M.\ edulis}$ by three divalent cations (as chlorides). $\triangle = Cu^{++}$, $O = Cd^{++}$, $\bullet = Zn^{++}$. Assays were performed on crude extracts as described in the text.

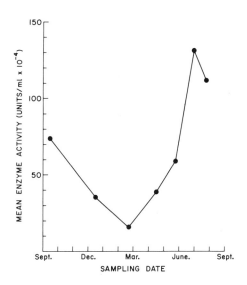

Figure 2. Seasonal variation of average leucine aminopeptidase activity in $\underline{M.\ edulis}$ during 1975-76, sampled from Shinnecock, Long Island, New York. Assays were performed on crude extracts from between 70 to 100 individuals in each sample. Standard deviations of the means are all less than 2.5×10^{-4} units/ml.

was highest, irrespective of the results of individual statistical tests.

These observations do not themselves provide a clue to the possible adaptive significance of variations in activity. They do, however, suggest that functional differences exist among the three alleles, that are apparent as differential activity among the six common phenotypes in the conditions of assay. These differences may be due either to functional variations of the individual alleles or to differences between the six phenotypes in the average concentration of enzyme protein, more likely the former (see below).

Functional differences between Lap alleles are further suggested by the relative inhibition of enzyme activity of the six phenotypes by millimolar quantities of certain divalent metal cations. The leucine aminopeptidase allozymes are extraordinarily sensitive to the presence of low concentrations of some divalent ions, a characteristic property of thiol proteinases (Barrett and Dingle, 1971). In a particlly purified preparation of mixed genotypes standard inhibition curves were determined for decrease in percent maximum activity in the presence of Cu^{++}, Cd^{++}, and Zn^{++}. Enzyme activity was inhibited over the range of 0.001 mM to 4.0 mM ion concentration (Fig.1). Enzyme activity was inhibited in much the same way by the three ions, but zinc appeared to have the strongest inhibitory effect. From the inhibition curve in Figure 1, two concentrations of zinc (0.04 and 0.10 mM), each producing partial inhibition, were selected as experimental concentrations in order to explore the possible differential inhibition of Lap phenotypes.

The relative inhibition of individual phenotypes was different in samples from Stony Brook and Shinnecock populations at both zinc concentrations (Table 3). In the Stony Brook sample, 0.04 mM zinc inhibited enzyme activity from between 64 to 47 percent of the controls, depending on the Lap phenotype. Although the magnitude of inhibition was greater at 0.10 mM zinc, the relative inhibition of Lap phenotypes was again unequal. At both zinc ion concentrations, the Lap-98/98 and Lap-98/96 phenotypes were inhibited less than the other three phenotypes tested. Lap-94/94 was not tested in this experiment since no individuals of this phenotype were represented in a collection of 74 individuals. In general, phenotypes with the Lap^{94} allele were inhibited to a greater extent than other phenotypes, though the inhibition of Lap-96/96 was equal. Similar results were obtained in a sample from Shinnecock (Table 3), at least at 0.04 mM zinc where overall inhibition was less: Lap-98/98

TABLE 3

Phenotype-dependent inhibition of leucine aminopeptidase (as percent remaining activity) in two population samples of *Mytilus edulis* by .04 and .10 mM Zn^{++} (as chloride)

	(Zn^{++})	N	98/98	98/96	Lap Phenotype 96/96	98/94	96/94	94/94	$F^\S$
Stony Brook									
	.04	74	64	60	47	47	51	--	21.6***
	.10	74	33	36	22	21	28	--	25.7***
Shinnecock									
	.04	75	83	81	68	73	74	74	2.9*
	.10	75	17	19	13	15	15	17	2.6*

§ One-Way Analysis of Variance on arc-sin transformed values; $*P < .05$, $***P < .001$

and Lap 98/96 were inhibited less than the other four tested phenotypes.

The same experimental concentrations of zinc ion do not produce the same overall magnitude of inhibition in samples from the two populations. The two populations exhibit differences in enzyme activity (see below), which are probably due to differences in the concentration of enzyme-protein. While differences in overall enzyme inhibition between populations could be due to differences in enzyme-protein concentration, between phenotype differences within each population could be either due to variations in the concentration of LAP enzyme or to allozyme properties.

Discussion

The possibility that the Lap locus is a target of selection, resulting in the marked genetic differentiation between populations, has been strengthened by the demonstration of differences in catalytic action between the six Lap phenotypes. The frequency of Lap^{94} is high in normal salinity environments and low in frequency in low salinity environments, in geographic areas south of Cape Cod. In more northern altitudes, Lap^{94} is low in frequency in high salinity habitats, thus implicating both temperature and salinity as affecting allele frequency distributions.

There are significant differences in leucine aminopeptidase activity between the individual Lap phenotypes in high and low salinity populations. These differences are maintained despite the substantial temporal variation in overall enzyme activity. Phenotypes of Lap^{94} always exhibited greatest enzyme activity. Hence, the allele with highest activity is highest in frequency in normal salinity populations which exhibit a generally high level of enzyme activity. In contrast, low salinity populations exhibit both low overall enzyme activity and highest frequency of the low activity Lap alleles. There appears to be both a phenotypic (enzyme activity) and genotypic (allele frequency) response to changes in salinity, though the actual mechanism (either enzymatically or selectively) of this response is not yet understood. Nevertheless, the average enzyme activity differences between phenotypes provide the phenotypic diversity upon which natural selection could act. In addition, the differential inhibition of the Lap phenotypes by zinc cations further emphasizes the differential functional characteristics of the alleles at this locus. At present, it is difficult to know whether or not the differential sensitivity of Lap phenotypes to cations has any physiological significance, though if so, it probably does not involve zinc. There are substantial changes in cellular ion profiles during salinity acclimation in bivalves (Robertson, 1964), and hence it is not unrealistic to suppose that some significant interaction between cellular ions and the LAP enzyme could occur. It is important to emphasize that the phenotype-dependent sensitivity is extremely clear cut, since 65% of the total variance in remaining enzyme activity can be attributed to between phenotype differences in ion sensitivity. This is a ten-fold greater contribution to phenotypes to activity variation than can be observed in comparisons of activities between phenotypes in natural populations. As of now, it seems unjustified to invoke any significant role for ions unless the observations with zinc can be extended to other more physiologically relevant ions. Appropriate experiments are currently in progress.

The activity and ion sensitivity results adequately demonstrate that the enzymological properties of the Lap alleles are not equivalent, though how these differences may be perceived by natural selection is not immediately obvious. The very dramatic correlation between within population allele frequencies and salinity together with salinity dependent activities at this locus, suggest a coherent picture whereby enzyme activity is somehow adaptively related to salinity variations. There is, however, no present evidence that enzyme activity, per se, is of particular significance in this species, nor that enzyme activity would be of particular importance as an adaptation to salinity variations in the environment. This critical question, involving the mechanism by which natural selection may act on this locus, cannot be adequately explored until we have a more complete understanding of the physiological role of this enzyme.

Enzymologically, the leucine aminopeptidase of Mytilus edulis is an aminopeptidase, suggesting that the enzyme functions in the metabolism of proteins, most obviously in the degradation of ingested protein materials within the intracellular lysosomal digestive system of the digestive gland. However, maximum seasonal activity does not correspond with maximum food availability, nor do enzyme levels respond to starvation and pulse feeding (Koehn and Bayne, unpublished). Also, while digestive gland material is rich in leucine aminopeptidase enzyme, LAP is ubiquitously distributed in non-digestive tissues. For example, nearly equal levels of enzyme activity can be found in the gill and palp, which function respectively as respiratory and mechanical organs for feeding. The ubiquitous distribution of LAP in non-digestive tissues, would suggest that the enzyme has a more general role in cellular protein metabolism. This might involve the mobilization of cellular protein as energy reserves and/or possibly functioning to "provision" the free amino acid pool for cellular osmotic accommodation. The very substantial temporal variation in enzyme activity may be important in deciphering the physiological role of this enzyme. Maximum activity in mid-summer corresponds to maximum environmental temperatures, a time when populations in all salinities in the Long Island area are experiencing temperature stress. A physiological consequence of high temperature stress is the inability to acclimate (i.e., maintain a temperature-independent metabolism - Widdows, 1972; Bayne, 1976). This could lead to the mobilization of protein reserves for energy, as metabolic rate would be, by and large, temperature-dependent. These possibilities cannot be experimentally approached until there exists a better understanding of the basic physiology of these populations.

Macro- and micro-geographic patterns of allele frequency variation have been used to infer the role of natural selection acting via the Lap locus (Koehn and Mitton, 1972; Koehn et al., 1976; Milkman and Koehn, 1977). The existence of phenotype-dependent functional characteristics of the LAP allozymes, as well as natural and experimental responses of these characteristics to salinity variations, constitute partial evidence that natural selection is occurring directly at the Lap locus. Nevertheless, to demonstrate unequivocably the selective maintenance of genetic variation at this locus requires a mechanism that explains how enzyme activity is ultimately reflected as differences in fitness. To identify this mechanism, we require a more thorough understanding of the general physiology of the Mytilus individuals which, collectively form these populations, of the particular function of leucine aminopeptidase in this physiology and finally of how Lap phenotypes differentially affect this physiology.

Summary

Genetic variation at the Leucine aminopeptidase (Lap) locus is phylogenetically widespread in marine bivalves. In populations of M. edulis on the east coast of North America, three common alleles occur: Lap^{98}, Lap^{96}, and Lap^{94}. Relative allele frequencies vary abruptly among certain geographic regions, most particulary the diminution of the frequency of Lap^{94} from 0.55 in coastal marine populations to 0.15 in estuarine populations.

Phenotype-dependent enzyme activity is described, both as it occurs in natural populations and as a response to the presence of certain divalent metal cations. Different overall levels of enzyme activity were detected between samples taken from two populations living in different ecological circumstances and differing in $\underline{Lap}^{94}$ frequency. There is high relative frequency of the high activity allele in populations exhibiting high overall enzyme activity in normal oceanic salinities, but lower frequency of $\underline{Lap}^{94}$ in populations of lower overall activity living in reduced salinity estuarine habitats. Additionally, there is significant temporal variation in enzyme activity in both populations.

The aminopeptidase activity of this enzyme would indicate that it is important in protein metabolism. Various possible physiological roles of the enzyme are discussed as well as how these may be important in aspects of protein metabolism.

Acknowledgements

The enormous number of enzyme assays that were performed during this study could not have been done without the dedicated assistance of Doris Wiener, Pat Gaffney, Dr. J. Peter W. Young, Dr. Norman Arnheim, Fred Immerman, Allen Lamb, David Innes, Walter Eanes, and Hans Lassen. Dr. Young performed many of the inhibition experiments. This work was supported by NSF research grant BMS 74-02522, USPHS research grant GM-21133, and USPHS Career Award GM-28963. This is contribution number 197 from the Program in Ecology and Evolution at the State University of New York at Stony Brook.

REFERENCES

Barrett, A.J. and Dingle, J.T. (eds.) 1971. *Tissue Proteinases*, North-Holland Publ. Co., Amsterdam.

Bayne, B.L. (ed.) 1976. *Marine Mussels: Their Ecology and Physiology*, Cambridge Univ. Press, Cambridge.

Frydenberg, O., Moller, D., Naevdal, G. and Sick, K. 1965. Haemoglobin polymorphism in Norwegian cod populations. Hereditas 53, 257-271.

Hjorth, J.P. and Simonsen, V. 1975. Genetics of *Zoarces* populations. VIII. Geographic variation common to the polymorphic loci *HbI* and *Est III*. Hereditas, 81, 173-184.

Johnson, G.B. 1976. Genetic polymorphism and enzyme function. In, *Molecular Evolution*. Ayala, F.J. (ed). Sinauer Associates, Inc., Sunderland, Mass.

Koehn, R.K. 1969. Esterase heterogeneity: dynamics of a polymorphism. Science 163, 943-944.

Koehn, R.K. and Mitton, J.B. 1972. Population genetics of marine pelecypods. I. Evolutionary strategy at an enzyme locus. Am. Nat. 106, 47-56.

Koehn, R.K., Milkman, R. and Mitton, J.B. 1976. Population genetics of marine pelecypods. IV. Selection, migration and genetic differentiation in the Blue Mussel, *Mytilus edulis*. Evolution 30, 2-32.

Koehn, R.K. and Rasmussen, D.I. 1967. Polymorphic and monomorphic serum esterase heterogeneity in catostomid fish populations. Biochem. Genet. 1, 131-144.

Levinton, J.S. and Koehn, R.K. 1976. Population genetics of mussels. In, *Marine Mussels: Their Ecology and Physiology*. Bayne, B.L. (ed). Cambridge Univ. Press, Cambridge.

Merritt, R.B. 1972. Geographic distribution and enzymatic properties of lactate dehydrogenase allozymes in the Fathead Minnow, *Pimephales promelas*. Am. Nat. 196, 173-184.

Milkman, R. and Koehn, R.K. 1977. Temporal variation in the relationship between size, numbers, and allele-frequency in a population of Mytilus edulis. Evolution (in press).

O'Gower, A.K. and Nicol, P.I. 1968. A latitudinal cline of haemoglobins in a bivalve mollusc. Heredity 23, 485-492.

Otto, K. 1971. Cathepsins B1 and B2. In, Tissue Proteinases. Barrett, A.J. and Dingle, J.T. (eds). North-Holland Publ. Co., Amsterdam.

Robertson, J. 1964. Osmotic and ionic regulation. In, Physiology of Mollusca. Vol 2 (Wilbur, K.M. and Yonge, C.M. (eds). Academic Press, New York.

Selander, R.K., Hunt, W.G. and Yang, S.Y. 1969. Protein polymorphism and genic heterozygosity in two European subspecies of the House Mouse. Evolution 23, 379-390.

Thorig, G.E.W., Schoone, A.A. and Scharloo, W. 1975. Variation between electrophoretically identifiable alleles at the alcohol deydrogenase locus in Drosophila melanogaster. Biochem. Genet. 13, 721-731.

Vigue, C.L. and Johnson, F.M. 1973. Isozyme variability in species of the genus Drosophila. VI. Frequency-Property-Environment relationships of allelic alcohol dehydrogenases in D. melanogaster. Biochem. Genet. 9, 213-228.

Widdows, J. 1972. Thermal acclimation by Mytilus edulis L. Ph.D. Thesis, University of Leicester, England.

Williams, G.C., Koehn, R.K. and Mitton, J.B. 1973. Genetic differentiation without isolation in the American Eel, Anguilla rostrata. Evolution 27, 192-204.

EXPERIMENTAL MORTALITY STUDIES AND ADAPTATION AT THE Lap LOCUS IN MYTILUS EDULIS

J. S. Levinton[1] and H. H. Lassen[2]

[1] Dept. of Ecol. and Evol. S.U.N.Y., Stony Brook, U.S.A.
and [2] Inst. of Ecol. and Gen., Univ. of Aarhus, Denmark

Introduction

A complete explanation of the forces that govern the distribution of allelic variation of allozyme polymorphism over space and time requires an integrated approach of population genetic, ecological, physiological and biochemical approaches. But most studies to date focusing upon selection on marine species, rely on the correlative approach of fitting distributional allozyme data to known environmental variation, such as the latitudinal temperature gradient. A typical approach is to observe a gradual change in allele frequency over space and find some change of environment over the same space (as in O'Gower and Nicol, 1968; Schopf and Gooch, 1971). This is unconvincing evidence for selection, as gradients of environmental variation are common along coastlines, and clines can be generated by population genetic forces excluding selection (see Endler, 1973 for discussion). More satisfying patterns indicating selection are different, concomitant, between-locus patterns of clinal variation for a given species along a coast (Christiansen and Frydenberg, 1974; Williams et al. 1973; Koehn et al 1976). Such differing between-locus patterns cannot be reconciled with any isolation model. Further convincing evidence is provided by allelic frequency changes in natural populations, following a local perturbation, consistent with predictions based on established environmentally correlated geographic allelic variation (e.g. Johnson, 1971). Further, the greater latitudinal variation in allele frequencies of east coast North American Mytilus edulis, relative to west coast Mytilus californianus is evidence for selection, given that dispersal is similar in both species and the east coast latitudinal temperature gradient is more pronounced than that of the west coast (Levinton and Suchanek, 1978).

Several studies show microgeographic variation in the tide zone or small scale geographic clinal variation, despite pelagic dispersal patterns of larvae, at the Lap and other loci of Mytilus edulis and M. californianus (see Levinton and Koehn, 1976; Balegot, 1971; Levinton and Fundiller, 1975; Koehn et al. 1973; Mitton et al. 1973). In particular a cline in the Lap^{94} allele of Mytilus edulis shows a smooth change from ca. 0.55 outside of Long Island Sound, New York, to 0.10 inside the Sound over a distance of only 30 km (Koehn et al. 1976; Lassen and Turano, unpublished). Within two weeks, less than 3 - 7 week pelagic larval lifespan of M. edulis larvae (Seed, 1976), floats released within the sound traverse the distance of the cline and travel at least 100 km along the coast (Paskausky and Murphy, 1976). Thus, this cline is maintained in the face of strong potential migration pressures. Furthermore, a similar cline appears in open marine-estuarine gradients in several other, much smaller, estuaries south of Cape Cod, Massachusetts, USA. The apparent suggestion of strong selective forces at the Lap locus has inspired efforts at biochemical characterization of kinetic differences between Mytilus Lap allozymes (see Koehn, this volume) and our attempts, described below, to understand where and how selection might occur in natural populations. If selection related to the locus is strong enough to cause selective mortality with respect to Lap phenotype then genetic differences must be manifested in physiological differences among individuals of different Lap genotypes. As selection should operate against those individuals less physiologically adapted to a given environment, between-genotype differences in physiological response should be measurable (Levinton and Fundiller, 1975). We have chosen the integral physiological responses of growth and mortality to test for between-genotype differential response at the Lap locus in Mytilus edulis. The scope for body growth is an integral of a mussel's ability to thrive under given ration, temperature and salinity conditions that incur a corresponding metabolic cost (Thompson and Bayne, 1974). Mortality is a traditional measure of physiological response to conditions (see Kinne, 1970) and is suited to the hypothesis that allele frequency changes occur in natural populations through selective mortality. We have also compared these physiological responses between estuarine (Long Island Sound) and open marine mussels exposed to the same conditions. Our working hypotheses are: (1) that estuarine and open marine populations have different physiological responses to the same stress; (2) between-genotype physiological differences result in differential growth; and (3) between-genotype physiological differences result in shift in genotypic and allele frequencies of populations subjected to experimental environmental change that induces mortality. An extension of hypothesis 3, in the present context, is that open marine $\underline{Lap}^{94}$ frequencies may be shifted towards estuarine frequencies by a low salinity shock (see Johnson and Powell, 1974, for a similar approach in Drosophila).

Materials and Methods

We employed estuarine (24-27 o/ooo mussels with $\underline{Lap}^{94}$ allele frequencies of ca. 0.1 collected from Flax Pond (Old field, New York) and Sunken Meadow Beach Park (east of North port, New York), and open marine (30-33 o/oo) populations (frequency of $\underline{Lap}^{94} \approx 0.5$) from rocks on the east side of Shinnecock Inlet, near Hampton Bays, New York. For comparative physiological response laboratory experiments, animals from both estuarine and open marine populations (referred to as Long Island Sound and Shinnecock below) were held in the laboratory for three weeks to one month at temperatures approximating field conditions and 32 o/oo salinity. For estuarine populations this represented an increase of only 7 o/oo over ambient and three weeks to a month is probably sufficient to ensure acclimation (B. L. Bayne, written communication). After this holding period animals were transferred to various salinity-temperature combinations and mortality was measured. To assess changes in $\underline{Lap}$ allele and genotype frequencies we kept populations of Shinnecock mussels for two weeks at 15°C, 32 o/oo and subjected them to shocks of low salinity (16 o/oo)-high temperature (25°C or 30°C), and normal salinity (32 o/oo)-high temperature. A random sample of the starting population was taken as a standard of comparison. Leucine aminopeptidase ($\underline{Lap}$: L-Leucyl-peptide-hydrolase; E.C. 3.4.1.1) phenotypes were separated and identified using horizontal starch block electrophoresis (Koehn and Mitton, 1972) and were classified according to the scheme published in Koehn et al. (1976).

To further estimate $\underline{Lap}$ phenotype dependent mortality and growth differences, we established subtidal transplant cages of wood and plastic mesh at Flax Pond (footbridge) and Shinnecock Inlet (Tully's Fish Market Dock). Animals were reciprocally transplanted with controls and mortality was estimated. Shinnecock and Flax Pond samples were transplanted to intertidal cages at Flax Pond and between-population growth and between-Lap phenotype growth was estimated. We have also monitored Lap phenotypes in natural populations at Shinnecock during 1975 and 1976. During January and February of 1976, we had the opportunity to monitor Lap frequencies before, during and after a mass mortality precipitated by a winter warm spell of greater than 12°C for a few days.

Results

Mortality. Open marine mussels collected from Shinnecock in November, 1974, were acclimated to 10°C, 32 o/oo and transferred to different temperature-salinity combinations for 48 hours. Mortality was estimated after a 24-hour return to the acclimation chamber. Table 1 shows the effect of high temperature on mortality as well as the enhancement of

TABLE 1

Percent mortality of experimental populations of M. edulis subjected to 48 hour shock, after 24 hour return to acclimation chamber (initial N=100, for each experimental condition).

Salinity	0 o/oo	8 o/oo	16 o/oo	24 o/oo	32 o/oo
Temp					
0°C	0	0	0	0	0
10°C	62	66	94	18	10
18°C	100	98	66	30	10
25°C	100	100	100	100	88
35°C	100	100	100	100	100

low salinity-induced mortality with increasing temperature. Mussels collected in June 1975 were similarly acclimated and survivorship was followed (Fig.1) at 3, 10 and 25°C at 16 o/oo. Mortality was negligible at 3°C, slight at 10°C and dramatic at 25°C. Thus low temperature-low salinity shock induces less mortality than high temperature-low salinity conditions.

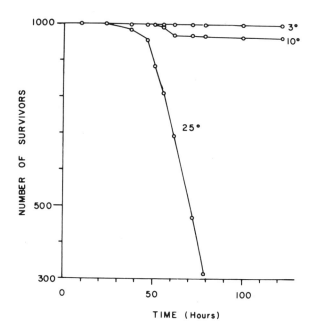

Figure 1. Mortality of identically acclimated mussels transferred to 16 o/oo at three temperatures.

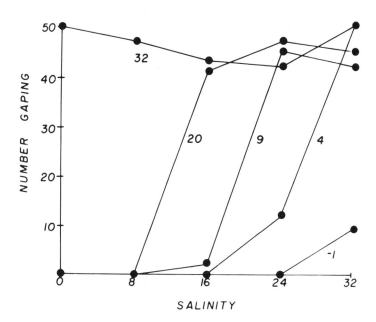

Figure 2. Number of Shinnecock mussels gaping at different temperature-salinity combinations after 24 hours (N equals 50 in all treatments).

Of interest is the negligible mortality of summer-collected mussels at 10°C, 16 o/oo after three days (ca. 5%) versus 94% mortality over the same conditions in late fall. This may be related to seasonal acclimatization as ambient water temperature was about 5°C in fall-collected mussels, and 20°C in the summer-collected samples.

The dynamics of this differential mortality response to salinity shock with changing temperature can be ascertained from the behavioural response of shell closure, followed by gaping under given temperature-salinity conditions. With a marked change in salinity the animal responds immediately by closing the shell valves, with subsequent reopening after a time, dependent upon the magnitude of the change (Gilles, 1972). We presume that oxygen comsumption resumes as the valves open (Bayne et al. 1976). Figure 2 shows the proportion of animals gaping after 24 h. at different temperature-salinity combinations. Note that low salinity depresses gaping at all test temperatures below 32°C. The increased frequency of gaping at a given salinity with increasing temperature may be due to the need to meet an increased metabolic demand with increasing temperature. A failure of physiological integration is probably the cause of similar gaping frequencies at 32°C over different salinities. At 8 and 0 o/oo, animals were noticeably swelled and most appeared to by dying.

Open marine mussels (Shinnecock) and Long Island Sound mussels (Sunken Meadow Beach) were collected on the same day (September, 1974) acclimated to 10°C, 32 o/oo and placed in aquaria of 0, 8, 16, 24 and 32 o/oo, all at 20°C. The test temperature was chosen because it is well within tolerance limits of mussels but is at the limit at which M. edulis can acclimate its routine respiration rate (Widdows, 1972, in Bayne et al. 1976). Figure 3 illustrates the results of two experiments showing no mortality for Long Island Sound mussels, but increased mortality with decreasing salinity for open marine mussels.

The same two populations were placed in the same range of test salinities at 32°C, a temperature above the generally accepted upper tolerance limit for Mytilus edulis of approximately 27 C (see Bayne et al. 1976). Figure 4 shows survivorship of the two populations at 8, 16, 24 and 32 o/oo (100% mortality occurred at 0 o/oo, 32°C within six hours). Note that at 32 o/oo the mortality curves for Shinnecock and Long Island Sound mussels almost overlap. With decreasing salinity the respective mortality curves increasingly diverge, open marine (Shinnecock) mussels dying at greater rates than

EXPERIMENTAL MORTALITY STUDIES AND ADAPTATION

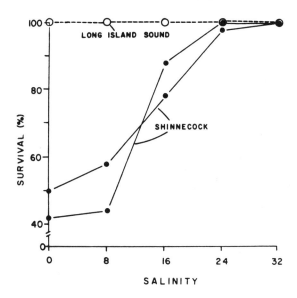

Figure 3. Survival of Shinnecock vs. Long Island Sound mussels, when exposed at 20°C to different salinities for 24 hours.

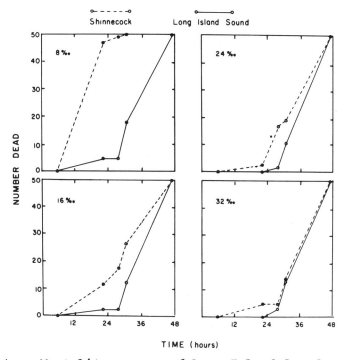

Figure 4. Mortality curves of Long Island Sound and Shinnecock mussels exposed to 32°C and differing salinities.

Long Island Sound mussels. The between-population difference in response is therefore explained by lowered salinity and not high temperature. Also note that the mortality curves for Long Island Sound populations are almost identical over the four test salinities shown (but mortality rate did increase in 0 o/oo). These results are consistent with hypothesis 1: that open marine and estuarine populations react differently to the same stress.

Transplants. In winter (1975/1976) mussels 10-20 mm in length were reciprocally transplanted in subtidal cages between Long Island Sound (Flax Pond) and open marine (Shinnecock) habitats with controls to test for mortality differences. Table 2 shows negligible mortality with no differences among treatments. The experiment was repeated for intertidal cages in summer with similar results. Therefore, although these two populations show different responses to mortality-inducing shock, the environmental differences between open marine and Long Island Sound waters do not cause mortality in transplants. We assume that after three months, successful acclimatization was possible.

Lap Phenotype-Specific Growth. Mussels 14-17 mm in length were transplanted from Shinnecock to intertidal cages in Flax Pond, Long Island Sound (24/4/76 - 6/9/76). Growth was measured with calipers and Lap phenotype identified for each individual. A similar analysis was done (9/6/76 - 6/9/76) for Flax Pond native mussels transplanted into Flax Pond intertidal cages. Table 3 shows mean weekly growth rates for Lap phenotypes for the two experiments. Between-Lap phenotype growth variation was estimated with a one-way analysis of variance for each experiment which showed no significant differences in growth. Therefore, hypothesis 2, that differences in Lap phenotype should be reflected in a differential integral physiological response such as growth, is not supported. The difference of growth rate between the two mussel populations in Long Island Sound is probably real because the Shinnecock experiment bracketed in time the Flax Pond mussel experiment and we observed increases of length of the Shinnecock mussels before 9/6/76. The explanation for the growth difference, however, cannot be inferred from our present information.

Lap Changes with Experimental Mortality. Table 4a shows the results of a series of mortality experiments testing for changes in Lap genotype and allele frequencies. Shinnecock mussels 10-25 mm long were employed in all experiments and low-salinity, high temperature was the usual test environment. Figure 5 shows a typical survivorship curve, with a period of

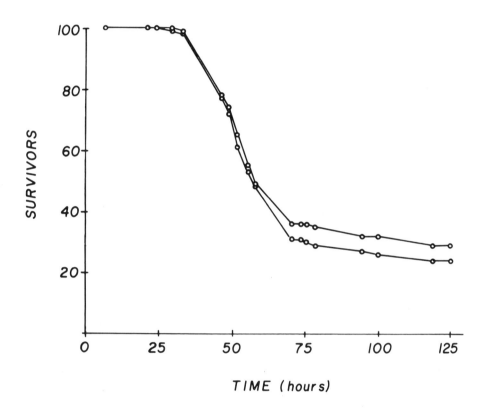

Figure 5. A typical survivorship curve (two replicates) for Lap shock mortality experiments.

TABLE 2

Transplant scheme and results (number of survivors) of subtidal transplant, with controls, between Shinnecock Inlet and Flax Pond.

Cage Number	9	10	11	12	13	14	15	16
Recruited From	Flax	Flax	Flax	Flax	Shinn	Shinn	Shinn	Shinn
Placed In	Flax	Shinn	Flax	Shinn	Flax	Shinn	Flax	Shinn
Initial Count 12/13/75	252	250	249	250	248	251	250	250
Survivors 2/27/76	251	250	246	247	243	247	244	243

TABLE 3

Among Lap phenotype growth rates (mm week^{-1}) of open marine (Shinnecock) and Long Island Sound (Flax Pond) Mytilus edulis transplanted to Long Island Sound cages (sample size in brackets). Analysis of variance shows no significant among-phenotype growth differences.

a. Mean Growth Rates

Lap phenotype	94/94	94/96	96/96	96/98	98/98	94/98
Long Island Sound	0.95 (16)	1.07 (63)	1.06 (151)	1.05 (217)	1.05 (237)	1.02 (79)
Open Marine	0.66 (300)	0.65 (162)	0.64 (45)	0.65 (80)	0.65 (85)	0.66 (231)

b. Analyses of Variance

	Source	df	SS	MS	F_s	
Long Island Sound	Phenotype	5	14.54	2.91	0.30	§
	Within	899	8843.27	9.84		
	Total	904	8857.81			
Open Marine	Phenotype	5	46.74	9.35	1.34	§
	Within	758	5279.95	6.97		
	Total	763	5326.69			

§ not significant

TABLE 4a

Results of genotype-dependent shock mortality experiments. Controls were randomly selected from the starting populations. All experimental populations were kept in the lab for two weeks at 32 o/oo, and ambient environmental temperature before shock.

		Temp °C	Salinity o/oo	N	94/94	94/96
Low initial frequency of Lap^{94} – Winter						
Control	A-0	--	--	191	0.178	0.283
Survivors	A-1	25	16	149	0.348	0.154
Survivors	A-2	25	16	156	0.308	0.173
Survivors	A-3	25	16	132	0.364	0.136
Control	B-0	--	--	157	0.312	0.178
Survivors	B-1	25	16	145	0.407	0.131
Survivors	B-2	25	16	129	0.341	0.209
Survivors	B-3	25	0	123	0.520	0.203
Survivors	B-4	25	16	150	0.300	0.187
Survivors	B-5	25	16	149	0.383	0.242
Survivors	B-6	25	16	173	0.225	0.410
Control	C-0	--	--	215	0.158	0.326
Survivors	C-1	30	16	319	0.248	0.273
Survivors	C-2	30	32	233	0.240	0.262
Other initial frequencies of Lap^{94}						
Control	1-0	--	--	380	0.318	0.179
Survivors	1-1	25	16	188	0.255	0.223
Control	2-0	--	--	223	0.363	0.188
Survivors	2-1	26	16	100	0.320	0.250
Survivors	2-2	26	32	245	0.298	0.196
Control	3-0	--	--	234	0.256	0.261
Survivors	3-1	25	16	125	0.204	0.314
Survivors	3-2	25	32	192	0.224	0.276

Table 4a (continued)

96/96	96/98	98/98	94/98	Lap^{94}	Lap^{94}	Lap^{94}
0.056	0.094	0.084	0.304	0.471	0.246	0.283
0.013	0.054	0.128	0.302	0.577	0.117	0.305
0.038	0.077	0.096	0.308	0.548	0.163	0.288
0.030	0.076	0.129	0.265	0.564	0.136	0.299
0.064	0.108	0.102	0.234	0.518	0.207	0.274
0.021	0.076	0.083	0.283	0.614	0.124	0.262
0.062	0.078	0.085	0.225	0.558	0.205	0.236
0.008	0.114	0.008	0.146	0.695	0.167	0.138
0.033	0.087	0.093	0.300	0.543	0.170	0.287
0.054	0.101	0.107	0.114	0.560	0.225	0.215
0.040	0.046	0.058	0.220	0.540	0.268	0.191
0.033	0.088	0.098	0.298	0.470	0.240	0.291
0.034	0.094	0.097	0.254	0.510	0.218	0.271
0.094	0.086	0.077	0.240	0.491	0.268	0.240
0.032	0.076	0.095	0.300	0.558	0.159	0.283
0.037	0.133	0.074	0.277	0.505	0.215	0.279
0.045	0.049	0.098	0.255	0.585	0.164	0.251
0.060	0.110	0.120	0.140	0.515	0.240	0.245
0.061	0.131	0.086	0.229	0.510	0.225	0.265
0.034	0.077	0.124	0.248	0.511	0.203	0.286
0.094	0.165	0.071	0.150	0.437	0.335	0.228
0.057	0.130	0.099	0.214	0.469	0.260	0.271

no mortality; followed by one of high mortality; ending with
a stabilization of the number of survivors (usually 25-35%).
We therefore stopped experiments at about 70% mortality.
Using a rows-by columns Chi-square contingency test, (Table 4b)
genotypic frequencies are seen to be significantly different
from controls in 9 of 16 experiments.   But a comparison of
controls of similar Lap^{94} starting frequency shows stronger
heterogeneity ($X^2 = 60.9$, $P < 0.005$ for controls A-O, B-O, C-O).
The preponderance of significant deviations may be no more
meaningful than among-sample variance.   The allele frequency
differences between survivors and controls were estimated by
standardizing the control to zero mean, deviations (d_i) of
survivors from controls should therefore be distributed with
zero mean and unit variance (see Christiansen et al., 1976)
and 95% of deviations (d_i) should be within the interval +2
to -2.   This is, however, not the case.   For Lap^{94}, 6 of
the 16 deviations fall outside this interval and a seventh is
equal to 1.9.   The probability, P, that 6 or more of 16
experiments would fall outside this interval can be estimated
by the binomial distribution where p, the probability of
falling outside is 0.05 and q equals 1-p.   The probability
then is:

$$P = \Sigma^{16}_{x=6} (x^{16})(0.05)^x(0.95)^{16-x}$$

This probability is much less than 0.001 for the Lap^{94} allele
 (a similar statement can be made for the Lap^{96} allele).
Compared with the grand mean of samples shown in Fig.6, 5 out
of 16 survivor samples show significant deviations of Lap^{94}
($p < 0.001$).

Although deviations in allele frequency are significant,
they are small, usually less than 0.1.   The only case of a
much larger deviation was one experiment at 25°C and 0 o/oo
where the frequency of Lap^{94} changed from 0.480 to 0.695.
However all deviations of Lap^{94} from controls were much less
than the difference in frequency found along the Long Island
Sound-open marine Lap^{94} cline.   In most low-salinity experiments we noted that LAP did not stain as well in gels and
that satellite bands were present.   This impression was confirmed by electrophoresing survivors of the 25°C, 0 o/oo
experiment on the same gels with controls held at 32 o/oo,
25°C and others kept in air.   LAP of animals in 0 o/oo
showed poorer staining and extensive development of satellite
bands, relative to mussels held in air and at 32 o/oo.

Allele frequencies of samples from natural populations
at Shinnecock during 1975 and 1976 are shown in Figure 6.

TABLE 4b

Difference between genotypic frequencies in control and survivors, (Table 4a) using χ^2 contingency test and deviation, d_i, of allele frequency.

Contrast	χ^2	d.f.	Significance	Lap^{94}	Lap^{96}	Lap^{98}
A-O x A-1	23.39	5	P < 0.005	+2.74	-4.26	+0.63
A-O x A-2	29.98	5	P < 0.005	+2.02	-3.68	+0.15
A-O x A-3	21.60	5	P < 0.005	+2.32	-3.43	+0.44
B-O x B-1	8.02	5	n.s.	+2.36	-2.51	-1.33
B-O x B-2	1.49	5	n.s.	+0.95	-0.01	-1.01
B-O x B-3	25.87	5	P < 0.005	+4.16	-1.16	-3.58
B-O x B-4	3.13	5	n.s.	+0.62	-1.13	+0.36
B-O x B-5	9.15	5	n.s.	+1.04	+0.55	-1.64
B-O x B-6	24.26	5	P < 0.005	+0.56	+1.93	-2.39
C-O x C-1	7.05	5	n.s.	+1.28	-0.84	-2.75
C-O x C-2	13.84	5	P < 0.025	+0.63	+0.96	-1.73
1-O x 1-1	8.16	5	n.s.	-1.69	+2.32	-0.14
2-O x 2-1	30.24	5	P < 0.005	-1.66	+2.28	-0.16
2-O x 2-2	11.09	5	P < 0.05	-2.30	+2.35	-0.49
3-O x 3-1	55.70	5	P < 0.005	-1.90	+3.91	-1.69
3-O x 3-2	5.90	5	n.s.	-1.22	+1.97	-0.49

If the mean allele frequency is p_0 (here the grand mean frequency of Lap^{94} over all samples), and all sample times and locations have the same allele frequency, then the deviation of a sample i,

$$d_i = (p_i - p_0)\, 2X_i/p_0 q_0$$

is approximately normally distributed with zero mean and unit variance (X_i is the number of individuals in sample i and $q_0 = 1 - p$; Christiansen et al., 1976). Two samples are more than 2 units below the mean and shock experiments using these populations as controls resulted in an increase of the frequency of Lap^{94} (see arrows in Figure 6). Populations within

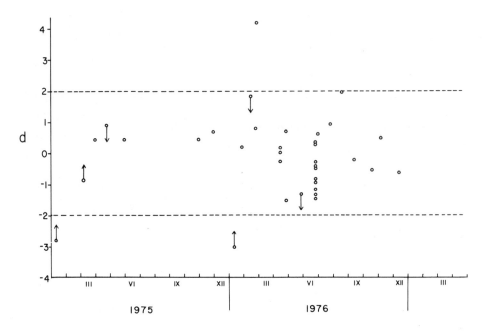

Figure 6. Seasonal variation at Shinnecock in the Lap^{94} allele standardized against the mean allele frequency for all samples. Samples with arrows are those employed as controls for shock experiments, with indicated change of Lap^{94}.

2 units of the mean yielded survivors with lowered Lap^{94} allele frequencies in shock experiments. Figure 7 shows the Lap^{94} deviation of survivors as a function of initial frequency of controls, pointing out the increase of Lap^{94} when the d_i of the initial population is less than 2.0. Large negative deviations of Lap^{94} frequency are associated with winter months,

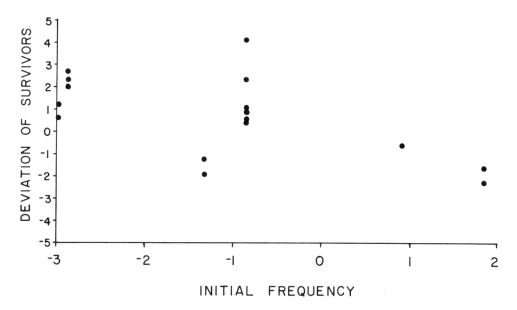

Figure 7. Standarized deviation of Lap⁹⁴ frequency in survivors, relative to that of controls. Frequency of controls are plotted in standard deviation units relative to the mean allele frequency for all Shinnecock samples in Figure 6.

but there is no obvious pattern of seasonal varation revealed in Figure 6. In both January,1975 and January 1976, Lap^{94} frequency does appear to increase from a yearly low, but frequencies of samples taken near this locality in January have higher allele frequencies. Therefore an overall seasonal pattern seems unlikely.

The increase observed from winter-low frequency controls

to survivors may also be explained by chance. As the controls show rather low Lap^{94} frequencies, any set of samples would be expected to have frequencies nearer the population mean resulting in a net increase.

Short term variation from January to February 1976 does coincide with laboratory mortality experiments. On January 13, the frequency of Lap^{94} was estimated at 0.470 (N = 233). Between January 14 and January 25, a period of several warm days over 12°C resulted in mass mortality of intertidal mussels (50 - 100% depending on locality) all over Long Island. We performed shock experiments on the 13 January (C1 and C2 in Table 4a) and found increases of Lap^{94} frequency (d_i < 2 in both cases). We also followed changes in the Shinnecock field locality with mortality and found that the frequency of Lap^{94} increased significantly (Figure 8). The last sample, taken on 26 February, was more than four standard deviations above the mean frequency (see Figure 6).

The data therefore verify hypothesis 3: that between-genotype physiological differences should be manifested in differential mortality of Lap genotypes or allele frequency changes. But the extension of the hypothesis, that shock experiments using low salinity should shift Lap^{94} frequencies from open marine to estuarine frequencies is not verified. The direction of change is complex and the magnitude too small.

Further evidence suggests that large allele frequency changes do not occur at the Lap locus. In June 1976 we discovered mussels 5 - 6 mm long in cages of much larger Shinnecock mussels that we had transplanted into Long Island Sound in April. Because Long Island Sound native mussels had not yet spawned and it is unlikely that larvae would stay for three weeks in the cages, these mussels were probably plantigrades (newly settled individuals) that settled among byssal threads the previous fall at Shinnecock and had now been transplanted to Long Island Sound for 2 - 3 months. We do not know how many, if any, died due to the transplant. But the Lap allele frequencies were identical to those of Shinnecock mussels, (Lap^{94} = 0.515, Lap^{96} = 0.254, Lap^{98} = 0.231, N = 132).

Discussion

A difference in mortality response by Long Island Sound and open marine mussels is clearly indicated by our data. The reason for this difference is less clear but is important to our understanding of the maintenance of the Lap polymor-

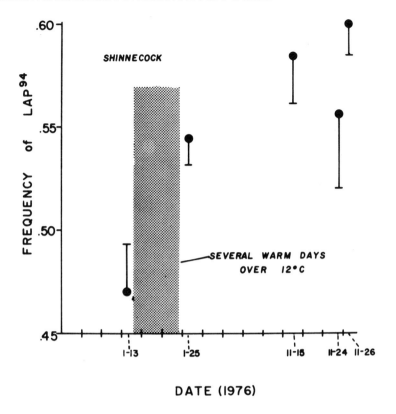

Figure 8. Change in Lap^{94} frequency at Shinnecock, before and after a warm spell in Januray, 1976.

phism. A large number of studies show that the respiration rate of <u>Mytilus edulis</u> is constant when measured in field-ambient salinities ranging from 5 - 30 o/oo (Remane and Schlieper,1971). Despite this, when the salinity is varied in the laboratory, above or below field ambient conditions, respiration is highest at the salinity in which the animal has lived in nature (Bayne <u>et</u> <u>al</u>. 1976). This picture is complicated by the possibility of long-term acclimatization

and the difficulty of acclimating invertebrates to laboratory
salinity changes. Theede (1965) was able to successfully
reverse the salinity tolerances of Baltic (15 o/oo) and North
Sea (30 o/oo) Mytilus edulis, through reciprocal transplantation and subsequent acclimatization. Therefore, the differences we have observed might disappear over a long-term laboratory
acclimation period.

Response differences similar to our results were found
for Mytilus edulis larvae. Bayne (1965) measured larval growth
in populations at different salinities. Larvae from an open
marine population in North Wales grew normally at 30-32 o/oo,
but growth was retarded at 24 o/oo, and ceased at 19 o/oo.
But larvae from a low salinity population in the Øresund
(Denmark, ca. 10 o/oo) grew at 14 o/oo. Thus larvae originating from open marine populations might have high mortality
when introduced into estuarine environments. Estuaries might
therefore be isolated from immigration of open marine Mytilus
edulis larvae. But estuarine larvae could probably successfully leave the estuary, given that adaptation for retention
within the estuary has not evolved (Wood and Hargis, 1971).

The possibility of genetic differentiation of the Long
Island Sound population of Mytilus edulis should not be excluded. Battaglia (1957) raised populations of the harpacticoid copepod Tisbe reticulata from low-salinity (Venice) and
open-marine salinity (Roscoff) populations for ten generations.
He found that the open-marine population was more sensitive
to environmental change than the Venice population. At
23.3 o/oo, fecundity of laboratory strains of the Roscoff
population was depressed by a factor of 14, relative to Roscoff
strains raised at 35 o/oo; and by a factor of nine, relative
to strains of Venice populations raised at 35 o/oo or 23.3 o/oo.

We have found no differential growth response of Lap
phenotypes of Long Island Sound or Shinnecock populations.
However, shock mortality experiments do shift genotypic and
allele frequencies to a significant degree, and in one case
a shock experiment paralleled an Lap^{94} change occurring in a
natural population. The largest change in Lap^{94} we observed
was with an extreme shock of 25°C, 0 o/OO. But, nevertheless,
the changes in allele frequency are usually less than 0.1 and
never more than 0.2. This difference is far less than the
frequency difference of 0.4 - 0.5 observed along the Long Island
Sound - open marine cline of 30 km.

The difference in direction of Lap frequency change, related to season, is unexplained but not unexpected. Many
seasonally related physiological asymmetries exist: tempera-

ture-energy budget relations, seasonal gametogenesis cycles, and food availability are just a few examples. In particular, Lap activity changes seasonally with a minimum in winter and a maximum in summer (Koehn, this volume). A specific example is relevant to our results. Newell and Kofoed (1977, and in preparation) show that the slipper shell Crepidula fornicata, when acclimated to low temperature, requires a much greater minimum maintenance concentration of food with increase in temperature than when acclimated to higher temperature. Winter animals which experience an increase in environmental temperature are, therefore, more starved than summer animals. Thus the change of temperature we observed in January, 1976, might have selected for genotypes coding for higher activity of an enzyme seemingly involved in digestion. Koehn (this volume) presents evidence that genotypes with the Lap^{94} allele (the one which increases after the mass mortality) have higher Lap activity than other genotypes.

Though smaller than the magnitude of the change in Lap^{99} frequency along the cline, the changes observed after natural mass mortality and in our experiments are consistent with other findings. Milkman and Koehn (1977) found evidence for a maximal change of only about 0.15 between years on the Long Island cline. Microgeographic variation at the Lap locus and juvenile-adult comparisons show fairly small changes in allele frequency (Balagot, 1971; Levinton and Fundiller, 1975; Levinton and Koehn, 1976). Thus a picture of modest changes in allele frequencies in adult populations, concomitant with mortality, is supported. With this result, how is the Long Island Sound cline maintained, in the face of apparent high migration rates? Several possibilities exist:

1. Selection among Lap genotypes occurs in the larval stage as larvae enter or leave the Sound, or just after metamorphosis. We can exclude the plantigrade stage and adults. Only experiments (now in progress) can verify this.

2. Migration is much less than we suppose. Long Island Sound has an estuarine flow and there is a net water loss from the Sound to adjacent open marine waters. Larvae may therefore leave the Sound and their density diluted in the outside marine waters with little larval influx into the Sound. Clines seen in much smaller estuaries (Koehn et al. 1976) would be difficult to regard as due to isolation.

3. Long Island Sound and adjacent open marine populations are different physiological races. The physiological difference may be due to long-term acclimatization or have a genetic basis. When open-marine larave enter the Sound their intolerance to

low salinity results in high mortality. Larvae leaving the Sound are rapidly diluted by the much larger open marine population. This effect causes isolation of Long Island Sound from adjacent open marine waters, allowing small changes in Lap^{94} frequency to accumulate over many generations in Long Island Sound. These small changes may be continuous, or episodic, accompanying times of high mortality. In effect, the mortality of larvae entering the Sound "protects" against the dilution of the Lap^{94} frequency in the Sound (0.1) by open marine immigration (0.55). This hypothesis is consistent with the studies on Mytilus larval tolerance (Bayne, 1965) and the establishment of physiological races based upon salinity differences (Battaglia, 1959). Experiments upon larvae are now in progress to test this hypothesis. We emphasize that with strong mortality as migrants enter or leave the Sound, a cline can be generated without selective mortality, among genotypes, at the Lap locus. Instead, populations with differing source Lap^{94} allele frequencies have differing mortality rates along the length of the cline, as the populations are mixed by currents. The Lap^{94} frequency difference between the Sound and adjacent open marine waters, however, would still have had to be generated by selection at the Lap locus or closely linked genes. Without some selection the small amount of gene exchange that must occur would be sufficient to eventually homogenize the difference.

Summary

Microgeographic intertidal variation and differences between open marine waters and estuaries suggest the possibility that selective mortality causes short term shifts in allele frequencies at the Lap locus in the mussel Mytilus edulis. A large change in allele frequency from the Long Island Sound estuary to adjacent open marine waters occurs in a cline of only 30 km distance. In such a short distance planktotrophic larval migration would be expected to homogenize this between-habitat genetic difference in the absence of strong selection.

We subjected Long Island Sound and open marine mussels to salinity and temperature shock, using standard dosage-mortality tests, to determine mortality patterns and allele frequency changes at the Lap locus. Long Island Sound mussels were more resistant to salinity shock than open marine mussels. The difference may be genetically based (physiological races) or due to long-term acclimatization, but mussels from the two habitats present different targets to selection. The effect of salinity shock on mortality increases with increasing temperature.

Temperature and salinity shock on open marine mussel populations produced survivors with allele frequencies only slightly differing (changes less than 0.1) from controls. The direction of frequency change appears to be seasonally dependent, but seasonal shifts in allele frequencies are not found in one locale studied over two years. One large increase in frequency of $\underline{Lap^{94}}$ observed in the laboratory coincided with a similar change in an open marine population in winter which suffered high mortality after a brief warm spell.

Transplants of mussels from open marine waters into Long Island Sound showed no mortality. Furthermore, juveniles less than 1 mm long accidentally transplanted into the sound showed frequencies characteristic of open marine conditions after several months. We conclude that unless selective mortality is extensive in pelagic larvae or in metamorphosing juveniles, small allele frequency shifts accompanying mortality are the usual event and isolation is more pronounced at the entrance of Long Island Sound than we now suspect. With such isolation, the small shifts observed in the lab could accumulate over generations, resulting in the differentiation of the Long Island Sound population. Changes observed in smaller estuaries would be enigmatic and difficult to reconcile with this model. If the two populations were physiological races or ecotypes, mortality could be high along the cline and $\underline{Lap}^{94}$ frequencies could change concomitantly with the mixing and subsequent high mortality of migrants of one ecotype into the optimum environment of the other. Some selection at the $\underline{Lap}$ locus would still be required to maintain the overall difference in allele frequencies.

Acknowledgements

We are very thankful to R. Hirschler and S. Brande who helped with most of the laboratory work. A. Bien, D. Innes and A. Lamb also assisted in the lab and in the field. B. L. Bayne provided helpful discussion and F. B. Christiansen aided our understanding of clinal distributions. We thank Niel Tully for providing his dock at Shinnecock Inlet and for selling us lobsters and scallops. Supported by NSF research grant BMS 74-02522 (to JSL) and grants from the Danish National Science Research Council and the University of Aarhus (to HHL). This is contribution number 205 from the Program in Ecology and Evolution at the State University of New York at Stony Brook.

REFERENCES

Balagot, B. P. 1971. Microgeographic variation at two biochemical loci in the Blue Mussel, Mytilus edulis. MA Thesis, State University of New York, Stony Brook.

Battaglia, B. 1959. Ecological differentiation and incipient intraspecific isolation in marine copepods. Ann. Biol. 33, 259-268.

Bayne, B. L. 1965. Growth and the delay of metamorphosis of the larvae of Mytilus edulis (L.). Ophelia 2, 1-47.

Bayne, B. L. 1976. The biology of mussel larvae. pp. 81-120. In, Marine Mussels, Their Ecology and Physiology. Bayne, B. L. (ed). Cambridge University Press, Cambridge.

Bayne, B. L., Thompson, R. J. and Widdows, J. 1976. Physiology: I. pp. 121-206. In, Marine Mussels, Their Ecology and Physiology. Bayne, B. L. (ed). Cambridge University Press, Cambridge.

Christiansen, F. B. and Frydenberg, O. 1974. Geographical patterns in four polymorphisms in Zoarces viviparus as evidence of selection. Genetics 77, 765-770.

Christiansen, F. B., Frydenberg, O., Hjorth, J. P. and Simonsen, V. 1976. Genetics of Zoarces populations. IX. Geographic variation at the three phosphoglucomutase loci. Hereditas 83, 245-256.

Endler, J. A. 1973. Gene flow and population differentiation. Science 179, 243-250.

Gilles, R. 1972. Osmoregulation in three molluscs: Acanthochitona discrepans (Brown), Glycymeris glycymeris (L.) and Mytilus edulis (L.). Biol. Bull. Woods Hole 142, 25-35.

Johnson, F. M. and Powell, A. 1974. The alcohol dehydrogenases of Drosophila melanogaster: frequency changes associated with head and cold shock. Proc. Nat. Acad. Sci. U.S.A. 71, 1783-1784.

Johnson, M. S. 1971. Adaptive lactate dehydrogenase variation in the Crested Blenny, Anoplarchus. Heredity 27, 205-226.

Kinne, O. 1970. Temperature: Animals-Invertebrates. pp. 407-514. In, Marine Ecology, V. 1, Environmental Factors, Part I. Kinne, O. (ed). Wiley, London.

Koehn, R. K. 1977. Biochemical aspects of genetic variation at the Lap locus in Mytilus edulis. This volume.

Koehn, R. K., Milkman, R. and Mitton, J. B. 1976. Population genetics of marine pelecypods. IV. Selection, migration and genetic differentiation in the Blue Mussel Mytilus edulis. Evolution 30, 2-32.

Koehn, R. K. and Mitton, J. B. 1972. Population genetics of marine pelecypods. I. Ecological heterogeneity and adaptive strategy at an enzyme locus. Am. Nat. 106, 47-56.

Koehn, R. K., Turano, F. J. and Mitton, J. B. 1973. Population genetics of marine pelecypods. II. Genetic differences in microhabitats of Modiolus demissus. Evolution 27, 100-105.

Levinton, J. S. and Fundiller, D. L. 1975. An ecological and physiological approach to the study of biochemical polymorphisms. pp. 165-178. In, Proceedings of the Ninth European Symposium on Marine Biology. Barnes, J. (ed). Aberdeen University Press, Aberdeen.

Levinton, J. S. and Koehn, R. K. 1976. Population genetics of mussels. pp. 357-384. In, Marine Mussels, Their Ecology and Physiology. Bayne, B. L. (ed). Cambridge University Press, Cambridge.

Levinton, J. S. and Suchanek, T. H. 1978. Geographic variation, niche breadth and genetic differentiation at different geographic scales in the mussels Mytilus edulis and Mytilus californianus. Mar. Biol. in press.

Milkman, R. and Koehn, R. K. 1977. Temporal variation in the relationship between size, numbers, and an allele frequency in a population of Mytilus edulis. Evolution 31, 103-115.

Mitton, J. B., Koehn, R. K. and Prout, T. 1973. Population genetics of marine pelecypods. III. Epistasis between functionally related isoenzymes of Mytilus edulis. Genetics 73, 487-496.

Newell, R. C. and Kofoed, L. H. 1977. The energetics of suspension-feeding in the gastropod *Crepidula fornicata* L. J. Mar. Biol. Ass. U. K. 57, in press.

O'Gower, A. K. and Nicol, P. I. 1968. A latitudinal cline of haemoglobins in a bivalve mollusc. Heredity 23, 485-492.

Paskausky, D. F. and Murphy, D. L. 1976. Seasonal variation of residual drift in Long Island Sound. Estuar. Coast. Mar. Sci. 4, 513-522.

Remane, A. and Schlieper, C. 1971. *Biology of Brackish Water*. Wiley-Interscience, New York.

Schopf, T. J. and Gooch, J. L. 1971. Gene frequencies in a marine ectoproct: a cline in natural populations related to sea temperature. Evolution 25, 286-289.

Seed, R. 1976. Ecology. pp. 13-65. In, Marine Mussels, *Their Ecology and Physiology*. Bayne, B. L. (ed). Cambridge University Press, Cambridge.

Theede, H. 1965. Vergleichende experimentelle Untersuchungen uber die zellulare Gefrierresistenz mariner Muscheln. Kieler Meersforschungen 21, 153-166.

Thompson, R. J. and Bayne, B. L. 1974. Some relationships between growth, metabolism and food in the mussel *Mytilus edulis*. Mar. Biol. 27, 317-326.

Widdows, J. 1972. Thermal acclimation by *Mytilus edulis* L. Ph.D. Thesis, University of Leicester, England.

Williams, G. C., Koehn, R. K. and Mitton, J. B. 1973. Genetic differentiation without isolation in the American Eel, *Anguilla rostrata*. Evolution 27, 192-204.

Wood, L. and Hargis, W. J. 1971. Transport of bivalve larvae in a tidal estuary. pp. 29-44. In, *Proceedings of the Fourth European Symposium on Marine Biology*. Crisp, D. J. (ed). Cambridge University Press, Cambridge.

SECTION 2

ECOLOGY, LIFE HISTORIES AND ADAPTIVE STRATEGIES

GENETIC CONSEQUENCES OF DIFFERENT REPRODUCTIVE STRATEGIES IN MARINE INVERTEBRATES

D. J. Crisp

N.E.R.C. Unit, Marine Science Laboratories, U.C.N.W.

Menai Bridge, Bangor, United Kingdom

Larval Dissemination

Benthic invertebrates are usually either sessile or slow moving in comparison with the rate of movement of the water masses above them. A few, such as the pectinids, cirolanid isopods and portunid crabs spend part of their adulthood swimming and therefore have greater mobility. For the rest the pelagic larva is the main instrument of dissemination.

Pelagic larvae which feed ("planktotrophic") might in theory survive indefinitely, though they usually reach the stage at which settlement is possible within two to three weeks, barely sufficient time to cross barriers of 50-100 kilometres (Crisp and Southward, 1953; Thorson, 1961). Scheltema (1971) has drawn attention to species whose pelagic larvae remain for months in the sea; such "teleplanic" larvae can migrate by currents over distances of thousands of kilometres. Pelagic larvae which neither feed nor absorb dissolved organics must be sustained on the remainder of the egg yolk ("lecithotrophic") and their potential time of survival will be limited by thermodynamic necessity. It will depend strictly on the rate of respiration and the proportion of non-essential energy reserves (Crisp, 1976b). Although a few non-feeding larvae can survive for many days, e.g. larval Rhizocephela, the majority that have been studied are littoral forms which lose the capacity to settle within a few hours (e.g. Spirorbis, Bugula etc.) The dissemination of such coastal species must rely mainly on the 6 hourly transit of tidal currents, so that to extend the larval life beyond this

time would be of little advantage and would waste energy
that might better be made available to the post-larva.
Further abbreviation of the larval phase, presumably in
exchange for higher survival of the post-larva, leads to
non-pelagic, oviparous or viviparous development. As
Thorson (1950) has demonstrated, the proportion of species
with non-pelagic development increases at higher latitudes.

Marine invertebrates characterised by pelagic dissemination have had to evolve mechanisms to reaggregate in order
to breed. Two variants of chemical recognition exist which
may be called gregarious and associative respectively, (Crisp,
1974). In the former, the larva seek out their own species
directly, while in the latter they seek the living surface
or exudate of some other animal, plant or micro-organism.
As an indirect result of associative settlement, larvae
originally dispersed over the area accumulate within a confined sector. In <u>Spirorbis</u> <u>spirorbis</u> and some barnacles,
both mechanisms appear to be combined (Crisp, 1965; Knight-
Jones, 1953; Williams, 1964). Gregarious and associative
settlement is presumably inherited, so that variants programmed to choose different target organisms would become
genetically isolated. This would account for the evolution
of sibling species associated with different substrata, e.g.
in the Spirorbidae, and for the development of epibionts,
e.g. epizoic barnacles, epiphytic Bryozoa etc.

Pelagic dissemination, even when countered by reaggregation, is wasteful of progeny, and its evolution must therefore
have some indirect consequence of high survival value. I
have argued that larval transfer is an essential process for
the survival of species which occupy isolated transient habitats subject to density independent catastrophes (Crisp, 1976a,
b). Many habitats occupied by marine invertebrates have this
property - not only the surfaces on which sessile epibionts
live but also the habitats of relatively immobile animals
dependent on isolated food sources, endangered by accumulations of predators, or limited by competitors for space;
adults living in all such situations require a disseminative
phase to break out of the trap. The dispersal of larvae
immediately following sexual union increases the diversity
of the propagules and their opportunity to colonise remote
habitats with characteristics different from those of the
parental environment (see Williams, 1975).

Larval transfer from remote populations must result in
a high rate of genetic exchange which might be advantageous
or otherwise. One would suspect that where <u>larval</u> <u>dissemination</u> is of the essence of survival - as for example, in a

pelagic epibiont whose offspring must find another host – then whatever the consequences in terms of greater genetic exchange the necessity of a pelagic phase must be accepted. On the other hand, where a variety of reproductive strategies coexist within the same habitat, as for example among littoral species of numerous Phyla (such as Gastropods, Bryozoa and Echinoderms), the long term genetic consequences of the adopted strategy would be one of the factors discriminating between the survival of competing species.

Spatial Genetic Diversity and Reproductive Strategy

At each zygote to zygote interval, genetic material will flow as a result of the movements of adult, larva and, in externally fertilizing species, of egg and sperm also. We know a little of the survival of invertebrate sperm once shed into the sea but, since adult invertebrates are generally aggregated, the majority of eggs are likely to be fertilised by sperm from neighbouring individuals unless there are mechanisms to prevent it. Thus pelagic larvae, when they exist, constitute the main propagules and in sessile organisms with internal fertilization, (e.g. barnacles), the only agents of gene flow. They play the same role as the wind-borne seeds of pollen and plants, being largely transported by capricious external forces. Wright (1931, 1943) considered an island model within a panmictic population and a continuous "isolation by distance" model, with short range dispersal. Both have analogies among marine invertebrates, the former resembling species with teleplanic larvae and the latter a population of periwinkles with short range dispersal.

If we define, by an imaginary line within a continuous population, a group or enclave then the probability that a new individual, replacing one which has disappeared, will have originated from within will obviously increase with the radius r of the enclave and decrease as a function of a central measure of the distance traversed between zygote and succeeding zygote. If diffusion laws apply, the root mean square distance, D, would be the appropriate statistic and will be referred to below as the "generation path". (Figure 1.) Figure 2 illustrates an environment with a randomly variable factor which operates for or against a particular element of the genotype. Three situations are shown. In 2a the generation path is larger than the scale of the spatial heterogeneities of the environment and so ensures effective panmixia over all such domains. Here enclaves, of radius r, lying within ecologically uniform domains, are smaller

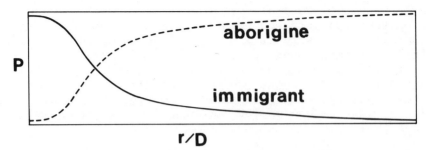

Figure 1. Type of relationship expected between the probability, P, (vertical axis) that a new individual within an enclave of dimension r has come from outside (immigrant) or from within (aborigine), and the root mean square generation path, D. The horizontal axis is the nondimensional number r/D.

than D, hence genetic immigration will be large enough to counter-balance selection at each generation. There may well be some selection during each generation, but at settlement the genotype will again be redistributed. Successful occupation of all domains would require a minimum genetic diversity while the panmixia intrinsic in the system would need either high fecundity to compensate for losses of ill-equipped individuals, or else the selection of genes that interact harmoniously within the genome (Mayr, 1963). Figure 2b shows the situation where D is of the order of r. Selection is only partially countered by immigration so that populations will develop distinctive but overlapping genotypes. Figure 2c illustrates the condition of minimum dissemination, where r greatly exceeds D, and the enclaves are effectively isolated. Selection is then virtually unaffected by immigration so that the enclaves may eventually acquire different subspecies or species. What this analysis indicates is that the significance of disseminative capacity measured by D to genetic variability depends not on D itself but rather on the relationship between D and the roughness or grain of the selecting factor in the environment.

Genetic and Environmental Diversity

I have assumed in the above discussion that a particular variable in the environment, represented by the vertical axis in Figure 2, was responsible for selection. Unfortunately there are not many instances in biology where we understand precisely the factors and how they operate in maintaining

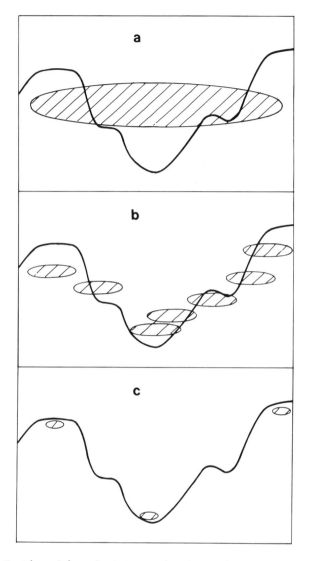

Figure 2 Relationships between the "roughness" of a spatially variable environmental factor resulting in post larval selection (vertical axis) and the generation path D. The hatched areas represent inbreeding groups of size in the order of D, and their displacement in the vertical axis represents variation in the genotype of the dispersed offspring resulting from selection.
    (a) Long larval life, D large compared with scale of environmental roughness.
    (b) D comparable with roughness.
    (c) Short larval life, D small compared with scale of environmental roughness.

genetic variability. Some examples are human sickle-cell anaemia (blood parasite incidence), melanism in Lepidoptera (background colour), copper tolerant grasses (soil copper level). There are few, if any, correspondingly clear cases among marine organisms. Hence perhaps there is a temptation among marine biologists to seek a general relationship between genetic variability as evidenced by the number of alleles at a sample of loci and the supposed diversity of the environment. The underlying assumption is that environments can be classified in some general way, being "stable", "predictable", "variable", "stressful" and so on, without specifying the factors referred to. I suggest this approach is likely to result in unrewarding debate. One cannot compare nor add together measures which are expressed in different units, for instance, temperature changes in space ($d\theta/dx$) with temperature changes in time ($d\theta/dt$) or the many physical and biotic variables. It would be like asking whether the weight of an elephant exceeded the noise of a jet engine! Nor is there any law of nature which stipulates that environments which have wide variations in one measure should have wide variations in others. Indeed, if one considers the examples of selection quoted above, they illustrate the irrelevance of any general attribute of variability within the environment.

Furthermore, in attempting to relate genetic and environmental diversity to one another, one has to have regard to the different potential polymorphism at various loci, the possibility that different groups of organisms have different potential for genetic variation as well as the subjectivity intrinsic in any assessment of diversity. Better by far to concentrate on understanding specific situations! Until we have such examples which include organisms with different disseminative capacities, one can only speculate. Species with pelagic larvae might be expected to need more varied genotypes to enable them to adapt to a greater variety of niches.

Quantitative Aspects of Larval Dispersal

Obviously the value of the generation path D depends on the longevity of the larvae and the hydrographic regime in which they are dispersed. There is sufficient information to assign an order of magnitude. Two approaches are possible: direct observation of larval movement and indirect estimates based on the movements of surface waters.

Since it is not feasible to mark and recapture larvae

in order to make direct observations, one must take advantage of special situations. For example, in the early stage of the invasion of a new species, the final stage larvae or new settlements can have arisen from no source other than a bridgehead population. This fact would allow one to trace the rate of diffusion. Thus between 1953-5 the rate of progress of <u>Elminius</u> along the North Wales coast was estimated at approximately 20-30 kilometres per year, representing probably the average movement in the plankton of a 10-15 day larva along a linear coast lacking large residual currents (Crisp, 1958). In a similar way, the failure of larvae to extend across regions inimical to their adults (Crisp and Southward, 1953) indicates that the separating distance exceeds D for most of the larvae.

The second approach is relatively easy for short lived larvae living, for example, in an estuary where currents are measurable. Such currents rarely exceed 2-4 knots, so that distances of the order of 5-15 kilometres are twice traversed in the 12 hour period of the semi-diurnal tide. In the open sea, conditions are complex, the scale of diffusion increasing with a linear scale of the area (Stommel, 1949). However, Okubo's (1971) summary of diffusion data allows predictions to be made for the magnitude of larval transfer in the absence of residual and tidal currents (Table 1). In practice, of course, tidal currents are likely to be as, or more, important to the movements of coastal species whose larvae remain in the water for less than a week.

TABLE 1

Oceanic Diffusion (based on Okubo, 1971).

Pelagic life	Log_{10} (probable distance transported in cms)	Order of magnitude	Whether likely to be exceeded by tidal currents
3 - 6h	4	100 m	Yes
1 - 2 days	5	1 Km	Yes
7 - 14 days	6	10 Km	?
14 days - 3 months	7	100 Km	No
1 year	8	1000 Km	No

Geographical Scale of Genetic Variation in Relation to Reproductive Strategy

There is good evidence that the length of the larval stage influences the scale of genetic variation in the way one would expect. To minimise extraneous factors one ought to consider a group of similar organisms such as the genus Littorina whose morphs have recently been well studied. Heller (1975a) distinguished four European species originally lumped as L. saxatilis, while Littorina obtusata is now at least two species, littoralis and mariae. As Table 2 shows, none of the above has pelagic larvae; all but one of the saxatilis group are viviparous, the others being oviparous with attached capsules. Past difficulties in separating these species suggest fairly recent speciation of the type illustrated in Figure 2c, for all can exist on the same shore, though occupying different niches. Moreover, within these species are colour morphs, especially in rudis, and littoralis, which are thought to be of cryptic value, as they are related to the background. Selection has been shown to operate over distances which are probably large in comparison with the generation path in L. rudis (Heller, 1975b).

The two species which release pelagic egg capsules and larvae, Littorina littorea and L. neritoides, appear uniform over large distances, having only a limited range of colour morphs. Specimens of L. littorea may be brick red or orange rather than brown, but these are isolated and exceptional individuals rather than populations. In the northern end of its range L. neritoides is shiny black, but in the Mediterranean blue-black, or sometimes diffusely striped with cream.

A study of isoenzyme distribution leads to a similar conclusion. Seven populations of L. saxatilis in the Cape Cod area exhibited significant variation at the phosphoglucose isomerase and leucyl alanine peptidase loci over distances of a few kilometres and independently of the separating distance (Snyder and Gooch, 1973). Over a wider geographical range but including the same area between Buzzard's Bay and Barnstaple, Berger (1973) found similarly large frequency variations in esterases in L. saxatilis but smaller differences in L. littorea. In the latter, Esterase IV was almost invariant, while the two alleles of Esterase III showed relatively small variations that may have been related to geography or water temperature. Gooch and Schopf (1970) and Schopf and Dutton (1976) demonstrated short range variations in allele frequencies in the bryozoa Bugula stolonifera and Schizoporella unicornis and in Schizoporella errata, all of

TABLE 2

N.W. European species of *Littorina* (Based on Heller, 1975a)

Species	Shore level (excluding pools)	Breeding	Colour morphs	Observed genetic variation	Range
rudis	MT-HWS	ovo-viviparous	Many	Large	Circum-polar
patula	MT-HW	"	several	?	?
neglecta	LW-HW	"	"	?	?
nigrolineata	MT-HW	oviparous	"	?	E. Atlantic
littoralis	LW-MT	oviparous	many	Large	E. & W. Atlantic
mariae	LW	oviparous	several	?	?
littorea	LW-MT	pelagic egg & larva	few	Clinal	E. Atlantic (W. Atlantic by immigration)
neritoides	MT-above HWS	pelagic egg & larva	few	?	Lusitanean

which have short-lived pelagic larvae. Not only were there
significant variations over distances of 10-15 kilometres,
which could be explained in terms of hydrographic isolation
(see Table 1), but there were also variations over distances
of a few metres, which must surely have indicated selection
during or after settlement. In contrast, Gooch et al.
(1972) found little genetic variation at the lactate dehydrogenase and "general protein" loci of Nassarius obsoletus,
either between widely separated populations or among size
classes. This species has a pelagic larva which can settle
in 10, or survive for more than 50 days, and probably exem-
plifies the situation shown in Fig 2a. Pesch (1974),
observed N-S allelic clines in allele frequencies in
Mercenaria mercenaria which also has pelagic larvae. Indeed,
in most species with long lived planktotrophic larvae,
appreciable genetic differences should require a distance of
separation at least the order of 100 Km. Balanus balanoides,
a well studied circumboreal barnacle, shows racial differences between the east and west Atlantic seaboards, both on
physiological (Crisp, 1964) and genetic criteria (Flowerdew
and Crisp, 1975), yet are fully interfertile (Crisp and
Flowerdew, unpublished observations). Moreover, the distribution of isozymes at an esterase locus were the same in
high and low water populations taken from the same locality,
as would be expected if shore level exercised no selection at
this locus. However, the distribution of alleles varied
greatly between seasons in a way very difficult to explain by
short term selection (Flowerdew and Crisp, 1976). In
Chlamys opercularis, a mobile pectinid with planktotrophic
larvae, Beaumont (private communication) has found that the
alleles of a myogen locus are almost uniform throughout the
Irish Sea, but with very different allele frequencies in
other sea areas to the north and south. As far as this
evidence goes, the geographical scale of genetic heterogeneity
seems to correlate well with hydrographic mixing processes.

There appear to be some exceptions to the generalisation
that taxa with pelagic larvae do not show local genetic
divergences - for example West Atlantic Mytilus edulis (Koehn,
1975) and the European M. edulis galloprovincialis complex,
though it is unprofitable to discuss the latter until we know
whether its component taxa are interfertile. Another which
deserves careful attention is the Balanus amphitrite complex,
each taxon of which may be a separate species, but in some
areas they may have been brought together by ship transport,

If limited larval dispersal allows greater intraspecific
variation, it ought in the long run to favour local speciation,

while conversely the species whose larvae spread freely should retain specific status over large areas. Thorson (1950), who has shown that pelagic larvae are uncommon in high latitudes, points out that species that retain a pelagic stage are highly successful over large geographical ranges, citing B. balanoides and Saxicava arctica as examples. However, there are also examples of widely distributed species both in space and time which lack pelagic stages, such as Littorina rudis, some species of Fucus and many interstitial animals. The correlation between geographical extent and reproductive strategy is therefore not entirely clear. Perhaps in the evolutionary time scale the exchange achieved by rafting, free floating or transport on sand grains and birds' feet (Boaden, 1964) are sufficient to maintain the integrity of conservative species.

Genetic Effects of High Fecundity

Pelagic dispersal is always associated with high fecundity to offset the mortality experienced during the larva's sojourn in the sea. Williams (1975) calls this situation the "Elm-Oyster model" referring to the close analogy between larvae and wind-borne seeds. He argues that intense competition in a multiple niche habitat will result from high fecundity and wide dissemination. Thus apart from compensating for non-selective pelagic mortality - though even resistance to predation and to starvation could conceivably be selected - there is little advantage to be gained by spawning so many copies of the same genotype that they compete within the space available only to a single individual. The premium on genetic diversity is therefore very great. As yet there are not enough data to judge whether the model is correct for marine organisms but the large genetic component of variability relating to size, vigour, fertility and tolerance is well established in cultivated plants e.g. Clausen and Hesey, 1958; Baker, 1969, Fripp and Caten, 1971, and is basic to successful plant breeding. Also, it is known that in plants heavy mortalities amongst seedlings are highly selective genetically (Harper, 1965a,b; Stebbins, 1970). The increased genetic variability must be associated with an increase in the variance of offspring viability, resulting in a higher proportion of malformed and poorly equipped individuals - the so-called genetic load. This too has been shown to occur in highly fecund plants and animals (Crumpacker, 1967). Conversely, given sufficient genetic exchange, highly fecund, genetically diverse species offer the prospect of producing individuals with superb combinations fitted for special niches.

The analogy between larvae and seeds needs experimental confirmation and it is clearly of great importance to the success of shellfish hatcheries. Experience shows that batches from matings of limited numbers of parents are most variable, some containing many malformed larvae, even after polyspermy has been eliminated (Gruffydd and Beaumont, 1970) though such variations have not as yet been related to genetic effects by inbreeding. Nor has much attention been given to the positive requirements for improving domestic strains. Indeed the work of Mathers, Wilkins and Walne (1974) on two loci in the hatchery stocks of <u>Crassostrea gigas</u> indicates that British hatcheries have for some time been derived from a single importation and probably from the offspring of only a pair of oysters! Not to have populated hatcheries initially from a wider sample of the gene pool may well have been a major genetic error.

Homozygote Excess in Marine Invertebrates

A significant, but not necessarily large, excess of homozygotes over Hardy-Weinberg expectations is commonly encountered in electrophoretic phenotypes of marine invertebrates. If not due to misscoring it might result from homozygote selection, both homozygotes being fitter than the heterozygote but occupying different niches. High fecundity and wide dissemination would presumably favour such a situation. Multiple niche selection in space is easy to envisage, but the same result could be achieved by differential selection in time. Thus animals homozygous for A' could be selected for in the larval stage and those homozygous for A'' in the post larval stage, leading to selection against the heterozygote $A' A''$.

Another factor which can give rise to apparent homozygote excess is the Wahlund effect. This is a purely mathematical result of mixing two populations each containing the same pair of alleles but at different frequencies and each population may individually conform to Hardy-Weinberg expectations. For more than two alleles the situation is complicated. Classically the populations would be geographically distinct, but would mix at the boundaries. However, they might also occupy distinct microhabitats with different selection pressures, and mix after pelagic redistribution. Another possibility is that each cohort might experience selection during the larval or early juvenile stage as well as in adulthood so that the age classes may show a different distribution of alleles. Such a difference has been found in <u>Chlamys opercularis</u> (Beaumont, private communication), where individuals of the 0 year class from Oban deviated only slightly from Hardy-Weinberg expectations, but the whole population of age classes showed a significant excess of homozygotes.

Unfortunately because we have so little data many possibilities exist; with a few ounces of fact we should be able to terminate several tons of theory!

Summary

Larval dissemination is necessary to the survival of many species with transient habitats. This must result in panmixia and, in the absence of selection, a uniform distribution of alleles over large areas. Conversely, species with non-dispersive strategies may show sharp differences in genotype over relatively small distances, together with a tendency to speciation. There is sufficient evidence in the physical oceanography literature to allow us to predict, from the length of the larval life, the order of magnitude of dispersal and the boundaries of panmictic populations.

High fecundity, which is needed to offset mortality in species with long-lived pelagic larvae, would be expected to give rise to wide genetic variability. As in the corresponding situation of opportunistic wild plants producing vast numbers of seeds, genetic variability is likely to have the disadvantage of a high genetic load, but compensated by the potential to produce combinations suited to specific situations - for example, in hatcheries and cultivation.

Homozygosity in excess of Hardy-Weinberg expectation is commonly encountered in marine species. Several possible causes - misscoring, null alleles, multiple niche selection and the combination of populations of mixed origin in space and time (Wahlund effect) might all contribute.

REFERENCES

Baker, R. J. 1969. Genotype-environment interactions in yield of wheat. Canad. J. Plant Science. 49, 743-571.

Berger, E. M. 1973. Gene enzyme variation in three sympatric species of Littorina. Biol. Bull. mar. biol. Lab. Woods Hole 145, 83-90.

Boaden, P. J. S. 1964. Grazing in the interstitial habitat: a review. In, Grazing in Terrestrial and Marine Environments. Crisp, D. J. (ed). Blackwell Oxford 299-303.

Clausen, J. and Hesey, W. M. 1958. Experimental studies on the nature of species. IV Genetic structure of ecological races. Carnegie Inst. Washington Publ. 615, 312 pp.

Crisp, D. J. 1958. The spread of Elminius modestus in North West Europe. J. mar. biol. Ass. U.K. 37, 483-520.

Crisp, D. J. 1964. Racial differences between North American and European forms of Balanus balanoides. J. mar. biol. Ass. U.K. 44, 33-45.

Crisp, D. J. 1965. Surface chemistry; a factor in the settlement of marine invertebrate larvae. In, Botanica Gothoburgensia III Proceedings 5th Marine Biological Symposium, Goteburg 1965. 51-63.

Crisp, D. J. 1974. Factors influencing the settlement of marine invertebrate larvae. In, Chemoreception in Marine Organisms. Grant, P. T. and Mackie, D. M. (eds). Academic Press. 177-205.

Crisp, D. J. 1976a. Energy relations of marine invertebrate larvae. Thalassia Jugoslavica 10, 103-120.

Crisp, D. J. 1976b. The role of the pelagic larva. In, Perspectives in Experimental Zoology. Spencer-Davies, P. (ed). Pergamon Press. 145-155.

Crisp, D. J. and Southward, A. J. 1953. Isolation of intertidal animals by sea barriers. Nature 172, 208.

Crumpacker, D. W. 1967. Genetic loads in Maize (Zea mais L.) and other crossfertilised plants and animals. Evol. biol. 1, 306-424.

Flowerdew, M. W. and Crisp, D. J. 1975. Esterase heterogeneity and an investigation into racial differences in the cirripede Balanus balanoides using acrylamide gel electrophoresis. Mar. Biol. 33, 33-39.

Flowerdew, M. W. and Crisp, D. J. 1976. Allelic esterase isozymes, their variation with season, position on the shore and stage of development in the cirripede Balanus balanoides. Mar. Biol. 35, 319-325.

Fripp, Y. J. and Caten, C. E. 1971. Genotype-environment interactions in Schizophyllum commune. Heredity 27, 393-407.

Gooch, J. L. and Schopf, T. J. M. 1970. Population genetics of marine species of the phylum Ectoprocta. Biol. Bull. mar. Biol. Lab. Woods Hole 138, 138-156.

Gooch, J. L., Smith, B. S. and Knupp, D. 1972. Regional survey of gene-frequencies in the mud snail Nassarius obsoletus. Biol. Bull, mar. biol. Lab. Woods Hole 142, 36-48.

Gruffydd, Ll. D. and Beaumont, A. R. 1970. Determination of the optimum concentration of eggs and spermatozoa for the production of normal larvae in Pecten maximus (Mollusca lamellibranchiata). Helgolander wiss. Meeresunters 20, 486-497.

Harper, J. L. 1965a. Establishment, aggression and co-habitation in weedy species. In, Genetics of Colonising Species. Baker, H. G. and Stebbins, G. L. (eds). Academic Press. 243-268.

Harper, J. L. 1965b. The nature and consequences of interference amongst plants. In, Proceedings 11th International Congress of Genetics, 2. Geerts, S. J. (ed). 465-482.

Heller, J. 1975a. The taxonomy of some British Littorina species with notes on their reproduction (Mollusca: Prosobranchia). Zool. J. Linn. Soc. 56, 131-151.

Heller, J. 1975b. Visual selection of shell colour in two littoral prosobranchia. Zool. J. Linn. Soc. 56, 153-170.

Knight-Jones, E. W. 1953. Gregariousness and some other aspects of the settling behaviour of Spirorbis. J. mar biol. Ass. U.K. 30, 201-202.

Koehn, R. K. 1975. Migration and population structure in the pelagically dispersing marine invertebrate Mytilus edulis. In, Isozymes, N. Genetics and Evolution. Markert, C. L. (ed). Academic Press. 945-959.

Mathers, N. F., Wilkins, N. P. and Walne, P. R. 1974. Phosphoglucose isomerase and esterase phenotypes in Crassostrea angulata and Crassostrea gigas (Bivalvia Ostreiche). Biochem. System and Ecology 2, 93-96.

Mayr, E. 1963. Animal Species and Evolution. Belknap Press, Harvard.

Okubo, A. 1971. Oceanic diffusion diagrams. Deep-Sea Res. 18, 789-802.

Pesch, G. 1974. Protein polymorphisms in the hard clams Mercenaria mercenaria and Mercenaria campechiensis. Biol. Bull. 146, 393-403.

Scheltema, R. S. 1971. The dispersal of the larvae of shoal-water benthic invertebrate species over long distances by ocean currents. In, Fourth European Marine Biology Symposium (Crisp, D. J. (ed). Cambridge University Press. 7 - 28.

Schopf, T. J. M. and Dutton, A. R. 1976. Parallel clines in morphologic and genetic differentiation in coastal zone marine invertebrates: the bryozoan Schizoporella errata. Palaeobiology 2, 255-264.

Snyder, T. P. and Gooch, J. L. 1973. Genetic differentiation in Littorina saxatilis (Gastropoda). Mar. Biol. 22, 177-182.

Stebbins, G. L. 1970. Variation and evolution in plants: Progress during the past twenty years. In, Essays in Evolution and Genetics in Honour of Theodosius Dobzhansky. Hecht, M. K. and Steere, W. C. (eds). Appleton-Century-Crofts. 173-208.

Stommel, H. 1949. Horizontal diffusion due to oceanic turbulence. J. Mar. Res. 8, 199-225.

Thorson, G. 1950. Reproductive and larval ecology of marine bottom invertebrates. Biol. Rev. 25, 1-45.

Thorson, G. 1961. Length of pelagic life in marine bottom invertebrates as related to larval transport by ocean currents. In, *Oceanography*. Am. Ass. Adv. Sci. Publ. No. 67, 455-474.

Williams, G. C. 1975. *Sex and Evolution*. Princeton University Press.

Williams, G. B. 1964. The effects of extracts of *Fucus serratus* in promoting settlement of larvae of *Spirorbis borealis* (Polychaeta). J. mar. biol. Ass. U.K. 44, 397-414.

Wright, S. 1931. Evolution in Mendelian populations. Genetics 16, 97-159.

Wright, S. 1943. Isolation by distance. Genetics 28, 114-138.

ECOLOGICAL, PHYSIOLOGICAL AND GENETIC ANALYSIS OF ACUTE OSMOTIC STRESS

R. W. Doyle

Department of Biology, Dalhousie University

Halifax, Nova Scotia, Canada

Introduction

Several papers in this symposium have dealt with fluctuating salinity as a source of stress in a population - that is, as an agent of natural selection. In the first part of this paper I will discuss the results of an artificial selection experiment involving the reaction of the estuarine amphipod <u>Gammarus lawrencianus</u> (Bousfield, 1973) to a sudden change in salinity. The experimental results indicate that selection for resistance to low salinities affects relative fitness in at least two ways, by changing survivorship and also by changing fecundity. The general conclusion of the first part of the paper is that it would be most unwise to make simple assumptions about the correlation between variation in physiological or biochemical traits involving salinity and variation in relative fitness. In order to assess correctly the correlation between physiological and fitness variation it will probably be necessary to trace all of the causal paths between the physiological variable and fitness. Beneficial paths to one component of fitness such as survivorship may be counterbalanced by detrimental ones to the same or to another component of fitness. The general outline of these paths can often be discovered in the physiological literature. However, physiology by itself is not enough because it is also necessary to construct a life table for the species in its natural environment with all the important relationships between physiology, mortality and fecundity known quantitatively. Furthermore, the quantitative ecological and physiological information must be of a statistical nature because what matters is the way <u>variation</u> in physiological phenotypes contributes to <u>variation</u> in relative fitness. Marine invertebrate physiologists have rarely designed their experiments to yield information about the covariance of traits in animal populations. In the second part of the paper I will describe the results of an experiment conducted with this objective in mind. The results will be interpreted using a statistical technique (path coefficients) which should in principle allow one to calculate the covariance of physiological traits with fitness through even the most complicated causal pathways.

Selection Experiments on G. lawrencianus. When <u>Gammarus lawrencianus</u> is abruptly transferred from its normal salinity into distilled water, rapid hydration occurs, during which the animals become increasingly sluggish and un-coordinated until eventually they are unable to right themselves. This vulnerable state is easily recognized and therefore the time taken to reach it, i.e. the "time-to-succumb", or TS as it will be called, can be measured with reasonable accuracy. The process ends in death if the animals are not returned to a more concentrated medium. While the behavioral changes occuring during the early stages of an "osmotic emergency" (Smith, 1963) are reversible in the laboratory, there can be little doubt that similar emergencies in more natural environments greatly increase the organisms' chances of being killed by a predator. Human predators - armed with wooden tongue depressors - find it relatively easy to capture these small amphipods during an osmotic emergency. The decrease in "time-to-capture", or TC, begins at once when the animals are subjected to the reduced salinity. The natural habitat of <u>G. lawrencianus</u> is the intertidal and subtidal zone, where it is fairly abundant along the North American coastline between Labrador and Maine (Bousfield, 1973; Steele and Steele, 1970). It is usually found in estuaries and other locations where the sea water is somewhat diluted by run-off. Sudden decreases in salinity accompanying tidal fluctuations and rainstorms are likely to render the amphipods more susceptible to predation by shorebirds, eels and other animals, especially during molt.

Experimental animals were collected from Easy Bay, Nova Scotia ($46°02'N$, $60°23'W$) and multiplied for three generations in the laboratory. The selection for TC, that is for resistance to osmotic emergencies, consisted of placing 66 adult males in a container of distilled water and removing them with a tongue depressor as they became sluggish enough to capture. The first 10 and the last 10 males were saved as breeding stock and mated with an equal number of similarly selected females to provide the "sensitive" and "resistant"

TABLE 1

Phenotypic data from selected lines. TC = time-to-capture, TS = time-to-succumb, TM = time-to-mature. TC and TS measured in minutes, TM in days; Log_e transformation used for TC and TS.

Generation	Trait	Sensitive Line $\pm \sigma$	Resistant Line $\pm \sigma$	Difference	Significance of difference
P	TC	-----	-----	3.1σ	-----
F_1	TC	$2.46 \pm .85$	$2.88 \pm .73$	$.53\sigma$	$p = .01$, df = 95
F_2	TS	$3.80 \pm .23$	$4.00 \pm .29$	$.77\sigma$	$p = .0053$, df = 53
F_2	TM	61.1 ± 9.4	69.9 ± 9.6	$.76\sigma$	$p = .031$, $U_{17,10} = 128$

lines. When the F_1 generation matured, random mating within lines produced F_2 offspring without further selection. TC was measured in the F_1 and TS in the F_2 generation. While the two measures are highly correlated, TC is more convenient when one is selecting individuals out of a crowded common container, and TS is more accurate when one is taking measurements on isolated individuals. In a second set of selection experiments presently underway, the rate of decrease of heartbeat is being used as a more reliable and less subjective measure of the response to salinity stress. While the F_1 animals were being raised it was noticed that the two lines appeared to be maturing at different times. Consequently, it was decided to quantify this difference in "time-to-mature", or TM, by setting up sensitive and resistant lines which were carefully matched by time of birth, population density and food supply. Four containers of newborn F_2 individuals were set up from each line. The results of these experiments are summarized in Table 1. TC in the resistant line is higher than in the sensitive line. Both

TC and TS were lognormally distributed in these experiments. With 15% selected from each end of the distribution the difference between the parental means is approximately 3.1σ on a logarithmic scale. The F_1 means diverged by approximately $.53\sigma$ giving a response/selection ratio of .17 , which can be taken as a conventional measure of "heritability". TS and TM are also greater in the resistant line. No differences in fecundity were observed. I have no satisfactory explanation for the peculiarly large correlated response, except to point out that TM is not normally distributed in this experiment and that the confidence limit is large. Obviously there is no indication of the kind of genetic interaction involved.

The correlation between TM and TS means that natural selection against less fit individuals will not necessarily improve the resistance of the gammarus populations to osmotic emergencies, even though artificial selection has apparently done so (Table 1). It would be possible for natural selection to produce intermediate or compromise values for both traits which would resist further evolutionary change, in large populations at least.

<u>Non-correlation of TS with fitness (at equilibrium)</u>. The method chosen for a preliminary investigation of the relationship between TS and relative fitness involves a representation of population growth called a Leslie matrix, which is familiar to population ecologists as an arrangement of mortality and fecundity information which allows one to calculate such population properties as growth rates and age distributions (Leslie, 1945, 1948; Pielou, 1969).

The principal eigenvalue of a Leslie matrix is the natural antilog of the intrinsic rate of increase, r, of populations, phenotypes or genotypes. The population with the highest r-value is the "fittest" in the sense that it will eventually supplant all the others, whether growth is resource limited or not, given the "compound interest" effects of simple population dynamics (Crow & Kimura, 1970). There are more complicated ecological models in which maximum r need not mean maximum fitness, but the present usage does seem to make sense in the case of <u>Gammarus</u> <u>lawrencianus</u> populations which expand very rapidly each spring and in general undergo very large fluctuations in size during the year. The effects of osmotic emergencies on the population can be represented by a survival probability, $p(1-m(1-a))$, which is incorporated into the Leslie matrix. In this formulation the frequency of osmotic emergencies is m, the probability of surviving an emergency is q and p is the

probability of surviving other sources of mortality over
the time interval. I assume that q is the expression of TC
or TS in a natural environment swarming with predators. TM
is the number (variable) of time intervals preceding reproductive maturity in the matrix. The optimal values of q
and TM - i.e. those resulting in highest fitness - are found
in the optimal Leslie matrix. The latter was obtained by
expanding the characteristic equation and using a Fibonacci
search procedure to locate the maximum r over the range of
q and TM. Whenever q < 1.0 and m > 0.0, osmotic emergencies
will exact a toll in each generation. The amount of this
selective mortality is $1-(1-m(1-q))^T$, where T is the length
of a generation (roughly TM + 6 days in the G. lawrencianus
example). Figure 1 illustrates various solutions of the
matrix in which the functional dependence of TM on q is

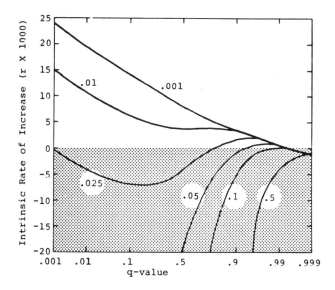

Figure 1 Population fitness (r) as a function of sensitivity
to osmotic emergencies (q). The numbers beside the lines
indicate the frequency of osmotic emergencies (m). At any
given m-value, natural selection moves the population toward
the q-value with the highest fitness. Populations falling
in the shaded region cannot indefinitely survive.
Abscissae are spaced on a probit scale (as on probability
paper).

assumed to be a simple dosage-response, by analogy with
quantitative genetic models of susceptibility to certain
diseases with a "threshold" of expression (Cavalli-Sforza &
Bodmer, 1971). In this functional relationship TM is
related directly and linearly to the probit transformation
of q. Qualitatively similar results are obtained if the
dependence of TM on q is a linear or concave function, or
if it is concave in the upper part as in the dosage-response.
The results will be qualitatively somewhat different if the
functional dependence is strongly convex, namely disruptive
selection with q=0 optimal at low and q=1 optimal at high
m-values. This sensitivity to the functional form of the
correlation between TC and TM makes a general treatment
difficult, except by the technique of path coefficients
described in the second part of this paper.

It can be seen in Figure 1 that in most cases the fittest
population (highest r) will have some residual susceptibility
to osmotic emergencies. In the specific example illustrated,
the parameters are derived from laboratory populations of
G. lawrencianus. (The length of the pre-reproductive period
varies from 39 to 59 days while q varies from 0 to 1 and the
value of p is .95/day. There are two reproductive episodes
separated by a 10-day interval, with 12 offspring in the
first and 18 in the second. The value of the survival probability is independent of age.)

At low (m = .001) or intermediate (m = .1) frequencies,
the fittest populations have q-values which result in 4.5%
and 8% of all deaths being caused by osmotic emergencies.
When the frequency of osmotic emergencies is sufficiently
high the optimum value of q does approach zero; however,
r is negative and so the population cannot sustain itself
indefinitely in this environment. The general conclusion
to be drawn is that there is no simple correlation between
overall fitness and a seemingly beneficial trait like
survivorship in the face of an osmotic emergency, when the
correlation is considered at the genetic level, or more
accurately at the level of evolutionary response to selection.
At equilbrium - supposing that each trait responds when the
other is selected, which has so far only been demonstrated
for selection on TS - the model population shows no linear
correlation at all between TS and fitness.

Physiological causes of TS variation. Placing a marine
animal into distilled water and observing the results is an
experiment that most biologists will have performed at some
time while obtaining their education - probably near the

beginning. The length of time an animal survives depends on the rates of several physiological processes, principally the influx of H_2O and loss of intra- and extra-cellular electrolytes. It also depends on the tolerance for increased internal pressure and/or blood dilution. Presumably both rates and tolerances can vary among individuals, but the relevant question is: how much do they vary? And how much does this variation contribute to variation in fitness?

In a preliminary attempt to answer this question a total of 37 adult G. lawrencianus were examined according to the following procedure. After being weighed and blotted dry, each amphipod was rinsed for 5 minutes in 4 changes of deionized water and transferred to a beaker containing 30 milliliters of deionized water. After a further 25 minutes the animal was removed from the beaker and incubated 1 minute in tritiated water (HTO) at a specific activity of 100 μCi/ml. Following a 45-second rinse the animal was placed in a second beaker containing 30 milliliters of deionized water where it was held until TS was reached. Throughout this procedure samples were removed periodically for analysis of Na^+ by atomic absorption spectrophotometry and analysis of HTO by scintillation counting. When TS was reached the animal was placed in a disposable plastic test tube to decompose for one week, following which a final Na^+ and HTO sample were taken.

This procedure allowed me to estimate the values of the following variables for each animal:

(1) W = wet weight.

(2) TS.

(3) C_{25} = Na^+ lost during the first 25 minutes (per mg. wet weight).

(4) NA LOSS = total Na^+ lost before TS (per mg. wet weight).

(5) NA CONC = total Na^+ lost from the animal during the 1-week experiment (per mg. wet weight).

(6) NA RATE = absolute rate of Na^+ loss during the first 25 minutes, computed as μg Na^+/mg wet weight/hour.

(7) K(NA) = specific rate of Na^+ loss computed over the first 25 minutes and computed as the first-

order specific rate constant

$$k_{(Na)} = \frac{1}{25} \ln \frac{NA\ CONC}{C_{25} - NA\ CONC}$$

(8) OSMOTIC INFLUX = defined as counts/minute/ mg wet weight taken up during the exposure to HTO and computed from the amount of HTO subsequently released with a correction for the loss during the 45-second rinse.

Unfortunately, the only ion loss which could be measured in this system was Na^+ because of the limiting sensitivity of the AA spectrophotometer.

A possible network of causal relationships among these variables is shown in Figure 2. The network reflects my interpretation of the literature on the physiological effects of salinity change on various species of Gammarus, based mainly on the publications of Lockwood (1962), Lockwood, et al.

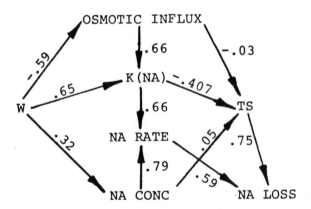

Figure 2 Path diagram of causal relationships among physiological variables affecting TS.

(1976), Potts & Parry (1964), Shaw & Sutcliffe (1961) and Sutcliffe (1971). A few examples will show how Figure 1 should be interpreted.

Variation in TS is assumed to be caused directly by variation in the osmotic influx rate (raising internal pressure), the specific rate constant for Na^+ loss (rate of blood dilution through excretory apparatus and integument) and the initial internal concentration of Na^+. Other contributary variables such as tolerance levels could not be measured directly and are not included in the diagram which omits arrows for "residuals". The osmotic influx rate also affects TS indirectly through its effect on K(NA); this can be interpreted as ion washout through the excretory apparatus. Variation in weight, or more probably some variable related to weight such as surface/volume ratio, influences the osmotic influx rate, K(NA) and possibly the initial Na^+ concentration.

The numbers beside the lines - called path coefficients - were estimated from the data on the 37 animals by multiple linear regression, i.e. regressing TS on OSMOTIC INFLUX, K(NA) and NA CONC, regressing K(NA) on OSMOTIC INFLUX and W, etc. The further steps in the conversion of regression coefficients into path coefficients (if any) are quite standard (reviewed in Wright, 1968 and Li, 1975) and need not be described here. The correlation or covariance of any of the variables with any other, including the fitness component TS can also be calculated from the path coefficients by following the rules established by the originator of this statistical technique, Sewall Wright. Furthermore, the correlations between the variables can be broken down into components representing the various pathways by which the variables are interconnected.

To give just one example of the kind of things one can discover through using this technique I will calculate the correlation between weight and TS, which in this case is merely the sum of the paths connecting the two variables:

$$r_{(calc)} = (-.03)(-.59) + (-.41)(.65) + (-.41)(.66)(-.59)$$
$$+ (.05)(.32)$$
$$= -.07.$$

This calculated correlation is not significantly different from the observed value of -.02. What is interesting about the connection between TS and weight, however, is the fact that it includes three pathways passing through OSMOTIC INFLUX and K(NA). Thus the negligible correlation between weight and TS does not mean that they are functionally unrelated. On the contrary; there are strong causal pathways connecting the two variables, but they have different signs and effectively cancel each other out. Since the change in TS which was observed in the selection experiments presumably involves a change in these pathways, the response of growth rate to selection for TS becomes less mysterious.

An obstacle to developing this line of thought any further lies in the fact that a system of linear relationships is fundamentally inappropriate in describing a phenomenon dominated by diffusion (at least, that is the present working hypothesis). I am not yet certain how great an obstacle this will eventually prove to be. A process involving first-order reaction kinetics can obviously be linearized in part with log transformations, and methods for dealing with certain kinds of non-linear relationships are available (Turner and Stevens, 1959). My next attempt at interpreting these data will be based on a log-transformed kinetic model.

Discussion and Perspective

The genetic component of this symposium on the "Genetics, Ecology and Evolution of Marine Organisms" has dealt with isoenzyme variation almost exclusively. In most instances the connection between isoenzyme genetics and the ecology of the populations has been made at a very high level of abstraction. For example, the amount of heterozygosity in the deep sea has been related to a generalized trophic stability by Ayala and Valentine, heterozygosity in Littorina spp. is discussed in terms of habitat overlap by Beardmore and the relationship between morphological and genetic variability in Thais is described by Campbell. The details of the functioning of each isoenzyme system is irrelevant to the purpose of these studies, which have as a major technical objective the identification of the maximum number of loci so that generalizations about the genome are affected as little as possible by sampling error (or a biassed sample of functional types). Koehn's work on the LAP locus in Mytilus edulis is an exception inasmuch as he is attempting to relate functional differences among gene products to variation in the ecological milieu in which populations find themselves. My contribution to the symposium is also related to the functional interpretation of variation at the genetic, biochemical, physiological and ecological levels.

A functional interpretation of genetic variation will

require ecological knowledge about the relative fitnesses of animals which differ in their physiology, as well as biochemical and physiological knowledge about the functioning of gene products. In the present paper I try to show that the causal relationships between fitness and the physiological response to salinity fluctuations involves complications at all levels. The most subtle analysis in one field may be obviated by a naive inter-relationship with one of the other fields, and unfortunately hardly any previous work in marine physiology, provides the variances and covariances required by evolutionary or population genetic theory. The methods of path coefficients, or structural equation methods in general, do seem suitable for this purpose and have been quite highly developed by econometricians (for introductions see Leser, 1969 or Land, 1973). In the present study, if data had been available on the maturation time of the animals used in the physiological experiment, a path coefficient approach might have been used throughout without recourse to computer simulation. It should be possible, given adequate data, to fit even the most complex biochemical, physiological and ecological data into a consistent conceptual structure. Anything less will almost certainly fail to produce an adequate explanation of gene frequencies based on the functional variation of the products.

Summary and Conclusions

Artificial selection for change in the resistance of _Gammarus_ _lawrencianus_ to an osmotic shock (distilled water) also altered the age at which sexual maturity is reached. Because of this correlation between resistance and maturation rate, populations will be maximally fit when both traits have intermediate values. Populations are predicted to remain susceptible despite continual selection in their natural environment, and there will be no linear correlation between variation in susceptibility and variation in relative fitness. Thus simple assumptions about what is a "good" physiological phenotype for an amphipod under salinity stress are unlikely to be valid.

In a second set of experiments a network of causal inter relationships involving permeability to Na^+ and H_2O is interpreted by the technique of path coefficient analysis, and a possible causal connection between growth (weight) and susceptibility is pointed out. It is suggested that path coefficients or some related technique may provide a consistent framework for relating biochemistry and physiology to evolutionary theory and population genetics.

REFERENCES

Bousfield, E. L. 1973. <u>Shallow-water Gammaridean Amphipods of New England</u>. Cornell Univ. Press., Ithaca, N.Y.

Cavalli-Sforza, L. L. and Bodmer, W. F. 1971. <u>The Genetics of Human Populations</u>. Freeman, San Francisco.

Crow, J. F. and Kimura, M. 1970. <u>An Introduction to Population Genetics Theory</u>. Harper & Row, New York.

Land, K. C. 1973. Identification, parameter estimation and hypothesis testing in recursive sociological models. In, <u>Structural Equation Models in the Social Sciences</u>. Goldberger, A. S. (ed). Harcourt Brace, New York. pp. 19-49.

Leser, C. E. V. 1969. <u>Econometric Techniques and Problems</u>. Griffin, London.

Leslie, P. H. 1945. The use of matrices in certain population mathematics. Biometrika 33, 183-212.

Leslie, P. H. 1948. Some further notes on the use of matrices in population mathematics. Biometrika 35, 213-245.

Li, C. C. 1975. <u>Path Analysis: A Primer</u>. Boxwood Press, Pacific Grove, Calif.

Lockwood, A. P. M. 1962. The osmoregulation of crustacea. Biol. Rev. 37, 257-305.

Lockwood, A. P. M., Croghan, P. C. and Sutcliffe, D. W. 1976. Sodium regulation and adaptation to dilute media in crustacea as exemplified by the isopod <u>Mesidotea entomon</u> and the amphipod <u>Gammarus duebeni</u>. In, <u>Perspectives in Experimental Biology, Vol. 1., Zoology</u>. Davis, S. (ed). Pergamon, London pp 93-106.

Pielou, E. C. 1969. <u>An Introduction to Mathematical Ecology</u>. Wiley, New York.

Potts, W. T. M. and Parry, G. 1964. <u>Osmotic and Ionic Regulation in Animals</u>. MacMillan, New York.

Shaw, J. and Sutcliffe, D. W. 1961. Studies on sodium balance in <u>Gammarus duebeni</u> Lilljeborg and <u>G. pulex</u> (L.). J. exp. Biol. 38, 1-15.

Smith, R. I. 1963. A comparison of salt-loss rate in three species of brackish-water nereid polychaetes. Biol. Bull. 125, 332-343.

Steele, D. H. and Steele, V. J. 1970. The biology of *Gammarus* (Crustacea; amphipods) in the northwestern Atlantic. IV. *Gammarus lawrencianus* Bousfield. Can. J. Zool. 48, 1261-1267.

Sutcliffe, D. W. 1971. Regulation of water and some ions in gammarids (amphipoda). I. *Gammarus duebeni* Lilljeborg from brackish water and fresh water. J. Exp. Biol. 55, 325-344.

Turner, M. E. and Stevens, C. D. 1959. The regression analysis of causal paths. Biometrics 15, 236-258.

Wright, S. 1968. *Evolution and the Genetics of Populations*. Vol. I. Univ. Chicago Press.

THE EVOLUTION OF COMPETING SPECIES

T. Fenchel, J.-O. Frier and S. Kolding

Dept. of Ecology and Genetics, University of Aarhus

Dk 8000 Aarhus C, Denmark

Introduction

The concepts of competitive exclusion and of niche differences in coexisting species constitute a central paradigm of ecology. A large number or investigations have been carried out in order to demonstrate differences in resource utilization of coexisting forms of differences in habitat selection among forms utilizing similar resources. More recently the formulation of the niche concept by Levins (1968) and by MacArthur and Levins (1967) has allowed quantitative predictions with respect to the limiting similarity of coexisting species which differ in their utilization of resource qualities along one "resource axis". Many field studies have since demonstrated the predictive value of this theory (see, e.g. MacArthur, 1972; May, 1973; and Christiansen and Fenchel, 1977).

From the present view point it is of interest to study the ecological and evolutionary mechanisms which have led to the differences between coexisting species observed in nature. There is much evidence to show that species interactions play a large role in the directional selection which leads to micro- and macroevolutionary changes (Bock, 1972). The present study discusses such changes brought about when closely related species coexist, i.e. "character displacement".

This term was originally coined by Brown and Wilson (1956) to describe the phenomenon that where two congeners coexist, they differ more from each other than where they occur allopatrically. It was recognized that two different mechanisms may be responsible for character displacement. If interspecific matings between two closely related, coexisting forms occur and if these matings are sterile or produce hybrids with lower fitness, then strong selection forces could be expected, leading to displaced mating periods, changes in sexual behaviour or other mechanisms which prevent cross-mating. This phenomenon is termed "reproductive character displacement". Character displacement could also be the result of competition for common resources or habitats. Morphological, behavioural or physiological changes might evolve which lead to differential resource utilization or habitat selection. We will here term this phenomenon "competitive character displacement". (The evolution of antagonistic mechanisms, e.g. interspecific aggression, ia another aspect of the evolution of competing species. The conditions for the evolution of such mechanisms are discussed in Christiansen and Fenchel (1977) but will not be treated here.) Theoretical models of the evolution of competitive character displacement in species competing for common resources and with Mendelian inheritance are discussed in Christiansen and Fenchel (1977) and Fenchel and Christiansen (1977).

A relatively large number of examples of both types of character displacement have been inferred from field observations. Several of these have been reviewed by Grant (1972) and by Pianka (1976). From these reviews, it is evident that many of the published examples of character displacement are not entirely convincing or have not been sufficiently analyzed. Thus, most cases have not been shown to be reproducible, viz. only one case of sympatry has been compared with cases of allopatry. These examples can often be explained as clines which have evolved and are maintained by reasons extrinsic to the interspecific relation. Furthermore, experimental and observational evidence clarifying the ecological significance of morphological or other differences is usually absent. Finally, evidence establishing the genetical basis of the observed character displacement is wanting.

The present paper reviews, in part studies in progress on three systems of congeneric marine species. Our purpose in these studies is eventually to achieve a thorough ecological and genetical analysis of the phenomenon of character displacement based on experimental evidence. Only some of the aspects of these studies can be given here; more detailed accounts are, or will be, published separately.

The Systems Studied

The systems of congeneric species studied comprise three marine and estuarine genera, viz. the prosobranch Hydrobia (with three species in the areas studied), the isopod Sphaeroma (with two species in the areas studied) and the amphipod genus Gammarus (with five species in the areas studied). The species belonging to these genera all show habitat selection with respect to salinity and within each of the genera the species more or less fill the entire environmental gradient from very dilute brackish water to fully marine conditions. They therefore show zonation patterns in estuaries, fjords and lagoons in European waters. Experimental evidence also shows that these species have a wider tolerance range than is usually realized in nature (Kinne, 1954; Fenchel, 1975a; Hylleberg, 1975; Frier, 1976). This indicates competitive interactions between the species. However, along the salinity gradients there is usually a more or less wide zone of overlap of the congeners and under some topographic conditions even three of the congeners may be found to coexist. Such distribution patterns, where in some places only one and in other places two or more of the congeners are found, may be understood as the interactions between competition and migration in a patchy environment (Fenchel, 1975a). Our main study areas are constituted by the complex estuaries and fjords along the east coast of Jutland and in particular the Limfjord. In these areas some 200 localities have been studied; these localities yield examples of both allopatric and sympatric populations of the species belonging to the three genera. In Gammarus and Hydrobia, populations from the central Baltic Sea (Stockholm Archipelago, Gotland, Tvarminne area in Finland) have also been studied. In the Baltic Sea the species are, to a much higher degree, found in sympatric populations. The zonation patterns characteristic of Danish estuaries and determined by salinity are absent in the Baltic, which is characterized by constant, low salinity.

Hydrobia. The hydrobias are small prosobranchs living mainly in, and on, shallow water sediments where typically they reach population densities of $1 - 3 \times 10^4$ m^{-2}. Three species are known from Danish waters, viz. H. ulvae, H. neglecta and H. ventrosa. The first species is mainly found in waters with the highest salinities, the last species mainly at low salinities and H. neglecta takes an intermediate position. Hydrobia neglecta is often absent between H. ulvae and H. ventrosa in salinity gradients in the Limfjord and it has not been found in the Baltic, where the two other species usually coexist at low salinites (c.6 o/oo) at which H. ulvae

is never found in Danish waters. Hydrobia ventrosa and
Hydrobia neglecta deposit egg capsules containing 1 (2) eggs
which develop directly into small juvenile snails. Hydrobia
ulvae deposit egg capsules containing many eggs which develop
into short lived planktonic larvae. All the species reproduce mainly during summer and they have a life span of 2 -3
years. Hybrids have never been observed and it has not been
possible to induce interspecific crossmatings in the laboratory (Lassen, 1978); we therefore believe that they never take
place. The three species ingest sediment particles and larger
diatoms. The diatoms and the microflora attached to sediment
particles constitute the food source of hydrobias. (For details on hydrobid biology and for references see Muus, 1967;
Fenchel, 1975a; Fenchel et al., 1975; Fenchel and Kofoed, 1976;
Kofoed, 1975; and Hylleberg, 1975,)

The relationship between H. ulvae and H. ventrosa has
been studied most intensively. Allopatric populations of
both species (and also of H. neglecta) consist of individuals
with about the same average shell length (c 3.5 mm). In
sympatric populations the species show a size difference,
H. ulvae always being the larger and the size ratio between
the shell lengths of the two species is usually in the range
1.3 - 1.4. This size ratio is maintained more or less
throughout the year. In the Baltic Sea this size difference
between the two species is always found (Fenchel, 1975b).
Collections of egg capsules from the field indicate that sympatric populations have shorter and displaced reproductive
periods whereas allopatric populations of both species reproduce throughout the summer. Laboratory studies, however,
have produced evidence for only a slight difference in the
reproductive periods of individuals deriving from sympatric
populations relative to individuals deriving from allopatric
ones (Fenchel, 1975b; Clarck and Lassen, personal communication).

A decrease in fitness, measured as individual growth rate
and as survival rate, as functions of density in laboratory
cultures, shows that the snails compete for food at natural
population densities. It has further been shown, that the
intensity of competition between two of the Hydrobia species
is identical to the intensity of intraspecific competition
when snails of the same size classes are used for the experiments. When snails of different size classes are used,
competition is relaxed. It has also been shown that the median
size of ingested food particles is a linear function of body
size and that the size difference found in the sympatric populations does indeed lead to a significant difference in the
size spectra of ingested food particles and thus a differen-

tial resource utilization by the two species. Finally it was shown that the size class of diatoms living in the sand is a function of the size class of snails grazing in the sediment (Fenchel, 1975b; Fenchel and Kofoed, 1976).

Thus, these studies have demonstrated that the character displacement found in sympatric populations leads to a relaxation in the competition for food resources and that the system conforms to the assumptions implied in the theoretical models predicting character displacement in species involved in exploitation competition (Christiansen and Fenchel, 1977).

The examples from the Limfjord are of special interest since they allow an estimate of the time scale involved in the microevolutionary changes observed in the sympatric populations. The Limfjord first became open to the North Sea at the beginning of the last century; prior to this it was a fjord system characterized by very low salinities. The present distribution patterns of the hydrobid snails and thus presumably the evolutionary changes in the local populations have thus taken place within the last about 150 years.

Sphaeroma. The isopod genus Sphaeroma is represented by two species in Danish waters, viz. S. hookeri and S. rugicauda of which the former is found in more brackish waters The species have a life span of one year with one annual generation and they reproduce in summer. The animals are mainly found in rock crevices which they leave only when feeding. The food consists of micro-algae which they obtain directly or by browsing on sand grains or other solid objects. The species show sexual dimorphism with respect to size, the males being the larger sex.

When allopatric populations of the two species are compared, it is seen that S. rugicauda generally attains a slightly larger size than does S. hookeri. In sympatric populations, however, this size difference is much greater, the size ratio being about 1.4 (Fig.1). Some evidence also indicates that the size difference between the sexes within each of the species is decreased in sympatric populations. This may illustrate the hypothesis of Schoener (1967) that sexual dimorphism in general signifies a niche expansion of a species and thus relaxes intraspecific competition. The finding of a decreased sexual dimorphism in sympatric populations of Sphaeroma is then in accordance with the prediction of Christiansen and Fenchel (1977) that the niche width is decreased by interspecific competition.

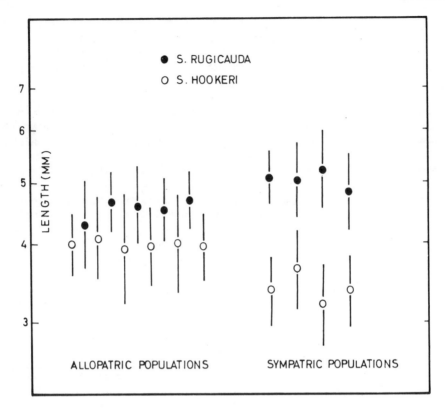

Figure 1. Average length (∓ 1 S.D.) of females of <u>Sphaeroma hookeri</u> and <u>S. rugicauda</u> in 11 allopatric and 4 sympatric populations.

Hutchinson (1959) predicted that in annual species which coexist on the basis of size difference, the larger species should reproduce first so that it remains the largest species throughout the life cycle. We could therefore predict that the reproductive periods would become displaced in coexisting populations of <u>Sphaeroma</u> spp. Allopatric populations reproduce almost simultaneously whereas in sympatric populations <u>S. rugicauda</u> reproduces about 14 days ahead of <u>S. hookeri</u> (Frier, in preparation; see also Fig.2).

The ecological interpretation of the character displacement in <u>Sphaeroma</u> is still under study. The species do not show any size-dependent particle size selection, either with respect to the food particles ingested or with respect to the particles on which they browse, which could explain the observed character displacement in terms of food resource competition. It has been possible to show, however, that there is a very clear size-dependent selection for the diameter of the crevices in which the animals are normally found hiding. We are, therefore, now investigating whether competition for hiding places in conjunction with predation (by fish) on free-swimming individuals may constitute the selective forces leading to the observed size differences in coexisting populations.

It is, however, also possible that the case of Sphaeroma represents an example of reproductive character displacement. According to Lejeuz (1966) cross matings between the two Sphaeroma species are possible in the laboratory, but that these result in a lowered fertility in succeeding generations. In nature hybridization has never been shown, although several attempts to find it have been made (Lejeuz, 1966; Møller, 1971), but precopulation between the species often takes place whenever conditions makes this possible. It is therefore possible that the displaced reproductive periods may be the result of selection against interspecific precopulations and that this again leads to the observed size differences as a consequence of different growth periods in the two species.

Gammarus. In the inner Danish waters and in the Baltic Sea five closely related species of the amphipod genus Gammarus are found, viz. G. duebeni, G. zaddachi, G. salinus, G. locusta, and G. oceanicus. In the Danish localities the species occur in the above mentioned order when passing from very dilute brackish water into fully marine conditions.

In salinity gradients the transition from one species to the other is often very sharp although some overlap does occur and G. oceanicus and G. locusta are found together in many of the Limfjord localities. However, all the species can complete their life cycle in a wide and overlapping salinity range in the laboratory (all the species grow and reproduce in the range 15 - 25 o/oo). The distribution patterns in the Danish localities therefore suggest that interspecific competition in these forms is important. With respect to the species pairs which do coexist to a larger or smaller extent (in particular G. oceanicus and G. locusta) studies have shown that the life cycles are displaced so that reproduction of the two species does not overlap and the size distribution overlap is minimal throughout the life-cycle (Fenchel and Kolding, in preparation). The displaced reproductive periods may exemplify reproductive character displacement since the species will perform sterile, interspecific precopulation in the laboratory (Kinne, 1954a, b). Also the displaced reproductive periods will lead to a succession in the utilization of habitats of the juvenile individuals (fine branched algae in very shallow water) by the different species. Under all circumstances, the species differences with respect

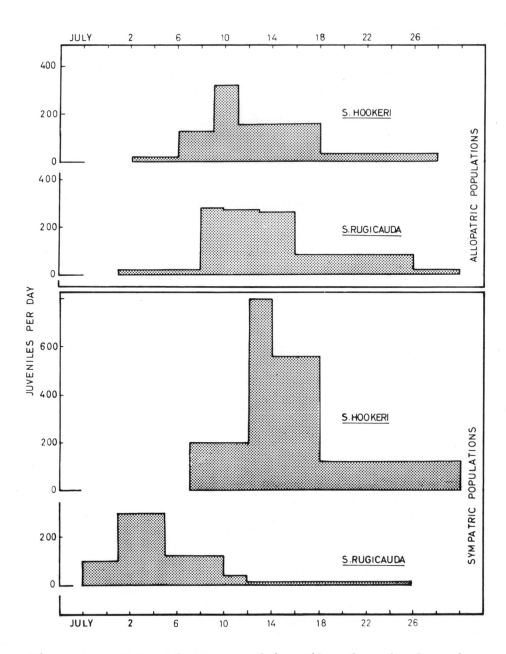

Figure 2. Dates of release of juveniles from females of Sphaeroma hookeri and S. rugicauda in two allopatric populations and in a sympatric one.

to their salinity preference and reproductive periods show
that species interactions have been important in the evolution
of the species after they had been formed and, either one
after the other or simultaneously, colonized Northern European
waters.

In the Baltic Sea the relations between the species are
different in some respects. Gammarus duebeni is largely
confined to more or less isolated marginal habitats, i.e.
rock pools where the other species are not found. The other
species seemingly coexist over wide areas in a rather constant
and low salinity of 6 - 7 o/oo. There is some indication of
habitat selection with respect to exposure and depth and often
only two of the species are common in any one locality.
Nevertheless the ability of the species to coexist in the
Baltic localities is striking compared with the situation in
Danish waters.

So far we have not been able to demonstrate morphological
or behavioural differences between the species in the Baltic
which could explain the differences in distribution patterns.
A detailed study of the life cycles of the Baltic populations
has been initiated. Preliminary results indicate that while
the life cycles of the species in the Baltic are basically
similar to those found in the Danish waters, the reproductive
periods seem to be shorter and more well defined in the Baltic
gammarids. It is not yet known whether these differences are
sufficient to explain the coexistence of the Gammarus species
in the Baltic localities.

General Discussion

Character displacement is, much like the principle of
competitive exclusion, an ecological and evolutionary phenomenon, the existence of which is hard to falsify. In fact
the main value of the concept may be that it has inspired
detailed observations and experiments. It does, however,
also open new possibilities for studying natural selection
in cases where definable selection forces are acting on the
species populations.

With respect to Hydrobia and Sphaeroma we have established
character displacement as a reproducible phenomenon, i.e.
it is found in all localities studied where sympatry of congeners occurs, but not in allopatric populations. The
hydrobias constitute an example of pure competition character
displacement and it has been possible to quantify the decrease
in competition intensity as the result of the change in a
morphological character, i.e. size. In Sphaeroma the charac-

ter displacement may be a response to the competition for
habitats, the result of selection against interspecific matings
or both factors may be significant. Both the Hydrobia and
the Sphaeroma examples shows that body size and temporal
aspects of the life cycle are quantitative characters which
may respond rapidly to selection.

The studies on Gammarus indicate at least that species
interactions have been important during the evolution of the
species which form a group covering the whole salinity range
in Northern European estuaries. The ability of the species
to coexist in the Baltic Sea in a uniform salinity and without
the same habitat selection remains to be explained. The
gammarids probably invaded the Baltic Sea about 5000 years
ago, but the present salinity conditions have only existed
for 2000 years (Segerstrale, 1957). Considering the time
scale involved in the microevolutionary changes in the
other genera studied, one could have expected more evident
changes in the populations of the species than the slight
changes in reproductive periods and perhaps in habitat selection. However, the gammarids may also examplify the
fact that the time scale of microevolutionary changes are
highly dependent on the available genetical variance and
on physiological constraints, factors which are very difficult to predict.

Summary

Interspecific interactions are believed to be an important factor in the evolution of species. Where closely related species coexist, microevolutionary changes may be
found which are absent where the species occur allopatrically.
Such "character displacements" may be due to different mechanisms. When cross matings between two congeners are sterile
or lead to hybrids with a low fitness, strong selection forces
may lead to changes in life cycles, behaviour or morphology
which prevent cross matings. Character displacement may also
be the result of interspecific competition for resources or
habitats. In sympatric populations selective forces may lead
to changes in morphology, life cycles or habitat selection
which decreases the intensity of interspecific competition.

The present contribution attempts to analyse such microevolutionary changes in sympatric and allopatric populations
of several species belonging to three different genera of
shallow water invertebrates. These comprise the mud snails
Hydrobia (3 species), the isopod genus Sphaeroma (2 species),
and the amphipod genus Gammarus (5 species). Field studies
and experimental work seem to establish the operation of all
the mechanisms referred to above.

REFERENCES

Bock, W.J. 1972. Species interactions and macroevolution. Evol. Biol. 5, 1-24.

Brown, J. and Wilson, E.O. 1956. Character displacement. Syst. Zool. 5, 49-64.

Christiansen, F.B. and Fenchel, T.M. 1977. Theories of Populations in Biological Communities. Springer-Verlag, Berlin.

Fenchel, T. 1975a. Factors determining the distribution patterns of mud snails (Hydrobiidae). Oecologia (Berl.) 20, 1-17.

Fenchel, T. 1975b. Character displacement and coexistence in mud snails (Hydrobiidae). Oecologia (Berl.) 20, 19-32.

Fenchel, T.M. and Christiansen, F.B. 1977. Selection and interspecific competition. In, Measuring Selection in Natural Populations. Christiansen, F. B. and Fenchel, T. M. (eds). Springer-Verlag, Berlin.

Fenchel, T. and Kofoed, L. H. 1976. Evidence for exploitative interspecific competition in mud snails (hydrobiidae). Oikos 27, 367-376.

Fenchel, T., Kofoed, L.H. and Lappalainen, A. 1975. Particle size-selection of two deposit feeders: the amphipod Corophium volutator and the prosobranch Hydrobia ulvae. Mar. Biol. 30, 119-128.

Frier, J. -O. 1976. Oxygen consumption in the isopods Sphaeroma hookeri Leach and S. rugicauda Leach. Ophelia 15, 193-203.

Grant, V. 1972. Convergent and divergent character displacement. Biol. J. Linn. Soc. 4, 39-68.

Hutchinson, G.E. 1959. Homage to Santa Rosalia or why are there so many kinds of animals. Am. Nat. 93, 145-159.

Hylleberg, J. 1975. The effect of salinity and temperature on egestion in mud snails (Gastropoda: Hydrobiidae). Oecologia (Berl.) 21, 279-289.

Kinne, O. 1954a. Die Gammarus-Arten der Kieler Bucht.

Zool. Jahrb. Abt. Syst. 82, 405-424.

Kinne, O. 1954b. Interspezifische Sterilpaarung als Konkurrenzökologischer Faktor bei Gammariden (Crustacea, Paracarida). Naturwiss. 41, 434.

Kofoed, L. H. 1975. The feeding biology of Hydrobia ventrosa (Montagu). I. The assimilation of different components of the food. J. Exp. Mar. Biol. Ecol. 19, 233-241.

Lassen, H. H. 1978. Electrophoretic enzyme patterns and breeding experiments in Danish mudsnails (Hydrobiidae). Submitted to Ophelia.

Lejeuz, R. 1966. Comparation morphologique, biologique et genetique de quelque especes du genre Sphaeroma (La treille) Isopodes Flabelliferes. Arch. Zool. Exp. Gen. 107, 469-667.

Levins, R. 1968. Toward an evolutionary theory of the niche. In, Evolution and Environment. Drake, E. T. (ed). New Haven, Conn. Yale Univ.

MacArthur, R. H. 1972. Geographical Ecology. Harper and Row, New York.

MacArthur, R. H. and Levins, R. 1967. The limiting similarity, convergence and divergence of coexisting species. Am. Nat. 101, 377-385.

May, R. M. 1973. Stability and complexity in model ecosystems. Princeton University Press, Princeton.

Muus, B. 1967. The fauna of Danish estuaries and lagoons. Medd. Danm. Fisk. Havunders. 5(1), 1-316.

Møller, B. 1971. Rapport over udført arbejde i forbindelse med en undersøgelse af amylase heterogenitet hos Sphaeroma hookeri (Leach) og Sphaeroma rugicauda (Leach). Unpublished Thesis, Copenhagen University.

Pianka, E. 1976. Competition and niche theory. In, Theoretical Ecology. May, R. M. (ed). Blackwell, Oxford.

Schoener, T. W. 1967. The ecological significance of sexual dimorphism in size in the lizard Anolis conspersus. Science 155, 474-477.

Segerstrale, S. G. 1957. Baltic Sea. Geol. Soc. America, Memoir 67, 751-800.

ON THE RELATIONSHIP BETWEEN DISPERSAL OF PELAGIC VELIGER LARVAE AND THE EVOLUTION OF MARINE PROSOBRANCH GASTROPODS

R. S. Scheltema

Woods Hole Oceanographic Institution

Woods Hole, Massachusetts U.S.A.

The concept of biological evolution as it is now generally understood includes two different phenomena. These are (1) the gradual transformation of species in time, a process sometimes also termed "gradualism" or "phyletic change", and (2) allopatric speciation, that is, the multiplication of species through reproductive isolation usually achieved by the spatial separation of populations over many generations.

The two processes, gradual transformation and species multiplication, I believe are closely coupled to one another and are related to the dispersal-capability of species. The relationship can be illustrated by a pictorial model (Fig.1).

Consider a hypothetical species, for example, a tropical prosobranch having a wide geographic range and pelagic larval development. This species is made up of a number of spatially separated populations (designated A through H, Fig.1). In the pictorial model, two dimensional space is represented along the horizontal axes; changes through time are shown on the vertical axes. Each disc denotes one generation and the differences in disc diameter designate variation in population size that occurs from place to place and from time to time. The adult of the species is regarded as essentially sedentary and hence populations will be isolated from one another; only the pelagic larval stages can move freely between populations.

Pelagic larval dispersal is denoted by the lines $\underline{d}$ between any two populations. In the case where there are no

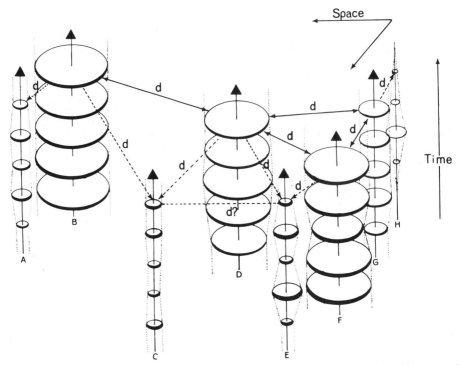

Figure 1. Pictorial model illustrating how larval dispersal provides genetic continuity between populations. See text for further explanation.

constraints to larval transport, the maximum possible number of unique interactions I will be shown by the equation:

$$I = \frac{(N-1)N}{2} \tag{1}$$

where N equals the total number of populations. On the pictorial model I is numerically equal to the maximum number of arrows that can be drawn between all populations of a species. This maximum number is, however, never actually realized.

The possibility for larval transport between any two points depends upon (1) the direction and velocity of the currents, (2) the distance between populations, and (3) the maximum duration of larval development for the species.

Often exchange in only one direction between populations is possible. In other instances, difference in net current direction at various depths may provide a mechanism for two-directional dispersal between populations. For example, it has been suggested that gastropod veliger larvae are transported in both directions across the tropical equatorial Atlantic Ocean, from east to west on the South Equatorial Current and west to east on the Equatorial Undercurrent (Scheltema, 1968, 1972b). To utilize such differences in horizontal current direction, larvae must be able to adjust their depth in the water column by their vertical swimming behaviour (vide, Scheltema, 1975; Sandifer, 1975).

The greatest distance that can be traversed by a pelagic larva depends upon the duration of its development. Certain gastropods, as well as the larvae of many other invertebrates, can delay settlement if a habitat suitable for post-larval survival is not available (e.g. Nassarius obsoletus, Scheltema, 1956, 1961; Phestilla sibogae, Hadfield and Karlson, 1969; Hadfield, 1977). A requirement for successful transport between populations is that the rate of passive dispersal, dependent upon the current velocity, will be great enough to carry pelagic larvae between two points within the maximum time that settlement can be delayed (Thorson, 1961).

The effective-larval-dispersal is the number of veligers that are dispersed to other populations and survive to reproduce. The frequency of effective-larval-dispersal $\underline{d}$ between two points is shown by the relationship:

$$\underline{d} = \underline{pt} \tag{2}$$

where $\underline{t}$ is the number of larvae available for transport and $\underline{p}$ is the probability of dispersal. The number of larvae available for transport ($\underline{t}$) depends upon (1) the size of the parent or "donor" population(s), $\underline{t}_p$, and (2) the fecundity of the species, $\underline{t}_f$ (i.e. $\underline{t} = \underline{t}_p \underline{t}_f$). The probability $\underline{p}$ for effective-larval-dispersal depends upon (1) the chance for survival of larvae during transport, viz. the survival coefficient (estimated to average no less than $2 \times \underline{t}_f^{-1}$, vide, Scheltema, 1972a) and (2) the likelihood that a current will carry larvae between two points, i.e. the drift coefficient, calculated from drift bottle data (vide, Scheltema, 1971b). For instance, the probability $\underline{p}$ that any particular larva will be transported across the tropical Atlantic is minute, estimated between 10^{-9} to 10^{-11}, but because the number of veligers available is very large, successful dispersal is probably much more frequent than might be supposed. It has been estimated that 1.3×10^{12} veligers of the gastropod Cymatium parthenopeum are transported across the North Atlantic each year (Scheltema, 1972a). Some of these larvae have successfully colonized the Azores and there are even scattered records of adults dredged off the Irish coast.

Small populations will produce relatively few pelagic larvae (i.e., $\underline{t}$ will be small) and can contribute only few emigrants in exchanges with larger populations (shown by the dotted lines between small and large populations, e.g. A and B; C and B, D, E etc. Fig.1). Conversely, pelagic larvae from large populations are not so likely to encounter small populations or "targets" ($\underline{vide}$, MacArthur and Wilson, 1967). Consequently, the value of $\underline{d}$ will usually be small in reciprocal exchanges to and from small populations.

Populations may differ not only in morphological but in physiological and biochemical characteristics. Geographic variation when found in a species (such as the one shown in the pictorial model, Fig.1) can have arisen theoretically from either (1) mutation, (2) random genetic drift, or (3) natural selection. The rates at which each of these three processes occur can differ temporally and spatially in each population.

Experimentally determined rates of mutation in natural populations have been shown to be quite low, between 10^{-5} to 10^{-6}. The number of generations required to replace a single allele by mutation pressure alone (i.e. in the absence of natural selection or genetic drift) is very large, more than two hundred thousand generations ($\underline{vide}$, Mettler and Gregg, 1969, pp. 80-83). Consequently, spontaneous mutation, although it introduces new genetic variation, is usually not regarded as important in the establishment of genetically determined geographic variation.

Genetic drift is probably largely confined to small isolated peripheral or island populations (e.g. A, C, E and H in Fig.1), where it operates under conditions of restricted immigration and low genetic variability. Such small peripheral and island populations may frequently become extinct (e.g., H in Fig.1).

Natural selection, usually regarded as the principal factor responsible for genetic variation between populations, may result from either (a) biological interactions such as competition and predation (e.g. Stanley, 1973) or from the interactions with the physical environment. Both biological and physical environmental factors may be aperiodic and unpredictable ($\underline{vide}$, Slobodkin and Sanders, 1969) or periodic and predictable (e.g. diurnal or seasonal). The notion of predictability, however, must be related to the average life-expectancy; events with a periodicity longer than the lifespan of an individual are unpredictable to members of that species (Grassle, 1972). Aperiodic or unpredictable events in the physical environment may cause extinction. For example, over geological time, unpredictable climatic and sea-level changes (either regression or transgression) can alter the geographic distribution of species with wide geographic ranges and even cause extinction among species with too restricted a distribution. There can also be a reciprocal interaction between the biological and physical environments ("feed back") and their impact upon any particular species. Although both the biological and physical factors responsible for natural selection may vary in intensity and periodicity from place to place and from time to time, stabilizing genetic mechanisms tend to counteract the disruptive effects of natural selection.

The immigration of larvae into a population is a source of new genetic variation that offsets the effects of natural selection, the principal source of geographic variation. Effective-larval-dispersal counteracts the effects of large-scale, unpredictable events by shifting and maintaining the geographic range of a species wherever and whenever favourable conditions exist. Extinction of the species is thereby avoided. By providing genetic continuity (i.e. gene flow) between spatially separated populations, pelagic larvae maintain a high genetic variability (i.e. eurytopy) within the species as a whole. Levins (1964) concluded that "the adaptive significance of gene flow [i.e., effective-larval-dispersal] is that it permits populations to respond under natural selection to long-term widespread fluctuations in the environment while damping the response to local ephemeral oscillation". At the same time effective-larval-dispersal tends to decrease geographic variability; Lewontin (1974) on theoretical grounds asserted that "even a migration rate as small as one individual in a thousand per generation is sufficient to prevent differentiation [arising from random drift] between populations of moderate size".

The importance of effective-larval-dispersal should now be evident from the pictorial model. Reproduction provides the continuity of genes over time (vertical arrows, Fig.1); larval transport effects gene flow through space (along d, Fig.1). Together the two processes, reproduction and dispersal, result in a flux of genes through time and space. While larval dispersal can maintain genetic continuity through space (vide, Gooch et al., 1972, for an example of gene flow between populations of the prosobranch gastropod Nassarius obsoletus), natural selection (and, to an unknown degree, genetic drift) can bring about geographic variation over short distances even in species with pelagic development (e.g.

vide, Struhsaker, 1968 for an instance of a case of differentiating over short distances in the prosobranch gastropod Littorina picta, a species with pelagic development).

The question can now be asked: What consequences do differences in dispersal capability (i.e., the magnitude of d in Fig.1, and I in equation 1) hold for species over long geological time? How do differences in the frequency of larval dispersal between populations affect the species as a whole over geological epochs?

Consider a continuum of dispersal capability. Near one end of such a continuum are species that have a teleplanic or "long distance" larva, at the other extreme are species with "direct development" and completely lacking a pelagic stage. Three cases can be used to amplify this point.

Species that have a teleplanic or "long distance" veliger larva (vide, Scheltema, 1971a, for definition) are forms that have a capacity for very widespread dispersal; their larvae can remain planktonic for many months, in some instances up to one year. It is predicted that because they have large geographic ranges they will be broadly adapted species with wide genetic diversity and therefore not readily susceptible to extinction and consequently will have a long fossil record. Even when some populations within the species range become extinct, such localities may become repopulated by immigrant larvae from remote regions when suitable conditions are again restored.

Is there evidence that species with teleplanic larvae existed in the geological past? Here only the Cenozoic will be considered. A comparison of the protoconch of fossil specimens with that of the larval shell of closely related contemporary species allows inferences to be made about the type of development in the geologic past. The subject is considered elsewhere in more detail (Scheltema, in press) so that only a few examples need be given here.

The protoconch of a well preserved specimen of Tritonium enode found by Sorgenfrei (1940) in the Klintinghoved Formation of the Danish Lower Miocene can be compared with the shell of a contemporary larva of Bursa granularis cubaniana taken from the plankton of the tropical Atlantic Ocean or with the protoconch of an adult specimen of the contemporary species Sassia remensa from the western Pacific. The likeness is readily apparent (Fig.2). Although placed by Sorgenfrei in a family of its own, Tritonium enode clearly belongs to the superfamily Cymatiacea and from the characters of the adult shell to the family Cymatiidae and probably the genus Sassia. Protoconchs

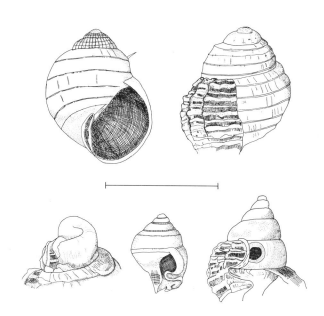

Figure 2. Fossil protoconchs and contemporary shells of Cymatiacea (above) and Muricacea (below) to illustrate method of inferring mode of development. Upper left: Teleplanic pelagic larva of <u>Bursa granularis cubaniana</u> from plankton of the tropical Atlantic Ocean, 18°00'N., 64°23'W.; Upper right: Fossil protoconch on juvenile <u>Tritonium</u> (= <u>Sassia</u>) <u>enode</u> from the Lower Miocene, Klintinghoved Formation, Denmark. This fossil species by comparison with the contemporary larva of <u>B. granularis cubaniana</u> at left is inferred to have had a teleplanic pelagic larval development; Lower centre: Teleplanic pelagic larva of <u>Thais haemastoma</u> from plankton of the tropical Atlantic Ocean, 7°20'S., 7°23'E.; Lower left: Protoconch on juvenile of the contemporary species <u>Thais emarginata</u> from the coast of California (redrawn from Scheltema, in press). The larval shell of this species with a known nonpelagic development differs conspicuously from the shell of <u>Thais haemastoma</u> to its right; Lower right: Fossil protoconch of the juvenile <u>Murex inornatus</u> from the Middle Miocene, Arnum Formation, Denmark (redrawn after Sorgenfrei, 1958) [= <u>Murex paucispinatus</u>, Lower Miocene Klintinghoved Formation, Denmark]. This fossil species by comparison with the larva of <u>Thais haemastoma</u> at lower centre and Thais emarginata at lower left is inferred to have had a teleplanic larval development. All specimens are illustrated at the same enlargement. Scale equals 2 mm.

in the genus Sassia are, however, unlike those of many Cymatiidae and by the reticulate pattern throughout the entire larval shell more nearly resemble the closely related family Bursidae (the latter family also included within the superfamily Cymatiacea). The exact systematic position of Tritonium enode in the Cymatiacea is for the present argument unimportant. The larval shell of Bursa granularis cubaniana (Fig.2) is large, over 2 mm in length typical of teleplanic veligers. Indeed, protoconchs of all pelagic Cymatiidae and Bursidae are very big. From what is known experimentally about growth rates of gastropod larvae (Scheltema, 1967; 1971b) it can be deduced that development of pelagic Cymatiacea veligers is very long, extending over many months. Extensive sampling in the tropical Atlantic Ocean has shown that the veligers of both the Bursidae and the Cymatiidae are widely dispersed and can be transported even across ocean basins (Fig.3). Concomitant with this dispersal capability, contemporary species of Cymatiacea have large geographic ranges, some extending over more than one major ocean. By inference it can therefore be concluded that Tritonium enode must also have had a long pelagic development and that it too must have been capable of wide dispersal on the surface waters of the Early Miocene Atlantic Ocean (vide, Berggren and Hollister, 1974, for Atlantic surface currents during the Miocene).

Evidence from both the Bursidae and the Cymatiidae supports the prediction of a long fossil record for species with teleplanic larvae. Woodring (1959) has shown, for example, that species complexes of Bursidae during the Middle Miocene of Central America differed very little from those found in the Caribbean and West Africa today. Indeed it can readily be argued that many Middle Miocene species differed from contemporary forms only subspecifically at most.

The protoconch of the fossil species Murex (= Thais?) inornatus from the Lower and Middle Miocene (Sorgenfrei, 1940, 1958) can be compared with the pelagic larva of Thais haemastoma (Fig.2). The similarity between the fossil protoconch and contemporary larva, viz. the high spire and sinusigerous lip, suggests a pelagic mode of development for the Miocene species. The correctness of this inference is further supported by a comparison of the fossil species with the protoconch of the related contemporary muricacean, Thais emarginata, known to have a nonpelagic development (Fig.2).

Veligers of contemporary Muricacea (belonging to the family Thaididae) are widely dispersed in the tropical and North Atlantic Ocean (Scheltema, 1971b, and in press) and

have been shown to be capable of being transported across
the Atlantic Basin. Again the inference is that the fossil
form must have had a capability for long-distance dispersal
similar to that of the contemporary species.

To summarize briefly, it is predicted that species of
gastropods with teleplanic or "long-distance" veliger larvae
will have a large geographic range and temporal longevity
of great duration. Direct comparison of fossil protoconchs
with the shells of contemporary larvae allow the recognition
of teleplanic gastropod veligers in the geological past.

The second case that can be considered along the continuum are those species having a pelagic development of restricted duration, extending from approximately three to six
weeks. These species, regarded as occupying a position near
the centre of the continuum, are typical of most temperate
and many tropical forms. The length of larval development
places constraints upon the distance of dispersal and such
larvae will not successfully cross biogeographic barriers
such as ocean basins. They may, however, have very large
geographic ranges along continental boundaries and throughout
shallow epicontinental seas. Such widespread distributions,
where they occur, result from the step-wise dispersal by
pelagic larvae over a large number of generations. Examples
may be found among Western Atlantic gastropod species that
range between Cape Cod and southern Brazil. Under some
conditions dispersal between populations may be infrequent
enough (or natural selection may be intensive enough) to
result in genetic isolation, particularly between peripheral
and island populations (in Fig.1, A, C, E and H). Isolation
between populations of estuarine species is likely to result
in geographical variation even over relatively short distances
(e.g. Crassostrea virginica between Chesapeake Bay and
Delaware Bay, Hillman, 1964).

Again, the direct comparison of protoconchs of fossil
specimens with those of closely related contemporary larval
shells makes it possible to identify larvae with a development of restricted duration in the plankton. For example,
a comparison of the larval shells of three contemporary species
of Nassarius grown in the laboratory (Scheltema, 1962;
Scheltema and Scheltema, 1965) with the protoconchs of fossils
from the same genus in the Klintinghoved and Arnum Formations,
Lower and Middle Miocene of Jutland (Sorgenfrei, 1940, 1958)
makes it possible to deduce that the Miocene species of
Nassarius had a pelagic development of approximately similar
duration, between 3 and 6 weeks.

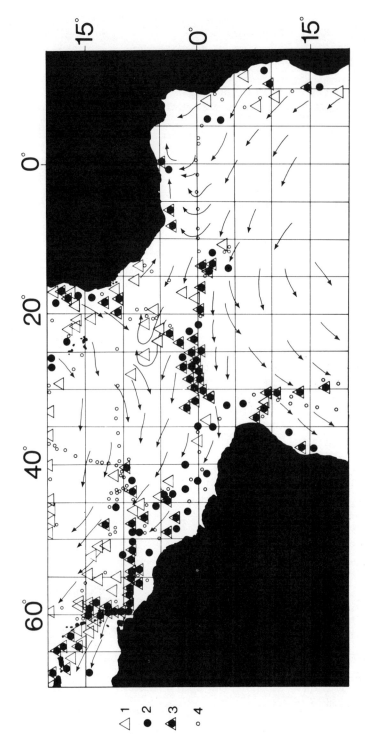

Figure 3. Distribution of teleplanic veliger larvae belonging to the prosobranch superfamily Cymatiacea and found in plankton samples made in the upper 100 meters of tropical Atlantic waters between 20° north latitude and 20° south latitude. The equatorial undercurrent may disperse veligers from west to east (i.e. South America to West Africa) along the equator (Scheltema, 1968) rather than from east to west as occurs in the surface currents. Symbols indicate: 1 Cymatiidae; 2 Bursidae; 3 locations where both Cymatiidae and Bursidae were found; 4 locations where Cymatiacea were absent.

There have been only few studies yet made of larval dispersal between estuaries or along coastal regions (vide, Scheltema, 1975, for brief review of this subject and also Sandifer, 1975, and Lough, 1976, for more recent work on coastal dispersal). However, from the examination of plankton samples made on the continental shelf of the eastern United States (unpublished data) it is already apparent that a great assortment of invertebrate larvae is transported along the coast.

Summarizing, it is predicted that species that have a planktonic development of restricted duration (i.e., three to six weeks) will have moderate to large geographical distributions ranging from a few hundred to many thousands of kilometers. There is the possibility for a widespread geographical range by "step-wise" larval dispersal over many generations. However, only restricted genetic exchange will occur between many widely separated populations of such species, the amount of gene flow depending upon the value of d between any two populations (Fig.1). Most genetic exchange will be restricted to neighbouring populations; the value of I (equation 1 above) will be less than in species with teleplanic larvae (e.g., in Fig.1 there may be continuous exchange between B and D but little direct larval exchange between B and F or G). Restricted gene flow to outlying distant populations can result in geographic variation (sometimes clines) or under certain circumstances lead to allopatric speciation.

Finally, at the other end of the continuum is the third case, species with a very short or completely lacking pelagic stage (i.e., lecithotrophic development). In this instance the species will have a small geographic range and dispersal will be largely restricted to the migration of adults. Exceptions will be minute species or egg capsules capable of being rafted on algae or floating debris. Nonpelagic development is almost the exclusive mode of reproduction among arctic gastropods (Thorson, 1950) and is also prevalent in most neogastropod families (Radwin and Chamberlin, 1973). Natural selection will act to narrow the genetic diversity of such species, particularly if they live in a homogeneous environment. Species lacking a dispersal stage will be particularly susceptible to climatic and sea-level changes. Under changing environmental conditions species with nonpelagic development must either evolve, adapting to alterations in their surroundings, or become extinct.

Stated briefly, it is predicted that marine gastropod species lacking pelagic larvae, in the absence of other modes of dispersal, will be both geographically and temporally restricted, that is, they will be limited both in range and species longevity.

Evidence to support this prediction comes from the genus Granulifusus with a nonpelagic development. This genus, found only in the western Pacific near Japan, includes five extinct species, one from the Late Miocene and Early Pliocene, two from the Pliocene, and two from the Pleistocene. There are also two contemporary species. Shuto (1974) has erected a lineage that shows a direct gradual transition of three recognized fossil species from the Late Miocene and Early Pliocene through the Pleistocene to the Recent. There are in addition, outside this direct lineage, one species restricted to the Pliocene and two other species found only in the Pleistocene. Thus all species belonging to this genus with nonpelagic development were restricted not only geographically but also temporally.

We return now to the original question, namely, "What consequences do differences in dispersal hold for species over long geological time"? According to the hypothesis presented here, it is predicted that species with a teleplanic larva will have a very wide geographic range, a large genotypic variability and will evolve only slowly over geological time. It may therefore be expected that they will have a long existence in the fossil record.

It is predicted that species with a pelagic larval stage of restricted duration (three to six weeks) will conform to the "punctuated equilibrium" model of Eldridge and Gould (1972, 1974) and will belong to lineages that have disjunct geographic and vertical distributions (so-called "breaks" in the fossil record). On the one hand patterns of allopatric speciation and extinction can be expected to result from differential natural selection; on the other extended geographic ranges can be explained by stepwise dispersal over many generations by means of pelagic larvae.

Finally, species lacking a pelagic veliger larva (or having a lecithotrophic larva of only a few days) will be predicted to fit the classical model of "gradualism" (vide, Hecht, 1974). They can be expected to belong to lineages that show gradual changes with time (i.e. in equation 1 the number of unique interactions I will approach O; in Fig.1, d will be very small). Thus prosobranch gastropod species with a nonpelagic development, it is expected, will have a short fossil history because they must either change with the environment (so-called "phyletic evolution") or become extinct.

Although only three points along a continuum have been selected, there are doubtlessly many species that will fall between these hypothetical forms (i.e. those that have intermediate values of I or d).

It is now possible to summarize the kind of evidence that will be required to test the hypothesis illustrated by the pictorial model (Fig.1). There are two aspects to this hypothesis and these can be put into the form of separate questions. The first question may be stated thus: What evidence from the fossil record supports the notion that dispersal-capability is related to species temporal longevity? The methods for obtaining information on the life history of fossil species are already suggested and are further elaborated elsewhere (Thorson, 1950; Shuto, 1974; Scheltema, in press). However the relationship between mode of reproduction and temporal longevity is known for only a relatively small number of species; among thirteen contemporary tropical prosobranchs demonstrated to have teleplanic larvae, 53% existed in their present form back into the Middle Miocene (approximately 13 million years ago), not evolving enough since that time to be distinguished beyond the subspecies level. Twenty-two species of prosobranch gastropods with pelagic larval development found in the Danish Lower Miocene, Klintinghoved Formation have a mean longevity of 13.5 million years. Species that represented families that almost exclusively have teleplanic larvae (i.e. Architectonicidae, Tonnidae, Cassidae, and Cymatiidae) had a greater mean longevity (18.8 million years) than that for the entire assemblage.

On the other hand the average longevity of seven species of Granulifusus with nonpelagic development was no more than three million years (vide, Shuto, 1974). Recently Hanson (1977) has reported within a single gastropod family from the Lower Tertiary of the Gulf of Mexico a remarkable correspondence between mode of development and species longevity. (These results have been submitted for publication.)

Although data bearing on the first question seem to support the proposed hypothesis, further evidence is needed before more definite conclusions can be formed. This would include additional information on the mode of reproduction and species longevity from a large variety of temperate and tropical fossil assemblages, particularly comparisons between species with a telepanic or "long-distance" pelagic larva and those with a nonpelagic larval development.

In testing the predictions of the hypothesis, other forms of dispersal should be taken into account, particularly among those forms known to lack a pelagic larval stage (Scheltema, 1977). For example certain taxa of Bryozoa known to have a very short pelagic development of only a few hours nevertheless were widely dispersed during the Tertiary by means of rafting (Cheetham, 1960). Dispersal by means other than by larval transport could obviously also affect the rate of evolution and species longevity, if the hypothesis proposed is correct.

The second question can be stated as follows: What evidence supports the proposed idea that allopatric speciation is connected with mode of reproduction and dispersal capability? One kind of evidence which should bear on this question is a comparison of mode of reproduction and dispersal between different individual lineages in the fossil record. It should be possible, for example, to make comparisons between a lineage with a predominately restricted development (three to six weeks) and ones that have teleplanic larvae or a nonpelagic development and to determine whether a relationship exists between species multiplication and mode of reproduction. For example, the genus Nassarius that has species with a three to six week pelagic development (and for which there is an extensive fossil record) might be compared with the genus Bursa with teleplanic larvae or the genus Granulifusus with nonpelagic development. The difficulty with such an approach is the relative lack of well described fossil marine gastropod lineages and as Sohl (1977) has observed, among gastropods "most studies have concentrated on the total assemblage of a small interval and have not emphasized the study of lineages over a large time span".

Another kind of evidence germane to this question can be derived from contemporary species by ascertaining the amount of genetically determined geographic variation (or incipient speciation) in relation to larval dispersal. For example, a study of the dispersal of teleplanic veliger larvae in the tropical and North Atlantic Ocean has given a strong indication that the frequency of larval transport (as measured by the number of occurrences in the plankton) is inversely related to the number of subspecies that have been assigned to a species by gastropod systematists (Scheltema, 1971b, 1972). Such a relationship between dispersal capability and geographic variation suggests that veliger larvae successfully bring about genetic exchange between populations and that a high capacity for dispersal inhibits or prevents geographical variation and allopatric speciation. Here difficulties of interpretation may arise, however, since the intensity of natural selection is not readily measured. For example, it is known that under conditions of very intense natural selection, morphological differentiation may arise over very short distances, even though a species may have a pelagic larval development. Such an instance has been shown in the intertidal prosobranch Littorina picta (Struhsaker, 1968).

Recent further insights into the relationship of larval dispersal of gastropod veligers to geographic variation are discussed and reviewed elsewhere (Gooch, 1975; Scheltema, 1975) and need not be reiterated here. This evidence from electrophoresis seems to support the view that pelagic dispersal between populations can in certain instances reduce genetic heterogeneity between populations (Gooch et al., 1972; Berger, 1972). In the absence of larval transport greater heterogeneity between populations seems to result.

What I have attempted to do in the present article is not to present data that will answer all the questions posed but rather to propose an hypothesis that will provoke further thought and research. Whatever future study eventually reveals, knowledge of the mode of reproduction of contemporary and fossil gastropods will provide new insights into the evolution of these benthic organisms.

Summary

The concept of evolution as it is now generally understood includes two different phenomena; these are (1) the gradual transformation of species in time, and (2) the multiplication of species by reproductive isolation (allopatric speciation). An hypothesis can be proposed that relates both these phenomena to the effective-larval-dispersal capability of a species. It is predicted that prosobranch gastropods with telepanic or "long distance" larvae will be broadly adapted species with large genotypic variability, broad geographic range and great temporal longevity. Species with a pelagic development of more restricted duration (e.g. three to six weeks) and hence with less dispersal-capability will show geographic variation and varying degrees of speciation (or species radiation). Disjunct vertical and geographical distribution will occur among such forms. Species completely lacking a pelagic dispersal stage will more nearly fit the classical concept of "gradualism"; they must either change with their environment (i.e. show "phyletic evolution") or become extinct. Such species can be expected to persist for only short intervals of geologic time.

It is proposed that natural selection is the principal factor responsible for temporal and spatial genotypic variation that eventually may lead to allopatric speciation.

Reproduction and larval dispersal on the other hand, provide genetic continuity, that is, a flux of genes through time and space.

Acknowledgements

I should like to thank my wife Amelie for her encouragement and critical reading of the manuscript. My research assistant Isabelle Williams has also made suggestions after carefully reading the entire text. The substance of this paper was first presented as a public lecture at Yale University in 1975. Much of this article was written while I was a guest at the Duke University Marine Laboratory at Beaufort, North Carolina, and I am grateful to Dr. John D. Costlow the director for his kind hospitality. I also wish to express my gratitude to the National Science Foundation for its continued support currently under grant OCE-73-00439. This is contribution No. 4029 from The Woods Hole Oceanographic Institution.

REFERENCES

Berger, E. M. 1973. Gene enzyme variation in three sympatric species of Littorina. Biol. Bull. mar. biol. Lab., Woods Hole, 145, 83-90.

Berggren, W. A. and Hollister, C. D. 1974. Paleogeography, paleobiogeography and the history of circulation in the Atlantic Ocean. pp. 126-186. In, Studies in Paleo-oceanography. Hay, W. W. (ed). Spec. Pub. No.20, Soc. Economic Paleontologists and Mineralogists.

Cheetham, A. H. 1960. Time, migration and continental drift. Bull. Amer. Petroleum Geologists, 44, 244-251.

Eldridge, N. and Gould, S. J. 1972. Punctuated equilibria: an alternative to phyletic gradualism. pp. 82-115. In, Models in Palaeobiology. Schopf, T.J.M. (ed). Freeman, Cooper, San Francisco.

Eldridge, N. and Gould, S. J. 1974. Morphological transformation, the fossil record, and the mechanism of evolution: a debate. Part II. Evol. Biol. 7, 303-308.

Gooch, J. L., Smith, B. S. and Knapp, D. 1972. Regional survey of gene frequencies in the mud snail Nassarius obsoletus. Biol. Bull. mar. biol. Lab., Woods Hole 142, 36-48.

Grassle, J. F. 1972. Species diversity, genetic diversity and environmental uncertainty. pp. 19-26. In, *Fifth European Marine Biological Symposium*. Battaglia, B. (ed). Piccin, Padua, Italy.

Hadfield, M. G. 1977. Chemical interactions in larval settling of a marine gastropod. pp. 403-413. In, *NATO Conference on Marine Natural Products*. Faulkner, D. J. and Fenical, W. A. (eds). Plenum Press, New York.

Hadfield, M. G. and Karlson, R. H. 1969. Externally induced metamorphosis in a marine gastropod. Amer. Zool. 9, 122. (Abstr. 317.)

Hanson, T. 1977. Larval ecology and species longevity in Lower Tertiary gastropods. Meeting of benthic ecologists, April 29 - May 1, 1977, held at Yale University, New Haven, Connecticut, unpublished.

Hecht, M. K. 1974. Morphological transformation, the fossil record, and the mechanism of evolution: a debate. Part I. Evol. Biol. 7, 295-303.

Hillman, R. E. 1964. Chromatographic evidence of intraspecific genetic differences in the eastern oyster, *Crassostrea virginica*. Systematic Zool. 13, 12-18.

Levins, R. 1964. The theory of fitness in a heterogeneous environment. IV. The adaptive significance of gene flow. Evolution 17, 635-638.

Lough, R. G. 1976. Larval dynamics of the dungeness crab, *Cancer magister*, off the central Oregon coast, 1970-71. U. S. Fish. Bull. 74, 353-375.

Lewontin, R. C. 1974. *The Genetic Basis of Evolutionary Change*. Columbia Univ. Press. New York.

MacArthur, R. H. and Wilson, E. O. 1967. *The Theory of Island Biogeography*. Princeton Univ. Press, Princeton, N. J.

Mettler, L. E. and Gregg, T. G. 1969. *Population Genetics and Evolution*. Prentice Hall, Englewood, N.J.

Radwin, G. E. and Chamberlin, J. L. 1973. Patterns of larval development in stenoglossan gastropods. Trans. San Diego Soc. Nat. History 17, 107-117.

Sandifer, P. A. 1975. The role of pelagic larvae in recruitment of populations of adult decapod crustaceans in the York River estuary and adjacent lower Chesapeake Bay, Virginia. Estuarine and Coastal Mar. Sci. 3, 269-279.

Scheltema, R. S. 1956. The effect of substrate on the length of existence of *Nassarius obsoletus*. Biol. Bull. Mar. Biol. Lab., Woods Hole 111, 312 (Abstr.)

Scheltema, R. S. 1961. Metamorphosis of veliger larvae of *Nassarius obsoletus* (Gastropoda) in response to bottom sediment. Biol. Bull. Mar. Biol. Lab., Woods Hole 120, 92-109.

Scheltema, R. S. 1962. Pelagic larvae of New England intertidal gastropods. I *Nassarius obsoletus* Say and *Nassarius vibex* Say. Trans. Amer. Microsc. Soc. 81, 1-11.

Scheltema, R. S. and Scheltema, A. H. 1965. Pelagic larvae of New England intertidal Gastropods III. *Nassarius trivittatus*. Hydrobiologia 25, 321-329.

Scheltema, R. S. 1967. The relationship of temperature to larval development of *Nassarius obsoletus* (Gastropoda). Biol. Bull. mar. biol. Lab., Woods Hole 132, 253-265.

Scheltema, R. S. 1968. Dispersal of larvae by equatorial ocean currents and its importance to the zoogeography of shoalwater tropical species. Nature 217, 1159-1162.

Scheltema, R. S. 1971a. The dispersal of the larvae of shoalwater benthic invertebrate species over long distances by ocean currents. pp. 7-28. In, *Fourth European Marine Biological Symposium*. Crisp, D. (ed). Cambridge Univ. Press, Cambridge, U.K.

Scheltema, R. S. 1971b. Larval dispersal as a means of genetic exchange between geographically separated populations of shoalwater benthic marine gastropods. Biol. Bull. mar. biol. Lab., Woods Hole, 140, 284-322.

Scheltema, R. S. 1972a. Dispersal of larvae as a means of genetic exchange between widely separated populations of shoalwater benthic invertebrate species. pp. 101-114. In, *Fifth European Marine Biological Symposium*. Battaglia, B. (ed). Piccin Editore, Padua, Italy.

Scheltema, R. S. 1972b. Eastward and westward dispersal across the tropical Atlantic Ocean of larvae belonging to the genus Bursa (Prosobranchia, Mesogastropoda, Bursidae). Intern. Rev. Gesamten. Hydrobiologie 57, 863-873.

Scheltema, R. S. 1975. The relationship of larval dispersal, gene flow and natural selection to geographic variation of benthic invertebrates in estuaries and along coastal regions. pp. 372-391. In, Estuarine Research, vol. I. Chemistry, Biology and the Estuarine System. Cronin, L. E. (ed). Academic Press, New York.

Scheltema, R. S. 1977. Dispersal of marine invertebrate organisms: paleobiogeographic and biostratigraphic implications. pp. 73-108. In, Concepts and Methods of Biostratigraphy. Kauffman, E. G. and Hazel, J. E. (eds). Dowden, Hutchinson, and Ross, Stroudsburg, Pennsylvania.

Scheltema, R. S. (In press.) Dispersal of pelagic larvae and the zoogeography of Tertiary benthic gastropods. In, Historical Biogeography, Plate Tectonics and the Changing Environment. Boucot, A. J. and Gray, J. (eds). Oregon State Univ. Press, Corvallis, Oregon. (In press.)

Shuto, T. 1974. Larval ecology of prosobranch gastropods and its bearing on biogeography and paleontology. Lethaia 7, 239-256.

Slobodkin, L. B. and Sanders, H. L. 1969. On the contribution of environmental predictability to species diversity. Brookhaven Symposia in Biology 22, 82-93.

Sohl, N. F. 1977. Utility of gastropods in biostratigraphy. pp. 519-539. In, Concepts and Methods of Biostratigraphy. Kauffman, E. G. and Hazel, E. J. (eds). Dowden, Hutchinson and Ross, Stroudsburg, Pennsylvania.

Sorgenfrei, T. 1940. Marint Nedre-Miocaen i Klintinghoved paa als- Et bidrag til løsing af Aquitanien-Spørgsmaalet. Danmarks Geologiske Undersøgelse. II Radkke Nr. 65,143 pp.

Sorgenfrei, T. 1958. Molluscan assemblages from the marine Middle Miocene of South Jutland and their environments. Danmarks Geologiske Undersøgelse. II Radkke Nr. 79, 2 Vols.

Stanley, S. M. 1973. Effects of competition on the rates of evolution, with special reference to bivalves and mammals. Syst. Zool. 22, 486-506.

Struhsaker, J. W. 1968. Selection mechanisms associated with interspecific shell variation in Littorina picta (Prosobranchia; Neogastropoda). Evolution 22, 459-480.

Thorson, G. 1950. Reproductive and larval ecology of marine bottom invertebrates. Biol. Rev. 25, 1-45.

Thorson, G. 1961. Length of pelagic life in marine invertebrates as related to larval transport by ocean currents. pp. 455-474. In, Oceanography. Sears, M. (ed). Pub. 67. Amer. Assoc. Advan. Sci., Washington, D.C.

Woodring, W.P. 1959. Geology and paleontology of the Canal Zone and adjoining parts of Panama. U.S. Geol. Surv. Prof. Paper 306B. 147-239.

ADAPTIVE STRATEGIES IN THE SEA

J. W. Valentine and F. J. Ayala

Depts. of Geology and Genetics

Univ. of California, Davis, California, U.S.A.

Introduction

In recent decades, much attention has been given by population and community ecologists to the explanation of patterns of species diversity or richness. It is now clear that there are many causes of diversity differences, and that explanations that account for local diversity patterns do not necessarily apply to large-scale global trends such as the latitudinal diversity gradient. Indeed the factors responsible for global patterns of diversity are still in dispute, despite the growing body of observational, experimental and theoretical results.

Environmental stability is the most generally cited factor in diversity regulation. However, the very definition of stability, the isolation of particular environmental parameters for which stability is significant with respect to diverstiy, and the elucidation of the mechanism that effects diversity control, are all subject to considerable disagreement. The aim of this paper is to show that in the sea, at least, there is a convergence of evidence and theory which eliminated many hypotheses of diversity regulation and which tends to support, if not corroborate, a particular model – the trophic resource model – composed of elements from several hypotheses.

Marine Species Diversity Patterns

In the sea there are three major adaptive zones or superzones which have such different conditions of life that they share a relatively small proportion of their adult taxa. These are the shelf benthic zone, the deep-sea benthic zone, and the pelagic zone. Diversity and stability patterns in these different environments have instructive similarities and differences.

The benthic communities of the continental shelves and of comparable depths around islands must contain nearly 90 percent of the world's marine invertebrate taxa (Valentine, 1973). The major diversity trend of the shelf faunas is a latitudinal gradient. Many tropical benthic communities contain five to ten times as many species as their high-latitude analogs, to judge from faunal accounts of the better-known large groups such as molluscs and echinoderms. As this diversity trend correlates only with factors that also vary systematically with latitude, a very large class of environmental variables which do not share this pattern are eliminated as having a major involvement with global diversity trends. Climatic stability, particularly temperature stability, is the most important variable thus eliminated (Valentine, 1972). Temperature is least stable (most seasonal) over mid-latitude shelves and very stable in high latitudes, yet diversity is intermediate and low, respectively, in these two geographic situations (Sverdrup, Johnson, and Fleming, 1942, p.130).

The seasonality of productivity is the most obvious correlate of the latitudinal diversity gradient, varying as it does with the seasonality of solar radiation. Longitudinal trends in marine shelf diversity are associated with trends in stability of the water column, where these can be documented, and thus with the seasonality of nutrient supplies (Valentine, 1972, 1973). Thus where seasonality in productivity appears to be greatest, owing to variation either in solar radiation or in nutrients, diversity within benthic communities is least.

It has been argued (for example by Fischer, 1960 and Sanders, 1969) that the sheer physical difficulties of life in progressively higher latitudes would progressively raise the extinction rates and thus create the diversity gradient. In fact there is little evidence that extinction rates are greater in higher latitudes. In a study of foraminiferan species, Stehli, Douglas, and Kafescioglu (1972) showed that the average age of species in high and low latitudes is very nearly the same, and indeed in an earlier study (Stehli, Douglas, and Newell, 1969) it was shown that the average ages of genera and families are significantly greater in high than in low latitudes. It thus seems highly unlikely that any diversity-independent factors, such as climatic ones, can regulate diversity trends on a long-term basis, although they might be responsible for local diversity patterns. On theoretical grounds it is more likely that diversity-dependent factors, such as space and trophic supplies, would be implicated in global diversity regulation (Valentine, 1972).

Pelagic ecosystems contain perhaps five percent of living marine invertebrate species as permanent inhabitants. These ecosystems appear to have global diversity patterns similar to those of the shelves (Ryther, 1969) in that the more diverse communities are tropical and of these the most diverse are Indo-Pacific. Shelf waters - the neritic environment - are more frequently perturbed, for example by episodes of upwelling, and support fewer pelagic species than the more stable offshore water masses. However, the more stable waters are, in general, nutrient poor and have lower though presumably more stable levels of productivity. All in all, pelagic diversity correlates with the general patterns of stability of the water column and of productivity insofar as they are known.

Deep-sea ecosystems, which contain five percent or more of the living marine invertebrate species, are poorly known. Judging from the pattern of diversity and endemism reported by Hessler and Sanders (1967), Sanders (1968), and Hessler and Jumars (1974), deep-sea forms tend to range widely geographically but narrowly in depth, being distributed in depth-restricted bands around the ocean basins. Endemism most often occurs, then, when sills between basins form barriers to species with depth ranges beneath the sill depth. Diversity patterns may be associated with patterns of productivity in the euphotic zone overlying the deep-sea bottoms. Beneath regions of stable productivities in the euphotic zones, such as beneath the centres of the major ocean gyres, diversity is high, although benthic biomass is evidently low there. The general stability attributed to the deep-sea environment suggests that trophic supplies should be relatively stable when compared with those of shallow waters. Certainly deep-sea benthic diversity is generally high when compared with diversity in shallow-water environments, such as muddy-bottom shelf regions, that are similar in habitat type and spatial heterogeneity to the deep-sea floor (Sanders, 1968).

Life History Strategies and Diversity

Adaptations to patterns of environmental variability have been termed strategies (Levins, 1968). Since species in stable and unstable environmental regimes often have different strategies, it is reasonable to examine strategy differences as a possible source of the diversity differences between communities from such contrasting regimes. It has been known for years that much of the difference between species from regions of high and low diversity is associated with reproduction and development, usually considered as adaptive to distinctive environmental regimes. One of the most-discussed strategy contrasts is between species with higher and those with lower reproductive potentials, termed r and K strategies respectively (see MacArthur and Wilson, 1967). Other characteristics attributed to r and K strategists vary among authorities; Pianka (1970) has assembled a rather long list which he believes to be usually associated with these strategies. For example, r strategists usually display rapid development, early reproduction, are semelparous and have relatively short lives, while K strategists display the opposite conditions. The r strategists are most commonly found in more uncertain or unstable environments, the K strategists in the more stable (Pianka, 1970, 1972). Thus these strategies are differentiated with respect to environments in which diversity differences occur.

Studies of natural populations indicate that r and K strategies can indeed represent a trade-off in response to environmental factors that affect the reproductive regimes and also adult mortalities (a recent review of many relevant findings is by Giesel, 1976). A good example is provided by Atlantic salmon populations. Those that live in longer streams which are not exploited by regular fishing are composed of individuals that tend to delay maturity and become relatively large and robust by the time of their first reproductive effort. They are thus able to cope with a difficult migration to their spawning grounds. These fish have relatively few offspring but are iteroparous. Thus the population has a K strategy. Conspecific populations in short streams that are heavily exploited by fishing, however, reproduce earlier and have more offspring, and a shorter life due to the fishery, and thus have r strategies. The K-selected individuals must be robust and long-lived as adults, while the r-selected individuals have little to gain from being robust adults since they stand a good chance of being fished, but instead mature rapidly to reproduce copiously before mortality. These two straegies have been nicely supported by computer studies of model populations with appropriate contrasting parameters (Schaffer and Elson, 1975 and references therein). Thus the r and K strategy trade-off occurs in nature and has a plausible adaptive explanation.

If r and K strategies are significantly associated with global patterns of diversity, then r strategists should be found in the less stable environments of low diversity, while K strategists should be most common in the more stable and diverse regions. In fact this pattern is not found. Tropical marine invertebrates from highly diverse shelf communities tend to have much higher reproductive potentials than those in high latitudes. Thorson (1950) and Mileikovsky (1971) have estimated that about 70 percent of tropical invertebrate species have planktonic larvae; some of these may live many months and can achieve very wide dispersal (Scheltema, 1971). At higher latitudes the percentage of planktonic larvae decreases and at very high latitudes there are very few. Instead, larvae are demersal or develop directly from large yolky eggs, and the average reproductive potential is greatly reduced when compared with low latitudes. Hence as far as reproductive potential itself is concerned, it is the tropical species in the more stable regimes that are r-selected, while arctic species in more unstable regimes are not.

As far as patterns of adult mortality are concerned, there are too few data as yet to compare adequately the longevity distributions within high and low latitude faunas. It is often assumed that high-latitude species live nearer the margin of their tolerance than do low-latidude forms, and are therefore particularly subject to catastrophic mortalities during severe environmental fluctuations (see for example Fischer, 1960). However, some benthic species in both high and low latitudes are known to live long lives (Frank, 1969); so for the present we can only say that there is no strong evidence that high-latitude benthic invertebrates do have higher adult mortalities than low-latitude ones. Support for an r and K strategy trade-off between unstable high latitudes and more stable low latitudes is thus lacking, from the standpoints of both reproduction and survivorship.

Life histories of species in the pelagic and deep-sea realms are not known for a wide spectrum of invertebrate taxa. In the pelagic realm, planktonic euphausiids (krill) are relatively well-studied (for example Brinton, 1962 and Mauchline and Fisher, 1969). Among krill, high-latitude species have higher reproductive potentials and live longer than their tropical counterparts (Mauchline and Fisher, 1969,

Chap. 10). For deep-sea invertebrates, there is evidence that reproductive potentials average lower but longevities higher than their shallow-water allies (Allen and Sanders, 1973; Schoener,, 1972; Scheltema, 1972; Turekian et al., 1975).

Such data as are available, then, suggest that r and K strategies can vary independently according to the exigencies of the environment, and neither appears to be consistently associated with highly stable or highly unstable environments, whether defined by climatic factors or by productivity. Figure 1 depicts the association of modal strategy components with major environmental regimes as indicated by available data, and contrasts these observations with the expectations based on an r and K strategy trade-off. It is not yet clear whether a larger fraction of the energy available to populations is channeled into reproduction in the high-latitude or low-latitude situations; perhaps the reproductive efforts are energetically similar in these cases (see Vance, 1973). Certainly the energy expended on reproduction is packaged differently - into numerous offspring in the tropics and fewer in high latitudes for shelf invertebrates, and evidently in the opposite sense for pelagic invertebrates - with corresponding differentials in reproductive potential. Reproductive success, however, is likely to be similar when normalized for environmental and biotic differences, which is precisely what adaptive strategies are expected to achieve. It is concluded that r and K selection are not functionally associated with regulation of diversity in marine ecosystems (in contrast to an earlier opinion, Valentine, 1971). Indeed if we are judging only from theoretical aspects of these strategies as they are observed in the shallow-water benthos, then we should conclude that the arctic shelves are the more and the tropical shelves the less stable environments, insofar as selection is concerned. The diversity gradients, however, suggest that the traditional view is more likely to be correct.

Niche Size Strategies, Competition, and Diversity

If diversity regulation is not mediated through lifehsitory strategies, then another much-discussed possibility is that it is involved with adaptive flexibility or adaptive specialization, sometimes conceptualized as niche breadth. The rationale for this notion is that populations in environments that are more unstable with respect to resources or tolerances require greater adaptive flexibility in order to cope with the environmental fluctuations. Their resource

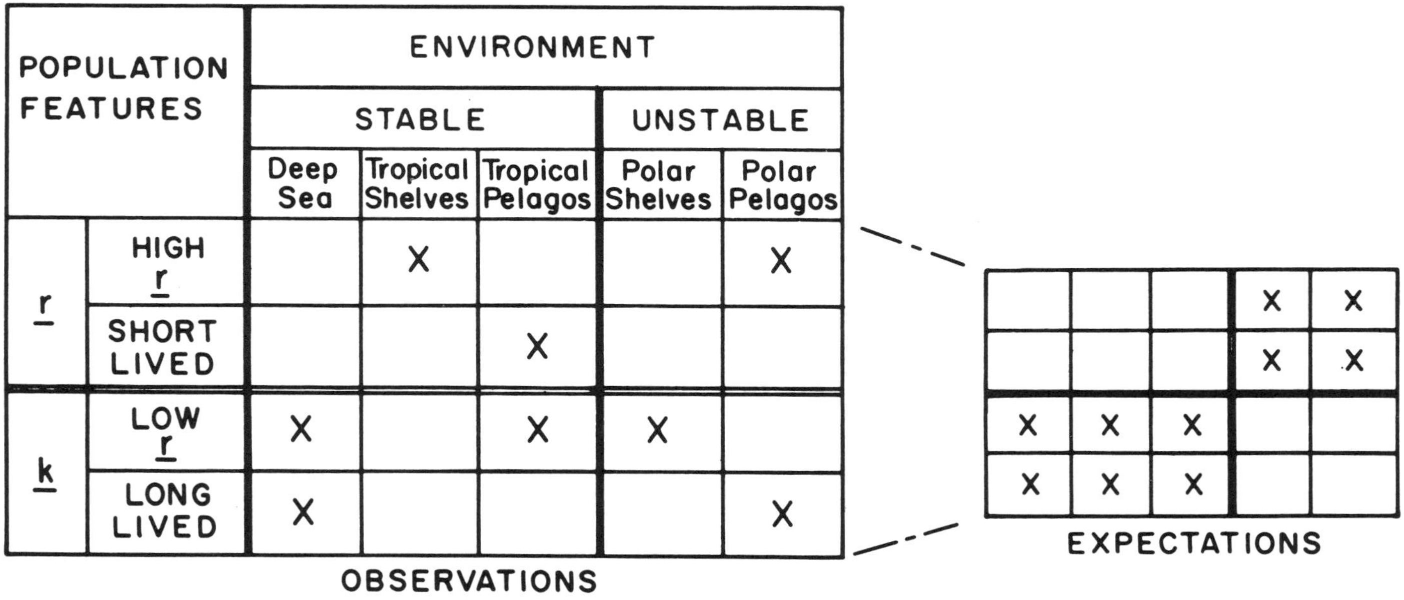

Figure 1. The broad trends observed in reproduction and longevity between populations inhabiting trophically stable and trophically unstable environments, contrasted with the trends expected if r- and K-strategy trade-offs are important between such regimes. The observed trends seem random with respect to expectations.

bases are therefore more generalized and the partitioning of niches less narrow than in more stable environments. More "niche space" - functional range - is required by each population, and therefore competition functions to limit the number of species within communities to a low level. In more stable environments each population may be more specialized or at least restricted through competition to a relatively narrow range of resources. More populations can therefore be packed into a given range of habitats to form more species - rich communities.

This expectation seems to accord with observations on actual niche widths in nature (for examples see Kohn, 1966; Arnaud, 1970; Abele, 1974). Although ecologically generalized species are found in the tropics as in high latitudes, those that are highly specialized are much more frequent in low latitudes and become more rare in more unstable environmental regimes. The narrowing of some niche sizes in stable regions has long been explained as an outcome of competition (Dobzhansky, 1950). However, it has not been verified that the reduction of diversity in unstable regions is a function of competition; this is difficult to test.

The concept of ecological grain (Levins and MacArthur, 1966; Levins, 1968) is useful to summarize this model. The environment may be regarded as a patchwork or mosaic of conditions; between patches, change in one or more environmental parameters and therefore in habitat conditions is more significant than within patches. Within a given major environmental realm, then, species may inhabit one or many patches depending upon their adaptive flexibility. Species that are virtually restricted to a single patch perceive the environment as coarse-grained, since the between-patch changes are limiting for them. Other species may range widely through the same patches, and perceive the environment as fine-grained, since the between-patch changes fall within their range of tolerances. To re-state the niche-breadth model in terms of the grain concept, then, communities from more stable environments will contain some species that have coarser-grained adaptations, and contain a higher proportion of coarse-grained species, than communities from less stable environments.

Community Stability, Diversity and Adaptive Strategies

The proposition that higher species diversities promote stability within communities, expressed in an elegant form by MacArthur (1955), has been contradicted by more recent work on model systems (May, 1972, 1974; Hubble, 1973).

May's findings are particularly relevant. In his models, oscillating subsystems somewhat analogous to natural populations are combined to form larger systems somewhat analogous to communities. The more complex the larger systems in terms of the number of subsystems and the richness of their interactions, the less stable the entire system becomes. This result seems quite general. The principle involved is that a system of oscillating subsystems cannot be stabilized by the addition of still more oscillating (unstable) subsystems or by an increase in oscillatory interactions; they can remain at about the same stability in some special cases, but more generally they are destablized. May has pointed out that natural selection may seek out the more stable configurations in a given environment to erect and maintain surprisingly complex ecosystems under appropriate circumstances.

Based on such models, an explanation may be proposed to explain relations between patterns of environmental stability (especially as defined by trophic resources), of adaptive strategies and of the diversity of invertebrate populations in marine communities. We assume that the various processes of speciation are quite powerful, capable of originating large numbers of species within relatively short periods of geological time, given appropriate environmental opportunities. The available environments, then, tend to be filled approximately to their capacities to support species. In environments wherein sunlight and nutrient supples are most constant, productivity can be most stable and animal population systems themselves can be the least oscillatory, owing to the stability of their trophic resources. Even where primary productivity is essentially absent, such as in the deep-sea, detrital trophic supplies may be reasonably constant, and thus populations may maintain relatively stable reproduction and growth regimes. Under such conditions, natural selection can regulate the dynamics of populations so that a relatively large number of different species can be accommodated together. As the number of species being supported on a given resource base grows, the niche size or functional range of the average species is narrowed, presumably through competition. The narrowing is accompanied by increased efficiency as specializations are developed.

In environments wherein the trophic resource base fluctuates. however, consuming populations also tend to fluctuate and thus the community structure is stable only at lower diversities. Niche sizes or functional ranges must be relatively broad in order to cope with the variable en-

vironmental conditions. Fitter individuals tend to be those
that are more broadly adapted, although they are not as
efficient in any specific environment as an appropriate
specialist. These generalists would tend to eat a wider
range of food items and to live in a wider range of habitat
conditions than the average species in stable regimes. Thus
communities of low diversity containing generalized, flexibly
adapted and thus fine-grained species would be present in
trophically unstable environments. In trophically stable
environments, communities could contain numerous specialized,
coarse-grained species, though fine-grained forms such as
high-level predators and perhaps scavengers, that need to
range widely, would also be present.

Adaptive Strategies and Genetic Variability

Most hypotheses that relate genetic and environmental
variability have predicted that genetic variabilities would
be higher in spatially more heterogeneous and temporally more
variable environments (see Levins, 1968; Grassle, 1973).
The theoretical prediction of greater genetic variability
in environments of greater spatial heterogeneity is supported
both by experiment (Powell, 1971; McDonald and Ayala, 1974)
and observation (for example, the classic studies of Cepaea
and Biston; for references see Hedrick et al., 1976).
Furthermore there is evidence that species that perceive a
given environment as more heterogeneous have greater genetic
variabilities than those that perceive the environment as
more homogeneous (Selander and Kaufman, 1973). The association of spatial and genetic variability seems well-established.

However, the prediction that high genetic variabilities
are associated with temporally variable environments is contradicted by the evidence from marine invertebrates. These
data are summarized in Table 1 and Figure 2, where low genetic
variabilities are shown to be commonly associated with temporally unstable environments, and high genetic variabilities
with temporally stable ones, when stability is measured by
global diversity patterns. Although this trend is clear,
moderately large ranges of genetic variabilities are found
within some diversity classes (as in class 3, Figure 2).
In these cases, species with relatively low genetic variabilities are those which have the finer-grained adaptations, so
far as we can tell: large vagile forms, commonly predators
or omnivores such as starfish or crabs. Similarly, fishes
tend to have lower genetic variabilities than invertebrates
from regions of similar diversities. The species with the

TABLE 1

Marine species for which estimates of genetic variability, based upon 15 or more loci, are used in Figure 2.

Species	Average Percent Heterozygosity	Source
Benthic Shelf Invertebrates		
Asterias vulgaris	1.1	Schopf & Murphy, 1973
Cancer magister	1.4	Hedgecock & Nelson in Valentine, 1976
Asterias forbesi	2.1	Schopf & Murphy, 1973
Liothyrella notorcadensis	3.9	Ayala et al., 1975a
Homarus americanus	3.9	Hedgecock & Nelson in Valentine, 1976
Hemigrapsus oregonesis	4.1	Hedgecock & Nelson, unpublished
Cancer gracilis	4.7	Hedgecock & Nelson, unpublished
Pachycheles rudis	4.7	Hedgecock & Nelson, unpublished
Crangon negricata	4.9	Hedgecock & Nelson in Valentine, 1976
Pagurus granosimanus	5.3	Hedgecock & Nelson, unpublished
Limulus polyphemus	5.7	Selander et al., 1970
Petrolisthes cinctipes	6.0	Hedgecock & Nelson, unpublished
Crangon franciscana	6.3	Hedgecock & Nelson, unpublished
Upogebia pugettensis	6.5	Hedgecock & Nelson in Valentine, 1976
Callianassa californiensis	8.2	Hedgecock & Nelson in Valentine, 1976
Phoronopsis viridis	9.4	Ayala et al., 1974
Emerita analoga	11.4	Hedgecock & Nelson, unpublished
Crassostrea virginica	12.0	W. W. Anderson in Valentine, 1976

Table 1 (continued)

Tridacna maxima	21.6	Ayala et al., 1973
		Campbell et al., 1975
Benthic Deep-Sea Invertebrates		
Asteroidea, four species	16.4	Ayala et al., 1975b
Frieleia halli	16.9	Valentine and Ayala, 1974
Ophiomusium lymani	17.0	Ayala & Valentine, 1974
Pelagic Invertebrates		
Euphausia superba	5.7	Ayala et al., 1975c
Euphausia mucronata	14.1	Valentine & Ayala, 1976
Euphausia distinguenda	21.3	Valentine & Ayala, 1976
Fish		
Trematomus borchgrevinki	0.5	Somero & Soule, 1974
Bathygobius ramosus	0.5	Somero & Soule, 1974
Sebastes caurinus	1.8	Johnson et al., 1973
Trematomus hansoni	2.5	Somero & Soule, 1974
Sebastes elongatus	3.2	Johnson et al., 1973
Trematomus bernacchii	3.3	Somero & Soule, 1974
Leuresthes tenuis	3.6	Somero & Soulé, 1974
Sebastes alutus	3.8	Johnson et al., 1973
Gibbonsia metzi	4.3	Somero & Soule, 1974
Gillichthys mirabilis	4.6	Somero & Soule, 1974
Abudefduf troschelii	5.0	Somero & Soule, 1974
Halichores sp.	5.7	Somero & Soule, 1974
Mugil cephalus	7.1	Somero & Soule, 1974
Amphiprion clarki	9.1	Somero & Soule, 1974
Dascyllus reticulatus	10.7	Somero & Soule, 1974
Mammal		
Mirounga leonina	3.0	McDermid et al., 1972

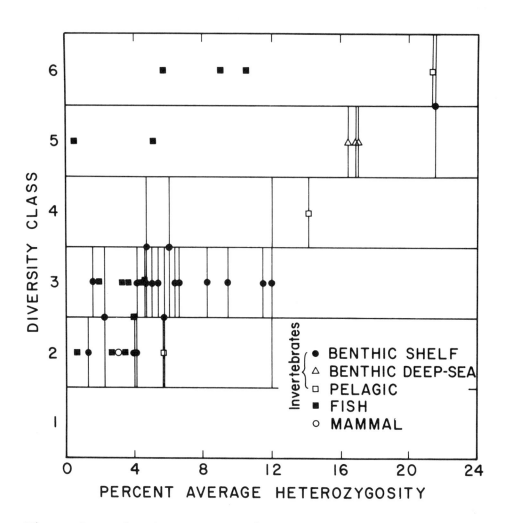

Figure 2. The observed relation between genetic variability of marine species (represented as the average percentage of heterozygosity per individual) and the diversity of the regions inhabited by each species. For invertebrates, the vertical bars indicate the total range of diversity classes with which the species are associated. For vertebrates, the symbol indicates the diversity class within which the samples were collected. The higher the number of the diversity class, the higher the diversity; regions of greater diversity exhibit greater trophic resource stability.

genetically higher variabilities are those with the coarser-grained adaptations: sessile forms such as oysters or burrowing phoronids. Since coarse-grained species are more common in stable environments, and since such species generally have proven to have higher genetic variabilities, then the latitudinal trend in genetic variability may well reflect the trend in the adaptive grain or species. Indeed the simplest way to summarise the observed patterns of genetic variability is that coarser-grained species have higher genetic variabilities than finer-grained species. The usual levels of genetic variabilities expected for marine invertebrates, fishes and mammals are depicted in Figure 3.

To explain this relationship, it has been hypothesized that the selective advantage of generalized versus specialized gene products varies with the environmental grain (Ayala et al., 1975a; Valentine, 1976). Figure 4 depicts a group of alleles at some locus, which have quite a range of activities and functions; some have broad functional ranges (those to the left) and others narrow ones. Alleles with broad functions have products that cleave more bonds and/or are active through greater ranges of pH, Eh, and other factors than alleles with narrow functions. In such cases, narrowly active enzymes have greater activities within their functional ranges than do broadly active enzymes. The more specialized alleles are more efficient than the less specialized when conditions are optimal for their functions, so to speak.

In environments that are fluctuating widely with respect to conditions that involve the activity of a given gene locus (such as environment I in Figure 4), the more flexible allele may be the fitter, since it maintains its function throughout the spectrum of conditions that the individual must face. More specialized alleles may be more efficient at times when the environment happens to permit them to function but they are useless at other times. In relatively stable environments (as environment II in Figure 4), however, appropriately specialized alleles would be fitter. Numerous specialized alleles can be retained in gene pools by various forms of balancing selection in relatively stable environments. They would enhance adaptation by accommodating individuals to local heterogeneities within the habitat range of the population.

In this view, patterns of diversity of populations and patterns of variability of genes are both outgrowths of selection for differences in functional ranges, of organisms on one hand and of gene products on the other. They thus

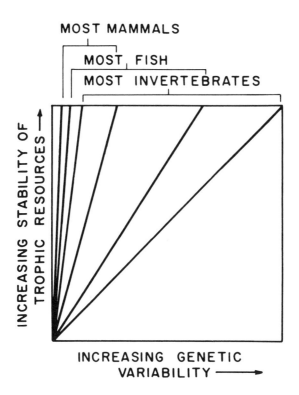

Figure 3. Theoretical relationship between genetic variability and trophic resource stability in species of marine mammals, fishes and invertebrates, based on expectations of the hypothesis that generalized alleles are found in species with fine-grained strategies and specialized alleles in species with coarse-grained strategies. In any given resource regime there is a range of strategies present, but increasingly coarse-grained strategies are found in more trophically stable environments. Mammals are modally finer-grained than invertebrates.

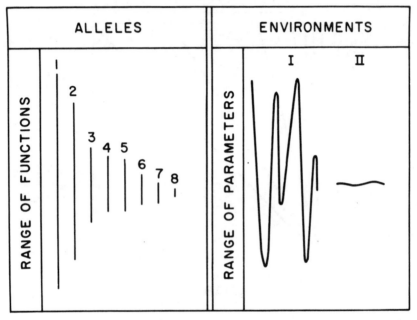

Figure 4. Functional range of alleles contrasted with the requirements of unstable and stable environmental regimes. Only the broadest allele (No.1) continues to produce a functional product over the entire range of environment I, while all the alleles (1-8) can function continuously in the more stable environment II.

represent separate levels of adaptation to a common environmental challenge, and indeed the functional flexibility or specialization of alleles underlies the functional flexibility or specialization of the phenotype they encode.

Summary

Strategies are adaptations to patterns of environmental variability. They are appropriately discussed in terms of the general components of fitness responsible for their evolution: reproduction, growth and maintenance. The trade-off in energy devoted to high early reproduction and growth on one hand, and to maintenance on the other, is well-documented for populations with high and low adult mortalities, respectively, and is plausibly explained by fitness differences associated with these life expectancy patterns. The extremes of such a trade-off can be characterized as r and K strategies.

On marine shelves, the gross environmental patterns do not favour such r and K trade-offs in a global scale. For example, in high latitudes the environment is harsh, but average survivorship is not known. Trophic resources are so highly seasonal and irregular that larvae are chiefly benthic with large yolky eggs and display direct development; relatively few are produced per individual. In low latitudes, by contrast, productivity is less seasonal and more regular, larvae are chiefly planktonic, and high reproductive and growth potentials are found, most likely as a tactic in competition for settling sites in a patchy environment. At the same time the adults face a milder environment and may maintain themselves for extended periods. Thus the r and K trade-off is not a dominant pattern; indeed, both strategies are commonly associated in the life histories of tropical species. Instead, the predominant adaptive strategy trade-off is clearly between different approaches to maintenance, in high latitudes most species are flexibly adapted, while in low latitudes many are narrowly adapted.

Patterns of species diversity within communities of marine organisms show a strong correlation with the modal functional ranges of the component populations; high diversities are associated with narrow functional ranges. Diversity patterns do not correlate with patterns of r and K strategy trade-offs.

Genetic variabilities are correlated with species diversities among marine animals; high variabilities are associated with high diversities. It is suggested that low genetic variability results from the higher fitness of the more functionally flexible gene products in harsh and seasonal environments. High genetic variability in specialized populations may act to maintain a somewhat flexible adaptive posture in a narrowly specialized genome; it is probably employed in fine-tuned adaptation to spatially differentiated microenvironments. This explanation can account for much of the pattern of genetic variability within regions of similar species diversities as well. Generally larger and more vagile forms, which must be more flexibly adapted, have lower genetic variabilities than smaller and more sedentary or sessile forms. It is likely that genetic variability is an adaptive strategy.

Acknowledgement

We are most grateful to Dr. Dennis Hedgecock, Bodega Marine Laboratory, for permission to use his extensive unpublished data on marine decapods.

REFERENCES

Abele, L. G. 1974. Species diversity of decapod crustacea in marine habitats. Ecology 55, 156-161.

Allen, J. A. and Sanders, H. C. 1973. Studies on deep-sea protobranchia (Bivalvia); the families Siluculidae and Lameltilidae. Bull. Mus. Comp. Zool. Harvard 145, 263-310.

Arnaud, P. M. 1970. Frequency and ecological significance of necrophagy among the benthic species of antarctic coastal waters. In, Antarctic Ecology. Holgate, M. W. (ed). 1, 259-266.

Ayala, F. J., Hedgecock, D., Zumwalt, G. S. and Valentine, J. W. 1973. Genetic variation in Tridacna maxima, an ecological analog of some unsuccessful evolutionary lineages. Evolution 27, 177-191.

Ayala, F. J. and Valentine, J. W. 1974. Genetic variability in the cosmopolitan deep-water ophiuran Ophiomusium lymani. Mar. Biol. 27, 51-57.

Ayala, F. J. and Valentine, J. W. 1978. Genetic variation of resource stability in marine invertebrates. This volume.

Ayala, F. J., Valentine, J. W., Barr, L. G. and Zumwalt, G. S. 1974. Genetic variability in a temperate intertidal phoronid, Phoronopsis viridis. Biochem. Genet. 11, 413-427.

Ayala, F. J., Valentine, J. W., DeLaca, T. E. and Zumwalt, G. S. 1975a. Genetic variability of the Antarctic brachiopod Liothyrella notorcadensis and its bearing on mass extinction hypotheses. J. Paleontology 49, 1-9.

Ayala, F. J., Valentine, J. W., Hedgecock, D. and Barr. L. G. 1975b. Deep-sea asteroids: high genetic variability in a stable environment. Evolution 29, 203-212.

Ayala, F. J., Valentine, J. W. and Zumwalt, G. S. 1975c. An electrophoretic study of the antarctic zooplankter Euphausia superba. Limnol. Oceanogr. 20, 635-640.

Brinton, E. 1962. The distribution of Pacific euphausiids. Bull. Scripps Inst. Oceanogr. Univ. California 8, 51-270.

Campbell, C. A., Valentine, J. W. and Ayala, F. J. 1975.
High genetic variability in a population of *Tridacna
maxima* from the Great Barrier Reef. Mar. Biol. 33,
341-345.

Dunbar, M. J. 1968. *Ecological Development in Polar Regions.*
Prentice Hall, Englewood Cliffs, New Jersey.

Fischer, A. G. 1960. Latitudinal variations in organic
diversity. Evolution 14, 64-81.

Frank, P. W. 1969. Growth rate and longevity of some
gastropod mollusks on the coral reef at Heron Island.
Oecologia 2, 232-250.

Giesel, J. T. 1976. Reproductive strategies as adaptations
to life in temporally heterogeneous environments.
Ann. Rev. Ecol. Syst. 7, 57-79.

Grassle, J. F. 1973. Variety in coral reef communities.
In, *Biology and Geology of Coral Reefs*. Jones, O. A.
and Endean, R. (eds). 2: 247-270. Academic Press,
New York and London.

Hedrick, P. W., Ginevan, M. E. and Ewing, E. P. 1976.
Genetic polymorphism in heterogeneous environments.
Ann. Rev. Ecol. Syst. 7, 1-32.

Hessler, R. R. and Jumars, P. A. 1974. Abyssal community
analysis from replicate box cores in the north central
Pacific. Deep-sea Res. 21, 185-209.

Hessler, R. R. and Sanders, H. C. 1967. Faunal diversity
in the deep-sea. Deep-sea Res. 14, 65-79.

Hubble, S. P. 1973. Populations and simple food webs as
energy filters: II Two-species systems. Am. Nat.
107, 122-151.

Johnson, A. G., Utter, F. M. and Hodgkins, H. O. 1973.
Estimate of genetic polymorphism and heterozygosity
in three species of rockfish (genus *Sebastes*).
Comp. Biochem. Physiol. 44B, 397-406.

Kohn, A. J. 1966. Food specialization in *Conus* in Hawaii
and California. Ecology 47, 1041-1043.

Levins, R. 1968. *Evolution in Changing Environments*.
Princeton Univ. Press, Princeton, New Jersey.

Levins, R. and MacArthur, R. 1966. The maintenance of genetic polymorphism in a spatially heterogeneous environment: variations on a theme by Howard Levene. Am. Nat. 100, 585-589.

MacArthur, R. H. 1955. Fluctuations of animal populations and a measure of community stability. Ecology 36, 533-536.

MacArthur, R. H. and Wilson, E. O. 1967. The Theory of Island Biogeography. Princeton Univ. Press, Princeton, New Jersey.

Mauchline, J. and Fisher, L. R. 1969. The biology of euphausiids. Adv. Mar. Biol. 7, 1-454.

May, R. M. 1974. Stability and Complexity in Model Ecosystems. 2nd edition (1st ed., 1972). Princeton Univ. Press, Princeton, New Jersey.

McDermid, E. M., Ananthakrishnan, R. and Agar, N. S. 1972. Electrophoretic investigation of plasma and red cell proteins and enzymes of Macquarie Island elephant seals. Anim. Blood Grps. Biochem. Genet. 3, 85-94.

McDonald, J. F. and Ayala, F. J. 1974. Genetic response to environmental heterogeneity. Nature 250, 572-574.

Mileikovsky, S A. 1971. Types of larval development in marine bottom invertebrates, their distribution and ecological significance: a re-evaluation. Mar. Bio. 10, 193-213.

Pianka, E. R. 1970. On r- and K-selection. Am. Nat. 104, 592-597.

Pianka, E. R. 1972. r and K selection or b and d selection? Am. Nat. 106, 581-588.

Powell, J. R. 1971. Genetic polymorphisms in varied environments. Science 174, 1035-1036.

Ryther, J. H. 1969. Photosynthesis and fish production in the sea. Science 166, 72-76.

Sanders, H. L. 1968. Marine benthic diversity: a comparative study. Am. Nat. 102, 243-282.

Sanders, H. L. 1969. Benthic marine diversity and the stability-time hypothesis. In, <u>Diversity and Stability in Ecological Systems</u>. Brookhaven Symp. Biology 22, 71-81.

Schaffer, W. M. and Elson, P. F. 1975. The adaptive significance of variations in life history among local populations of Atlantic salmon in North America. Ecology 56, 577-590.

Scheltema, R. S. 1971. Larval dispersal as a means of genetic exchange between geographically separated populations of shallow-water benthic marine gastropods. Biol. Bull. 140, 284-322.

Scheltema, R. S. 1972. Reproduction and dispersal of bottom-dwelling deep-sea invertebrates: a speculative summary. Barobiol. Exper. Biol. Deep Sea, 58-66.

Schoener, A. 1972. Fecundity and possible mode of development of some deep-sea ophiuroids. Limnol, Oceanogr. 17, 193-199.

Schopf, T. J. M. and Murphy, L. S. 1973. Protein polymorphism of the hybridizing seastars <u>Asterias forbesi</u> and Asterias vulgaris and implications for their evolution. Biol. Bull. 145, 589-597.

Selander, R. K. and Kaufman, D. W. 1973. Genic variability and strategies of adaptation in animals. Proc. Nat. Acad. Sci. USA 70, 1875-1877.

Selander, R. K., Yang, S. Y., Lewontin, R. C. and Johnson, W. E. 1970. Genetic variation in the horseshoe crab (<u>Limulus polyphemus</u>), a phylogenetic "relic". Evolution 24, 402-419.

Somero, G. N. and Soule, M. 1974. Genetic variation in marine fishes as a test of the niche-variation hypothesis. Nature 249, 670-672.

Stehli, F. G., Douglas, R. G. and Kafescioglu, I. A. 1972. Models for the evolution of planktonic foraminifera. In, <u>Models in Paleobiology</u>. Schopf, T. J. M. (ed). pp. 116-128. Freeman and Cooper, San Francisco.

Stehli, F. G., Douglas, R. G. and Newell, N. D. 1969. Generation and maintenance of gradients in taxonomic diversity. Science 164, 947-949.

Sverdrup, H. U., Johnson, M. W. and Fleming, R. H. 1942. The Oceans. Prentice-Hall, New York.

Thorson, G. 1950. Reproductive and larval ecology of marine bottom invertebrates. Biol. Rev. 25, 1-45.

Turekian, K. K., Cochran, J. K., Kharker, D. P., Cerrato, R. M., Vaisnys, J. R., Sanders, H. C., Grassle, J. F. and Allen, J. A. 1975. Slow growth rate of a deep-sea clam determined by ^{228}Ra chronology. Proc. Acad. Nat. Sci. USA 72, 2829-2832.

Valentine, J. W. 1971. Resource supply and species diversity patterns. Lethaia 4, 51-61.

Valentine, J. W. 1972. Conceptual models of ecosystem evolution. In, Models in Paleobiology. Schopf, T.J.M. (ed). pp 192-215. Freeman and Co., San Francisco.

Valentine, J. W. 1973. Evolutionary Paeleocology of the Marine Biosphere. Prentice-Hall, Englewood Cliffs, New Jersey.

Valentine, J. W. 1976. Genetic strategies of adaptation. In, Molecular Evolution. Ayala, F. J. (ed). pp. 78-94. Sinauer Associates, Sunderland, Massachusetts.

Valentine, J. W. and Ayala, F. J. 1975. Genetic variation in Frieleia halli, a deep-sea brachiopod. Deep-Sea Res. 22, 37-44.

Valentine, J. W. and Ayala, F. J. 1976. Genetic variability in Krill. Proc. Nat. Acad. Sci. USA 73, 658-660.

Vance, R. R. 1973. On reproductive strategies in marine benthic invertebrates. Am. Nat. 107, 339-352.

LIFE HISTORIES AND GENETIC VARIATION IN MARINE INVERTEBRATES

J. F. Grassle & J. P. Grassle

Woods Hole Oceanographic Institution and

Marine Biological Laboratory, Woods Hole, Mass., U.S.A.

Introduction

Life history features that relate to dispersal and population size are the principal concerns in studies of population structure and genetic variability. We have been studying a group of sibling species of polychaete worms of the genus Capitella, which were formerly thought to be the same species (Grassle and Grassle, 1976). Each of six species from shallow-water environments in the New England region has a somewhat different dispersal ability which relates to ability to respond to disturbed environments (Grassle and Grassle, 1977). Species with the best dispersal as measured by length of larval life (Ia and III) have seasonal reproduction and an intermediate ability to increase following environmental disturbance (Table 1). These two species are genetically the least diverse with the most common allele having a frequency of 0.95 or more in 7 out of 8 and 5 out of 6 respectively of the loci studied (Table 2). These species also show little differentiation with distance. Species I, which responds most rapidly at all seasons by sharp increases in numbers (most opportunistic), has an intermediate degree of genetic variation and shows some spatial division into subpopulations. This is the species that increased dramatically in number following an oil spill and increased to densities of 250,000 per m^2 within a month in field experiments (Grassle and Grassle, 1974, 1977). A deep-sea species, not included in previous descriptions of this sibling group, has the slowest population response and the highest genetic variability. The individuals studied

TABLE 1

Comparison of response to disturbance, dispersal ability, and genetic variability (measured as percent electrophoretic loci polymorphic) in 6 species of Capitella.

Capitella species	Response to disturbance	Dispersal ability measured as length of larval life	% loci polymorphic[1]
IIIa	-	none	50
I	less than 1 month	several hours	50
II	-	6-24 hours	25
III	seasonal (winter)	$\geq$ 2 weeks	20[2]
Ia	seasonal (winter)	2 months	12.5
deep-sea	several months	unknown	62.5

[1] Loci where the common allele has a frequency $\leq$ 0.95. At least several hundred of each of these species have been run except the deep-sea species where 44 individuals were studied.

[2] Three out of 8 loci active in all other shallow-water Capitella species show no activity in species III.

TABLE 2

Loci polymorphic[2] in six species of Capitella

Species	IIIa	I	II	III	Ia	Deep-Sea
Loci[1]						
Phi	xx	xx	o	x	x	xx
Xdh	xx	xx	xx	xx	xx	--
Pgm	xx	o	o	na	o	xx
Idh-1	o	o	x	na	x	o
Idh-2	xx	xx	x	na	?o	o
Mdh-1	o	o	o	o	o	--
Mdh-2	x	xx	xx	x	x	--
αGpdh	o	o	o	o	o	--
G6pdh	--	--	--	--	--	xx
To	--	--	--	--	--	xx
Hk	--	--	--	--	--	o
Lap	--	--	--	--	--	xx

[1] Phi = phosphohexose isomerase; Xdh = xanthine dehydrogenase; Pgm = phosphoglucomutase; Idh = isocitrate dehydrogenase; Mdh = malate dehydrogenase; αGpdh = - glycerophosphate dehydrogenase; G6pdh = glucose-6-phosphate dehydrogenase; To = superoxide dismutase; Hk = hexokinase; Lap = leucine aminopeptidase.

[2] xx = polymorphic (common allele with frequency ≤ 0.95)
x = polymorphic (common allele with frequency 0.95-0.99)
o = monomorphic
na = no activity

were collected from wood panels placed on the bottom at 3644 m for nine months (Turner, 1973). Other in situ experiments in the deep sea show that members of either the same species or a closely related sibling species have rates of colonization of unoccupied substrate measured in years.

Data in the literature have led to a number of generalizations which may be summarized (in modified form) as follows: 1. Species with better individual regulatory ability (including physiological tolerances, mobility and behavioural adaptations) have lower genetic variability (Selander and Kaufman, 1973). 2. Species in environments where temporal change is less variable and severe have higher genetic variability (Ayala et al., 1975a; Valentine, 1976). 3. Species with greater dispersal ability tend to be less divided into subpopulations and have lower genetic variability (Gooch, 1975a). These generalizations must be viewed as hypotheses that have not been adequately tested. In each case direct evidence for a cause and effect relationship is not available and the second and third generalizations, in particular, depend on a few isolated examples. Of even greater concern is the possibility that allelic variation at loci assayed electrophoretically may bear little relationship to the variability in the genome as a whole. Much of the variation is concealed and the enzymes that stain in interpretable patterns are not necessarily a random representation of the genome. Despite these reservations these generalizations are useful and we consider each of them separately.

The inverse relationship between individual flexibility and genetic plasticity

Selander and Kaufman (1973) and Selander (1976) have reviewed the data on genetic variability in invertebrates and vertebrates. The data show that vertebrates are less variable than invertebrates and that fish are the most variable of the vertebrates. The behavioural and physiological regulatory ability of vertebrates enables each individual to adapt to a wide range of environmental change. In invertebrates variability among individuals plays a greater role in the adaptive process. The terms coarse-grained and fine-grained have been applied to these different modes of adaptation. We avoid using these terms since they have been used to compare within-phenotype as well as between-phenotype environmental grain (MacArthur and Wilson, 1967). In invertebrates adaptation to environmental fluctuations that occur frequently relative to the length of the life cycle is at the level of the individual. For example, in the estuarine crab, Rithropanopeus harrisii, no genetic variation has been detected at 10 loci (Gooch, 1975b). The daily fluctuations in the environment encountered by each individual may be as great as any encountered in the remainder of the life cycle. Changes that occur with a frequency of the same order of magnitude as the life cycle involve population adjustments, and short-term selection may be a significant component of adaptation (Hutchinson, 1965 pp. 60-63, Grassle and Grassle, 1974, 1977). Capitella species I is an example of a species with a very opportunistic life history involving high mortality and short-term selection. The Ia and III Capitella species are more like Rithropanopeus with low variability and little population differentiation. Elsewhere in this volume Battaglia and Bisol discuss a similar phenomenon using the contrasting examples of the copepods Tigriopus and Tisbe. Tigriopus has greater individual flexibility and a much longer generation time than Tisbe. In a particularly well-studied terrestrial example, the sea oat, Avena barbata, has greater individual phenotypic plasticity whereas genetic variation between individuals plays a greater role in adaptation in Avena fatua (Marshall and Allard, 1970).

The inverse relationship between genetic plasticity and individual flexibility has frequently been discussed in the genetics literature (Gause, 1942, Thoday, 1953, Battaglia, 1964). The only new aspect to this is that more is known about the evolution of life histories. Genetic plasticity may be more important in opportunistic species adapted for life in short-lived unpredictable habitats. A definition of predictability was presented by one of us at an earlier symposium on marine genetics held in Venice: "Changes causing environments to be less predictable differ greatly from the mean in extent and occur sporadically and without autocorrelation within the life span of an individual. Changes included within the physiological tolerance of every individual, i.e., those which do not result in mortality or reduced viability do not affect predictability for the species considered" (Grassle, 1972). Species for which the environment is relatively unpredictable may be termed opportunistic or r-selected (Grassle and Grassle, 1975). Relatively opportunistic species are characterized by rapid rates of population increase and high mortality as adaptations to ephemeral habitats. The degree of opportunism of a species may be expressed in terms of generation time and the length of time the habitat is suitable for reproduction (Southwood, 1977). Many benthic invertebrates produce large numbers of offspring which must colonize new habitats as old habitats deteriorate. In these populations intense selection may occur each generation and only individuals with genetically variable offspring have a high probability of contributing to the next generation.

Environmental Stability and Genetic Variability

Species diversity. The hypothesis that the deep sea is characterized by species with low genetic variability (Grassle, 1967) has been shown to be false for a number of species (Gooch and Schopf, 1972; Ayala and Valentine, 1974; Valentine and Ayala, 1975; Ayala et al., 1975b; Murphy et al., 1976). Ayala et al., (1975a) and Valentine (1976) have explained these results by the trophic stability hypothesis. In simplest form, the hypothesis predicts correlation between seasonality of productivity and low levels of genetic variability (Valentine and Ayala, 1976). Like the environmental predictability-genetic variability hypothesis, the trophic stability hypothesis was first used to explain gradients of species diversity (Valentine, 1971). As an explanation of gradients of species diversity it is a special case of the environmental predictability - species diversity relationship (Sanders, 1969, Grassle, 1972). The trophic stability hypothesis is particularly applicable in instances where fluctuations in winds and currents result in seasonal nutrient upwelling of variable timing and extent from year to year. The upwelling of nutrient rich water results in blooms of one or a few species of phytoplankton. The opportunistic response of these phytoplankton species is similar to the population response of a variety of other taxa to environmental perturbation.

The trophic stability theory obviously cannot explain gradients in diversity of primary producers such as trees (Whittaker, 1969), nor can it account for the local diversity gradients of coral reefs (Loya, 1972) or estuaries (Sanders et al., 1965; Boesch, 1974). A more general environmental stability theory may, however, be applicable both to species diversity and to genetic variation. To estimate the magnitude and temporal scale of variation relevant to a particular taxon the concept of predictability is needed. As discussed in the previous section, if an environmental change is predictable for a species, adaptation is primarily at the level of the individual so that every individual can adjust to the changes that occur. If environmental changes are unpredictable only a portion of the population can adapt to change, and short-term selection may be a significant component of adaptation.

Genetic Variability and Population size

The original prediction of low genetic variability in the deep sea (Grassle, 1972) was based on (1) the expectation of small average population size in communities with high diversity and (2) the lack of temporal heterogeneity of the environment. Species confined to species-rich areas have reduced dispersal ability and small population size (MacArthur and Wilson, 1967, p. 159; Diamond, 1975). Small populations would be expected to have high rates of random loss of alleles as well as selection for a reduced number of genotypes through increased ecological specialization (more restricted habitat as a result of competition and predation). An empirical relationship between population size and electrophoretic variation in marine invertebrates has been suggested by Soule (1976). Species with large population size as well as broad distribution are less likely to experience genetic bottlenecks. Species with broad distributions and long-distance larvae found primarily in tropical areas, show the greatest persistence through geological time (Jackson, 1974; Scheltema in press).

Studies of a variety of groups including rodents, lizards, fish and insects show reduction of variability in small populations (Laing et al., 1976; Soulé, 1976; Selander, 1976 and references therein). These examples are from tropical or subterranean populations where the environment would be relatively homogeneous. Species with such small population sizes could not survive in temporally hetergeneous environments (Grassle, 1972). On the other hand temporally homogeneous environments may contain species with broad distribution and large effective population size. The few conspecific populations with low variability and low similarity studied are all from temporally homogeneous environments (Laing et al., 1976).

The deep-sea species such as ophiuroids, asteroids, brachiopods and decapods and the single tropical marine invertebrate studied (Tridacna maxima) (Ayala et al., 1973; Campbell et al., 1975) differ from the majority of marine species in their relatively large adult size and broad range of distribution. The larvae of Tridacna maxima spend 11 days in the plankton and, like many bivalves, a very large number of eggs are released each spawning ($10^6 - 10^7$) (Jameson, 1976). The best studied deep-sea species are ophiuroids (Gooch and Schopf, 1972; Ayala and Valentine, 1974; Murphy et al., 1976). Ophiuroids are characterized by good dispersal and unusually broad distributions vertically and horizontally in comparison with other major deep-sea taxa (Sanders and Grassle, 1971; Hendler, 1975). The effective population size of

these species is likely to be much larger than for the majority
of deep-sea species with relatively limited dispersal ability.
The high genetic variability in the deep-sea Capitella is
expected since it is broadly distributed and has a very patchy
distribution.   The deep-sea asteroids found to have relatively
high genetic variability (Ayala et al., 1975; Murphy et al.,
1976) also have relatively broad distributions (Cherbonnier
and Sibuet, 1972; Ayala et al., 1975b).   The deep-sea
brachiopod Frieleia halli (Valentine and Ayala, 1975) is also
highly variable genetically, but is unusual in being found
both at shelf and bathyal depths and in having seasonal
reproduction (Rokop, 1974).   The highly diverse taxa in the
deep sea such as polychaetes, gastropods, and peracarid crustacea have the narrowest depth distributions and show the
greatest endemicity to ocean basins (Sanders and Grassle, 1971).
The species in these groups with narrow distributions have
not been studied.   The relative absence of physical barriers
in the ocean may mean that very few of the species are sufficiently restricted spatially to adapt to a uniform environment.

   The strongest evidence for the trophic stability theory
comes from a study of species of a single genus of planktonic
krill from upwelling areas in the Pacific (Valentine and Ayala,
1976).   The tropical upwelling species (Euphausia distinguenda)
has the greatest heterozygosity followed by the Peru upwelling
species (Euphausia mucronata).   The species from the Atlantic
convergence (Euphausia superba) shows the least variability.
The low variability of the Antarctic species, E. superba, is
said to depend on the seasonality of productivity rather than
environmental variation since the physical environment is
most variable in the Peru upwelling region.   Euphausia superba
is 50-60 mm in length whereas E. mucronata is 17 mm and E.
distinguenda is 8.0-8.5 mm long (Brinton, 1962).   The large
size of E. superba suggests that it may have greater individual
regulatory ability than the other two species.   Since E.
superba lives 2+ years (Mauchline and Fisher, 1969) the year
to year variation in timing and amount of productivity may be
more important than seasonality which occurs within the life
span of an individual.   Without more information on the life
histories of these populations and the environments in which
they live it is not possible to say which species lives in
the more predictable environment.   Euphausia distinguenda
had a much broader distribution than E. mucronata, with populations in the Indian Ocean and Red Sea in addition to the
Pacific (Brinton, 1962).   This complicates any simple comparison of the levels of genetic variability in these two
species.

Biotic Spatial Heterogeneity and Genetic Variability

The population size-genetic variability relationship discussed by Soulé (1976) relates as much to the selective effects of habitat heterogeneity as to random loss of alleles. The most likely explanation for high genetic variability in widely distributed tropical and deep-sea species is the biotic spatial heterogeneity of these relatively species-rich environments. This may be trophic resource specialization as suggested by Ayala et al., (1975. p7) or spatial differences in interactions with other species. Although temporal heterogeneity is low in these environments spatial heterogeneity is greater and would be an even more effective means for maintaining polymorphism (Hedrick et al., 1976). In the coral reef giant clam, Tridacna maxima, there is a color polymorphism in the mantle which may be maintained by predators feeding predominantly on the most common juvenile color morph. The predation regime may vary from one part of a reef to another. The algal symbionts that are the principal source of food for Tridacna may also vary from one part of a population to another. Even though there is little separation into subpopulations each adult may be selected from a million offspring. The sporadic, and patchy recruitment that occurs on coral reefs (Frank, 1969) is yet another source of heterogeneity.

Spatial heterogeneity in the deep sea is less obvious. However, Jumars (1976) indicates that patchiness on the scale of individual organisms is likely to be important in structuring deep-sea communities. Disturbance on scales of centimeters may be a source of spatial heterogeneity for unusually long periods owing to the low rates of recolonization (Grassle, 1977). For Tridacna, and for Ophiomusium and other deep-sea species, each individual encounters a different suite of competitors, predators, symbionts, pathogens or food items. Each propagule of the species is a colonist, and each entering colonist encounters different biotic regimes in each local environment. Under such circumstances it is advantageous for the larvae to be genetically variable since the local biotic regime selects for different genotypes in different local habitats. The metaphor used by Williams (1975), in discussing the advantages of sexual reproduction, that a variety of lottery tickets are better than many copies of a single lottery ticket in colonizing new areas, is applicable. One example, is the electrophoretic variation among subpopulations of Drosophila species found on different larval substrates (Richardson, Smouse and Richardson, 1977).

Even though discrete subpopulations may not exist, spatial differentiation results from selection in each microhabitat following larval settlement. If microhabitat differences are on a scale of individuals each survivor would be adapted to a somewhat different suite of local conditions. Ayala and Valentine (1974) and Valentine and Ayala (1975) found that trawl samples of the deep-sea species, Ophiomusium lymani and Frieleia halli, about a kilometer apart show little genetic differentiation. Even more remarkable is the lack of differentiation between Tridacna populations separated by thousands of kilometers of ocean (Campbell et al., 1975). This supports the suggestion that the relevant scale of heterogeneity may be within what might normally be thought of as a single population. We still have little idea of the effectiveness of dispersal on this small scale and the extent to which genetic heterogeneity persists from generation to generation.

The Hawaiian drosophilids are perhaps the best studied group of organisms with high species diversity living under relatively constant temporal conditions. Although the populations are confined to islands and might be expected to have relatively small population size, a number of the species are as variable as other Drosophila populations (Ayala, 1975). As has been suggested for coral reefs and the deep sea, highly localized spatial heterogeneity in the biotic regime may be the means by which this variation is maintained. In Drosophila setosimentum there is considerable variation within and between populations from all five volcanoes of Hawaii (Carson and Johnson, 1975). At least some of these populations may have been isolated for many generations. In other Hawaiian drosophilids, D. silvestris and D. heteroneura, allozymic similarity is great both within and between species, and similarity coefficients between populations within species are of the same order of magnitude as those between the two species (Sene and Carson, 1977). Average individual heterozygosity in these species is lower (0.083 and 0.089) than usual for Drosophila (0.157 - Powell, 1975) (Sene and Carson, 1977). It may be that the more constant parts of tropical environments will contain species with a broader range of heterozygosities than other environments. The mean for all tropical zone invertebrates (0.109 $\pm$ 0.009) is somewhat less than for temperate zone invertebrates (0.132 $\pm$ 0.012) (Powell, 1975).

Species with Greater Dispersal Ability show less Population Differentiation

Gooch (1975a) has reviewed the literature on genetic differentiation in marine invertebrates. Species with good dispersal show relatively little differentiation over distances of 1000 km or more for at least some loci. The species studied are all very abundant temperate estuarine and shallow coastal species. A planktonic larval stage enables a species to find suitable sites for settlement and to escape from deteriorating habitats. Lack of population differentiation may indicate greater individual regulatory ability including the behavioural adaptations needed for habitat selection at the time of larval settlement. In contrast, species such as the highly opportunistic and less well-dispersed Capitella type I are more divided into subpopulations because short-term selection is likely to be a major component of adaptation.

There may be a dichotomy between saturation dispersal, resulting from the periodic release of very large numbers of larvae of species with intermediate lengths of larval life (cf. Schram, 1968, 1970), and the more limited gene flow (Endler, 1973) in species with very long-lived larvae (Scheltema, 1971) and more restrictive conditions for larval settlement. This dichotomy would be analogous to that between wind-dispersed plants and species that rely on animal dispersal of seeds. Lack of differentiation between populations may be confined to the few species with very abundant planktonic larvae and wind-dispersed plants. In tropical rain forests wind dispersal is uncommon and the more efficient seed dispersal by animals may allow exploitation of rare patchily distributed resources (Ashton, 1969; Stern and Roche, 1974; Regal, 1977). Ashton (1969) studied the Brunei rain forest where only 1 of 760 species was wind pollinated. He believes that final occupancy of a space by adult trees results from selection as a result of a complex series of biotic interactions as well as chance settlement. This is the same sort of natural selection process that may occur in temporally constant marine habitats. We do not know whether this selection can occur within generations or whether a genetic mosaic on a scale of individuals persists through many generations. In temperate regions catastrophic changes in the environment (both biotic and abiotic) may result in subdivision into large discrete subpopulations such as occurs with Capitella sp. I. In species with very high rates of dispersal there is little separation into sub-populations, more adaptation at the level of the individual, and somewhat lower genetic variability.

Sexual Reproduction

Much of the discussion on the adaptive value of genetic variability in populations applies equally to the advantages of sexual reproduction. Williams' (1975) models mostly use high fecundity plants and marine invertebrates as examples. Groups such as rotifers, aphids, weevils and some earthworms that live in the most temporally variable habitats, are characterized by a high proportion of asexually reproducing species. Asexually reproducing terrestrial arthropods with closely related sexual relatives tend to be distributed in areas characterized by abiotic disturbance (Glesener and Tilman, in press). In such habitats the environment may be so variable that only a few genotypes are selected for and adaptation is mostly at the level of the individual. The uncertainty of the biotic environment appears to be a major factor selecting for sexual reproduction (Glesener and Tilman, in press). In such groups as corals, bryozoa and some asteroids on coral reefs asexual reproduction alternates with occasional sexual reproduction preceding the phase of the life history when new areas are colonized. It is likely that some of the syllid polychaetes so common on coral reefs reproduce only asexually. These species may be confined to local areas where the environment is so constant that genetic variation is not advantageous. At one environmental extreme asexual reproduction is advantageous in replicating an individual tolerant of most local temporal changes in the environment. These offspring must be broadcast over an area larger than the spatial extent of catastrophic changes. At the opposite extreme, in more constant environments, asexual reproduction is advantageous in replicating an individual which can hold a small piece of ground against competitors and predators.

Ecologists are just beginning to classify the range of relationships between life histories and habitats (Southwood, 1977). Much more information is needed on short-term selection and dispersal. A classification of life histories and population structure will eventually provide the matrix for understanding genetic variation.

Summary

The proportion of polymorphic loci in six sibling species of <u>Capitella</u> (Polychaeta) is related to a combination of population response time following disturbance and dispersal ability measured as length of larval life. The literature on protein polymorphisms in marine invertebrates is reviewed in the context of three generalizations: 1. Species with better individual regulatory ability (including physiological tolerances, mobility and behavioural adaptations) have lower genetic variability. 2. Species in environments where temporal change is less variable and severe have greater genetic variability. 3. Species with greater dispersal ability tend to be less divided into subpopulations and have lower genetic variability. The high genetic variability of species in tropical and deep-sea environments is best explained in terms of biotic spatial heterogeneity. Further studies of genetic variation should be accompanied by analyses of life histories and population structure.

Acknowledgment

This research was supported by N.S.F. grants DEB76-20336 and OCE76-21968. This paper is contribution No. 4010 from the Woods Hole Oceanographic Institution.

REFERENCES

Ashton, P. S. 1969. Speciation among tropical forest trees: some deductions in the light of recent evidence. Biol. J. Linn. Soc 1, 155-196.

Ayala, F. J. 1975. Genetic differentiation during the speciation process. Evol. Biol. 8, 1-78.

Ayala, F. J. and Valentine, J. W. 1974. Genetic variability in a cosmopolitan deep-water ophiuran Ophiomusium lymani. Mar. Biol. 27, 51-57.

Ayala, F. J., Hedgecock, D., Zumwalt, C.S. and Valentine, J. W. 1973. Genetic variation in Tridacna maxima, an ecological analog of some unsuccessful evolutionary lineages. Evolution 27, 177-193.

Ayala, F. J., DeLaca, T.E. and Zumwalt, G. S. 1975a. Genetic variability of the Antarctic brachiopod Liothyrella notorcadensis and its bearing on mass extinction hypothesis. J. Paleont. 49, 1-9.

Ayala, F. J., Hedgecock, D. and Barr, L. G. 1975b. Deep-sea asteroids: high genetic variability in a stable environment. Evolution 29, 203-212.

Battaglia, B. 1964. Advances and problems of ecological genetics in marine animals. pp. 451-463. In, Genetics Today. Proc. XI Int. Congr. Genetics, Pergamon Press.

Boesch, D. F. 1974. Diversity, stability and response to human disturbance in estuarine ecosystems. Proc. Intern. Congr. Ecol. 1, 109-114.

Brinton, E. 1962. The distribution of Pacific euphausiids. Bull. Scripps Inst. Oceangr. 8, 51-269.

Campbell, C. A., Valentine, J. W. and Ayala, F. J. 1975. High genetic variability in a population of Tridacna maxima from the Great Barrier Reef. Mar. Biol. 33, 341-345.

Carson, H. L. and Johnson, W. E. 1975. Genetic variation in Hawaiian Drosophila I. Chromosome and allozyme polymorphism in D. setosimentum and D. ochrobasis from the Island of Hawaii. Evolution 29, 11-23.

Cherbonnier, G. and Sibuet, M. 1972. Résultats scientifiques de la campagne Noratlante: Asterides et Ophiurides. Bull. Mus. Hist. nat. Paris, 3e ser., n° 102, Zoologie 76, 1333-1394.

Diamond, J. M. 1975. Assembly of species communities. pp. 342-444. In, Ecology and Evolution of Communites. Cody, M. L. and Diamond, J. M. (eds). Harvard Univ. Press, Cambridge.

Endler, J. A. 1973. Gene flow and population differentiation. Science 179, 243-250.

Frank, P. W. 1969. Growth rates and longevity of some gastropod mollusks on the coral reef at Heron Island. Oecologia 2, 232-250.

Gause, G. F. 1942. The relation of adaptability to adaptation. Quart. Rev. Biol. 17, 99-114.

Glesener, R. and Tilman, D. 1977. Sexuality and the components of environmental uncertainty: A comparative approach. American Naturalist (In press).

Gooch, J. L. 1975a. Mechanisms of evolution and population genetics. pp 351-409. In, Marine Ecology. A Comprehensive, Integrated Treatise on Life in Oceans and Coastal Waters. Vol II pt. 1. Kinne, O. (ed). Wiley, London and New York.

Gooch, J. L. 1975b. Some current problems in marine genetics. pp. 85-107. In, The Ecology of Fouling Communities. Costlow, J. D. (ed). U.S. Office of Naval Research.

Gooch, J. L. and Schopf, T. J. M. 1972. Genetic variability in the deep sea: relation to environmental uncertainty. Evolution 26, 545-552.

Grassle, J. F. 1967. Influence of environmental variation on species diversity in benthic communities of the continental shelf and slope. Ph.D. Dissertation, Duke Univ. 68-2723, University Microfilms, Inc., Ann Arbor, Michigan.

Grassle, J. F. 1972. Species diversity, genetic variability and environmental uncertainty. pp 19-26. In, *Fifth European Marine Biology Symposium*. Battaglia, B. (ed). Piccin Editore, Padua.

Grassle, J. F. 1977. Slow recolonization of deep-sea sediment. Nature, 265, 618-619.

Grassle, J. F. and Grassle, J. P. 1974. Opportunistic life histories and genetic systems in marine benthic polychaetes. J. Mar. Res. 32, 253-284.

Grassle, J. F. and Grassle, J. P. 1977. Temporal adaptations in sibling species of *Capitella*. pp. 177-189. In, *Ecology of Marine Benthos*. Coull, B. C. (ed). Univ. of South Carolina Press, Columbia.

Grassle, J. P. and Grassle, J. F. 1976. Sibling species in the marine pollution indicator, *Capitella capitata* (Polychaeta). Science 192, 567-569.

Hedrick, P. W., Ginevan, M. E. and Ewing, E. P. 1976. Genetic polymorphism in heterogeneous environments. Ann. Rev. Ecol. Sys. 7, 1-32.

Hendler, G. 1975. Adaptational significance of the patterns of ophiuroid development. Amer. Zool. 15, 691-715.

Hutchinson, G. E. 1965. The Ecological Theatre and the Evolutionary Play. Yale Univ. Press, New Haven.

Jackson, J. B. C. 1974. Biogeographic consequences of eurytopy and stenotopy among marine bivalves and their evolutionary significance. Amer. Nat. 108, 541-560.

Jameson, S. C. 1976. Early life history of the giant clams *Tridacna crocea* Lamarck, *Tridacna maxima* (Roding), and *Hippopus hippopus* (Linnaeus). Pacific Science 30, 219-233.

Jumars, P. A. 1976. Deep-sea diversity: does it have a characteristic scale. J. Mar. Res. 34, 217-246.

Laing, C., Carmody, G. R. and Peck, S. B. 1976. Population genetics and evolutionary biology of the cave beetle *Ptomaphagus hirtus*. Evolution 30, 484-498.

Loya, Y. 1972. Community structure and species diversity of hermatypic corals at Eilat, Red Sea. Mar. Biol. 13, 100-123.

MacArthur, R. and Wilson, E. O. 1967. The Theory of Island Biogeography. Princeton Univ. Press, Princeton.

Marshall, D. R. and Allard, R. W. 1970. Isozyme polymorphisms in natural populations of *Avena fatua* and *Avena barbata*. Heredity 25, 373-382.

Mauchline, J. and Fisher, L. R. 1969. The biology of euphausiids. Advances in Mar. Biol. 7, 1-454.

Murphy, L. S., Rowe, G. T. and Haedrich, R. L. 1976. Genetic variability in deep-sea echinoderms. Deep-Sea Res. 23, 339-348.

Powell, J. R. 1975. Protein variation in natural populations of animals. Evol. Biol. 8, 79-119.

Regal, P. J. 1977. Ecology and evolution of flowering plant dominance. Science 196, 622-629.

Richardson, R. H., Smouse, P. E. and Richardson, M. E. 1977. Patterns of molecular variation. II. Associations of electrophoretic mobility and larval substrate within species of the *Drosophila mulleri* complex. Genetics 85, 141-154.

Rokop, G. J. 1974. Reproductive patterns in the deep-sea benthos. Science 186, 743-745.

Sanders, H. L. 1969. Benthic marine diversity and the stability-time hypothesis. Brookhaven Symp. Biol, 22, 71-80.

Sanders, H. L. and Grassle, J. F. 1971. The interactions of diversity, distribution and mode of reproduction among major groupings of the deep-sea benthos. pp. 260-262. In, The World Ocean, Proc. Joint Oceangr. Assembly. Uda, M. (ed). Japan Soc. Promotion Sci., Tokyo.

Sanders, H. L., Mangelsdorf, P. C. (Jr.) and Hampson, G. R. 1965. Salinity and faunal distribution in the Pocasset

River, Massachusetts. Limnol. Oceanogr., 10 (Suppl.), R216-R229.

Scheltema, R. S. 1978. Dispersal of marine invertebrate organisms: paleobiogeographic and biostratigraphic implications. In, Concepts and Methods of Biostratigraphy. Kauffman, E. G. and Hazel, J. E. (eds). Dowden, Hutchinson, and Ross, Stroudsburg, Pa. (in press.)

Scheltema, R. S. 1971. Larval dispersal as a means of genetic exchange between geographically separated populations of shallow-water benthic marine gastropods. Biol. Bull. 140, 284-322.

Schram, T. A. 1968. Studies on the meroplankton in the inner Oslofjord. I. Composition of the plankton at Nakkholmen during a whole year. Ophelia 5, 221-243.

Schram, T. A. 1970. Studies on the meroplankton in the Inner Oslofjord. II. Regional differences and seasonal changes in the specific distribution of larvae. Nytt Mag. Zool. 18, 1-22.

Selander, R. K. 1976. Genic variation in natural populations. pp. 21-45. In, Molecular Evolution. Ayala, F. J. (ed). Sinauer, Sunderland, Mass.

Selander, R. K. and Kaufman, D. W. 1973. Genic variability and strategies of adaptation in animals. Proc. Natl. Acad. Sci., U.S.A. 70, 1875-1877.

Sene, F. M. and Carson, H. L. 1977. Genetic variation in Hawaiian Drosophila. IV. Allozymic similarity between D. silvestris and D. heteroneura from the island of Hawaii. Genetics 86, 187-198.

Soulé, M. 1976. Allozyme variation: its determinants in space and time. pp. 60-77. In, Molecular Evolution. Ayala, F. J. (ed). Sinauer, Sunderland, Mass.

Southwood, T. R. E. 1977. Habitat, the templet for ecological strategies? J. Anim. Ecol 46, 337-365.

Stern, K. and Roche, L. 1974. Genetics of Forest Ecosystems. Springer-Verlag, New York.

Thoday, J. M. 1953. Components of fitness. Symp. Soc. Exp. Biol. 7, 96-113.

Turner, R. D. 1973. Wood-boring bivalves, opportunistic species in the deep sea. Science 180, 1377-1379.

Valentine, J. W. 1971. Resource supply and species diversity patterns. Lethaia 4, 51-61.

Valentine, J. W. 1976. Genetic strategies of adaptation, pp. 78-94. In, Molecular Evolution. Ayala, F. J. (ed). Sinauer, Sunderland, Mass.

Valentine, J. W. and Ayala, F. J. 1975. Genetic variation in Frieleia halli, a deep-sea brachiopod. Deep-Sea Res. 22, 37-44.

Valentine, J. W. and Ayala, F. J. 1976. Genetic variability in krill. Proc. Nat. Acad. Sci. U.S.A. 73, 658-660.

Whittaker, R. H. 1969. Evolution of diversity in plant communities. Brookhaven Symp. Biol. 22, 178-196.

Williams, G. C. 1975. Sex and Evolution. Princeton Univ. Press. Princeton, N. J.

IS THE MARINE LATITUDINAL DIVERSITY GRADIENT MERELY ANOTHER EXAMPLE OF THE SPECIES AREA CURVE?

T. J. M. Schopf, J.B. Fisher and C.A.F. Smith III

Department of Geophysical Sciences

University of Chicago, Chicago, U.S.A.

Introduction

The purpose of this paper is to examine the relationship between faunal diversity and area for modern marine faunal provinces of the world's continental shelves. The rationale leading to this study is that each faunal province is, to a significant extent, a habitat island with a large number of endemic species, and with a physical environment more similar within the province than between provinces. Where data are available, adjacent faunal provinces differ by 20 to 40% of their species (see Schopf, 1978), and are delineated by physiographic barriers, such as passage around a major cape, or from one major current system to another. If faunal provinces can be treated as habitat islands, then an examination of their diversity might follow the theory of island biogeography which has been applied to such diverse islands as "true" islands (MacArthur and Wilson, 1967), coral heads as habitat islands for decapod crustacea (Abele, 1976), plant species as habitat islands for insect species (Lawton and Schroder, 1977), etc. Since faunal provinces are of different size (in particular, tropical provinces are by far the largest) our idea was that the latitudinal trend toward lower diversity in temperate and boreal regions (Fischer, 1960; Stehli, 1968) might simply be a reflection of island biogeographic relations.

The study was also undertaken for strong paleontological reasons. As one of us (Schopf, 1973, 1976) has argued, changes in faunal diversity over geological time may be a function of the number of faunal provinces existing at any given time, as modified by the changes in habitable area within each province, or even within all provinces on a world-wide scale. Thus a major goal is to examine the extent to which changes in faunal diversity on marine continental shelves may be related to changes in habitable area.

Materials and Methods

On the basis of the location of marine faunal communities, Ekman (1953) divided the oceans into eleven major regions. We used this basic organization together with information on water current patterns (Sverdrup et al., 1942), salinity and temperature (Fairbridge et al., 1966), the shape and extent of the shelf (National Geographic Atlas of the World, 1963), and previous experience (Schopf, 1978) to further divide these regions into thirty-two faunal provinces, as shown in Figure 1 and listed in Table 1. The weakest link in this part of the problem is the grouping of the polar regions into a single faunal province. The shelf area of the polar region is immense, much of it ice covered, and it seems that one might equally as well subdivide the Arctic into several smaller provinces to take this into account (but this is not attempted here).

The National Geographic Atlas of the World outlines the continental shelf (depth of 200 m) for all regions of the world of concern to us. We used data from several different projections, (but never a Mercator), each of which has a scale independent of latitude for the whole of the map area (and thus is effectively an equal area projection). By placing the atlas on a light table, we were able to trace on graph paper the outline of the shelf for each province. The tracings were carefully cut out and their masses were determined on a standard analytical balance. As a standard of comparison, large sections of graph paper were also cut out with dimensions from the scale of each map, and their masses were determined. By converting the mass of the standard to square kilometers, a simple calculation gave the shelf area for each province. The estimate thus obtained for total area of the continental shelf is 27.50×10^6 km^2. This is close to the values of 27.12×10^6 km^2 of Menard and Smith (1966) and of 27.49×10^6 km^2 of Kossinna (1921) (each of whome used different procedures) and indicates the general validity of our measurements. An estimate of the average latitude of each province was made, taking into account the centre of gravity of area of the continental shelf of each province.

The bivalve data (from Stehli, personal communication;

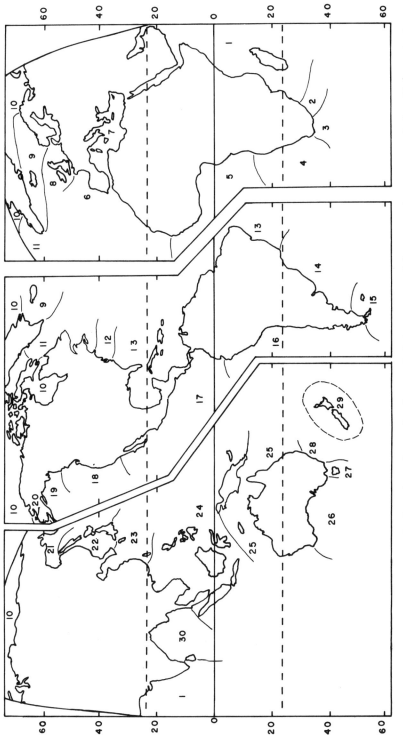

Figure 1. Map of world showing boundaries of biogeographic provinces used in this study. The seaward boundary in each instance is taken as the 200 m line, approximately the boundary of the continental shelf and continental slope. Numbers refer to provinces listed in Table 1.

TABLE 1
Data on faunal provinces and diversity

Faunal Region	Province Number	LAT°	Faunal Province	Area ($km^2 \times 10^5$)	Bryozoan Species
Tropical	1	18°	Ethiopian	12.8	100
	5	10°	Tropical West African	3.8	180
	13	20°	Western Atlantic Tropical	17.9	280
	17	18°	Mexican	3.5	520
	24	5°	Malayan	40.3	640
	25	15°	Papuan	20.1	380
	30	16°	Indian	4.5	135
	32	21°	Hawaiian	0.098	80
[Subtotal]				103.0	
Warm Temperate	2	33°	Temperate East African	0.2	
	6	30°	Temperate East Atlantic	2.1	
	7	37.5°	Mediterranean	6.2	300
	12	40°	Virginian	1.8	120
	14	38°	Argentinean	9.6	200
	16	28°	Chilean	2.1	200
	23	31°	Chinean	11.2	640
	26	35°	South Australian	3.3	295
	28	33°	Eastern Australian	0.32	
[Subtotal]				36.8	
Boreal, Anti-Boreal	3	36°	South African	0.91	175
	4	26°	West		

Table 1 (continued)

Bivalve Species	Bivalve Genera	Bivalve Families	Foram. Genera	Foram. Families	Coral Genera
443	108	52	98	42	60
263	91	45	28	17	8
302	111	55	186	51	27
557	124	54	83	37	8
1033	116	52	127	46	69
671	131	66	89	37	61
			34	20	41
					13
					11
285	87	49	54	29	2
230	83	50	107	41	
			68	31	
			93	38	
			24	17	
			68	35	63
314	100	56	55	25	
			61	28	
338	115	55			

Table 1 (continued)

			African	1.6	
	8	55°	Eastern North Atlantic Boreal	10.7	325
	9	65°	Norwegian	3.6	
	11	47.5°	Labradorean	10.9	100
	15	50°	Falkland	4.1	
	18	45°	American	4.7	
	19	57°	Aleutian	2.6	200
	20	62°	Alaskan	11.8	
	22	46°	Western North Pacific Boreal	2.9	200
	27	40°	Tasmanian	1.7	
	29	41°	New Zealand	2.3	
[Subtotal]				57.8	
Polar	10	70°	Arctic	56.4	200
	21	59°	Okhotsk	5.0	
	31	70°	Antarctic	16.0	320
[Subtotal]				77.4	
TOTAL				275.0	

Table 1 (contined)

182	80	44	89	34
107	44	32	11	9
100	50	32	65	32
178	72	43	46	27
203	88	44	86	35
			37	23
			39	24
411	116	50	92	35
279	98	51		
320	95	46	58	24
94	49	31	88	29
			18	11
81	31	24		

data used in Stehli et al., 1967) represent a compilation of 39 monographic studies taken by various groups over the past century. For each collection, the name of each of the bivalve families was listed and the number of genera and species was given for each family. Each monographic study was assigned to a faunal province. For those provinces which contained more than one collection site, we used the number of genera and species from the largest collection. The bryozoan data (Schopf, 1978) represent 23 collections or estimates of the number of bryozoan species found at various sites. These sites were placed in our faunal provinces. The foraminiferal data (from Stehli, personal communication; data used in Durazzi and Stehli, 1972) consisted of 104 collections for which each family and genus was identified by name. This enabled us to add up foraminiferal taxa from different stations when a province contained more than one station. The coral data (from Stehli, personal communication; data used in Stehli and Wells, 1971) consisted of a series of base maps in which each map outlines the localities where a given coral genus occurs. The number of genera in each province was determined by comparison of these base maps to the boundaries of our faunal provinces.

The number of taxa found in each faunal province was graphed as a function of the area of the province on a log-log plot. This was done also for 4-point graphs, with provinces grouped into (1) tropics, (2) temperate, (3) boreal and anti-boreal, and (4) world totals. By the least squares method, a best fit curve was plotted for species (S) versus area (A), yielding an equation of the form $S = kA^z$, where k (the intercept) and z are fitted constants. Taxonomic diversity was also plotted as a function of latitude for each faunal province, and best-fit lines for these graphs were calculated by the least-squares method. In addition, the residuals from each of the best-fit curves for the area and latitude analyses were graphed as a function of, respectively, latitude and area, following the method used by Raup (1976) and Sepkoski (1976). Plots of taxonomic "density" versus latitude were made by taking the taxa per unit area, using the "z-value" for the taxa obtained in the taxa-area plots. This was done in order to see if taxonomic richness and taxonomic density behaved in the same way with respect to latitude.

Results

Figure 2 shows the relationship between area and latitude of faunal provinces. The tropical provinces (8) are, on average, twice as large as either the temperate (9) or boreal

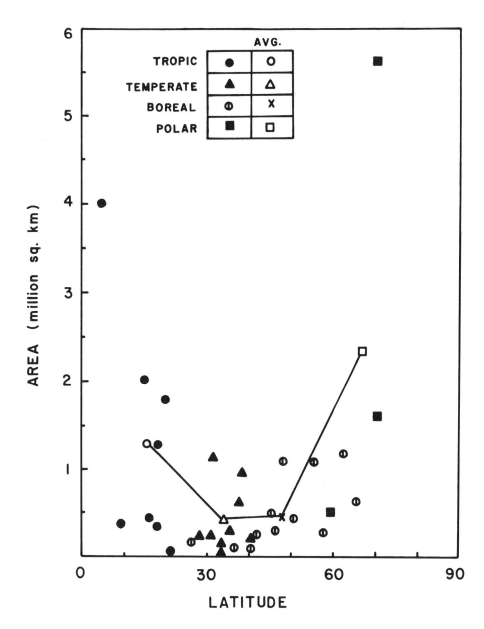

Figure 2. Relationship between area of faunal province and latitude. Note that the tropical provinces are approximately twice as large as temperate and boreal provinces. The arctic province is likely overestimated in size owing to the inclusion of all of the Arctic in a single province.

(12) provinces. The average polar province is largest of all owing to the inclusion of all of the various subdivisions of the Arctic single province. This region seems least well known of all major areas, and therefore the Arctic province data are most suspect of those used in this paper.

We report in Tables 2 - 4 the results of the analyses for taxa versus latitude, taxa versus area, and density versus latitude for bryozoa and bivalve species, foraminfera, coral and bivalve genera, and foraminifera and bivalve families. Further results will be reported only in terms of taxonomic diversity since no new information seems to be gained by considering taxonomic density.

Area is either significantly ($P = 0.05$ level) or highly significantly ($P = 0.01$ level) correlated with (1) bryozoan species, (2) foraminiferal genera, (3) coral genera, and (4) all genera (Figure 3). Latitude is significantly or highly significantly correlated with (1) bivalve species diversity, (2) total species diversity (Figure 4), (3) bivalve genera, (4) foraminiferal families, (5) bivalve families, and (6) all families (Figure 5). Thus of the 10 comparisons, 4 are best accounted for by taxa versus area, and 6 by taxa versus latitude.

Analyses of the residuals either from the plots of taxa versus area, or from taxa versus latitude (as explained in Methods), did not yield any further significant correlations. This appears to suggest that after the initial correlation is made (whatever it happens to be), the residual variation is not strongly associated with either area or latitude as a major secondary factor.

Discussion

The biological basis for suspecting that the habitat area of any environmental type is a good predictor of diversity is that (1) there seems to be a limit on the number of species which can be supported by the energy flow into a given region; and that (2) if that limit operates in the same fashion in different geographic regions (but of similar ecologic history), then the size of the habitable area should predict the number of taxa living there. In the cases at hand, area is a significant predictor of diversity in 4 of the 10 comparisons. The "area" used here is <u>total</u> area of a faunal province, and this includes more than simply the habitable area for any given taxon. Infaunal bivalves, for example, would appear to be

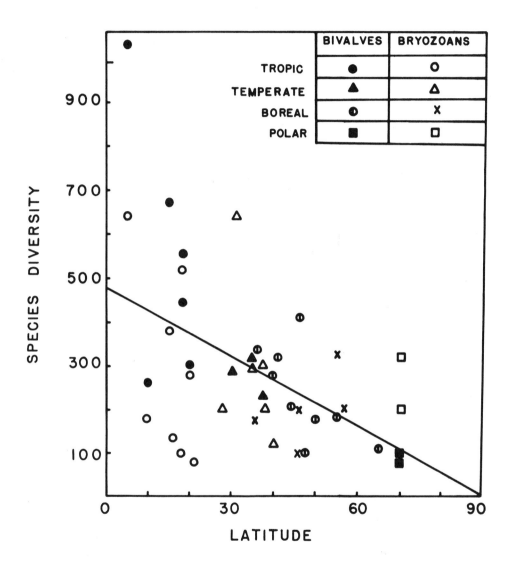

Figure 3. Species diversity as a function of latitude for bryozoa and bivalves. The least-squares best fit is Y = -5.29X + 481.8; r = 0.51**; n = 41.

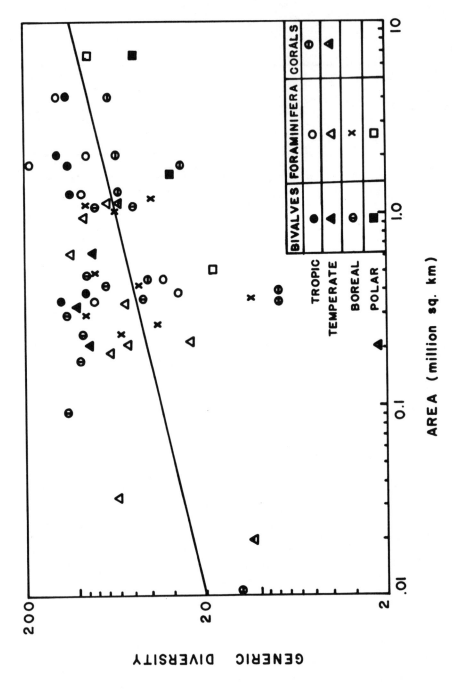

Figure 4. Generic diversity as a function of area for foraminifera, corals, and bivalves. The least square best fit is Y = 2.04X$^{.25}$; r = 0.37**; n = 57.

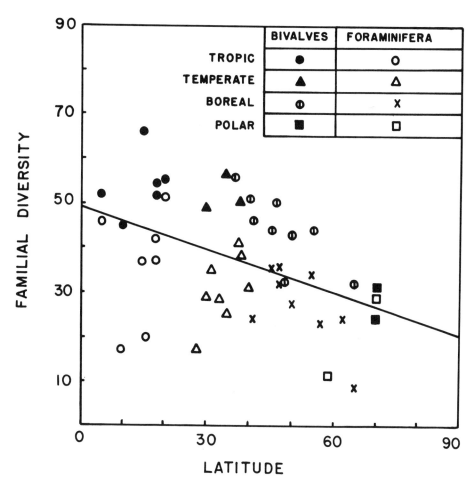

Figure 5. Familial diversity as a function of latitude for foraminifera and bivalves. The least-squares best fit is Y = -0.31X + 48.9; r = 0.44**; n = 46.

TABLE 2

Diversity as a function of area, and diversity and density as a function of latitude for bryozoan and bivalve species. Significant relationships (0.05) shown by *; highly significant relationships (0.01) shown by **. Note that for species as a whole, the best correlation is with latitude.

SPECIES

	BRYOZOA	BIVALVES	ALL
Diversity as a function of area	$Y = 14.74X^{.21}$ $r = 0.49*$ $n = 21$	$Y = 453.33X^{-.04}$ $r = 0.00$ $n = 20$	$Y = 50.57X^{.12}$ $r = 0.24$ $n = 41$
Diversity as a function of latitude	$Y = -1.92X + 331.55$ $r = 0.22$ $n = 21$	$Y = -9.04X + 660.31$ $r = 0.77**$ $n = 20$	$Y = -5.29X + 481.80$ $r = 0.51**$ $n = 41$
Density as a function of latitude	$Y = -0.10X + 19.32$ $r = 0.24$ $n = 21$	$Y = -16.17X + 1,158$ $r = 0.75**$ $n = 20$	$Y = -0.67X + 77.01$ $r = 0.39*$ $n = 41$

TABLE 3

Diversity as a function of area, and diversity and density as a function of Latitude for foraminiferal, coral and bivalve genera. Significant relationships (0.05) shown by *; highly significant relationships (0.01) shown by **. Note that for genera as a whole, the best correlation is with area.

GENERA

	FORAMINIFERA	CORALS	BIVALVES	ALL
Diversity as a function of area	$Y = 2.57X^{.24}$ $r = 0.41^*$ $n = 26$	$Y = 0.21X^{.35}$ $r = 0.58^*$ $n = 11$	$Y = 279.41X^{-.09}$ $r = 0.24$ $n = 20$	$Y = 2.04X^{.25}$ $r = 0.37^{**}$ $n = 57$
Diversity as a function of latitude	$Y = -0.73X + 96.58$ $r = 0.35$ $n = 26$	$Y = -0.98X + 52.41$ $r = 0.33$ $n = 11$	$Y = -1.18X + 133.90$ $r = 0.80^{**}$ $n = 20$	$Y = -0.41X + 83.45$ $r = 0.20$ $n = 57$
Density as a function of latitude	$Y = -0.02X + 3.68$ $r = 0.32$ $n = 26$	$Y = 0.001X + 0.28$ $r = 0.055$ $n = 11$	$Y = -4.26X + 458$ $r = 0.83^{**}$ $n = 20$	$Y = -0.01X + 2.75$ $r = 0.10$ $n = 57$

TABLE 4

Diversity as a function of area, and diversity and density as a function of latitude for foraminiferal and bivalve families. Significant relationships (0.05) shown by *; highly significant relationships (0.01) shown by **. Note that for families as a whole, the best correlation is with latitude.

FAMILIES

	FORAMINIFERA	BIVALVES	ALL
Diversity as a function of area	$Y = 4.38X^{.14}$ $r = 0.37 \quad n = 26$	$Y = 101.55X^{-.06}$ $r = 0.26 \quad n = 20$	$Y = 14.52X^{.06}$ $r = 0.17 \quad n = 46$
Diversity as a function of latitude	$Y = -0.24X + 38.77$ $r = 0.41* \quad n = 26$	$Y = -0.41X + 62.00$ $r = 0.77** \quad n = 20$	$Y = -0.31X + 48.91$ $r = 0.44** \quad n = 46$
Density as a function of latitude	$Y = -0.03X + 5.82$ $r = 0.40* \quad n = 26$	$Y = -0.97X + 140.0$ $r = 0.79** \quad n = 20$	$Y = -0.13X + 21.70$ $r = 0.41** \quad n = 46$

underrepresented in the Antarctic where bottom sediments are preponderately coarse, and sponge-bryozoan-brachiopod groups (i.e. epifaunal taxa) are well represented (Bushnell and Hedgpeth, 1969; Hedgpeth, 1971; Bullivant, 1967). The further discrimination of total area into habitable area, and non-habitable area for such faunal province would be a significant extension of the present study. Since paleontological applications of the area-species model may generally preclude division of marine faunal provinces into habitable and uninhabitable regions for particular taxa, it was not attempted in this first examination of the topic.

Paleontological uses of the area-species relationship in studies of marine faunal extinction have used world-wide summations of data (not province by province data) (Schopf, 1974, 1978; Simberloff, 1972). All tropical diversity and tropical area can be summed to yield a single "tropical" point; similarly for temperate, boreal and "world wide" parts. Such a summation might be expected to reduce the effect of one part of any region being unduly well or poorly studied. Correlation coefficients (r) for these plots are 0.84 and 0.97 for bryozoan and bivalve species, 0.90 and 0.94 for foraminiferal and bivalve genera, and 0.73 and 0.97 for foraminiferal and bivalve families. In addition, as theory predicts, the "z-value" falls as taxonomic rank increases, with values of 0.34 and 0.48 for species, 0.18 and 0.31 for genera, and 0.06 and 0.10 for families. Thus if a paleontologist were concerned with data on a world-wide basis, the results of the present study suggest that grouping of data into large areal units is a reasonable way to proceed.

The study may also have implications for considerations of Pleistocene diversity. Owing to the removal of ocean water to ice caps, sea level was lowered some 130 m during the Pleistocene Epoch (Milliman and Emery, 1968). The slope of the continental slope (often about 2°) is greater than that of the continental shelf (much less than 1°) and thus some compression occurred at the seaward edge of faunal provinces within the photic zone. This reduction in habitable area would very likely have had an effect on marine species diversity and the topic of marine Pleistocene extinctions should be further examined in a quantitative way.

If area _per se_ is important in predicting diversity, then for the latitudinal diversity gradient one might well ask what limits (for each province) the origination rate, and the extinction rate. Rates of speciation are seemingly _independent_ of latitude (Stehli _et al._, 1972), but rates of generic origination are markedly higher nearer the tropics

(Stehli and Wells, 1971; Durazzi and Stehli, 1972). The greater longevity of genera in the Arctic is perhaps owing to fewer Arctic species lineages which need to be grouped into genera than exist in the tropics. If so, although the turnover rates of species are the same from equator to pole, the much larger range of opportunities for diversification in the tropics automatically leads to a higher turnover rate of genera. The larger number of opportunities may be related to the greater number of environmental aspects which are relatively uniform (and with respect to which organisms can specialize) in the tropics (and deep-sea) (Schopf, 1972; 1973; 1977).

Conclusions

We conclude that area is sometimes shown to be a significant factor in predicting marine taxonomic diversity and that therefore its use as a significant factor in understanding major periods of evolutionary divergence and/or extinction may be correct. Latitude remains, however, a better statistical predictor of diversity than area in a small majority (6 of 10) of the cases we considered. Habitable area alone provides a reasonable ecological explanation for limitations on numbers of individuals, and through that to numbers of taxa (Preston, 1962), for similar ecological conditions (Simberloff, 1972). We know of no similar biological reason for suspecting that the well-known latitudinal diversity gradient has any property in, and of, itself which would allow for the prediction of diversity. We are cautiously optimistic that when habitable area (instead of merely total area) can be estimated for each faunal province that such estimates will help to account for the latitudinal diversity gradient.

Summary

1. The latitudinal diversity gradient is perhaps the foremost generalization in marine biogeography. It is therefore important to try to determine its cause.

2. The continental shelves of the world were divided into 32 faunal provinces.

3. The area of each province was carefully measured from equal area maps of the shelf. Tropical provinces (8) are, on the average, more than twice as large as temperate (9) and boreal (12) provinces. If the Arctic is a single faunal province, then polar provinces (3) are largest of all.

4. Faunal diversity was determined from as many of the 32 faunal provinces as was possible using species (Bryozoans and Bivalves), genera (Foraminifera, Corals and Bivalves), and families (Foraminifera and Bivalves). As far as we are aware, these are the only groups for which data are readily retrievable for nearly world-wide coverage.

5. Significant or highly significant correlations are found for <u>area</u> with respect to Bryozoan species, Foraminiferal genera, Coral genera, and all species grouped together, Significant or highly significant correlations are found for <u>latitude</u> with respect to Bivalve species, all species grouped together, Bivalve genera, Foraminiferal families, Bivalve families, and all families grouped together. Thus of the 10 comparisons, 4 are best explained by area, and 6 by latitude.

6. Not all of the area of a faunal province is equally available to any particular taxon (for example, most infaunal bivalves must necessarily be absent from coarse sediments). This systematic error probably decreases the correlation between area and diversity.

7. If area is significant in predicting diversity of faunal provinces, then one might wish to ask what limits the origination rate, and the extinction rate, in the equilibrium of a faunal province.

REFERENCES

Abele, L. G. 1976. Comparative species richness in fluctuating and constant environments: coral-associated decapod Crustaceans, Science 192, 461-463.

Bullivant, J. S. 1967. Ecology of the Ross Sea Benthos. In, <u>The Fauna of the Ross Sea</u>. Part 5. General accounts, station lists and benthic ecology. Bullivant, J. S. and Dearborn, J. H. pp. 49-75. New Zealand Dep. Sci. Industr. Res. Bull. 176.

Bushnell, V. C. and Hedgpeth, J. W. 1969. Antarctic Map Folio Series, Folio 11. American Geographical Society, New York.

Durazzi, J. T. and Stehli, F. G. 1972. Average generic age, the planetary temperature gradient, and pole location. Syst. Zool. 21, 384-389.

Ekman, S. 1953. <u>Zoogeography of the Sea</u>. Sidgwick and Jackson Ltd. pp. 417.

Fairbridge, R. W. ed. 1966. <u>The Encyclopedia of Oceanography</u>. Reinhold, New York.

Fischer, A. G. 1960. Latitudinal variations in organic diversity. Evolution 14, 64-81.

Hedgpeth, J. W. 1971. Perspectives of benthic ecology in Antarctica. In, <u>Research in the Antarctic</u>. pp. 93-136. Am. Assoc. Adv. Science.

Kossinna, E. 1921. Die Tiefen des Weltmeeres. Inst. Meereskunde, Veroff., Geogr.-naturwiss. 9, 1-70.

Lawton, J. H. and Schroder, D. 1977. Effects of plant type, size of geographical range and taxonomic isolation on number of insect species associated with British plants. Nature 265, 137-140.

MacArthur, R. H. and Wilson, E. O. 1967. <u>The Theory of Island Biogeography</u>. Princeton University Press. Princeton, New Jersey.

Menard, H. W. and Smith, S. M. 1966. Hypsometry of ocean basin provinces. J. Geophys. Res. 71, 4305-4325.

Milliman, J. and Emery, K.O. 1968. Sea levels during the past 35,000 years. Science 162, 1121-1123.

National Geographic Atlas of the World. 1963. Melville Bell Grosvener, Editor-in-Chief, National Geographic Society, Washington D.C.

Newell, N. 1971. An outline history of tropical organic reefs. Am. Mus. Novit. No. 2465, pp. 37.

Preston, F. W. 1962. The canonical distribution of commonness and rarity: Part I and Part II. Ecology 43, 185-215, 410-432.

Raup. D. M. 1976. Species diversity in the Phanerozoic: an interpretation. Paleobiology 2, 289-297.

Schopf, T. J. M. 1972. Varieties of paleobiologic experience. In, Models in Paleobiology. Schopf, T. J. M. (ed). pp. 8-25. Freeman-Cooper & Co., Sam Francisco, California.

Schopf, T. J. M. 1973. Ergonomics of polymorphism: its relation to the colony as the unit of natural selection in species of the Phylum Ectoprocta. In, Animal Colonies. Boardman, R. S., Cheetham, A. H. and Oliver, W. A. (Jr.), (eds). pp. 247-294. Dowden, Hutchinson, & Ross, Stroudsburg, Pennsylvania.

Schopf, T. J. M. 1974. Permo-Triassic extinctions: relation to sea-floor spreading. J. Geol. 82, 129-143.

Schopf, T. J. M. 1977. Patterns and themes of evolution among the Bryozoa. In, Patterns of Evolution. Hallam, A. (ed). Elsevier Pub. Co. pp. 159-207.

Schopf, T. J. M. 1978 The role of biogeographic provinces in regulating marine faunal diversity thru geologic time. In, Historial Biogeography, Plate Tectonics and the Changing Environment. Gray, J. and Boucot, A. J. (eds). Biology Colloquium, Oregon State University. (In press)

Sepkoski, J. J., Jr. 1976. Species diversity in the Phanerozoic: species-area effects. Paleobiology 2, 298-303.

Simberloff, D. 1972. Models in biogeography. In, Models in Paleobiology. Schopf, T. J. M. (ed). pp. 160-191. Freeman, Cooper & Co., San Francisco, California.

Stehli, F. G. 1968. Taxonomic diversity gradients in pole location: the Recent model. In, Evolution and Environment. Drake, E. T. (ed). pp. 163-227. Yale University Press, New Haven, Connecticut.

Stehli, F. G., McAlester, A. L. and Helsley, C. E. 1967. Taxonomic diversity of Recent Bivalves and some implications for geology. Geol. Soc. Am. Bull. 78, 455-466.

Stehli, F. G. and Wells, J. W. 1971. Diversity and age patterns in hermatypic corals. Syst. Zool. 20, 115-126.

Stehli, F. G., Douglas, R. G. and Kafescioglu, I. A. 1972. Models for the evolution of Planktonic Foraminifera. In, Models in Paleobiology. Schopf, T. J. M. (ed). pp. 116-129. Freeman, Cooper & Co., San Francisco, California.

Sverdrup, H. U., Johnson, M. W. and Fleming R. H. 1942. The Oceans. Prentice-Hall, Inc.

SECTION 3

GENETIC VARIATION AND TAXONOMY

GENETIC VARIABILITY IN THE EUROPEAN EEL, ANGUILLA ANGUILLA L

E. Rodinó and A. Comparini

Istituto di Biologia Animale

Università di Padova, Italy

Introduction

The peculiar life cycle of the European eel, Anguilla anguilla L., has aroused the interest and curiosity of biologists for a long time.

Briefly (for a more complete treatment of eel biology, the works of Bertin, 1956; Deelder, 1973; Tesch, 1973 may be consulted), A. anguilla, the European fresh water eel, is a migratory catadromous fish. At the right season (on Italian coasts mainly during the winter months or in early spring) the young unpigmented individuals, called glass-eels or elvers, move to inshore waters, enter lagoons or ponds of brackish waters, or swim up the rivers in great numbers often reaching very far inland. In inland waters, elvers become pigmented and grow as "yellow eels", becoming "silver eels" after a number of years which may be ten or more. At this stage the pigmentation becomes silvery on the ventral side and there are other morphological and physiological modifications. The silver eel then begins the trip down the rivers back to the sea to reproduce. The breeding journey has no return and for a long time the problems of the migration and reproduction of the eel remained a mystery.

Many, but not all, of these problems were solved with the discovery of the eel's breeding area, through the careful and brilliant work of the Danish biologist Schmidt (1912; 1923; 1925) before and after World War I. According to his observations, all the European eels (see Fig.1 for the very

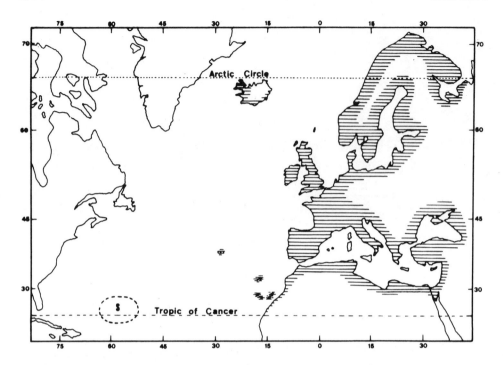

Figure 1. Anguilla anguilla. Geographic distribution (indicated by shading) along the European coasts and localization of the spawning area, S (modified from Schmidt, 1923 and Tesch, 1973).

extensive area of distribution of this species) migrate to reproduce in only one spawning area situated in the western Atlantic in the region of the Sargasso Sea. Very small larvae of $\underline{A}$. anguilla were not caught elsewhere and the spawned adults probably also perish in this area.

This spawning zone is contiguous with, and partially overlaps, that of the other Atlantic eel species: the American eel Anguilla rostrata Le Sueur. The two species are very similar but they can be distinguished on the basis of some meristic characters such as the average numbers of vertebrae (Ege, 1939); the distribution areas are also clearly separated: $\underline{A}$. rostrata is found along the East coast of North and Central America from the South-west of Greenland to the Guyanas (Tesch, 1973).

The eel larvae, or leptocephali, are carried passively by the currents from the Sargasso Sea towards the coasts; the larvae begin to metamorphose over the continental shelves, near the 1000 m. line, becoming elvers. The pelagic stage is completed in about three years by the European eel leptocephali, while American eels complete their larval cycle in only one year.

Not everyone was completely satisfied by this theory, because many problems remained, (for example, the need for European eels coming from the extremes of the distribution area to travel more than 4000 miles to reach the spawning zone); however, the most elaborated challenge to Schmidt's theory came, many years later, from Tucker (1959). In this author's opinion, European eels do not return to the remote breeding area but perish in their continental waters. Thus only American eels contribute to the breeding pool of the Atlantic eel population. A. anguilla and A. rostrata must both originate from the same gene pool and therefore not distinct species and their distinguishing characters are environmentally determined.

This hypothesis received much attention during the following years (D'Ancona, 1959, 1960; Jones, 1959; Deelder, 1960; Bruun, 1963). More recently, some evidence has accumulated in favour of the existence of two species with separate gene pools. Thus in the American eel, there is a haemoglobin polymorphism, while the European eel is monomorphic (Sick et al., 1967) and there are very significant differences in the frequency of malate dehydrogenase allozymes between samples of American and European eels (De Ligny and Pantelouris, 1973). Ohno et al. (1973) showed that the karyotype of A. rostrata is different from that reported for A. anguilla by Chiarelli et al. (1969).

Tucker's hypothesis was not the only challenge to the classic Schmidt theory. Many years before, some Italian biologists had questioned the assumption that all European eels, in particular those from the Mediterranean, make a long journey to reproduce in the Sargasso Sea; they maintained instead their belief in the presence of spawning areas in the Mediterranean (Grassi, 1914, 1919; Mazzarelli, 1914; Sanzo, 1928).

There was no real evidence in favour of this hypothesis; in fact young larval stages of eel were never caught in the Mediterranean. Sanzo (1928), however, pointed out that the investigations in this sea were not carried out as intensively as in the Atlantic ocean, and that failure to catch the younger stages is not definite proof of their total absence.

There were, moreover, some observations not in accord with the Schmidt theory: first of all, a clear demonstration

of the migration of large numbers of silver eels through
the Straits of Gibraltar towards the Atlantic Ocean was (and
still is) lacking (Ekman, 1932).   Secondly, Grassi (1914,
1919) was able to collect in the Messina strait and from the
Ionian Sea, small leptocephali 50 - 60 mm in length.   As
some of these were collected during the winter months it is
difficult to account for the presence of such animals so far
from the supposed breeding area.   The same author also reported biometrical and morphological differences in the
structure of the last ipural bone, between samples of elvers
from the North Tyrrhenian Sea (Pisa and Leghorn) and other
Mediterranean and Atlantic samples (which were reasonably
homogeneous).   He made no statistical analysis, but an evaluation of his published data shows that the observed differences are often significant.   However, such differences
might be caused by environmental effects.

A critical demonstration of genetic differences between
samples of Mediterranean eels, or of differences from Atlantic
eels, would strongly support the Grassi hypothesis of the presence of eel breeding grounds in the Mediterranean.

The Grassi hypothesis has received some support in recent
years from the discovery of differences in the phenotypic distribution of transferrin variants between samples of Mediterranean eels and between these and eels from the Atlantic
coast (Drilhon and Fine, 1971).   This evidence has, however,
been criticised on methodological grounds (Koehn, 1972).
Moreover, the use of adults for this kind of study leaves open
the possibility that the differences observed are the result
of differential selection.   There is, after all, good evidence
from fishes and other organisms, that local environmental conditions may cause differential survival of genotypes (Koehn
et al., 1971): This might be particularly relevant to the eel
which is distributed over a wide geographic area and, as an
elver, may meet a wide range of very different environments.
Genetic differences between adult and juvenile samples of the
same locality have indeed been observed by Williams et al.
(1973) in A. rostrata.

In order to find indirect evidence of the presence of
breeding areas for A. anguilla in the Mediterranean, we began
to search for differences in the frequencies of alleles at
several loci in samples of eels caught in various localities
of the Mediterranean Sea (Fig.2).   Our samples came mostly
from the Tyrrhenian coast of Italy, where the eel runs are most
abundant and particularly from Marina di Pisa at the mouth
of Arno river in Tuscany (taken as a reference station).   We

examined samples of elvers caught in consecutive years and at different seasons, to look for possible genetic heterogeneity. We obtained also one Atlantic sample from the shore of Wales near Swansea.

For consistency in assessing the genetic composition of migrating eel stocks, we examined, when possible, only samples of elvers at the beginning of their ascent of the river i.e. "recruits" following the Williams et al. (1973) terminology. In some cases, it was not possible to obtain elvers in sufficient numbers (Sardinia, Goro in the Adriatic Sea); these samples consisted mostly or in part of young yellow eels not more than 20 cm long, they were "resident" eels but not for more than a year. This was also the case for the Marseille sample. This fact does not seem to have had any important effect on our results.

Results

Our research began in the winter of 1975 with an examination of three elver samples from the Tyrrhenian Sea (Comparini et al., 1975). The chosen stations were: Marina di Pisa, Castiglione and Orbetello on the Tuscany coast and about 100 Km apart.

We first tried the enzyme systems found to be polymorphic by Williams et al. (1973) in A rostrata: Esterases, Malate dehydrogenase (MDH), Phosphoglucose isomerase (PGI) and Sorbitol dehydrogenase (SDH). All these systems are also variable in A. anguilla. We found it convenient to classify our samples at the four loci which could be most clearly and consistently scored: Mdh-2, Pgi-1, Pgi-2 and Sdh which are all polymorphic. Each of the first three loci has one allele which is much commoner than the others. The Sdh locus has two common alleles and some very rare alleles. The alleles are indicated by letters (a, b, c and so on) in the order of their respective anodic mobility. The detailed electrophoretic methods and the genetic interpretation of the complex electrophoretic patterns are reported in full elsewhere (Comparini et al., 1975, 1977).

The three samples showed considerable similarity in frequencies at all four loci and the differences were not significant. The observed zygotic distributions were also well in accord with Hardy-Weinberg expectations at all the loci, apart from slight but not significant homozygote excess at the Sdh locus. The results were interpreted as an indication that all the samples were probably part of one large,

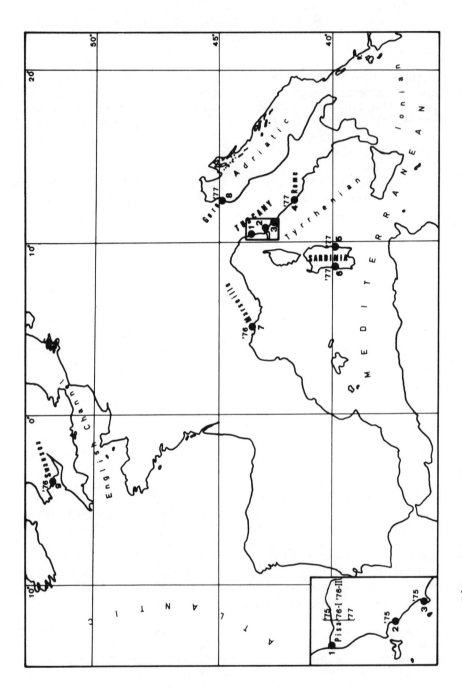

Figure 2. *Anguilla anguilla.* Localities of sampling indicating the year of capture.

genetically homogeneous population (Comparini et al., 1975).

The pooled results of the first three samples (under the heading, Pisa '75) are shown in Table 1, together with the results from samples from Marina di Pisa in two successive years. For each locus, the sample size, number of alleles, frequency of alleles and observed and expected heterozygote frequencies are given.

In addition to the four loci considered initially, the later samples were also scored for one of the Lactate dehydrogenase (Ldh) loci. Three Ldh loci are active in A. anguilla as in other Anguilliformes (Markert et al., 1975), but only one of them, is polymorphic with two alleles. The electrophoretic conditions for Ldh are the same as for Mdh (Continuous Tris-Citrate buffer, pH 7.5) and the staining scheme followed Shaw and Prasad (1970). In the winter of 1976 we caught and analyzed two samples of elvers (in early January and mid March respectively) to look for possible heterogeneity between elvers migrating up the Arno river at the beginning and at the end of the run.

Differences among the Pisa samples are very slight and not significant (see Table 2 (A)). This would indicate a substantial genetic homogeneity and stability of gene composition of the waves of elvers migrating to this shore in successive years. The Hardy-Weinberg expectations were followed closely in almost every case; a non-significant heterozygote deficit at the Sdh locus was observed in the last two samples.

The study was then broadened to include other localities and in Table 3 we report the results of analysis of five Mediterranean samples and one Atlantic sample (see Fig.2), between the autumn of 1976 and the spring of 1977. Two of the samples were from the central Tyrrhenian Sea (from the eastern coast of Sardinia and from Fiumicino, near Rome, at the mouth of the Tevere river). The northern Adriatic sample was from Goro on the Po river delta. The other Mediterranean samples came from the western coast of Sardinia and from the southern French coast, at the mouth of the Rhone river. The last sample was composed of Atlantic elvers and very young eels from a small river near Swansea (Wales).

It is interesting to note that again all the samples, including the Atlantic one, show the same kind of variability with very similar allele frequencies at all loci (Table 3). In no case is the contingency chi-square significant (Table 2 (B)). Further comparison of the pooled Pisa samples and all

TABLE 1

Anguilla anguilla. Observed numbers (N) of alleles and frequencies at 5 loci (Mdh-2, Pgi-1, Pgi-2, Sdh, Ldh-2) in different samples from Pisa. The observed and expected (in parentheses) frequencies of heterozygous individuals and the sample sizes (n) are also given for each locus.

Sample		Pisa '75		Pisa '76-I		Pisa '76-II		Pisa '77	
n		257		118		104		110	
Locus	Allele	N.	Freq.	N.	Freq.	N.	Freq.	N.	Freq.
Mdh-2	a	18	0.035	10	0.042	9	0.043	9	0.041
	b	24	0.047	9	0.038	7	0.034	10	0.046
	c	404	0.786	189	0.801	167	0.803	180	0.818
	d	12	0.023	6	0.025	8	0.038	4	0.018
	e	54	0.105	20	0.085	14	0.067	16	0.073
others		2	0.004	2	0.008	3	0.014	1	0.005
Heterozygosity		0.366 (0.367)		0.331 (0.347)		0.356 (0.346)		0.309 (0.321)	
n		239		109		96		127	
Pgi-1	a	50	0.105	21	0.096	14	0.073	22	0.087
	b	385	0.805	172	0.789	160	0.833	209	0.823
	c	41	0.086	22	0.101	17	0.089	20	0.079
others		2	0.004	3	0.014	1	0.005	3	0.012

Table 1 (continued)

Heterozygosity			0.331 (0.334)		0.312 (0.358)		0.323 (0.293)		0.331 (0.309)	
n			239		109		105		127	
Pgi-2	b		454	0.950	202	0.927	196	0.933	240	0.945
	c		22	0.046	12	0.055	11	0.052	12	0.047
	others		2	0.004	4	0.018	3	0.014	2	0.008
Heterozygosity			0.092 (0.095)		0.128 (0.137)		0.133 (0.127)		0.110 (0.105)	
n			239		98		75		127	
Sdh	b		238	0.498	101	0.515	74	0.493	141	0.555
	c		232	0.485	92	0.469	75	0.500	111	0.437
	others		8	0.017	3	0.015	1	0.007	2	0.008
Heterozygosity			0.481 (0.517)		0.531 (0.515)		0.427 (0.507)		0.449 (0.501)	
n					118		113		110	
Ldh-2	a		no		216	0.915	207	0.916	209	0.950
	b		data		20	0.085	19	0.084	11	0.050
Heterozygosity					0.136 (0.156)		0.150 (0.154)		0.100 (0.095)	

TABLE 2

<u>Anguilla anguilla</u>. Summary of results of comparisons of allelic distributions at five enzyme loci. Contingency tables, obtained by pooling rare alleles with the common one of most similar mobility, were used. The comparisons are as follows: (A) between the samples from Pisa (Table 1); (B) between the other samples (Table 3); (C) between the Pisa samples (pooled) and all the others (pooled); (D) between the sample from Swansea and all the Mediterranean ones (pooled).

Comparisons:	(A)		(B)		(C)		(D)	
	df	χ^2	df	χ^2	df	χ^2	df	χ^2
Mdh-2	12	6.24	20	15.58	4	2.95	4	5.56
Pgi-1	6	2.21	10	6.53	2	4.42	2	1.39
Pgi-2	3	1.12	5	3.72	1	0.00	1	0.61
Sdh	3	2.62	5	9.16	1	0.75	1	0.05
Ldh-2	2	2.60	5	6.20	1	2.17	1	4.13*

*Statistically significant, $P < 0.05$

TABLE 3

Anguilla anguilla. Observed numbers of alleles and frequencies at 5 loci in samples from different localities of the Mediterranean (Rome, Sardinia E, Sardinia W, Marseille, Goro), and in one sample from the Atlantic coast (Swansea). (H = heterozygosity.)

Locality		Rome		Sardinia E	
n		105		33	
Locus	Allele	N.	Freq.	N.	Freq.
Mdh-2	a	7	0.033	4	0.061
	b	12	0.057	3	0.045
	c	165	0.786	50	0.758
	d	3	0.014	3	0.045
	e	20	0.095	6	0.091
	others	3	0.014	-	0.000
H		0.390	(0.368)	0.424	(0.409)
n		105		33	
Pgi-1	a	13	0.062	4	0.061
	b	177	0.843	56	0.848
	c	20	0.095	6	0.091
	others	-	0.000	-	0.000
H		0.286	(0.276)	0.303	(0.269)
n		105		33	
Pgi-2	b	200	0.952	61	0.924
	c	7	0.033	4	0.061
	others	3	0.014	1	0.015
H		0.095	(0.092)	0.152	(0.142)

Table 3 (continued)

		n	105		27	
Sdh	b		123	0.586	32	0.593
	c		84	0.400	22	0.407
	others		3	0.014	–	0.000
H			0.476	(0.496)	0.444	(0.483)

		n	95		33	
Ldh-2	a		171	0.900	57	0.864
	b		19	0.100	9	0.136
H			0.200	(0.180)	0.273	(0.235)

Locality		Sardinia W		Marseille	
n		77		79	
Mdh-2	a	3	0.020	9	0.057
	b	5	0.032	11	0.070
	c	131	0.851	121	0.766
	d	2	0.013	5	0.032
	e	12	0.078	11	0.070
	others	1	0.007	1	0.006
H		0.272	(0.268)	0.405	(0.399)

n		77		78	
Pgi-1	a	18	0.117	10	0.064
	b	125	0.812	128	0.821
	c	11	0.071	17	0.109
	others	–	0.000	1	0.006
H		0.325	(0.322)	0.346	(0.310)

Table 3 (continued)

	n		77		78	
Pgi-2	b		144	0.935	146	0.936
	c		10	0.065	8	0.051
	others		-	0.000	2	0.013
H			0.130	(0.122)	0.128	(0.121)
	n		77		74	
Sdh	b		77	0.500	67	0.453
	c		75	0.487	80	0.540
	others		2	0.013	1	0.007
H			0.494	(0.513)	0.450	(0.503)
	n		77		79	
Ldh-2	a		144	0.935	147	0.930
	b		10	0.065	11	0.070
H			0.130	(0.122)	0.139	(0.130)

Locality			Goro		Swansea	
	n		93		103	
Mdh-1	a		7	0.038	7	0.034
	b		6	0.032	14	0.068
	c		150	0.806	155	0.752
	d		4	0.022	6	0.029
	e		17	0.091	20	0.097
	others		2	0.011	4	0.019
H			0.366	(0.339)	0.437	(0.418)

Table 3 (continued)

		n		93		101
Pgi-1	a		11	0.059	12	0.060
	b		156	0.839	172	0.851
	c		18	0.097	16	0.079
	others		1	0.005	2	0.010
H			0.269	(0.283)	0.257	(0.266)
		n		93		101
Pgi-2	b		175	0.941	191	0.945
	c		10	0.054	8	0.040
	others		1	0.005	3	0.015
H			0.108	(0.112)	0.099	(0.105)
		n		93		91
Sdh	b		103	0.554	98	0.538
	c		81	0.435	83	0.456
	others		2	0.011	1	0.006
H			0.398	(0.504)	0.538	(0.503)
		n		93		102
Ldh-2	a		170	0.914	179	0.877
	b		16	0.086	25	0.123
H			0.172	(0.157)	0.225	(0.216)

the other pooled samples, (Table 2 (C)), shows no indication of the heterogeneity hypothesized by Grassi (1914, 1919) on the basis of the differences in the last ipural bone of elvers.

The Mediterranean samples were also scored for the 6-phosphogluconate dehydrogenase (6-Pgd) locus; this locus shows a high level of variability with six alleles, three of them common and three rare. The best electrophoretic conditions for this enzyme are the same as for Ldh. The results are reported in Table 4 and again here, all the samples were very similar with no significant differences.

The results, taken together, suggest an overall genetic homogeneity among Atlantic and Mediterranean eels, confirming and extending the results obtained by de Ligny and Pantelouris (1973) for the Mdh locus.

If the Atlantic sample is compared with the pooled Mediterranean samples, the differences are again not significant (Table 2 (D)) except for the Ldh-2 locus. In this case a chi-square value is obtained of 4.13, just significant at the 0.05 level.

It is interesting to note that, even if there is no significant heterogeneity among the Mediterranean samples, there nevertheless seems to be a pattern of geographic variation at the Ldh-2 locus. Figure 3 shows a map of the Mediterranean Sea with the observed frequencies at the b allele in different localities. The frequency of this allele drops from 0.123 in the Atlantic to about 0.100 in the central Tyrrhenian Sea and then 0.073 (pooled Pisa samples), 0.070 and 0.065 gradually going to the North and then to the West and to the South.

A geographic gradient seems also to be present at the Sdh locus. The frequency of the b allele (see Fig.3) appears to decrease gradually from about 0.59 in the central Tyrrhenian to 0.514 (pooled Pisa samples) and then to 0.500 and 0.453 respectively at the W. Sardinia and Marseille samples.

Patterns of variation, on a much wider geographic scale on the American coasts, were observed, and discussed, by Williams et al. (1973) in A. rostrata. More extensive sampling on the Mediterranean coasts will be needed, however, to confirm these observations that seem to indicate a genetic differentiation in the Mediterranean eels.

Another possible indication of genetic heterogeneity is derived from the small but persistent homozygote excess at

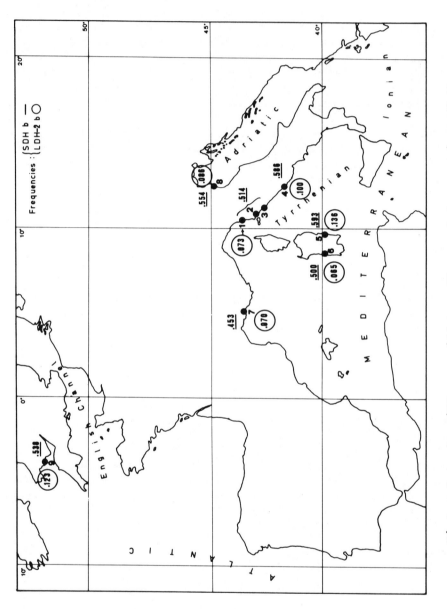

Figure 3. *Anguilla anguilla*. Observed frequencies of the b allele at the Sdh and Ldh-2 loci. Localities are numbered as in Figure 2. Pisa frequencies are for the pooled samples.

TABLE 4

Anguilla anguilla. Observed numbers and frequencies of alleles at the 6-Pgd locus in different samples from the Mediterranean Sea, and homogeneity χ^2 test.

Localities n		Pisa '77 61		Rome 75		Sardinia E 30		Sardinia W 73		Goro 93	
Locus	Allele	N.	Freq.	N.	Freq.	N.	Freq.	N.	Freq.	N.	Freq.
6-Pgd	b	29	0.238	30	0.200	15	0.250	29	0.199	34	0.183
	d	56	0.459	72	0.480	24	0.400	67	0.459	93	0.500
	f	33	0.270	46	0.307	19	0.317	46	0.315	53	0.285
	others	4	0.033	2	0.013	2	0.033	4	0.027	6	0.032
Heterozygosity		0.705 (0.659)		0.613 (0.635)		0.600 (0.676)		0.644 (0.650)		0.656 (0.634)	

$\chi^2_8 = 3.99$, $P > 0.80$ (Contingency table obtained by pooling rare alleles with the common one of most similar mobility)

the Sdh locus. In general the genotypic distributions for
the other loci, are in good accord with Hardy-Weinberg expectations, as demonstrated by the small differences between the
observed and expected heterozygote frequencies (Tables 1, 2
and 4). At the Sdh locus, however, a heterozygote deficit is
observed in eight of the nine Mediterranean samples. In
Table 5, the observed and expected zygotic distribution, for
this locus, are reported in full; the rare alleles were pooled
with the allele of most similar electrophoretic mobility for
statistical purposes (Li and Horvitz, 1953). Only in the case
of the Goro sample is the chi-square value significant at the
0.05 level, but if all the Mediterranean samples are pooled
together, the observed homozygote excess becomes significant
at the 0.01 level of probability.

The problem of homozygote excess is frequently reported
and discussed in the literature regarding the genetic variation of organisms (for marine Teleosts see Møller, 1968; for
invertebrates: Gooch and Schopf, 1970; Boyer, 1974; Tracey
et al., 1975 and many others). Excluding, for obvious reasons,
the possibility of persistent inbreeding in our case, there are
three other possible causes for this situation: (1) the
presence of a silent or "null" allele; (2) selection against
heterozygotes at the Sdh locus during the early and juvenile
stages; (3) the sampling together of individuals originating
from different populations with different gene frequencies
(the Wahlund effect).

At the moment we have no evidence in favour of the first
hypothesis. The second hypothesis, whose general implications
are discussed by Manwell and Baker (1969), cannot be excluded
although we would expect to see a stronger effect of selection
on the samples of older individuals. This is seen in the
Goro sample but not in those from Sardinia and Marseille.
Any selection effects, however, will be dependent on local environment and a different response in different localities is
not very surprising.

Thirdly, the Wahlund effect hypothesis, if true, would
require the presence of at least two eel populations in the
Mediterranean Sea differing considerably in allelic frequencies
at this locus only. One of these populations would then be
composed of Atlantic eels coming from the Sargasso Sea.
We would then expect to observe a significant genetic difference, at the Sdh locus, between our Atlantic sample and some of
those caught in the Mediterranean. In our opinion the existing data are not sufficiently clear to support this hypothesis;
while there is heterogeneity at the Sdh locus among the
Mediterranean samples, it is not so large as to become significant, and the Atlantic sample has a frequency of the Sdh

TABLE 5

Anguilla anguilla. Comparison of genotypic distributions at the Sdh locus with the Hardy-Weinberg expectations, and chi-square for tests of goodness-of-fit of observed to expected values. Rare alleles were pooled with the common one of most similar electrophoretic mobility.

Samples	genotypes	bb	bc	cc	N	χ^2
Pisa '75		64 (60.75)	113 (119.49)	62 (58.75)	239	0.71
Pisa '76-I		26 (27.06)	51 (48.87)	21 (22.06)	98	0.19
Pisa '76-II		22 (18.75)	31 (37.50)	22 (18.75)	75	2.25
Pisa '77		44 (40.25)	55 (62.49)	28 (24.25)	127	1.83
Rome		39 (37.80)	48 (50.40)	18 (16.80)	105	0.24
Sardinia-E		10 (9.48)	12 (13.03)	5 (4.48)	27	0.17
Sardinia-W		20 (19.75)	38 (38.49)	19 (18.75)	77	0.01
Marseille		17 (15.17)	33 (36.67)	24 (22.17)	74	0.74
Goro		34 (29.08)	36 (45.85)	23 (18.08)	93	4.29*
Total (Mediterranean)		276 (256.55)	417 (455.90)	222 (202.55)	915	6.66**
Swansea		25 (26.38)	48 (45.24)	18 (19.38)	91	0.34

*Statistically significant, P < 0.05
**Statistically significant, P < 0.01

b allele very close to the average of the pooled Mediterranean samples (0.538 against 0.522).

Everything considered, we may conclude that at the moment, even with the evidence of some genetic heterogeneity among Mediterranean eels, we have no definite proof of the existence of local races of eels in this sea.

Genetic Variation in Anguilla anguilla and Conger conger

To obtain further information on the genetic structure of the Mediterranean eel population, we have estimated the degree of genetic variation in a sample of Tyrrhenian elvers from Pisa. Table 6 lists the enzymes assayed (only those that were clearly resolved), the buffer system used and the best tissue source for each locus.

The results for the 20 loci we could reliably classify are given in Table 7. For each locus the number of alleles, the sample size, the observed and expected heterozygote frequencies and whether or not the locus is polymorphic is given. We have considered a locus not to be polymorphic when the frequency of observed heterozygotes was, on average, less than one in every 50 sampled individuals (frequency of the common allele higher than 0.990).

The proportion of polymorphic loci (65%) and the average heterozygosities (> 0.18) are a clear indication of the presence in this species of a very high level of genetic variation, particularly so for a vertebrate.

Comparing this result with published data on genetic variability in other Teleosts, it is interesting to note that only in one other species, the Cyprinodont Fundulus heteroclitus (Mitton and Koehn, 1975), has a similarly high level of genetic variation been observed. In the review of Powell (1975), only 5 of the 36 Teleost species for which at least ten loci were considered, showed a level of heterozygosity higher than 0.1. The mean value for the 36 species was 0.057. Selander (1976) reports an average proportion of 0.078 heterozygous loci per individual and 0.306 polymorphic loci per population in 14 fish species.

Other recently reported data on Teleosts (Utter et al., 1973; Kornfield and Koehn, 1975; Avise and Ayala, 1976; Johnson and Utter, 1976; Siebenaller, this Volume; Ward and Galleguillos, this Volume) give data on a further 39 species. The mean heterozygosities range from 0.00 (for 13 loci) in the

Cichlid <u>Ciclhasoma</u> <u>cyanoguttatum</u> (Kornfield and Koehn, 1975) and in 4 of 15 species (14 to 16 loci) studied by Johnson and Utter (1976), to 0.10 for 46 loci in the flatfish <u>Pleuronectes platessa</u> (Ward and Galleguillos, this volume). The average heterozygosity estimate, for all 39 species, is around 0.04.

All these reported values are very low in comparison with our results in <u>A</u>. <u>anguilla</u>, even though the estimates are based on data obtained with different techniques and in species often not closely related.

Anguilliformes are a group of Teleosts of great phylogenetic and comparative interest. The superorder Elopomorpha, of which they are a part, branched very early from the phyletic line leading to the higher Teleosts (Greenwood <u>et al</u>., 1966; Olson, 1971); among modern representatives of this group, Anguilliformes represent a specialized order, which however retains several primitive peculiarities (Romer, 1966).

For a better evaluation of the high level of genetic varation found in <u>A</u>. <u>anguilla</u>, a comparison with a related species would be very useful. We therefore examined a sample of another species of the Anguilliformes, the conger eel <u>Conger</u> <u>conger</u> <u>L</u>. (Congridae). This species is strictly marine, and has a biological cycle similar to that of the eel, but is more localized. The conger leptocephali migrate passively from the breeding areas to shallow waters near the coasts (Schmidt, 1911), where the juvenile individuals live and grow. The adults then migrate to reproduce in spawning areas situated in the open sea, over great depths, and like common eels, they do not return. According to Schmidt (1931), this species spawns in several areas both in the Atlantic Ocean and in the Mediterranean.

We obtained one conger eel sample of more than 50 individuals from the Tyrrhenian Sea. The results obtained from assay of 19 loci are reported in Table 8. The degree of genetic variation in <u>C</u>. <u>conger</u> (0.316 polymorphic loci and 0.076 observed mean heterozygosity) seems more or less average for a Teleost species. It is interesting to note that almost all the variability is limited, in this species, to only three loci (<u>Est-1</u> and <u>3</u>, <u>Idh-1</u>).

In Table 9 we summarize our results in the two species, considering first all of the loci tested and then the 15 homologous loci scored in both species (there were problems in establishing homologies for Esterase loci because of the differences in the electrophoretic patterns). The degree of polymorphism, mean numbers of alleles and average heterozygosities

TABLE 6

Anguilla anguilla. Enzymes assayed, number of loci scored and buffer systems used. The best tissue source for each locus are also reported. Electrophoresis was performed in 11.5% starch gel (for Esterases only, a 5.5% Bis-acrylamide gel was used).

Enzyme (E.C. No.)	Abbreviation	Buffer system(*)	Number of loci	Best tissue for locus (")
Adenosine deaminase (3.5.4.4)	Ada	A	1	l
Esterases (3.1.1)	Est	C	3	l
-Glycerophosphate dehydrogenase (1.1.1.8)	α-Gpd	A	1	m
Glucose-6-phosphate dehydrogenase (1.1.1.49)	G6pd	B	1	l
Isocitrate dehydrogenase (1.1.1.42)	Idh	A	2	l(1); m(2)
Lactate dehydrogenase (1.1.1.27)	Ldh	A	3	e,h(1); m(2); k(3)
Malate dehydrogenase (1.1.1.37)	Mdh	A	3*	e,h,m
6-Phosphogluconate dehydrogenase (1.1.1.44)	6-Pgd	A, B	1	l

Table 6 (continued)

Phosphoglucose isomerase (5.3.1.9)	Pgi	B	2	h,l(1); m(2)
Phosphoglucomutase (2.7.5.1)	Pgm	A, B	2	m
Sorbitol dehydrogenase (1.1.1.14)	Sdh	B	1	1

(*) A = Tris-citrate, pH 7.5 buffer (Comparini et al., 1975); B = Lithium hydroxide-borate, Tris-citrate, pH 8.2 discontinuous buffer system (Shaw and Prasad, 1970); C = Tris-citrate 0.025 M, pH 7.2 continuous buffer system for vertical acrylamide gel slab electrophoresis (method of Akroyd, 1967, modified).

* The activity of the m MDH (mitochondrial MDH) locus demonstrated following the method described by Whitt (1976).

(") Symbols for tissues are: e = eye, h = heart, k = kidney, l = liver, m = skeletal muscle. The numbers in parentheses correspond to the locus. Best tissues are those with the clearest resolution (not necessarily with the highest activity).

In elvers, we used commonly the post-branchial segment including several tissues (Comparini et al., 1977). Best tissues are also the same, for the loci scored, in C. conger.

TABLE 7

Anguilla anguilla. Genetic variation in a sample of elvers from the Tyrrhenian Sea.

Locus	No. of alleles	Sample size	Frequency of heterozygotes Observed	Frequency of heterozygotes Expected	P.99	P.95 (*)
Ada	3	50	0.300	0.266	+	+
Est-1	11	72	0.806	0.846	+	+
Est-2	3	72	0.028	0.028	+	–
Est-3	3	72	0.125	0.143	+	+
α-Gpd	1	50	0.000	0.000	–	–
G6pd	1	50	0.000	0.000	–	–
Idh-1	4	46	0.109	0.104	+	+
Idh-2	2	30	0.033	0.033	+	–
Ldh-1	1	110	0.000	0.000	–	–
Ldh-2	2	110	0.100	0.095	+	+
Ldh-3	1	110	0.000	0.000	–	–
sMdh-1	2	110	0.018	0.018	–	–
sMdh-2	6	110	0.309	0.321	+	+
mMdh	2	20	0.200	0.255	+	+
6-Pgd	6	61	0.705	0.659	+	+
Pgi-1	4	127	0.331	0.309	+	+

Table 7 (continued)

Pgi-2	3	127	0.110	0.105	+	+		
Pgm-1	1	35	0.000	0.000	–	–		
Pgm-2	1	35	0.000	0.000	–	–		
Sdh	3	127	0.449	0.501	+	+		
Loci 20	3.00 ± 0.54		0.181 ± 0.053	0.184 ± 0.054	13	11		
	All./locus		Average heterozygosity		(0.650)	(0.550)		

(*) A locus is considered polymorphic (+) when the frequency of the most common allele is, respectively, $\leqslant 0.990$ (P.99) or $\leqslant 0.950$ (P.95).

are in every case very much higher in A. anguilla than in C. conger.

This comparison confirms that the European eel, a species unique among other Teleosts in the peculiarities of its biological cycle, stands out also in its high level of genetic variation.

Discussion

Since the first demonstration of the presence of large amounts of protein polymorphism in natural populations (Lewontin and Hubby, 1966), a number of hypotheses have been put forth to try to explain the biological significance of this kind of polymorphism: whether it has an adaptive significance and is therefore under selective control, or whether it is selectively neutral and can simply be accounted for by the drift of neutral, or nearly neutral, mutations (Kimura and Ohta, 1971).

Our observations in A. anguilla probably do not support the hypothesis (Ayala and Valentine, 1974; Valentine, 1976; Valentine and Ayala, this volume) that presumes a positive correlation between genetic variability and the environmental and trophic resource supply stability, as exemplified in many deep-sea and tropical marine organisms. The European eel is characterized, as a species, by great temporal and spatial heterogeneity in its habitat, very long larval life with great dispersal capabilities in a very extended area, and by a peculiar ability of juvenile and adult individuals to survive and grow in very different and often extreme environmental conditions.

The presence of a high level of genetic variation in this species is perhaps more in accord with the theory (Levins, 1968) that organisms living in heterogeneous or variable environments should have more genetic polymorphism than those living in more stable conditions. This theory also predicts that different species may experience or perceive their environment as "fine-grained" or as "coarse-grained". For species of the former kind, which in general are large in size, mobile and provided with homeostatic control, the best adaptive strategy would probably be a low level of genetic variation. For organisms of the latter type, mostly small invertebrates, the opposite would be true. This theory is supported by the demonstration (Selander and Kaufman, 1973) that populations of vertebrates have, on average, a very much lower level of genetic variation, than populations of invertebrates.

Whether the eel, unlike the majority of the other vertebrates, perceives its environment as "coarse-grained", will not be discussed here, for it is liable to be an "a posteriori" approach. Of more general interest is the comparison between A. anguilla and C. conger which can be interpreted as a direct test, as proposed by Somero and Soulé (1974), of the so called "niche-variation hypothesis" which states that a high genetic variability is an adaptive strategy in a temporally and spatially heterogeneous habitat. Our results, unlike those obtained by the above mentioned authors, seem to be consistent with this hypothesis.

The same results, however, can be interpreted also from a neutralist point of view: as pointed out by Somero and Soule (1974), different amounts of polymorphism among species, may reflect differences in population size, since the number of neutral alleles that can be maintained in a population is, in theory, proportional to the effective population size N_e (Kimura and Ohta, 1971). Soulé (1976) has shown that there is, indeed, a direct correlation between heterozygosity and estimated population size in many animal species. To estimate the effective population size of a species is not an easy task (see discussion in Soulé, 1976); A. anguilla, however, most closely approaches a species with effectively almost complete panmixia. If Schmidt's theory is true, and there is a single breeding area, N_e in this species must correspond to the total number of adult breeding individuals. In A. anguilla this number must be very great and probably larger than in many other Teleost species. Certainly N_e is much smaller in C. conger. at least in the Tyrrhenian population we have sampled.

Thus it is apparent that, even without discussing the other possible hypotheses, both selectionist and neutralist arguments can be used in the interpretation of the significance of the high degree of genetic variation present in A. anguilla. It is not easy to choose between the two possibilities; they may both play a part in determining levels of variation. In this species, indeed, the large effective population is made up of the adults coming to the only breeding zone from a number of different micro-habitats spread out in a wide area of distribution. In those micro-habitats, in which the individuals have lived for many years, different genotypes have probably been selected. Then, every year, large numbers of silver eels migrate to the breeding area to re-establish the common genetic pool, thus maintaining the high genetic variability of the species.

The solution of the problem may also be approached experimentally: it should be possible to test whether samples

TABLE 8

Conger conger. Genetic variation in a sample from the Tyrrhenian Sea.

Locus	No. of alleles	Sample size	Frequency of heterozygotes Observed	Expected	P.99	P.95
Est-1	2	20	0.300	0.320	+	+
Est-2	2	40	0.075	0.072	+	–
Est-3	3	40	0.475	00.508	+	+
Est-4	1	40	0.000	0.000	–	–
α-Gpd	1	30	0.000	0.000	–	–
G6pd	3	40	0.075	0.073	+	–
Idh-1	4	46	0.457	0.507	+	+
Idh-2	1	30	0.000	0.000	–	–
Ldh-1	1	43	0.000	0.000	–	–
Ldh-2	1	43	0.000	0.000	–	–
sMdh-1	1	43	0.000	0.000	–	–
sMdh-2	1	43	0.000	0.000	–	–
mMdh	1	43	0.000	0.000	–	–
6-Pgd	1	20	0.000	0.000	–	–
Pgi-1	2	58	0.017	0.017	–	–
Pgi-2	2	58	0.017	0.017	–	–

Table 8 (continued)

Pgm-1	1	35	0.000	0.000	–	–	–
Pgm-2	1	35	0.000	0.000	–	–	–
Sdh	3	55	0.036	0.036	+	–	–
Loci 19	1.68 ± 0.22		0.076 ± 0.035	0.082 ± 0.038	6		3
	All./locus		Average heterozygosity		(0.316)		(0.158)

TABLE 9

Anguilla anguilla and Conger conger. Summary of genetic variation for all the tested loci (A) and for the homologous loci tested in both species (B)

	(A)		(B)	
	A. anguilla	C. conger	A. anguilla	C. conger
Number of loci	20	19	15	15
Average number of genes sampled per locus	152.4 $\pm$ 16.4	80.2 $\pm$ 4.8	153.1 $\pm$ 21.	82.9 $\pm$ 5.5
Average number of alleles per locus	3.00 $\pm$ 0.54	1.68 $\pm$ 0.22	2.60 $\pm$ 0.45	1.60 $\pm$ 0.25
Frequency of polymorphic loci (P.99)	0.650	0.316	0.600	0.200
Frequency of polymorphic loci (P.95)	0.550	0.158	0.533	0.067
Average heterozygosity per locus: observed	0.181 $\pm$ 0.053	0.076 $\pm$ 0.035	0.158 $\pm$ 0.054	0.040 $\pm$ 0.030
expected	0.184 $\pm$ 0.054	0.082 $\pm$ 0.038	0.160 $\pm$ 0.053	0.043 $\pm$ 0.034

of yellow and silver eels, caught in different environments or from different parts of the distribution area, show the same degree of variability and similar gene frequencies, and to compare the results with those obtained from samples of elvers. It should be remembered that there are, indeed, indications of genetic heterogeneity, among different adult samples and also between "residents" and "recruits", in **A. rostrata** (Williams et al., 1973). Other indications of genetic differentiation, among adult samples of the European eel, were also reported by Pantelouris et al. (1970) and, as indicated above, by Drilhon and Fine (1971).

We hope that our future work, confirming or refuting these indications, will help to solve this interesting problem. At present, it would seem that the high level of genetic variation found in **A. anguilla** is consistent with what is known about the peculiar life cycle of this species.

Summary

In order to test the old hypothesis that different populations or races of eels (**Anguilla anguilla**) occur in the Mediterranean Sea, several samples of elvers or young yellow eels have been analyzed by means of starch gel electrophoresis. The elvers were captured in different localities, and in successive years, in the Mediterranean; one sample came from the Atlantic Ocean (coast of Wales).

The results seem to confirm the genetic homogeneity of European eels; the Atlantic sample was not significantly different from the Mediterranean ones. Some geographical variation was, however, observed among the Mediterranean samples. On the whole, at the present time, there is no definite evidence for the presence of different races within Mediterranean eels.

We have also studied the level of genetic variation on a sample of elvers from the Mediterranean. 13 out of 20 loci assayed (65%) were found to be polymorphic. The observed average heterozygosity per locus was 0.181.

For comparative purposes we have measured the genetic variation in a sample of another species of Anguilliformes, the Conger eel **Conger conger**. This species is strictly marine with a biological cycle which is similar, but much more localized, to that of the eel. 6 out of 19 loci were found to be polymorphic (31.6%). The observed mean heterozygosity per locus was 0.076.

The possible significance of the level of genetic variation found in A. anguilla is discussed in relation to what is known about the peculiar life cycle of this species.

Acknowledgments

This research was supported by the Italian National Research Council (C.N.R.) through funds granted to the Institute of Marine Biology of Venice. Many thanks are due to those who assisted us in obtaining samples, particularly Professor M. Cottiglia (University of Cagliari); Drs. V. Scali and M. Masetti (University of Pisa); Dr. R. Rossi (University of Ferrara); Dr. R. D. Ward (University College of Swansea U.K.).

REFERENCES

Akroyd, P. 1967. Acrylamide gel slab electrophoresis in a simple glass cell for improved resolution and comparison of serum protein. Anal. Biochem. 19, 399-410.

Avise, J. C. and Ayala, F. J. 1976. Genetic differentiation in speciose versus depauperate phylads: evidence from the California Minnows. Evolution 30, 46-58.

Ayala, F. J. and Valentine, J. W. 1974. Genetic variability in the cosmopolitan deep-water ophiuran Ophiomusium lymani. Mar. Biol. 27, 51-57.

Bertin, L. 1956. Eels. A Biological Study. London, Cleaver-Hume Press.

Bruun, A. F. 1963. The breeding of the North Atlantic freshwater-eels. Adv. mar. Biol. 1, 137-169.

Boyer, J. F. 1974. Clinal and size-dependent variation at the LAP locus in Mytilus edulis. Biol. Bull. 147, 535-549.

Chiarelli, B., Ferrantelli, O. and Cucchi, C. 1969. The caryotype of some Teleostea fish obtained by tissue culture in vitro. Experientia 25, 426-427.

Comparini, A., Rizzotti, M., Nardella, M. and Rodinó, E. 1975. Ricerche elettroforetiche sulla variabilità genetica di Anguilla anguilla. Boll. Zool. 42, 283-288.

Comparini, A., Rizzotti, M. and Rodinó, E. 1977. Genetic control and variability of phosphoglucose isomerase (Pgi) in eels from the Atlantic Ocean and the Mediterranean Sea. Mar. Biol. 43, 109-116.

D'Ancona, U. 1959. Old and new solutions to the eel problem. Nature 183, 1405.

D'Ancona, U. 1960. The life cycle of the Atlantic eel. Symp. zool. Soc. Lond. 1, 61-75.

Deelder, C. L. 1960. The Atlantic eel problem. Nature 185, 589-591.

Deelder, C. L. 1973. Exposé synoptique des données biologiques sur l'anguille, Anguilla anguilla (Linnaeus) 1758. (French. Transl.) Synop. FAO sur les Peches, n.80, pp. var.

De Ligny, W. and Pantelouris, E. M. 1973. Origin of the European eel. Nature 246, 518-519.

Drilhon, A. and Fine, J. M. 1971. Les groupes de transferrines dans le genre Anguilla L. Rapp. P.-V. Reun. Cons. perm. int. Explor. Mer. 161, 122-125.

Ege, V. 1939. A revision of the genus Anguilla Shaw. A systematic, phylogenetic and geographical study. Dana Rep. 16, pp. 256.

Ekman, S. 1932. Prinzipielles über die Wanderungen und die tiergeographische Stellung des Europäischen Aaales. Zoogeographica 1, 85-106.

Gooch, J. L. and Schopf, T. J. M. 1970. Population genetics of marine species of the phylum Ectoprocta. Biol. Bull. 138, 138-156.

Grassi, B. 1914. Quel che si sa e quel che non si sa intorno alla storia naturale dell'Anguilla. Memorie R. Com. talassogr. Ital. 37, 1-50.

Grassi, B. 1919. Nuove ricerche sulla storia naturale dell'Anguilla. Memorie R. Com. talassogr. Ital. 67, 1-141.

Greenwood, P. H., Rosen, D. E., Weitzman, S. H. and Myers, G. S. 1966. Phyletic studies of Teleostean fishes with a provisional classification of living forms. Bull. Am. Mus. Nat. Hist. 131, 339-456.

Johnson, A. G. and Utter, F. M. 1976. Electrophoretic variation in intertidal and subtidal organisms in Puget Sound, Washington. Anim. Blood Grps. Biochem. Genet. 7, 3-14.

Jones, J. W. 1959. Eel migration. Nature 184, 1281.

Kimura, M. and Ohta, T. 1971. Protein polymorphism as a phase of molecular evolution. Nature 229, 467-469.

Koehn, R. K. 1972. Genetic variation in the eel: a critique. Mar. Biol. 14, 179-181.

Koehn, R. K., Perez, J. E. and Merritt, R. B. 1971. Esterase enzyme function and genetical structure of populations of the freshwater fish, Notropis stramineus. Am. Nat. 105, 51-69.

Kornfield, I. L. and Koehn, R. K. 1975. Genetic variation and speciation in New World Cichlids. Evolution 29, 427-437.

Levins, R. 1968. Evolution in Changing Environments. Princeton Univ. Press, Princeton N.Y.

Lewontin, R. C. and Hubby, J. L. 1966. A molecular approach to the study of genic heterozygosity in natural populations. II. Amount of variation and degree of heterozygosity in natural populations of Drosophila pseudoobscura. Genetics 54, 595-609.

Li, C. C. and Horwitz, D. G. 1953. Some methods of estimating the inbreeding coefficent. Amer. J. Hum. Genet. 5, 107-117.

Manwell, C. and Baker, C. M. A. 1969. Hybrid proteins, heterosis and the origin of species. I. Unusual variation of polychaete Hyalinoecia "nothing dehydrogenase" and of a quail Coturnix erythrocyte enzymes. Comp. Biochem. Physiol. 28, 1007-1028.

Markert, C. L., Shaklee, J. B. and Whitt, G. S. 1975. Evolution of a gene. Science 189, 102-114.

Mazzarelli, G. 1914. Note critiche sulla biologia dell'-Anguilla. Riv. Pesca Idrob. Anno IX (XVI), 2, 49-59.

Mitton, J. B. and Koehn, R. K. 1975. Genetic organization and adaptive response of allozymes to ecological variables in Fundulus heteroclitus. Genetics 79, 97-111.

Møller, D. 1968. Genetic diversity in spawning cod along the Norwegian coast. Hereditas 60, 1-32.

Ohno, S., Christian L., Romero, M., Dofuku, R. and Ivey, C. 1973. On the question of American eel, Anguilla rostrata, versus European eel, Anguilla anguilla. Experientia 29, 891.

Olson, E. C. 1971. Vertebrate Paleozoology. John Wiley & Sons ed., New York.

Pantelouris, E. M., Arnason, A. and Tesch, F. W. 1970. Genetic variation in the eel. II. Transferrins, haemoglobins and esterases in the eastern North Atlantic. Possible interpretations of phenotypic frequency differences. Genet. Res. 16, 277-284.

Powell, J. R. 1975. Protein variation in natural populations of animals. Evol. Biol. 8, 79-119.

Romer, A. L. 1966. Vertebrate Paleontology. Univ. Chicago Press.

Sanzo, L. 1928. Biologia dell'Anguilla. In, Atti del Convegno di Biologia marina applicata alla Pesca. (Messina, Giugno 1928), Fasc. VII, 1-16.

Schmidt, J. 1911. Biology of the eel-fishes especially of the conger. Nature 86, 61-63.

Schmidt, J. 1912. Danish researches in the Atlantic and Mediterranean on the life-history of the freshwater-eel (Anguilla vulgaris, Turt.). With notes on other species. Internat. Revue Hydrobiol. Hydrograph. 5, 317-342.

Schmidt, J. 1923. Breeding places and migrations of the eel. Nature 111, 51-54.

Schmidt, J. 1925. The breeding places of the eel. Smithsonian Rpt. for 1924, 279-316.

Schmidt, J. 1931. Eels and conger eels of the North Atlantic. Nature 128, 602-604.

Selander, R. K. 1976. Genic variation in natural populations. In, Molecular Evolution. Ayala, F. J. (ed). Sinauer Associates, Inc., Sunderland, Mass., pp. 21-45.

Selander, R. K. and Kaufman, D. W. 1973. Genic variability and strategies of adaptation in animals. Proc. Natl. Acad. Sci. U.S.A. 70, 1875-1877.

Shaw, C. R. and Prasad, R. 1970. Starch gel electrophoresis of enzymes - a compilation of recipes. Biochem. Genet. 4, 297-320.

Sick, K., Bahn, B., Frydenberg, O. Nielsen, J. T. and Von Wettstein, D. 1967. Haemoglobin polymorphism of the American freshwater eel Anguilla. Nature 214, 1141-1142.

Somero, G. N. and Soulé, M. 1974. Genetic variation in marine fishes as a test of the niche-variation hypothesis. Nature 249, 570-572.

Soulé, M. 1976. Allozyme variation: its determinants in space and time. In, Molecular Evolution. Ayala, F. J. (ed). Sinauer Associates, I c., Sunderland, Mass. 60-77.

Tesch, F. W. 1973. Der Aal. Biologie und Fischerei. Hamburg und Berlin, Verlag Paul Parey.

Tracey, M. L., Bellet, N. F. and Graven, C. D. 1975. Excess allozyme homozygosity and breeding population structure in the mussel Mytilus californianus. Mar. Biol. 32, 303-311.

Tucker, D. W. 1959. A new solution to the Atlantic eel problem. Nature 183, 495-501.

Utter, F. M., Allendorf, F. W. and Hodgins, H. O. 1973. Genetic variability and relationships in pacific salmon and related trout based on protein variations. Syst. Zool. 22, 257-270.

Valentine, J. W. 1976. Genetic strategies of adaptation. In, Molecular Evolution. Ayala, F. J. (ed). Sinauer Associates, Inc., Sunderland, Mass. 77-94.

Whitt, G. S. 1970. Genetic variation of supernatant and mitochondrial malate dehydrogenase isozymes in the Teleost Fundulus heteroclitus. Experientia 26, 734-736.

Williams, G. C., Koehn, R. K. and Mitton, J. B. 1973. Genetic differentiation without isolation in the American eel, Anguilla rostrata. Evolution 27, 192-204.

TAXONOMY INTERSPECIFIC VARIATION AND GENETIC DISTANCE IN THE PHYLUM BRYOZOA

J. P. Thorpe[1,2], J. A. Beardmore[1] and J. S. Ryland[2]

Departments of [1]Genetics and [2]Zoology

University College of Swansea, Swansea, U.K.

Introduction

Since the original studies using electrophoretic techniques to examine levels of genetic variation at enzyme loci (Lewontin and Hubby, 1966; Hubby and Lewontin, 1966; Harris, 1966), data of this sort have become available in a wide range of animal species. However the principal rationale underlying many more recent studies has lain in attempts to support or to disprove hypotheses concerning the reasons for the existence of the observed variation. A great deal of time and effort has thus been devoted by many geneticists to investigations bearing upon the selectionist-neutralist controversy. Such work is, of course, absolutely essential for better understanding of both adaptation and evolutionary change and it has, consequently, tended to dominate the study of electrophoretically resolvable variation.

A field of great relevance for studies of enzyme variation, but which has received comparatively little attention, is that of taxonomy. This neglect is particularly unfortunate because taxonomic studies, although perhaps less controversial, are likely to provide a major area for the use of electrophoresis well into the future and possibly long after the selectionist-neutralist controversy has been resolved.

Undoubtedly the most obvious role for electrophoresis in conjunction with systematics is in the identification and description of species. In most cases where the status of a 'species' is in dispute, an electrophoretic investigation of

enzyme loci is likely not only to produce an answer less open
to argument than one based on conventional morphological
criteria, but it is able to produce results more quickly.
The utility of electrophoresis for distinguishing species is
well known to geneticists and is also becoming obvious to some
taxonomists.   A less common use in systematics lies in attempts to delineate taxa above the level of species.

Close similarities between taxonomic schemes based on
interspecific enzyme variation and those based on conventional
morphological characters have led various authors to imply
that enzyme studies could be used as a basis for establishing
taxonomic relationships between species.   (Avise, 1974, 1976;
Avise, Smith and Selander, 1974a; Johnson and Selander, 1971;
Selander et al., 1971; Smith, Selander and Johnson, 1973,
Avise et al., (1974b) states "The range of electrophoretic
similarities between most congeners is ideal for making precise interspecies comparisons.   Electrophoretic data no
doubt offers a great degree of reliability...".

It is easy to accept precise figures for genetic identity
between species and on the basis of these to revise or sweep
aside existing taxonomies.   However, it should be remembered
that conventional systematics is in one sense also a branch of
genetics and is based on the study of combined effects of a
large and unknown number of genes with environmental modification.

To base interspecific taxonomy solely on electrophoresis
is tempting because of the relative ease with which quantifiab-
le data may be obtained.   However to do so one must make various assumptions which are open to question.   It is potentially
dangerous to assume that information about expected levels of
interspecific genetic identity corresponding to varying levels
of taxonomic similarity and based almost entirely on studies
of vertebrates and Drosophila can be extrapolated to cover,
for example, marine lower invertebrates.   This problem is
however insoluble while such information is lacking in the
group concerned and all one can do is to proceed with caution.
Tentative taxonomic conclusions based on interspecific genetic
identities may well be justified in groups where information
is available.

One is also confronted with almost philosophical questions.
Bearing in mind that taxonomy is in any case largely artificial
and is intended to be useful, to what extent is one justified
in a revision of the taxonomic status quo based solely on
electrophoretic data?   This applies especially to interspecific
groupings at the generic and subgeneric level where morphologi-

cally based taxonomy is likely to prove more useful even if not necessarily more correct. Another complication is that taxonomists generally happily recognise morphologically based genera whilst accepting that they differ greatly in the amount of variation which they encompass.

From this one arrives at problems of definitions of taxa. There is as yet no totally acceptable definition of the most fundamental taxon, the species, whilst very few people have even made the attempt to define a genus. From discussions with taxonomists it would appear that the primary aim in constructing a genus is to have a group of species such that each is morphologically more similar to one or more other species within the genus than to any species in any other genus. Thus to most taxonomists a genus does not encompass any preconceived amount of genetic or morphological diversity.

A further problem in the use of enzyme variation for interspecific taxonomy is the substantial errors necessarily involved in almost any estimate of genetic identity, similarity or distance between two species. Even the calculation of errors is not simple.

In this article we discuss measures of genetic similarity and suggest a method for the estimation of confidence limits for such measures. The results of an electrophoretic survey of fourteen bryozoan species in four genera are reported and are compared with data from other taxa in an attempt to assess the potential contribution of electrophoretic analysis to taxonomy at the level of species and genera.

Measures of Interspecific Variation

Various measures of genetic distance, similarity or identity are availabe (Cavalli-Sforza, 1969; Hedrick, 1971; Latter, 1972; Rogers, 1972; Nei, 1972) all of which are likely to give closely correlated results (Avise, 1974). The relative merits and disadvantages of all of these measures are reviewed by Li and Nei (1975). For all except that of Nei (1972) no measure of standard deviation or confidence limits of any kind is available, but variances are in all cases expected to be large (Li and Nei, 1975). The two commonly used measures are similarity (S) and distance of Rogers (1972) and identity (I) and distance of Nei (1972). Both are fairly time-consuming to compute, but that of Nei (1972) is probably slightly easier and is now the method in general use. It also has other major advantages, it gives a measure of genetic distance D ($= - \log_e I$) which is intended to indicate net codon differences and time over which evolutionary divergence has taken place between two populations. Also there are published methods (Nei and Roychoudhury, 1974; Li and Nei, 1975) for calculating errors in estimates of Nei's (1972) genetic distance. For taxonomic purposes precise figures for net codon differences are of little significance though they are obviously of importance in evolutionary genetics. Unfortunately the calculation of standard deviation of D is not simple and the measure is not generally included by authors of published estimates of interspecific genetic distance. A computer programme for its estimation is available (Nei and Roychoudhury, 1974). However, under most conditions, standard deviations of Nei's D are likely to be extremely large. Li and Nei (1975) state that "the standard deviations of heterozygosity and genetic distance are so large that a large number of loci must be used in estimating the average heterozygosity and genetic distance per locus". Using two theoretical pairs of speciating populations Li and Nei (1975) were able to show that to keep the standard deviation of genetic distance below 50% of the value of D over most of the range of D a minimum of 64 genetic loci must be compared in the two species. As most authors have used of the order of 20 loci in making estimates of genetic distance, it is obvious that very substantial errors must be involved.

For the taxonomist the disadvantages of Nei's measure of genetic distance may be summarised as follows:-

(1) Considerable calculation is involved.

(2) Estimates of net codon differences are not particularly useful.

(3) Errors involved are very large and not easy to calculate.

(4) Populations used must be totally outbreeding and in Hardy-Weinberg equilibrium.

For taxonomic purposes a measure of genetic similarity can be obtained simply by taking the arithmetic mean over all pairs of loci of the sums of overlapping allele frequencies at each pair of loci.

i.e. Similarity (s) for a single locus with j alleles

$$= \sum \left[\left(\frac{f1x + f1y}{2}\right) - \left(\frac{|f1x - f1y|}{2}\right)\right] + \left[\left(\frac{f2x + f2y}{2}\right)\right.$$

$$\left. - \left(\frac{|f2x - f2y|}{2}\right)\right] + \cdots \left[\left(\frac{fjx + fjy}{2}\right) - \left(\frac{|fjx - fjy|}{2}\right)\right]$$

where fjX = frequency of allele j in population X.

fjY = frequency of allele j in population Y.

Mean Similarity (S) for n loci

$$= \sum s_1 + s_2 + s_3 \cdots + s_n$$

where s_n = Similarity of nth locus.

From this confidence limits can then be calculated. Nei and Roychoudhury (1974) and Li and Nei (1975) have shown that compared to interlocus sampling errors the intralocus errors are so small that they can safely be regarded as negligible.

If all n loci were monomorphic with A of them identical between populations X and Y and B monomorphic and different:-

$$S = \frac{A}{(A + B)}$$

therefore the confidence limits of S may be set by calculating the cumulative binomial probability of obtaining the observed or any more extreme value of S given that the true value of S is some other value. The two assumed values of S (one higher and one lower) giving cumulative probabilities of obtaining the observed value of S of 0.01 will set the 99% confidence limits. Approximate standard deviations may be estimated by calculating 68% confidence limits.

For monomorphic populations this gives a reasonably accurate estimate of the confidence limits of S. For polymorphic loci with overlapping allelic distributions the deviation of s from S is likely to be less than that of monomorphic loci. How much less will depend upon the means of all the values of s above and of those below the estimate of S. Intuitively it is to be expected that most values of s for overlapping polymorphic loci will be close to the extreme values of 1 or 0. This is confirmed by Ayala (1975) using data from the Drosophila willistoni group. Using the I of Nei (1972) he concludes that a minority of loci have values of I falling in the range 0.05 - 0.95. It would seem therefore that, for polymorphic loci with overlapping allele frequencies, the mean deviation of s from S would be very similar to that of monomorphic loci. Since the percentage of polymorphic loci cannot itself be assessed from the data without large sampling errors and is likely, for most species, to be around 30%, the effects of overlapping alleles at polymorphic loci on estimated confidence limits of S will be extremely small and can therefore safely be ignored. Typically this might lead to an error of about 2% in setting the confidence limits. This is likely, however, to be of the same order as the under estimation of confidence limits caused by ignoring the effect of intralocus sampling error (Nei and Roychoudhury, 1974; Li and Nei, 1975). The two would therefore tend to counteract each other.

An estimate of genetic distance (D) can be made using:-

$$D = (1 - S)$$

The confidence limits or standard errors for D can then be estimated as for S.

It should be noted that in the special case where S = 1 a confidence limit can be set using:-

$$C = \sqrt[n]{p}$$

where C = value of S having probability p (i.e. 100(1 -p)% confidence limit) and n = number of loci used.

Similarly, if S = 0 :-

$$C = 1 - (\sqrt[n]{p})$$

An empirical comparison of errors involved suggests that the simple methods suggested here produce significantly smaller estimates of standard error or confidence limits than those of Nei and Roychoudhury (1974). This of course can make a substantial difference to the level of significance when comparing estimates of genetic distance between species.

The measurement of genetic similarity of two species as the mean over all loci of the sum of overlapping allele frequencies at each locus also eliminates other disadvantages of Nei's (1972) genetic identity. His I is non-linear with respect to allele frequencies and therefore not so easily visualised. Figure 1 shows values of S (overlapping alleles) and I for a locus (A) in two theoretical populations X and Y. In one plot of I the locus is monomorphic in population X being fixed for allele A1. In population Y two alleles A1 and A2 are present, their frequencies being varied from 0 to 1 and

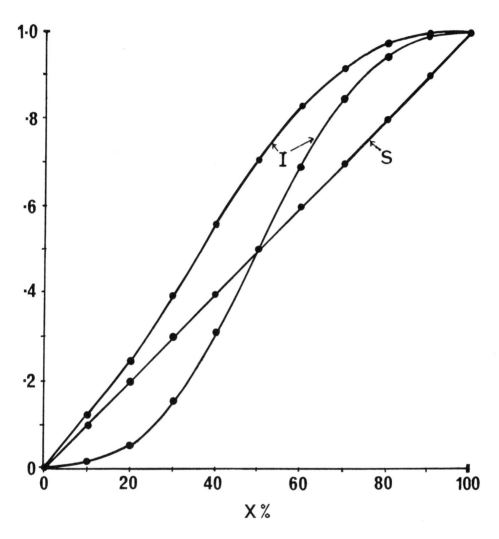

Figure 1. A comparison of values of S and I with changing frequencies of common alleles (X%) (For full explanation, see text).

1 to 0 respectively by increments of 0.1, x represents the frequency of allele A1 in population Y.

In the other plot the locus concerned is assumed to have three alleles A1, A2 and A3 of which A1 and A2 are found in population X, and A2 and A3 in population Y. It has also been assumed for convenience that the frequency of allele A2 (which is plotted as x) is variable between 0 and 1 in both populations X and Y, but whatever this frequency it is always the same in X as in Y. The frequency of allele A2 is then increased in increments of 0.1.

Intuitively it is to be expected that if the frequency of allele A1 in population Y is not varied and the number of the other alleles in population Y increases (X remaining fixed for allele A1) then the measure of similarity of the two populations should, if anything, decrease. With S based on overlapping alleles there is obviously no variation but, surprisingly, as the number of alleles increases, Nei's I tends rapidly towards unity. For example if frequencies of allele A1 in populations X and Y are 1.0 and 0.2 respectively S is constant at 0.2 irrespective of the number of other alleles in population Y. However, if the remaining 0.8 is divided equally amongst all other alleles in population Y then I with one other allele is 0.243, but increases to 0.447 with four other alleles and to 0.577 with eight others. If the number of other alleles is increased to 80 each with a frequency of 0.01 then I = 0.913.

It would seem, therefore, that for taxonomic purposes the use of a simple method of estimating genetic similarity based on overlap of allele frequencies has some advantages over Nei's I. Obviously the two are closely correlated but exactly how closely depends upon the data used. Both measures will give the same or very similar values close to 0 or 1 but for most data I is typically about 0.1 higher than S for polymorphic loci giving values of about 0.4 - 0.6. If monomorphic loci only are used I will be equal to S. Overall, unless an unusually high proportion of loci are polymorphic, values of I and S are likely to be similar with I perhaps slightly higher in the middle of the range.

In the present work interspecific comparisons of genetic similarity in bryozoans showed almost no cases of partially overlapping allele frequencies. For nearly all individual loci the value of s obtained was 1 or 0. In this case the value of S obtained would be identical to that of Nei's (1972) I.

One problem with all published measures of genetic similarity and also with that proposed here is their inability to deal with situations where very similar patterns are obtained between species (i.e. where bands on gels have similar spacing and intensity). Only bands of the same absolute mobility are considered to indicate any similarity. Examples of such patterns are those for G6PDH in pupfishes (Cyprinodon) (Fig.2) (from Turner,1974) and aminopeptidase in some species of the Bryozoan genus Alcyonidium (Fig.3). There is no immediately obvious solution to this problem.

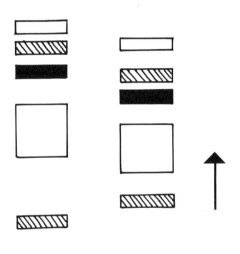

1   2

Figure 2. G6PDH in species of Cyprinodon (From Turner, 1974). Similarities of patterns are far greater than measures of genetic similarity would suggest.

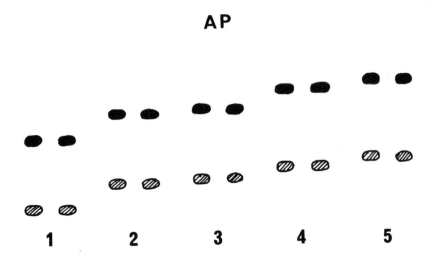

Figure 3. Aminopeptidase gel phenotypes in five species of bryozoan of the genus Alcyonidium, two samples of each. The drawing is only an approximate indication of relative mobilities and is not a picture of an actual gel. The line indicates the point of origin. Because of taxonomic difficulties the species are not named. All loci illustrated are either completely or effectively monomorphic. Rare variants in some species indicate that in each case there are probably two separate genetic loci involved.

However, the simple arithmetic measure of interspecific genetic similarity detailed above, although bearing no readily discernable relationship to net codon differences or to evolutionary time, could have some advantages for taxonomy. Both S and D are on linear scales from 1 - 0 and 0 - 1 and are therefore perhaps more easily visualised than non-linear or log scales. The other obvious advantages are that both are extremely rapid to calculate and also that with the aid of a calculator programmed for binomial probabilities, estimates of confidence limits or approximate standard errors may readily be ascertained. The estimation of errors is of paramount importance before any taxonomic use can be made of measures of interspecific distance or similarity, but methods by which it may be quickly and readily performed are not available for published measures of interspecific variation.

Similarities between systematic schemes based on genetic identity and those based on conventional morphology would appear to have led authors to overstate the potential usefulness of the former. Even with the large errors existing in estimates of genetic identity it is to be expected that most comparisons between pairs of species would, by chance, show a good correlation. However unless the number of loci used is substantially larger than that used in almost all samples published to date, little reliance can be placed upon any particular estimate of genetic distance. It should also be noted that, since for practical purposes almost the entire error depends upon the total number of different loci used, the error involved cannot be estimated by calculating genetic identities between different pairs of populations of the same two species, based on the same group of genetic loci in each case. Standard errors of such data would be based largely on genuine variation between conspecific populations and partly on intralocus sampling errors, although the latter have been shown to be small (Nei and Roychoudhury, 1974; Li and Nei, 1975).

Genetic Variation Between Species

Such work as has been published concerning interspecific enzyme variation is largely confined to arthropods (as represented by the genus *Drosophila*) and sundry species of vertebrates.

The first major investigation of interspecific enzyme variation was that of Hubby and Throckmorton (1968) using twenty-seven species of *Drosophila*. They found congeneric non-sibling species to differ at between 71% and 92.1% of their loci (mean 81.7%) whilst sibling species differed at between 14.3% and 77.5% (mean 47.6%).

A large scale survey over several years by Ayala and various co-workers (largely summarised by Ayala *et al.*, 1974) of the *Drosophila willistoni* group has shown a mean genetic identity, I, (Nei, 1972) in conspecific populations of 0.970, of 0.795 for semispecies, of 0.517 for sibling species and 0.352 for non sibling species.

In vertebrates many papers are available on interspecific

similarity. Avise and Smith (1974) find similarities, S (Rogers, 1972), of 0.37-0.79 between 10 teleost species of the genus Lepomis. A similar range of 0.31-0.89 is found in kangaroo rats (11 species) of the genus Dipodomys (Johnson and Selander, 1971). These have a range in conspecific populations of 0.92-1.0.

Gorman and Kim (1976) found 12 congeneric species (assuming A. marmoratus and A. lividus to be conspecific) of lizards (Anolis) to have genetic identities (I) of 0.57-0.94 and conspecific populations to have identities of 0.92-0.997. However work on four different species of the same genus (Webster, Selander and Yang, 1972) produced congeneric similarities in the range 0.16-0.29 and only 0.69-0.82 for conspecific populations.

Several other vertebrate genera, for example teleosts, Menidia (Johnson, 1975; frogs, Hyla (Case, Haneline and Smith, 1975); mice, Peromyscus (Kilpatrick and Zimmerman, 1975; Avise, Smith and Selander, 1974a; Avise et al., 1974b), Mus (Selander, Hunt and Yang, 1969), Thomomys (Patton, Selander and Smith, 1972) and Geomys (Selander et al., 1975); rats, Sigmodon (Johnson et al., 1972), Dipodomys (Johnson and Selander, 1971) and Rattus (Patton, Yang and Myers, 1975) have been found to give genetic identities for conspecific or interspecific populations similar to those already quoted.

Information about interspecific variation in invertebrates other than Drosophila is in short supply and not always within the range which might be predicted. Genetic identity between various populations of two alleged species of the snail Cerion was found to be 0.9895-0.9999. On the basis of this information Gould, Woodruff and Martin (1974) concluded that only one species was involved. A similarity of only 0.217 between two populations of the millipede Scoterpes copei led Laing, Carmody and Peck (1976) to conclude that speciation had taken place. Snyder and Gooch (1973) found conspecific genetic identities of 0.62-0.97 for the winkle Littorina saxatilis and 0.87-0.99 for another marine gastropod Nassarius obsoletus.

In a review covering most of the work then available on interspecific variation in animals, Avise, Smith and Ayala (1975) concluded that genetic similarities for most pairs of congeneric species lay in the range 0.30-0.80 with a mean of 0.50-0.60. For conspecific populations the values were far higher, usually higher than 0.90.

A similar range of values has also been for conspecific and interspecific genetic variation in plant species (Clegg and Allard, 1972; Gottlieb, 1973a,b, 1974). Babbel and Selander (1974) surveyed variation in two plant genera. Lupinus texensis was found to have a genetic identity between conspecific populations of 0.90-0.98 and L. subcarnosus of 0.93-1.0. Between these species the identity was 0.35. Hymenopappus scabiosaeus and H. artemisiaefolius had conspecific I values of 0.97 and 0.94 respectively with an interspecific identity of 0.896.

A survey of the literature suggests that a large proportion of interspecific congeneric genetic identities falls in the range 0.25-0.85 whilst conspecific values are generally above 0.90. These conclusions are in close agreement with those of Avise, Smith and Ayala, (1975).

Few authors give information about intergeneric genetic identities. Calculations from the data of Johnson and Selander (1971) give an I of 0.16 between two genera (Dipodomys and Perognathus) of myomorph rodents. The data of King and Wilson (1975) give an identity of 0.538 between chimpanzee (Pan) and man (Homo) which leads them to question the existing taxonomy. The genetic identities for all six possible pairs of four related genera of deep sea asteroids were found to be 0.000, 0,002, 0,002, 0,009, 0,076 and 0,257 (Ayala et al., 1975). Avise, Smith and Ayala (1975) find a most surprisingly high intergeneric genetic identity of 0.948 between the freshwater teleosts Lavinia exilicauda and Hesperoleucus symmetricus. Since the two species could not be distinguished at 23 of 24 loci it would seem most unlikely that they are correctly placed in different genera. If this latter example is then disregarded it appears that intergeneric genetic identities mainly fall in the range 0-0.3 or thereabouts.

In order to attempt to assess the potential utility of electrophoretic assays for interspecific taxonomy, estimates of genetic similarity and distance have been made for various pairs of bryozoan species using starch gel electrophoresis. Details of the practical aspects of this work are published elsewhere (Thorpe, 1977; Thorpe and Beardmore (in preparation)).

Comparisons were made between three bryozoan species of the cheilostome genus Schizomavella, nine species of the ctenostome genus Alcyonidium, and also between the two cheilostome genera Membranipora and Electra using the species

M. membranacea and E. pilosa. In Bryozoa the predominantly low levels of intraspecific polymorphism substantially ease the problems of interspecific comparisons. Results for Schizomavella and Alcyonidium are given in Tables 1 and 2 respectively. Taxonomic difficulties make it impossible to

TABLE 1

Schizomavella. Genetic distance (D) above diagonal, Genetic similarity (S) (or identity I) below diagonal. Figures in brackets are 99% confidence limits for each measurement. All estimates are based on a sample of nine loci.

	S. linearis	S. hastata	S. auriculata
S. linearis	1.	0.000 (0.401)	1.0 (0.599)
S. hastata	1.0 (0.599)	-	1.0 (0.599)
S. auriculata	0.000 (0.401)	0.000 (0.401)	-

give specific names to all the species of Alcyonidium. Four of these are therefore distinguished using letters (a) - (d), (Taxonomy based on morphology is difficult in Alcyonidium since the absence of a mineralized skeleton deprives the taxonomist of characters conventionally utilised in other Bryozoa). It should also be noted that the A. albidum used is that of Prenant and Bobin (1956) and not that of Hincks (1880).

Based on a sample of nineteen loci the genetic similarity between Electra pilosa and Membranipora membranacea was estimated as 0.279. The 99% confidence limits of this value are approximately 0.10-0.57. The equivalent value of I (Nei, 1972) is 0.284.

Implications of Interspecific Variation in Bryozoans

It was predicted that the genetic similarity for most pairs of congeneric species should fall in the range 0.25-0.85. This is clearly not the case for the thirty-six

TABLE 2

Genetic similarities between species pairs in *Alcyonidium*. Figures in brackets are approximate 68% confidence limits.

	A. gel.	A. sp.a	A. hirs.	A. sp.b	A. sp.c	A. sp.d	A. pol.	A. var.
A. albidum	0.016 (0.00– 0.31)	0.000 (0.00– 0.21)	0.000 (0.00– 0.21)	0.000 (0.00– 0.24)	0.000 (0.00– 0.21)	0.000 (0.00– 0.21)	0.000 (0.00– 0.21)	0.000 (0.00– 0.21)
A. gelatinosum		0.417 (0.29– 0.69)	0.000 (0.00– 0.17)	0.000 (0.00– 0.24)	0.000 (0.00– 0.17)	0.000 (0.00– 0.18)	0.000 (0.00– 0.09)	0.000 (0.00– 0.15)
A. sp. (a)			0.000 (0.00– 0.11)	0.000 (0.00– 0.24)	0.000 (0.00– 0.17)	0.000 (0.00– 0.18)	0.000 (0.00– 0.10)	0.000 (0.00– 0.15)
A. hirsutum				0.600 (0.39– 0.77)	0.000 (0.00– 0.17)	0.571 (0.39– 0.73)	0.000 (0.00– 0.12)	0.000 (0.00– 0.15)
A. sp. (b)					0.000 (0.00– 0.24)	0.400 (0.23– 0.61)	0.000 (0.00– 0.24)	0.000 (0.00– 0.24)
A. sp. (c)						0.000 (0.00– 0.18)	0.000 (0.00– 0.18)	0.000 (0.00– 0.18)
A. sp. (d)							0.000 (0.00– 0.18)	0.000 (0.00– 0.18)
A. polyoum								0.222 (0.13– 0.39)

measures of similarity for the genus <u>Alcyonidium</u> of which only four (11.1%) fall within this range. Thirty species pairs showed no similarity at all and the mean of 0.06218 is well outside the predicted range.

In <u>Schizomavella</u> none of the three figures falls within the range. However there is some evidence (to be published elsewhere) to suggest that <u>S</u>. <u>linearis</u> and <u>S</u>. <u>hastata</u> may be the same species. In this case only one measure of genetic similarity would be involved and this could hardly be regarded as an adequate sample size from which to draw any conclusions.

In any case the overall fit of interspecific genetic distances within the two genera to the predicted range is negligible. In the genus <u>Alcyonidium</u> the mean genetic similarity is highly significantly less than 0.25 ($P \ll 0.001$), although because of the interlocus sampling errors involved no single figure differs significantly (at $P < 0.01$) from the expected range.

In the intergeneric comparison of <u>Membranipora</u> and <u>Electra</u>, the observed result ($S = 0.279$) is within the expected range of 0 - 0.3. Although only a single result and subject to large errors, this value does suggest that <u>Electra</u> <u>pilosa</u> and <u>Membranipora</u> <u>membranacea</u> are more closely related than many of the species pairs within the genus <u>Alcyonidium</u>.

This apparently anomalous situation is however quite acceptable taxonomically since, as has already been pointed out, to many taxonomists a genus implies no particular predetermined amount of variation even within a single group. It therefore presents no problem if the genus <u>Alcyonidium</u> encompasses a greater range of genetic variation than the genera <u>Electra</u> and <u>Membranipora</u> combined. In the past such situations were inevitable but whether they are desirable is open to debate. The study of enzyme variation by electrophoretic and allied methods provides a means by which attempts may be made to quantify intrageneric genetic variation.

The results reported here suggest that whilst in the future it may be possible, using genetic distance, to suggest consistent criteria for the definition of genera, attempts to do this should be limited to groups for which a substantial body of published data is available.

Summary

The great value of electrophoretic methods for progress in taxonomy is underlined. Information available on the extent of genetic differences between conspecific populations, between congeneric species and between species of related genera is briefly reviewed. These differences can be quantified by a variety of measures of genetic identity, similarity or distance. An examination of the sampling errors involved in making estimates of genetic similarity shows that the errors depend almost entirely upon the number of loci examined and are likely to be substantial. An additional problem is that these errors are also extremely difficult to estimate with any accuracy.

A simplified measure of genetic distance is presented by the use in which it is possible to estimate confidence limits relatively easily.

Expected levels of genetic similarity between congeneric species are compared with estimates derived from two genera of bryozoans (nine and three species respectively). The fit in both cases is found to be poor. However, the value for two species from different genera falls within the predicted range. The consequences of such findings for taxonomy are briefly discussed.

Acknowledgements

We wish to thank the Natural Environment Research Council for financial support which enabled this work to be carried out.

REFERENCES

Avise, J. C. 1974. Systematic value of electrophoretic data. Syst. Zool. 23, 465-481.

Avise, J. C. 1976. Genetic differentiation during speciation. In, *Molecular Evolution*. Ayala, F. J. (ed). pp. 106-122. Sinauer Associates, Massachusetts.

Avise, J. C. and Smith, M. H. 1974. Biochemical genetics of sunfish. II. Genic similarity between hybridising species. Am. Nat. 108, 458-472.

Avise, J. C., Smith, M. H. and Ayala, F. J. 1975. Adaptive differentiation with little genic change between two native California minnows. Evolution 29, 411-426.

Avise, J. C., Smith, M. H. and Selander, R. K. 1974a. Biochemical polymorphism and systematics in the genus *Peromyscus*. VI. The boylii species group. J. Mammal. 55, 751-763.

Avise, J. C., Smith, M. H., Selander, R. K., Lawlor, T. E. and Ramsey, P. F. 1974b. Biochemical polymorphism and systematics in the genus *Peromyscus*. V. Insular and mainland species of the subgenus *Haplomylomys*. Syst. Zool. 23, 226-238.

Ayala, F. J. 1975. Genetic differentiation during the speciation process. Evol. Biol. 8, 1-78.

Ayala, F. J., Tracey, M. L., Hedgecock, D. and Richmond, R. C. 1974. Genetic differentiation during the speciation process in *Drosophila*. Evolution 28, 576-592.

Ayala, F. J., Valentine, J. W., Hedgecock, D. and Barr, L. G. 1975. Deep sea asteroids: high genetic variability in a stable environment. Evolution 29, 203-212.

Babbel, G. R. and Selander, R. K. 1974. Genetic variability in edaphically restricted and widespread plant species. Evolution 28, 619-630.

Case, S. M., Haneline, P. G. and Smith, M. F. 1975. Protein variation in several species of *Hyla*. Syst. Zool. 24, 281-295.

Cavalli-Sforza, L. L. 1969. Human diversity. In, *Proceedings of the XIIth International Congress of Genetics*

(Tokyo), 3. 405-416.

Clegg, M. T. and Allard, R. W. 1972. Patterns of genetic differentiation in the slender wild oat species Avena barbata. Proc. Natl. Acad. Sci. U.S.A. 69, 1820-1824.

Gorman, G. C. and Kim, Y. J. 1976. Anolis lizards of the Eastern Carribean: A case of study in evolution. II. Genetic relationships and genetic variation in the bimaculatus group. Syst. Zool. 24, 62-77.

Gottlieb, L. D. 1973a. Genetic differentiation sympatric speciation and the origin of a diploid species of Stephanomeria. Am. J. Bot. 60, 545-553.

Gottlieb, L. D. 1973b. Enzyme differentiation and phylogeny in Clarkia franciscana, C. rubicunda and C. amoena. Evolution 27, 205-214.

Gottlieb, L. D. 1974. Genetic confirmation of the origin of Clarkia lingulata. Evolution 28, 244-250.

Gould, S. J., Woodruff, D. S. and Martin, J. P. 1974. Genetics and morphometrics of Cerion at Pongo Carpet: a new systematic approach to this enigmatic land snail. Syst. Zool. 23, 518-535.

Harris, H. 1966. Enzyme polymorphisms in man. Proc. Roy. Soc. B. 164, 298-310.

Hedrick, P. W. 1971. A new approach to measuring genetic similarity. Evolution 25, 276-280.

Hincks, T. 1880. British Marine Polyzoa. Van Voorst, London.

Hubby, J. L. and Lewontin, R. C. 1966. A molecular approach to the study of genic heterozygosity in natural populations. I. The number of alleles at different loci in Drosophila pseudoobscura. Genetics 54, 577-594.

Hubby, J. L. and Throckmorton, L. H. 1968. Protein differences in Drosophila IV. A study of sibling species. Am. Nat. 102, 193-205.

Johnson, M. S. 1975. Biochemical genetics of the atherinid genus Menidia. Copeia 1975, 662-691.

Johnson, W. E. and Selander, R. K. 1971. Protein variation

and systematics in Kangaroo rats (genus Dipodomys). Syst. Zool. 20, 377-405.

Johnson, W. E., Selander, R. K., Smith, M. H. and Kim. Y. J. 1972. Biochemical genetics of sibling species of the cotton rat (Sigmodon). Univ. Texas Publ. 7213, 297-305.

Kilpatrick, C. W. and Zimmerman, E. G. 1975. Genetic variation and systematics of four species of mice of the Peromyscus boylii species group. Syst. Zool. 24, 143-162.

King, M. C. and Wilson, A. C. 1975. Evolution at two levels. Molecular similarities and biological differences between humans and chaimpanzees. Science 188, 107-116.

Laing, C. D., Carmody, G. R. and Peck, S. B. 1976. How common are sibling species in cave inhabiting invertebrates? Am. Nat. 110, 184-189.

Latter, B. D. H. 1972. Selection in finite populations with multiple alleles. III. Genetic divergence with centripetal selection and mutation. Genetics 70, 475-490.

Lewontin, R. C. and Hubby, J. L 1966. A molecular approach to the study of genic heterozygosity in natural populations II. Amount of variation and degree of heterozygosity in natural populations of Drosophila pseudoobscura. Genetics 54, 595-609.

Li, W. and Nei, M. 1975. Drift variances of heterozygosity and genetic distance in transient states. Genet. Res. 25, 229-248.

Nei, M. 1972. Genetic distance between natural populations. Am. Nat. 106, 283-292.

Nei, M. and Roychoudhury, A. K. 1974. Sampling variances of heterozygosity and genetic distance. Genetics 76, 379-390.

Patton, J. L., Selander, R. K. and Smith, A. H. 1972. Genic variation in hybridising populations of pocket gophers (genus Thomomys). Syst. Zool. 21, 263-275.

Patton, J. L., Yang, S. Y. and Myers, P. 1975. Genetic and morphologic divergence among introduced rat populations (Rattus) of the Galapagos archipelago, Ecuador.

Syst. Zool. 24, 296-310.

Prenant, M. and Bobin, G. 1956. Faune de France (Vol 60 (1)). Lechavelier, Paris.

Rogers, J. S. 1972. Measures of genetic similarity and genetic distance. Univ. Texas Publ. 7213, 145-153.

Selander, R. K., Hunt, W. G. and Yang, S. Y. 1969. Protein polymorphism and genic heterozygosity in two European subspecies of the house mouse. Evolution 23, 379-390.

Selander, R. K., Kaufman, D. W., Baker, R. J. and Williams, S. L. 1975. Genic and chromosomal differentiation in pocket gophers of the Geomys bursarius group. Evolution 28, 557-564.

Selander, R. K., Smith, M. H., Yang, S. Y., Johnson, W. E. and Gentry, J. B. 1971. Biochemical polymorphism and systematics in the genus Peromyscus. I. Variation in the old-field mouse (Peromyscus polionotus). Univ. Texas Publ. 7103, 49-90.

Smith, M. H., Selander, R. K. and Johnson, W. E. 1973. Biochemical polymorphism and systematics in the genus Peromyscus III Variation in the Florida deer mouse (Peromyscus floridanus), a Pleistocene relic. J. Mammal. 54, 1-13.

Snyder, T. P. and Gooch, J. L. 1973. Genetic differentiation in Littorina saxatilis (Gastropoda). Mar. Biol. 22, 177-182.

Thorpe, J. P. 1977. Unpublished thesis, University of Wales.

Turner, B. J. 1974. Genetic divergence of Death Valley pupfish species: biochemical versus morphological evidence. Evolution 28, 281-294.

Webster, T. P., Selander, R. K. and Yang, S. Y. 1972. Genetic variability and similarity in the Anolis lizards of Bimini. Evolution 26, 523-535.

THE SYSTEMATICS AND EVOLUTION OF MYTILUS GALLOPROVINCIALIS Lmk

R. Seed

Department of Zoology

University College of North Wales, Bangor, U.K.

Introduction

Marine mussels are widely distributed and their economic and ecological importance has made them the subject of considerable research. The Family Mytilidae includes several genera though the systematics of this group are still surprisingly confused due to the lumping together of species of apparently different origins on the basis of superficial similarities brought about through convergence.

Since first described by Lamarck the systematics of Mytilus galloprovincialis have been extensively discussed. Whilst regarded by some authorities as a subspecies of the larger M. edulis complex, by others it is considered to be a distinct species (see Lubet, 1973). It occurs principally in the Mediterranean, Adriatic and Black Seas, but extends northwards as far as N. France, Ireland and S.W. Britain. M. edulis, by contrast, ranges from the Arctic to N. Africa but is largely absent from the Mediterranean.

This paper compares these mytilids from numerous diverse habitats throughout Europe and considers the systematics and evolution of M. galloprovincialis in the light of such information.

Differences between M. galloprovincialis and M. edulis

Overall shell morphology: Mytilus with its inequivalve,

wedge-shaped shell is ideally suited to a gregarious, epifaunal way of life. Overall shape in mytilids, however, is subject to considerable phenotypic variation (Plate 1) and even distantly related mussels from similar habitats can appear superficially similar.

Densely packed mussels are proportionately more elongate with higher length to height ratios than mussels from less crowded conditions (Table 1b). They also frequently orientate such that their posterior margins are directed away from the substratum (Fig.1). Contact with surrounding mussels thus reduces growth in shell height and the emphasis here is on increase in length in an attempt to elevate the posterior feeding currents above neighbouring conspecifics. Environmental control of shape (and growth) can be demonstrated experimentally by transplanting mussels from one habitat to another (Fig.2).

Shape also varies with size (=age). Mussels become progressively wider and more elongate (Table 1) and old individuals are often incurved giving the shell a pronounced hooked appearance (Seed, 1968). Maximum width in younger mussels occurs approximately midway between dorsal and ventral margins but in old animals is displaced ventrally so that in cross section the under-surface of the shell appears exceedingly flattened (Fig.1).

These trends in morphology are exhibited by both M. edulis and M. galloprovincialis resulting in considerable convergence. Care is therefore required when using overall appearance in mytilid taxonomy. However, differences between these mussels are sometimes quite marked especially in low level populations protected from wave action. Typically, M. galloprovincialis is more pointed with a taller more gently rounded shell and a steeper angle between ventral and ligamentary margins than M. edulis of comparable size; M. galloprovincialis shells are also much flatter ventrally (Fig.3). At high tidal levels, where many older mussels occur, and more especially in the high density populations typical of exposed shores, shell characteristics of these mussels can merge until identification on external morphology becomes difficult or even impossible.

Mantle edge pigmentation and shell colour. Mantle colour is typically purple-violet in M. galloprovincialis and yellow-brown in M. edulis. Intermediate shades do occur and colour in M. edulis from N.E. England varies from grey to deep purple. Most Mediterranean specimens are deeply pigmented and M. edulis grown in strong light seems to deposit pigment suggesting that pigmentation may be protective. Analysis of

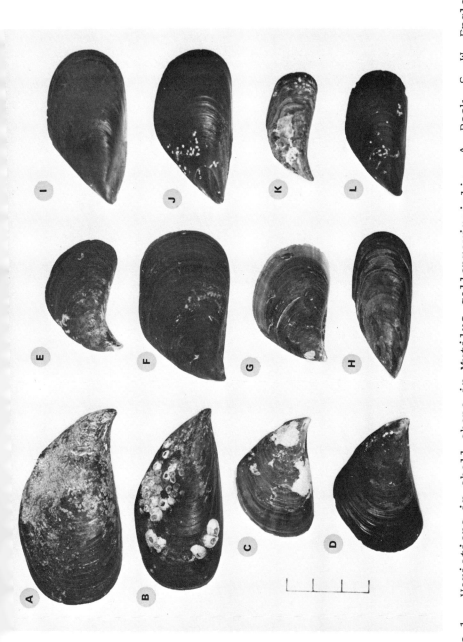

Plate 1 Variations in shell shape in Mytilus galloprovincialis. A. Rock, S. W. England. B. Newquay, S. W. England. C. N. Africa. D. Newquay, S. W. England. E. Castletownbere, S. W. Ireland. F. Marseille. G. Arcachon. H. Kilkee, W. Ireland. I. Marseille. J. Naples. K. Ardgroom, S. W. Ireland. L. Naples.

TABLE 1

Changes in shell proportions of Mytilus according to age and density.

	n	Density	Trends in shell ratios Small → Large*		Approx. age and size of largest mussels	
			Length:Height	Height:Width	yrs.	cm.
a) Exposed shore (Yorkshire) M. edulis						
High level	864	Moderate	1.82 → 2.16	1.31 → 0.98	17+	3.5
Low level	180	High	1.89 → 2.17	2.12 → 1.18	4-5	3.5
b) Sheltered shore (Yorkshire) M. edulis						
High level	980	Moderate	1.67 → 2.27	1.54 → 0.88	17+	5.5
Low level	720	High	1.92 → 2.20	1.63 → 1.12	9-10	5.5
Low level	204	Low	1.59 → 1.90	1.80 → 1.25	11-12	6.5
c) Rock (Cornwall)						
M. edulis	125	Low	1.47 → 2.09	—	—	9.0
M. gallopro-vincialis	125	Low	1.54 → 2.23	—	—	9.0

*The range indicated by arrows implies the trend from small to large mussels. Large mussels are usually longer and wider. Small mussels measured 1.0cm. in length; large mussels were largest specimens in the sample.

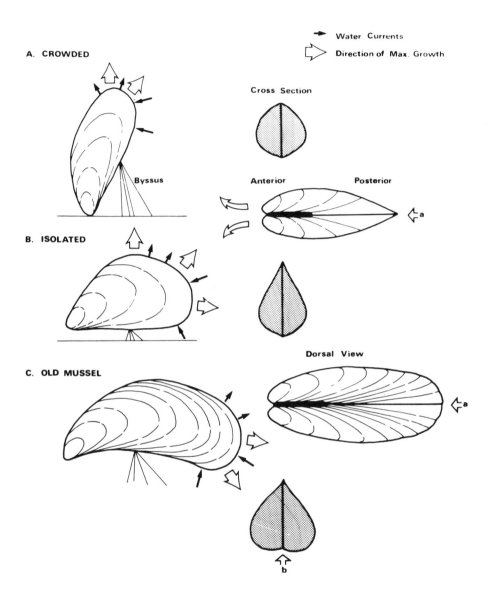

Figure 1. Effects on crowding and old age on shell shape in Mytilus.

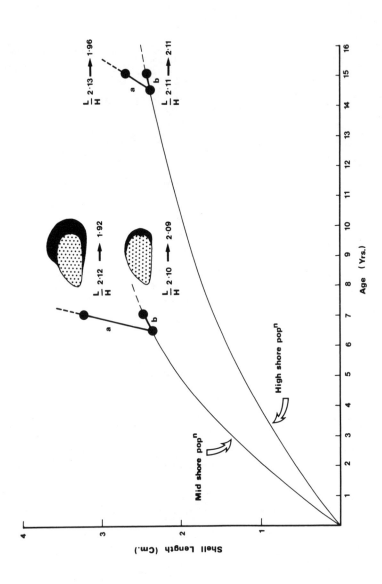

Figure 2. Changes in shell shape and growth over one growing season in two exposed shore populations of M. edulis after transplantation to cages at low levels in a sheltered estuary. a = experimental animals, b = controls, L:H = length/height.

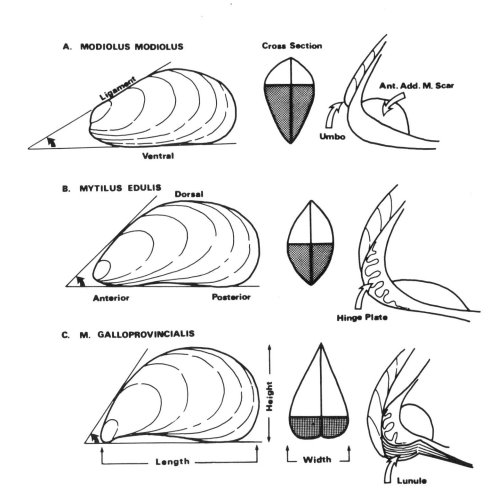

Figure 3. Shell characteristics of A) Modiolus modiolus B) Mytilus edulis C) M. galloprovincialis.

mantle pigments, however, has failed to reveal any marked differences (Seed, 1971). Shell colouration and raying are of limited taxonomic value especially on exposed shores where shells may be badly eroded.

Infection with Pinnotheres; growth. Mussels at Rock in S.W. England differ significantly in their infection with Pinnotheres pisum. M. edulis is consistently more heavily infected though no seasonal trends are evident (Fig.4). Subtle differences must therefore exist between these mussels but precisely what these might be is at present unclear. Infested mussels are in poorer condition and have heavier shells than non-infested mussels of similar size (Seed, 1969). Gill damage caused by the parasite impairs filtering efficiency and probably leads to poorer growth. It is significant therefore that the largest mussels at Rock are M. galloprovincialis whilst M. edulis of similar length have heavier shells (Lewis & Seed, 1969).

Growth experiments conducted on the N.E. coast of England using Rock mussels showed that M. edulis grew much faster than M. galloprovincialis under the experimental regime though this might be reversed at the higher temperatures prevailing in S. W. England. These experiments suggest that basic physiological differences exist between these mussels (see also Lubet & Chappuis, 1966).

Protein electrophoresis. The potentials of biochemical taxonomy have been recognised for some time and electrophoresis has facilitated the separation of several closely related species where identification has otherwise proved stubborn. Myogens from Rock mussels revealed slight but consistent differences when separated on polyacrylamide gels (Seed, 1971). More recently, however, several polymorphic enzyme systems in Irish mussel populations have shown considerable variability (Murdock et al., 1975) suggesting that identification using biochemical techniques may not always be possible throughout the entire geographical range of these mytilids. Since genetic variability in Mytilus is discussed elsewhere in this volume nothing further will be said here concerning this important field of research which should ultimately provide us with a clearer understanding of the selective forces operating on mussel populations.

Adductor muscle scars; hinge plate. (Plate 2) Anterior (and posterior) scars in M. galloprovincialis are consistently smaller than in M. edulis (Lewis & Seed, 1969; Seed, 1972; 1974). Figure 5 shows the frequency distribution of mean scar length to shell length ratios in mussels from

SYSTEMATICS AND EVOLUTION OF *M. galloprovincialis* Lmk

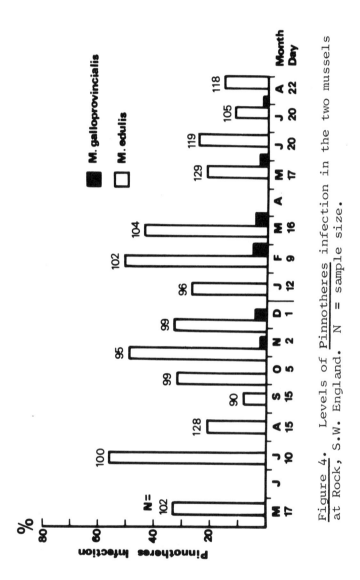

Figure 4. Levels of <u>Pinnotheres</u> infection in the two mussels at Rock, S.W. England. N = sample size.

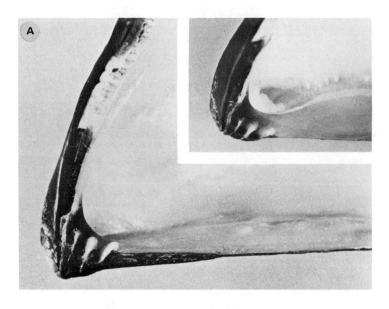

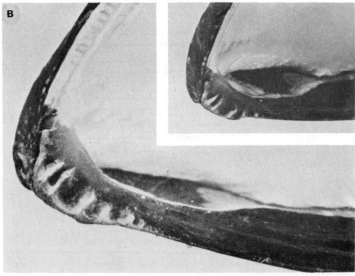

Plate 2   The anterior end of type specimens of A. *M. galloprovincialis* and B. *M. edulis*. Note differences in hinge plate and anterior adductor muscle scars. Both specimens are 5.8cm in shell length. (From Lewis & Seed, 1969).

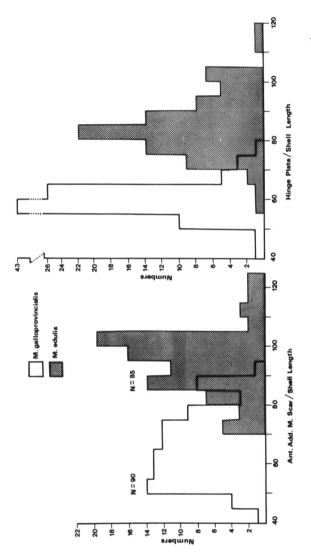

Figure 5. Size frequency distributions of the mean adductor/length and hinge plate/length ratios ($\times 10^3$) of M. edulis and M. galloprovincialis from numerous sites in Europe. N = No. of populations sampled.

numerous sites along the Atlantic, Channel and Mediterranean coasts. Although some overlap does occur, the trend to larger scars in M. edulis is quite pronounced. The adaptive significance of this is unclear though the early evolution of M. galloprovincialis in the relatively tideless less waveswept conditions of the Mediterranean may have been contributory. M. diegensis, which occurs in sheltered Californian bays also has small adductors and is similar in other respects to M. galloprovincialis.

Typically the hinge plate in M. edulis is a gently curving structure whereas in M. galloprovincialis it describes a tighter arc with its posterior end more clearly delimited from the adjacent ventral valve margin (Fig.3). Variations in the position of the lunule, hinge plate and ventral shell margin relative to the umbones modify the degree of pointedness of the anterior end to which reference has earlier been made. Apart from differences in its shape and position the hinge plate is again consistently smaller in M. galloprovincialis (Fig.5).

Reproduction. The annual reproductive cycles of the Rock populations are illustrated in Figure 6. Several differences are immediately apparent. Spawning in M. edulis starts in March-April, when sea temperature is about $9^{\circ}C$ and continues until early July. M. galloprovincialis, however, does not spawn until August when temperatures are maximal. The cyclical pattern in this mussel, here approaching its northern limits, is also less pronounced than in M. edulis. These differences are discussed in detail elsewhere (Seed, 1971) but it is concluded that extensive hybridisation between these populations seems unlikely. Cross fertilisation can, however, be induced and normal veligers have been successfully reared in the laboratory.

Discussion

The degree of variation in M. galloprovincialis and M. edulis throughout much of Europe is now well documented. What then is the status of M. galloprovincialis? To answer this question we must first define the term "species". This has proved surprisingly difficult and most definitions have their limitations. The morphological species, defined on the basis of character discontinuity is rather subjective and of little value for cryptic species. The biological species stresses lack of hybridisation and whilst permitting a more precise definition it presents problems for asexual, apomictic and fossil species. The solution may be to rec-

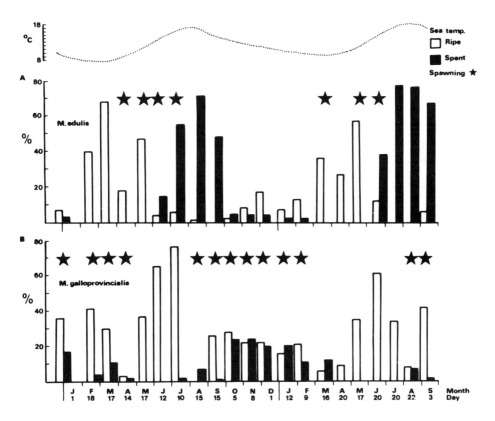

Figure 6. Reproductive cycles of A) <u>M. edulis</u> and B) <u>M. galloprovincialis</u> at Rock, S.W. England. Asterisks denote periods of above average spawning activity.

ognise several types of species (Ross, 1974) - the concept
applying to one type not necessarily applying to others.

Differences between *M. galloprovincialis* and *M. edulis*
seem to be sufficiently pronounced in certain localities to
justify their separation as distinct morphological species.
Furthermore, at Rock comparatively little gene flow appears
to occur between the two populations by virtue of their
different breeding cycles. Interbreeding can, however,
occur and the degree of hybridisation may vary annually.
These mussels tend to be most distinctive in protected habitats
(harbours, estuaries) where local temperature and tidal conditions might lead to enhanced reproductive isolation when
compared with outer coast populations living in waters which
are more thoroughly mixed. Mussels exhibiting intermediate
characteristics are encountered and these show an interesting
geographic distribution (Fig.7). They abound in W. Ireland
and Brittany, regions which are similar climatically. On
the Channel coasts and further south in the Bay of Biscay
the two mussels are generally more distinctive. Such trends
may reflect seasonal variations in spawning periods. At
Rock, for example, interbreeding is largely precluded and
at Arcachon breeding cycles are also known to differ (Lubet,
1959). Genetic isolation perhaps only occurs to any great
extend in these mussels towards the edges of their respective
distributions where spawning periods may be broadly out of
phase. Elsewhere, spawning periods might progressively
overlap resulting in a higher incidence of hybrids. At
present this remains speculative though garter snakes in
N. America can form good isolated species in parts of their
range whilst freely interbreeding elsewhere. Ricklefs
(1973) suggests that far from straining the biological species
concept this simply emphasises the problems of extending the
concept geographically. Populations are not uniform throughout their range and there is no reason to believe that the
degree of reproductive isolation between populations must be
the same wherever they coexist.

Barsotti and Meluzzi (1968) consider that *M. galloprovincialis* has only recently evolved from the *M. edulis* stocks
originally present both on Atlantic and Mediterranean coasts
(Fig.8). The warmer conditions which developed in the
Mediterranean and the reduced contact between the Atlantic
and Mediterranean favoured the differentiation of these stocks -
a process which is probably still in progress.

Many of the differences in shell characteristics between
M. edulis and *M. galloprovincialis* are extensions of those
changes occurring between the more primitive modiolids and

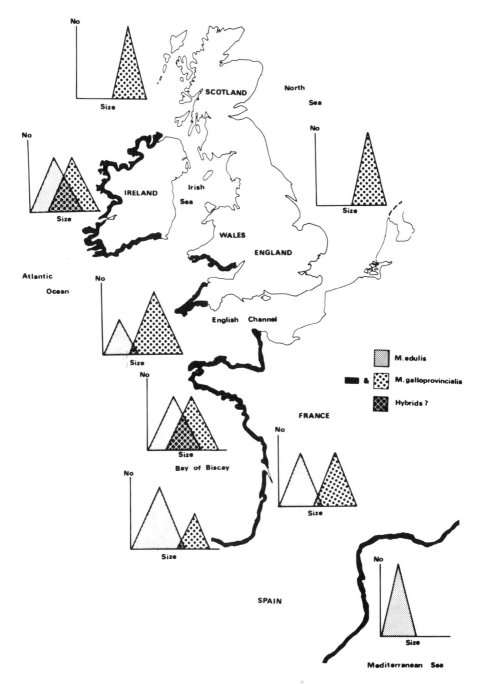

Figure 7. Schematic size frequency distributions of anterior adductor scar/shell length ratios of M. edulis and M. galloprovincialis from several regions in Europe. Note relative frequencies of intermediate forms (= hybrids?). Geographical distribution of M. galloprovincialis indicated by thick black line.

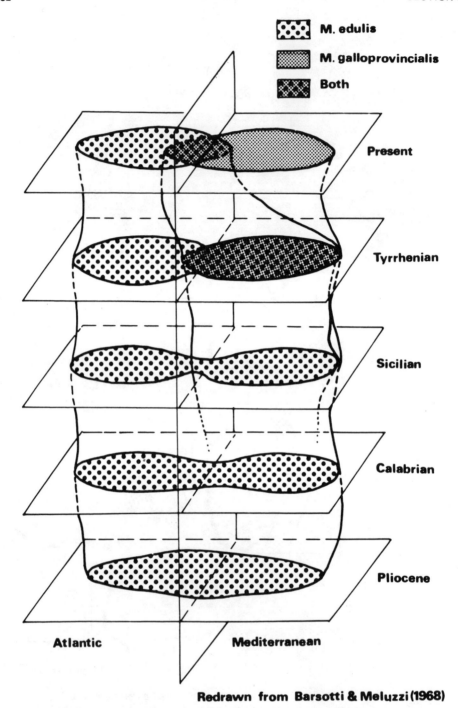

Figure 8. Probable phylogeny of <u>Mytilus edulis</u> L. and <u>M. galloprovincialis</u> Lmk.

the more recent mytilids (Fig.9). The Modiolus-Mytilus lineage represents a transition from an endobyssate to an epibyssate life style (Stanley, 1972). Changes accompanying this transition are specialisations for stronger byssal attachment and physical stabilisation (see also Yonge, 1953; Yonge & Campbell, 1968). During this transition the umbones become more pointed and terminal, the anterior adductor is reduced in size and the shell assumes a more wedge-shaped profile. Byssus retractor muscles move forward to provide a powerful downward force for more effective anchorage. The development of a hooked or recurved shell is also considered to be a specialisation for epifaunal life. In endobyssate forms maximum shell width lies at or dorsal to the mid dorso-ventral axis but in epibyssate species it moves ventrally to provide a broader base for attachment (Fig.10). Many of these characteristics associated with epifaunal life are also particularly obvious in M. galloprovincialis. They also become increasingly pronounced in larger (= older) specimens of both M. edulis and M. galloprovincialis suggesting that these features do indeed have an adaptive (and survival) value. Increase in length to height ratios with age could also be adaptive since elongate shells have more surface area in contact with the substratum thereby enhancing stability even further.

Stasek (1966) suggests that posterior expansion may have evolved in tropical waters to increase food intake. M. galloprovincialis, which evolved in warmer seas, consistently has taller more triangular shells than M. edulis. However, this trend is also seen in both mytilids in protected habitats where feeding conditions are generally more favourable. This probably reflects the less crowded conditions in such habitats. Nevertheless, elongate (i.e. less tall) shells do probably provide less resistance and could therefore prove advantageous in exposed situations.

Thus whilst presenting an interesting taxonomic problem, the relationship between M. edulis and M. galloprovincialis is of further interest in that it provides additional information concerning the possible evolutionary trends within this group. To understand the process of adaptation and evolution of these mussels more fully it is essential that we know more about their ecological-genetical relationships over their entire range. This field of research is interdisciplinary and the results multidirectional, systematics is vitally important to the ecologist whilst ecology has much to contribute to systematics.

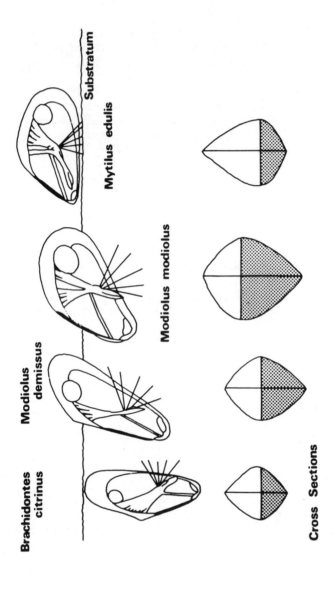

Figure 9. Grades of evolution in Recent Mytilidae.

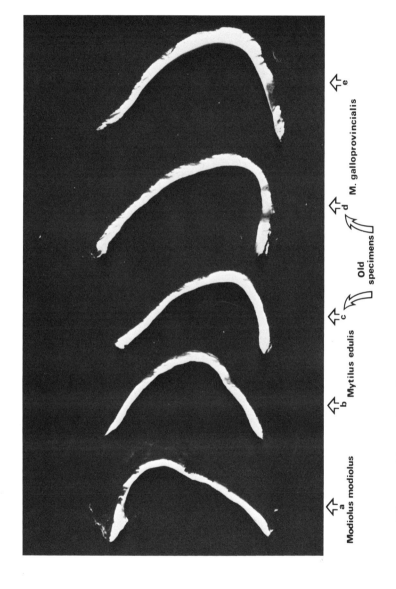

Figure 10. Sections through the shells of <u>Modiolus</u> (endobyssate) and <u>Mytilus</u> (epibyssate). Note the similarity between c and e. The ventral shell margin in <u>Mytilus</u> becomes progressively flatter with increasing age (c and d).

Summary

This paper examines the systematics and evolution of _Mytilus galloprovincialis_ in the light of the differences known to exist between this mussel and _M. edulis_. Both are exceedingly variable and individuals from similar habitats exhibit substantial convergence.

In certain localities reproductive isolation has resulted in the evolution of two quite distinctive mussels. Elsewhere, however, it appears that varying degrees of interbreeding may occur and the concomitant mixing of individual characteristics can then make the separation of these mussels difficult or even impossible.

M. galloprovincialis is of relatively recent Mediterranean origin but whether it is best regarded as a true species or a subspecies of _M. edulis_ perhaps depends not only on which definition of the species is used but also on which part of its geographical range is being considered. The systematics of these mytilids emphasises the problems inherent in obtaining a satisfactory and practical species definition, especially when the concept is extended geographically.

Differences in shell characteristics between _M. edulis_ and _M. galloprovincialis_ are also considered in relation to the changes which are thought to have accompanied the transition from a more primitive endobyssate to a more recent epibyssate way of life within the Mytilidae.

REFERENCES

Barsotti, G. & Meluzzi, C. 1968. Osservazioni su *Mytilus edulis* L. e *M. galloprovincialis* Lmk. Conchiglie 4, 50-58.

Lewis, J.R. & Seed, R. 1969. Morphological variations in *Mytilus* from south-west England in relation to the occurrence of *M. galloprovincialis* Lmk. Cahiers de Biologie Marine 10, 231-253.

Lubet, P. 1959. Recherches sur le cycle sexuel et l'émission des gamètes chez les Mytilidae et les Pectinidae (Moll. Bivalves). Revue des travaux de l'Office (scientifique et technologique) des Pêches maritimes 23, 387-548.

Lubet, P. 1973. Exposé synoptique des données biologique sur la moule *Mytilus galloprovincialis* (Lamarck 1819). Synopsis FAO sur les pêches. No. 88. (SAST-Moule, 3, 16(10), 028, 08, pag.var.) FAO, Rome.

Lubet, P. & Chappuis, J. 1966. Etude du débit palleal et de la filtration par une méthode directe chez *Mytilus edulis* et *M. galloprovincialis*. Bulletin de la Société linnéenne de Normandie 10, 210-216.

Murdock, E.A., Ferguson, A. & Seed, R. 1975. Geographical variation in leucine aminopeptidase in *Mytilus edulis* L. from the Irish coasts. Journal of experimental marine Biology and Ecology 19, 33-41.

Ricklefs, R.E. 1973. *Ecology*. Thos. Nelson & Sons Ltd.

Ross, H.H. 1974. *Biological Systematics*. Addison Wesley Publishing Co. Incorporated.

Seed, R. 1968. Factors influencing shell shape in the mussel *Mytilus edulis*. Journal of the Marine Biological Association, U.K. 48, 561-584.

Seed, R. 1969. The incidence of the pea-crab *Pinnotheres pisum* in the two types of *Mytilus* (Mollusca:Bivalvia) from Padstow, south-west England. Journal of Zoology 158, 413-420.

Seed, R. 1971. A physiological and biochemical approach to the taxonomy of *Mytilus edulis* L. and *M. galloprovincialis* Lmk. Cahiers de Biologie Marine 12, 291-322.

Seed, R. 1972. Morphological variations in *Mytilus* from the French coasts in relation to the occurrence and distribution of *M. galloprovincialis* Lmk. Cahiers de Biologie Marine 13, 357-384.

Seed, R. 1974. Morphological variations in *Mytilus* from the Irish coasts in relation to the occurrence and distribution of *M. galloprovincialis* Lmk. Cahiers de Biologie Marine 15, 1-25.

Stanley, S.M. 1972. Functional morphology and evolution of byssally attached bivalve molluscs. Journal of Palaeontology 46, 165-212.

Stasek, C.R. 1966. Views on the comparative anatomy of the bivalved molluscs. Malacologia 5, 67-68.

Yonge, C.M. 1953. The monomyarian condition in the Lamellibranchia. Transactions of the Royal Society of Edinburgh 62, 443-478.

Yonge, C.M. & Campbell, J.I. 1968. On the heteromyarian condition in the Bivalvia with special reference to *Dreissena polymorpha* and certain Mytilacea. Transactions of the Royal Society of Edinburgh 68, 21-43.

GENETIC AIDS TO THE STUDY OF CLOSELY RELATED TAXA OF THE GENUS MYTILUS

D. O. F. Skibinski, J.A. Beardmore and M. Ahmad

Dept. of Genetics, University College of Swansea

Swansea, United Kingdom

Introduction

The great value of gel electrophoresis as a tool in experimental studies of evolutionary processes is now well accepted by geneticists. Its use in taxonomy is perhaps less well appreciated though the taxonomic importance of allozyme data was recognised by Hubby and Throckmorton in 1965. The advantages and disadvantages of electrophoretic data in systematics have been summarised by Avise (1974) and elsewhere in this volume Thorpe et al. discuss an application of this approach.

Good evidence to satisfy the essential biological criterion of specific distinction can be obtained if individuals in a population are classified into two morphological types or putative species which are then shown to have non-overlapping allele frequency distributions at a number of loci. The absence of heterozygotes for different alleles each of which is common in one of the two types, but uncommon in the other, can be interpreted as evidence against hybridisation.

The geographic distributions of, and the morphological and physiological differences between, the two mussels Mytilus edulis and Mytilus galloprovincialis have been described by Lewis and Seed (1969) and Seed (this volume). The amount of phenotypic overlap in morphology was found to vary among sympatric populations in West Ireland, South West England and Brittany, possibly reflecting different levels of hybridisation. The study of such closely related or incipient species is of obvious interest in relation to theories of speciation, and in particular to the importance of the contact zone in the development of reproductive isolation between populations which have diverged allopatrically.

Ahmad and Beardmore (1976) have shown that M. edulis and M. galloprovincialis differ significantly in allele frequency distributions at three polymorphic loci. The greatest differences were found for the Aminopeptidase (Ap) and Phosophohexose isomerase (Phi) loci. Skibinski et al. (1978) showed further very large differences at two other loci, Esterase-D (Est-D) and Leucine aminopeptidase-1 (Lap-1), in two populations in South West England and presented evidence for the existence of hybrid individuals.

This paper describes a genetic analysis of populations showing varying degrees of morphological overlap.

Materials and Methods

The method of preparing starch gels and mussel extracts, and the staining schedules are described fully in Ahmad et al. (1977). The main populations investigated were at Croyde in North Devon, Robin Hood's Bay near Whitby in North East England, and King's Dock in Swansea, South Wales. Mussels were collected from the mid-littoral zone in the first two localities. The King's Dock population is constantly submerged and mussels were scraped from the sides of small jetties. At Croyde the substrate is rock and the population extends from high to low tide levels. At Robin Hood's Bay, the mussels are situated on flat scars of rock running at right angles to the shore line; two sites (RHB I and RHB II) 300-350 m apart and both confined to the mid-littoral zone were sampled. There was no intentional bias in the choice of mussels at any sampling site, though as individuals less than one cm in length give lower enzyme activity, these were rejected.

A morphological analysis was made of samples from Croyde, RHB I, and King's Dock using the characters which Lewis and Seed (1969) found most useful in separating the two forms. All mussels collected from RHB II were, morphologically, typical M. edulis. The characters used are the overall shell contours, the relationship of the umbo to the lunule, the size and shape of the hinge plate, the size and shape of the anterior adductor muscle scar, the degree of inrolling of the ventral edge of the shell, and the extent of pigmentation in the mantle. For each character each mussel was given a score according to the degree of resemblance to M. edulis or M. galloprovincialis. Usually a score of 0 was given if the individual was M. edulis - like, a score of 2 for M. galloprovincialis - like and a score of 1 for intermediates. For each mussel the scores for the different characters were then summed in an unweighted way. Typical M. edulis individuals thus have low, and typical M. galloprovincialis individuals high, total scores on the resulting index. This method of analysis involving a degree of subjective judgement is possibly less satisfactory than a strictly quantitative approach. Certain important aspects of the discriminatory characters are, however, difficult to quantify.

Within all three populations there was no indication of a pattern (for example, clumping) in the spatial distributions of the two forms within the small areas (about 30 cm^2) from which samples of mussels were removed.

Results

For some time we have been studying geographic variation at six polymorphic loci (Ap, Phi, Lap-1, Lap-2, Phosphoglucomutase (Pgm) and Est-D) in British mussel populations. With three exceptions, no clearly significant geographic heterogeneity in allele frequencies or in any other genetic variable has yet been observed between populations extending from Liverpool in the North West, along the coasts of Wales and southern and eastern England, to just south of the Scottish border. This situation contrasts to some extent with that on the East coast of North America, where Koehn et al. (1976) have observed sharp clines at the Lap-2 locus, sometimes over quite small distances, and to that in Ireland where Murdock et al. (1975) have observed considerable geographic heterogeneity at the Lap-1 locus. The three exceptions to the geographic homogeneity are the South West England populations, RHB I, and King's Dock, where the allele frequencies at a number of loci and particularly at Lap-1 and Est-D show large differences from the fairly uniform values elsewhere. In each of these localities M. galloprovincialis - like individuals have been observed. M. galloprovincialis could have become established in the North East of England, which is certainly outside its normally accepted range (Seed, this volume), by a single migratory event: the close proximity of the Robin Hood's Bay Marine Laboratory may be relevant here, although Lewis and Seed (1969) have observed high frequencies of mussels with deeply pigmented red-brown mantles low on the shore on Filey Brigg which is further south along the coast.

At Croyde and RHB I typical M. edulis and M. galloprovincialis individuals could be identified, the main difference between the populations being in the mantle colour of the M. galloprovincialis types which was dark purple or violet at Croyde and dark red brown at RHB I. Intermediates and individuals showing all combinations of M. edulis and M. galloprovincialis characters were present. For all characters except mantle colour and adductor scar shape and size, M. edulis and M. galloprovincialis types could be recognised unequivocally at King's Dock and the majority of individuals resembled M. galloprovincialis more closely. But most individuals had the large adductor scar and the light coloured mantle characteristic of M. edulis in Britain. Taking all characters together, the King's Dock mussels were thus somewhat intermediate with no individuals like the extreme M. edulis and M. galloprovincialis found in the other two populations. Quantification of the morphological variation within and between populations is certainly desirable and will be attempted in future studies.

Table 1 gives allele frequencies at the Est-D and Lap-1 loci for Croyde, RHB I, RHB II, and King's Dock together with corresponding values for Ilfracombe (North Devon) and Mumbles (Swansea Bay). Some of the data for Croyde in this and other tables are taken from Skibinski et al. (1978). The allele frequencies at Mumbles may be taken as representative of British M. edulis; the population at Ilfracombe near Croyde in North Devon seemed to consist solely of M. galloprovincialis. Also given are allele frequencies for the extremes of the morphology score distributions (E stands for the M. edulis-like extreme, G for the M. galloprovincialis-like extreme) for the Croyde, RHB I and King's Dock populations. At both loci the allele frequencies for these three populations are intermediate between Mumbles M. edulis and Ilfracombe M. galloprovincialis. At the Est-D locus allele 4 is at high frequency in M. edulis while allele 2 is at high frequency in M. galloprovincialis. Lap-1 alleles 2 and 3 are at high frequency in M. edulis, alleles 5 and 6 at high frequency in M. galloprovincialis. At Croyde the E and G allele distributions are very different with little overlap and are similar to the Mumbles and Ilfracombe distributions respectively. At RHB I the differences between the E and G groups are in the same direction as at Croyde but are less marked. The E group, which consisted of individuals which were typical M. edulis, is not as extreme in terms of the frequency of the most common M. edulis alleles as either Croyde E or Mumbles and many in-

TABLE 1

Allele frequencies at the Est-D and Lap-1 loci. N is number of individuals typed. n is the number of individuals picked at the morphological extremes for typing. Allele 7 has highest and allele 1 lowest anodal mobility. E - M. edulis-like individuals; G = M. galloprovincialis-like individuals

Est	N	Allele 1	2	3	4	5	6	7
Croyde	762	0.013	0.324	0.006	0.633	0.007	0.017	–
RHB I	166	0.021	0.425	0.006	0.539	0.003	0.006	–
RHB II	119	0.020	0.040	–	0.890	–	0.050	–
King's Dock	205	0.005	0.356	0.002	0.602	0.002	0.032	–
Mumbles	1000	0.006	0.027	0.001	0.937	0.005	0.023	0.002
Ilfracombe	34	–	0.868	0.015	0.118	–	–	–
	n/N							
Croyde E	166/790	0.006	0.018	0.003	0.955	0.003	0.015	–
Croyde G	37/790	–	0.932	0.041	0.027	–	–	–
RHB I E	43/166	–	0.209	0.012	0.767	–	0.012	–
RHB I G	36/166	0.084	0.556	0.014	0.347	–	–	–
King's Dock E	32/205	–	0.406	0.016	0.516	–	0.063	–
King's Dock G	35/205	0.029	0.386	–	0.571	–	0.014	–

Lap-1	N	1	2	3	4	5	6	7
Croyde	762	0.011	0.146	0.507	0.035	0.106	0.186	0.007
RHB I	166	0.012	0.102	0.377	0.039	0.205	0.253	0.012
RHB II	119	–	0.280	0.660	0.040	0.010	–	–
King's Dock	205	0.002	0.146	0.339	0.042	0.268	0.193	0.010
Mumbles	1000	0.018	0.196	0.726	0.040	0.017	0.004	–
Ilfracombe	34	–	–	–	0.074	0.456	0.456	0.015
	n/N							
Croyde E	166/790	0.015	0.191	0.733	0.045	0.015	–	–
Croyde G	37/790	–	–	0.068	0.041	0.297	0.541	0.054
RHB I E	43/166	0.012	0.151	0.593	0.023	0.093	0.116	0.012
RHB I G	36/166	0.014	0.042	0.236	0.042	0.278	0.375	0.014
King's Dock E	32/205	–	0.125	0.344	0.031	0.328	0.172	–
King's Dock G	35/205	–	0.186	0.414	0.014	0.186	0.186	0.014

dividuals had characteristic M. galloprovincialis genotypes. At King's Dock there is no significant difference in allele frequency distribution between the morphological extremes. The allele frequencies at RHB II are more similar to Mumbles M. edulis and to other sampled populations of M. edulis in the North East of England than to RHB I and RHB I E. Our data indicate that the difference between these two RHB populations is seen in all size classes above 1 cm.

At Croyde, where the difference between the morphological extremes is most marked, alleles at both loci can be classified according to their relative frequencies in groups E and G. Est-D alleles 1, 2 and 3 and Lap-1 alleles 5, 6 and 7 are relatively high frequency in M. galloprovincialis and will be called G alleles. At each locus these G alleles can be combined to form a compound allele (called G_{Est} at Est-D, and G_{Lap} at Lap-1). Similarly the remaining alleles at each locus will be called E alleles and can be combined to form a second compound allele (E_{Est} at Est-D and E_{Lap} at Lap-1). Further analysis is performed using these compound alleles. Table 2 gives observed and expected numbers of E_{Est}/G_{Est} and E_{Lap}/G_{Lap}

TABLE 2

Observed and expected numbers of heterozygotes at Est-D and Lap-1. F is Wright's fixation index.

Est-D

	N	Observed	Expected	F	$F\sqrt{N}$	P
Croyde	762	136	343	0.6037	16.66	<0.001
RHB I	165	64	82	0.2159	2.77	<0.01
King's Dock	205	103	95	-0.0859	-1.23	n.s.

Lap-1

	N	Observed	Expected	F	$F\sqrt{N}$	P
Croyde	762	127	321	0.6041	16.68	<0.001
RHB I	165	52	82	0.3669	4.71	<0.001
King's Dock	205	95	102	0.0700	1.00	n.s.

heterozygotes in random samples. F, which is Wright's (1951) fixation index, is used to measure deviation from Hardy-Weinberg expectation. F equals $(4ac-b^2)/(2a+b)(2c+b)$ where a and c are the observed numbers in the two different homozygote classes and b is the observed number of heterozygotes. In a population in Hardy-Weinberg equilibrium $F = 0$, +ve and -ve values of F indicate deficiencies and excesses of heterozygotes respectively. The sampling variance of F has been derived by Brown (1970) and, in a test of the null hypothesis $F = 0$, $F\sqrt{N}$ is normally distributed with unit variance. At both loci the Croyde and RHB I populations show significant deviations from Hardy-Weinberg expectation with heterozygote deficits. There are no significant deviations in the King's Dock sample though Est-D shows a non-significant excess of heterozygotes. At both loci there are directional changes in the value of F from high values at Croyde through intermediate values at RHB I to low values around zero at King's Dock.

The coefficient of linkage disequilibrium (Lewontin and Kojima, 1960) between non-alleles E_{Est} and E_{Lap} is given by

$$D = g_{EE}\, g_{GG} - g_{EG}\, g_{GE}$$

where g_{EE}, g_{EG}, g_{GE} and g_{GG} are the gametic frequencies for the two loci. D cannot be calculated for the array of two locus genotype frequencies since coupling and repulsion heterozygotes cannot be distinguished by the electrophoretic techniques used. A value may be estimated by maximum likelihood if random mating is assumed (Hill, 1974), but this assumption is inappropriate here in view of the large deviations from Hardy-Weinberg expectations. The numbers of individuals in the nine different two locus genotypic classes are given in rows 1 to 3 of Table 3. The value of D would lie between +0.164 ($P < 0.001$ for test of null hypothesis that there is no disequilibrium) and +0.123 ($P < 0.001$) at Croyde, between + 0.091 ($P < 0.001$) and + 0.012 (n.s.) at RHB I, and between + 0.063 ($P < 0.001$) and - 0.054 ($P < 0.001$) at King's Dock. So D is positive and must be significantly different from zero at Croyde but might, or might not, be significant at RHB I and King's Dock.

Although D cannot be measured easily, it is possible to test for, and measure, the strength of association between genotypes at the two loci. In a 2 x 2 contingency table this can be done using the product moment correlation coefficient (R). Thus for the table:

	Y_1	Y_2	
X_1	g_1	g_2	$n_1 = g_1 + g_2$
X_2	g_3	g_4	$n_2 = g_3 + g_4$
	$m_1 = g_1 + g_3$	$m_2 = g_2 + g_4$	N

$$R = \frac{g_1 g_4 - g_2 g_3}{\sqrt{n_1 \times n_2 \times m_1 \times m_2}}$$

For each population, R. has been calculated for the homozygote classes alone (data in rows 3 to 6 of Table 3) and for the combined homozygotes with heterozygotes (rows 7 to 9). Also shown in Table 3 are chi-squared and probability values for tests of independence in the 2 x 2 tables. The first test shows a very large and highly significant excess of the E_{Est}/E_{Est} E_{Lap}/E_{Lap} and G_{Est}/G_{Est} G_{Lap}/G_{Lap} double homozygotes at Croyde. At RHB I this effect remains but R is smaller indicating a looser association. At King's Dock there is no association and R is close to zero. The second test shows a significant excess of the E_{Est}/G_{Est} E_{Lap}/G_{Lap} double heterozygotes at Croyde. At RHB I, R is smaller and the chi-squared value is significant at the 0.05 level only if Yates' correction is not used. At King's Dock there is no association and R is close to zero.

A statistical comparison of the strengths of association can be made by comparing $(g_1 + g_4)$ and $(g_2 + g_3)$ between two populations in a further 2 x 2 contingency table, which is legitimate provided that column and row frequency values are the same in the two original contingency tables from which the data is taken. Although this conditions is not met perfectly some comparisons have been made. For the data in rows 4 to 6 of Table 3, Croyde versus RHB I gives $X_1^2 = 76.681$ ($P < 0.001$) and RHB I versus King's Dock gives $X_1^2 = 3.358$ ($P > 0.05$); ($X^2 = 4.081$, $P < 0.05$ without Yates correction): for the data in rows 7 to 9, Croyde versus RHB I gives $X_1^2 = 32.871$ ($P < 0.001$) and RHB I versus King's Dock gives $X_1^2 = 4.014$ ($P < 0.05$). This analysis provides evidence that the directional changes in strength of association (as measured by R), from high values at Croyde through intermediate values at RHB I to low values at King's Dock, are statistically significant.

TABLE 3

Numbers of individuals in different two locus genotypic classes for Est-D and Lap-1, and R, χ^2 and P values. Expected numbers are given in parentheses.

Population	Row	Est-D→ E/E Lap-1→ E/E	E/E G/G	G/G E/E	G/G G/G	E/G E/E	E/G G/G	E/E E/G	G/G E/G	E/G E/G	R	χ^2	P
Croyde	1	403 (267)	6 (94)	8 (119)	144 (42)	58 (84)	16 (30)	24 (72)	41 (32)	62 (23)	—	—	—
RHB I	2	36 (22)	10 (18)	10 (16)	19 (13)	16 (24)	22 (20)	13 (19)	13 (13)	26 (20)	—	—	—
King's Dock	3	26 (24)	19 (19)	5 (7)	5 (5)	30 (31)	25 (25)	34 (37)	13 (11)	48 (48)	—	—	—
Croyde	4	403 (230)	6 (109)	8 (111)	144 (41)	—	—	—	—	—	0.937	487.379	<0.001
RHB I	5	36 (28)	10 (18)	10 (18)	19 (11)	—	—	—	—	—	0.438	12.587	<0.001
King's Dock	6	26 (25)	19 (20)	5 (6)	5 (4)	—	—	—	—	—	0.060	0.009	n.s.
Croyde	7	561 (522)			74 (113)		65 (104)			62 (23)	0.362	97.181	<0.001
RHB I	8	75 (69)			38 (44)		26 (32)			26 (20)	0.156	3.360*	n.s.
King's Dock	9	55 (55)			55 (55)		47 (47)			48 (48)	0.005	0.004	n.s.

* $\chi^2 = 4.020$, $P < 0.05$ without Yates' correction.

Data on allozyme variation at other loci are of importance for the interpretation of the results described for Est-D and Lap-1. There are allele frequency differences between typical M. edulis and M. galloprovincialis from South West England at the Ap and Phi loci. Some representative data are given in Table 4. For Ap RHB I is similar to M. edulis, while King's Dock has a high frequency of the faster alleles and is more similar to M. galloprovincialis. For Phi, RHB I is roughly intermediate while King's Dock shows some differences from both M. edulis and M. galloprovincialis. Allele 3 has a higher frequency and allele 4 a much lower frequency than in both M. edulis and M. galloprovincialis. For Pgm the allele frequency distributions are similar for M. edulis, M. galloprovincialis and RHB I. At King's Dock a most striking observation is the high frequency of Pgm allele 6 which is at very low frequency in both M. edulis and M. galloprovincialis. Ap and Phi allele frequencies for the morphological extremes at RHB I show a difference with the E and G group values tending to lie in the direction of typical M. edulis and typical M. galloprovincialis respectively. (E versus G comparisons of allele frequency distributions give $\chi^2_2 = 13.124$ (P < 0.01) for Phi and $\chi^2_1 = 2.581$ (n.s.) for Ap.) At King's Dock there are no significant differences in allele frequency between the morphological extremes for Phi although there is a significant difference for Ap ($\chi^2_3 = 8.002$, P < 0.05).

Discussion

At Croyde, the evidence for the existence of two distinct gene pools between which there is restriction on gene flow can be summarised as follows:

(1) There are large differences in allele frequencies between M. edulis and M. galloprovincialis at two loci and significant, though smaller, differences at other loci.

(2) There are large deficits of heterozygotes compared with Hardy-Weinberg expectations at the Est-D and Lap-1 loci.

(3) There is an excess of E_{Est}/E_{Est} E_{Lap}/E_{Lap} and G_{Est}/E_{Est} G_{Lap}/G_{Lap} double homozygotes compared with the expectations based on the genotype frequencies at each locus.

Each of these effects is predicted for a mixture of two populations having differences in allele frequency at allozyme loci. The excess of E_{Est}/G_{Est} E_{Lap}/G_{Lap} double heterozygotes suggests that F_1 hybrids are being produced in the population.

At RHB I heterozygote deficits at the Est-D and Lap-1 loci are smaller than at Croyde. The allele frequency differences between the morphological extremes are smaller; in particular a number of G_{Est}/G_{Est} G_{Lap}/G_{Lap} animals have typical M. edulis morphology. The excess of double homozygotes is again highly significant though less marked and there is a non-significant excess of the double heterozygote.

At King's Dock allele frequencies at the Est-D and Lap-1 loci are roughly intermediate between M. edulis and M. galloprovincialis as is general morphology. There are no associations of morphology with Est-D, Lap-1 and Phi alleles although there is a significant association between morphology and allele frequency at the Ap locus.

The results can be summarised by saying that the extent of genetic mixing is low at Croyde, intermediate at RHB I, and high at King's Dock.

Barsotti and Meluzzi (1968) have suggested that M. galloprovincialis is a recent Mediterranean derivative of M. edulis. The three populations could therefore represent different levels of hybridisation and introgression between M. edulis and M. galloprovincialis immigrants. The morphological differences among the populations are consistent with the genetic data.

Seed (this volume) noted a higher frequency of mussels exhibiting intermediate characters in populations in West Ireland and Brittany than in South West England. These populations probably exhibit greater genetic mixing than occurs at Croyde and may be more similar to RHB I.

In nature, hybrid zones between closely related species are not uncommon (Mayr, 1963; Remington, 1968). Usually mating is at random within populations in the hybrid zone and every possible combination of characters of the two species is found. Avise and Smith (1974), in a study of two morphologically distinct subspecies of blue gill (Lepomis macrochirus), found no evidence, within the hybrid zone, of association of genotypes at two discriminatory allozyme loci or of deviations from Hardy-Weinberg expectations. The situation in Mytilus is clearly different with variation between sites in the amount of genetic and morphological mixture. Increased mixing could reflect an increased period of association between M. edulis and M. galloprovincialis though it is likely that the first British M. galloprovincialis immigrants arrived in the South West. Differences in the extent of mixing could

TABLE 4

Allele frequencies at Phi, Ap and Pgm. (Data from Padstow, Croyde and Ilfracombe are pooled to give representative M. edulis and M. galloprovincialis values from S. W. England.)

Phi	N	1	2	3	Allele 4	5	6	7
Mumbles	1000	0.014	0.036	0.234	0.055	0.596	0.052	0.014
S.W. England edulis	60	0.008	0.042	0.242	0.042	0.642	0.008	0.017
S.W. England gall.	165	0.015	0.073	0.433	0.300	0.145	0.033	–
RHB I	136	0.015	0.051	0.320	0.195	0.393	0.022	0.004
King's Dock	100	0.015	0.065	0.665	0.020	0.230	0.005	–

Ap	N	1	2	3	4	5	6	7
Mumbles	1000	0.011	0.015	0.770	0.140	0.056	0.008	–
S.W. England edulis	132	0.004	0.019	0.701	0.167	0.106	0.004	–
S.W. England gall.	161	–	0.019	0.512	0.311	0.134	0.025	–
RHB I	136	0.004	0.022	0.735	0.210	0.022	0.007	–
King's Dock	100	0.015	0.015	0.350	0.255	0.235	0.130	–

Pgm	N	1	2	3	4	5	6	7
Mumbles	1000	0.007	0.099	0.652	0.230	0.013	0.001	–
S.W. England edulis	129	–	0.112	0.678	0.194	0.016	–	–
S.W. England gall.	151	–	0.139	0.566	0.288	0.003	0.003	–
RHB I	136	0.011	0.121	0.603	0.261	0.004	–	–
King's Dock	150	–	0.077	0.490	0.320	0.013	0.100	–

Phi	n/N	1	2	3	4	5	6	7
RHB I E	35/136	–	0.029	0.371	0.071	0.529	–	–

Table 4 (continued)

	n/N	1	2	3	4	5	6	7
RHB I G	29/136	0.017	0.086	0.259	0.310	0.310	0.017	–
King's Dock E	28/100	0.036	0.071	0.696	0.018	0.175	–	–
King's Dock G	31/100	–	0.065	0.629	0.032	0.258	0.016	–

Ap	n/N	1	2	3	4	5	6	7
RHB I E	35/136	–	0.029	0.800	0.171	–	–	–
RHB I G	29/136	0.017	0.017	0.655	0.293	–	0.017	–
King's Dock E	28/100	–	–	0.482	0.179	0.196	0.143	–
King's Dock G	31/100	0.016	0.016	0.258	0.306	0.323	0.081	–

also be caused by large differences in the proportions of
parental types; this cannot apply here. It is more likely
that the different levels of mixing reflect different equilibrium points between introgression and selection against introgressed genes. Differences in spawning time probably influence the rate of introgression (Seed, this volume), and
the existence of large allele frequency differences between
RHB I and RHB II is not easy to explain without involving
differential mortality, especially in view of the known high
dispersal ability of Mytilus.

There is much evidence pointing to the temporal stability
of hybrid zones; a well-known example is of the crows Corvus
corvus corone and C. corvus corax (Meise, 1928) where a narrow
hybrid zone has apparently not broadened appreciably during
the last 5,000 years. A report (Donovan, 1802) of
'Mediterranean', mussels presumably M. galloprovincialis in
South West England suggests that there has been hybridisation
between M. edulis and M. galloprovincialis for at least 150
years. At present natural selection might, however, be
strengthening the reproductive barrier and reducing the amount
of hybridisation. Such temporal changes should be detectable
by examining the extent of genetic mixing in successive year
classes within populations.

If an introgressed population is thought of as a simple
mixture the frequency of an allele will depend solely on its
frequency in the two parental populations and their relative
contributions. In the introgressed King's Dock population,
our observations suggest that a complex situation exists in
which any or all of selection, founder effects and drift
after immigration of M. galloprovincialis into existing M.
edulis and subsequent isolation from nearby open shore populations of M. edulis, might have operated. Thus Pgm allele
6 is at higher frequency, and Phi allele 4 at lower frequency,
in the Dock than in both M. edulis and M. galloprovincialis,
and allele frequencies at Est-D are somewhat closer to M.
edulis while allele frequencies at the Ap locus and the Phi 3
and 5 allele frequencies are closer to M. galloprovincialis.
No mixture of typical M. edulis and M. galloprovincialis
could produce the King's Dock allele frequencies at all loci
simultaneously. With simple mixing the ratio of the frequency
of allele E between two populations should be the same at the
Est-D and Lap-1 loci provided that $E_{Lap} = E_{Est}$ in both M.
edulis and M. galloprovincialis. This condition does hold
approximately. The frequencies of allele E at Croyde, RHB I
and King's Dock are shown in Table 5. Pairwise tests in
2 x 2 contingency tables between populations give a significant

TABLE 5

Frequency of the compound E allele at Est-D and Lap-1.

	Croyde	RHB I	King's Dock
Est-D	0.657	0.552	0.637
Lap-1	0.699	0.533	0.529

Croyde versus RHB I $\chi^2_1 = 0.779$, n.s.
Croyde versus King's Dock $\chi^2_1 = 6.077$, $P < 0.05$
RHB I versus King's Dock $\chi^2_1 = 1.321$, n.s.

value of chi-squared for the Croyde versus King's Dock comparison only. This suggests that there are differences among the populations in the relative extent of introgression at the two loci. Hunt and Selander (1973) found differences among loci in the extent of allelic introgression in the hybrid zone between semispecies of house mice (Mus musculus), while Avise and Smith (1974) found equal levels of introgression at two allozyme loci between sunfish subspecies. The Mytilus data indicates a more complex situation.

A hypothesis that contrasts with the idea of introgression is that the M. galloprovincialis characters and allele frequencies have arisen by selection on an M. edulis base population within the populations in which they are observed. At Croyde and RHB I disruptive selection would then be implicated because of the evidence of two interbreeding yet distinct gene pools. At King's Dock disruptive selection could be in the very early stages or directional selection could be in operation. This hypothesis does imply considerable genetic and morphological convergence towards an M. galloprovincialis-like phenotype. It is a problem to guess what selective forces could be common to South West England, to King's Dock where there is considerable heat input and pollution, and to part of one fairly exposed bay in North East England and we are inclined to reject this idea. However perhaps climatic changes since the evolution of M. galloprovincialis in the Mediterranean now favour a similar evolutionary process in other geographic regions. The hypothesis is, moreover, consistent with the evidence, based largely on the results of laboratory experiments with Drosophila, that disruptive selection can cause divergence, reproductive isolation and sympatric speciation (see Thoday, 1972, for review). The existence of genetic and morphological differences among 'galloprovincialis' types from different geographic regions so large as to defy explanation in terms of founder effects or drift or selection after immigration would corroborate this hypothesis. The data included in this paper show no major differences, though a more definite answer might be found after extensive sampling throughout Europe.

Summary

This paper considers allozyme variation at five polymorphic loci (Lap-1, Est-D, Phi, Pgm and Ap) in a number of British Mytilus populations where individuals resembling M. edulis and M. galloprovincialis occur sympatrically.

In a sample from Croyde in South West England the morphological extremes differ greatly in the allele frequency distribution at the Est-D and Lap-1 loci; there is a large deficit of heterozygotes and a strong association between genotypes at the two loci. These results provide evidence that the mussel population consists of two distinct gene pools between which there is a considerable barrier to gene flow. At Robin Hood's Bay in North East England similar results are obtained though here the genetic and morphological associations are weaker. At King's Dock in Swansea, South Wales the allele frequencies at the Est-D and Lap-1 loci are roughly intermediate between typical M. edulis and M. galloprovincialis from South West England and there are no striking genetic or morphological associations.

Allele frequency differences between typical M. edulis and typical M. galloprovincialis are also found at the Phi and Ap loci.

It appears probable that the three populations exhibit increasing levels of hybridisation and introgression between M. edulis and M. galloprovincialis in the order Croyde $<$ RHB I $<$ King's Dock. At King's Dock there is clear evidence of differences in the extent of introgression among alleles at the Est-D, Lap-1, Pgm and Phi loci. These differences can be attributed to either natural selection, a founder effect or genetic drift.

Acknowledgements

This work was supported by grant GR3/2452 from the

N. E. R. C. The authors gratefully acknowledge various forms of help given by Dr. B. Bayne. D. O. F. Skibinski received an award under the Younger Research Workers Exchange scheme of the British Council.

REFERENCES

Ahmad, M. and Beardmore, J.A. 1976. Genetic evidence that the "Padstow Mussel" is Mytilus galloprovincialis. Mar. Biol. 35, 139-147.

Ahmad, M., Skibinski, D. O. F. and Beardmore, J. A. 1977. An estimate of the amount of genetic variation in the common mussel Mytilus edulis. Bioch. Genet. 15, 833-846.

Avise, J. C. 1974. Systematic value of electrophoretic data. Syst. Zool. 23, 465-481.

Avise, J. C. and Smith, M. H. 1974. Biochemical genetics of sunfish. I. Geographic variation and subspecific intergradation in the bluegill, Lepomis macrochirus. Evolution 28, 42-56.

Barsotti, G. and Meluzzi, C. 1968. Osservazioni su Mytilus edulis L. e M. galloprovincialis Lmk. Conchiglie 4, 50-58.

Brown, A. H. D. 1970. The estimation of Wright's Fixation Index from genotypic frequencies. Genetica 41, 399-406.

Donovan, E. 1802. The Natural History of British Shells. F. and C. Rivington, London.

Hill, W. G. 1974. Estimation of linkage disequilibrium in randomly mating populations. Heredity 33, 229-239.

Hubby, J. L. and Throckmorton, L. H. 1965. Protein Differences in Drosophila. II. Comparative species genetics and evolutionary problems. Genetics 52, 203-215.

Hunt, W. G. and Selander, R. K. 1973. Biochemical genetics of hybridization in European house mice (Mus musculus). Heredity 31, 11-33.

Koehn, R. K., Milkman, R. and Mitton, J. B. 1976. Population genetics of marine Pelecypods. IV. Selection migration and genetic differentiation in the blue mussel Mytilus edulis. Evolution 30, 2-32.

Lewis, J. R. and Seed, R. 1969. Morphological variations in *Mytilus* from South-West England in relation to the occurrence of *M. galloprovincialis* (Lamarck). Cah. Biol. Mar. 10, 231-253.

Lewontin, R. C. and Kojima, K. 1960. The evolutionary dynamics of complex polymorphisms. Evolution 14, 458-472.

Mayr, E. 1963. *Animal Species and Evolution*. Harvard Univ. Press, Cambridge.

Meise, W. 1928. Die Verbreitung der Aaskrake (Formankreis *Corvus corone* L.). J. Orn. Lpz. 76, 1-203.

Murdock, E. A., Ferguson, A. and Seed, R. 1975. Geographical variation in leucine aminopeptidase in *Mytilus edulis* L. from the Irish coasts. J. Exp. Mar. Biol. and Ecol. 19, 33-41.

Remington, C. L. 1968. Sature-zones of hybrid interaction between recently joined biotas. Evol. Biol. 2, 321-428.

Seed, R. 1971. A physiological and biochemical approach to the taxonomy of *Mytilus edulis* L. and *M. galloprovincialis* Lmk. Cah. Biol. Mar. 12, 291-322.

Skibinski, D. O. F., Ahmad, M. and Beardmore, J.A. 1978. Genetic evidence for naturally occurring hybrids between *Mytilus edulis* and *Mytilus galloprovincialis*. Evolution, June, in press.

Thoday, J. M. 1972. Disruptive selection. Proc. R. Soc. Lond. B. 182, 109-143.

Wright, S. 1951. The genetical structure of populations. Ann. Eugenics 15, 323-354.

CHROMOSOME EVOLUTION IN SOME MARINE INVERTEBRATES

D. Colombera and I. Lazzaretto-Colombera

Istituto di Biologia Animale

Universita di Padova, Italy

Introduction

Although there has been considerable development of cytotaxonomic studies in the last twenty years, marine invertebrates have usually been disregarded by karyologists. For instance, after the first researches in the early 1900's, only a few authors examined echinoderms and tunicates and out of the thousands of harpacticoid copepods only five species had been studied from a karyological point of view (Ar-Rushdi, 1963; Bocquet, 1951; Goswami and Goswami, 1972).

In fact, it is only as a result of our research on the chromosomes of ascidians, echinoderms and harpacticoids, initally undertaken to investigate problems of genetics, chordate origins and speciation, that a preliminary cytotaxonomy can be presented for the above mentioned groups.

Copepoda. Out of the seven orders into which copepods are classified, only the Calanoida, Harpacticoida and Cyclopoida have been studied chromosomally. Karyological data are very scanty or absent for the small orders Notodelphoida, Monstrilloida, Caligoida and Lerneopodoida, which contain only parasitic forms.

On the whole copepods are characterised by the presence of a distal parallel pairing of homologues during late prophase - (Fig.1) and metaphase-I of oocytes (Fig. 2). It has been suggested that this pairing might be important in the course of chromosome evolution since it might hinder the forma-

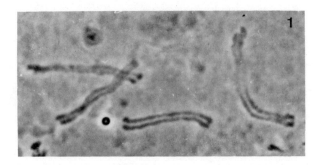

Figure 1.   Late prophase-I chromosome in oocyte of Tisbe holothuriae (Colombera et al., 1975).

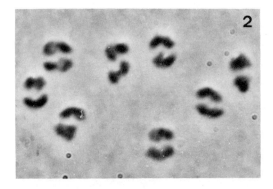

Figure 2.   Metaphase-I chromosomes in oocyte of Tisbe biminiensis (Lazzaretto-Colombera et al., 1976).

tion of chiasmata (Heberer, 1938) and influence the differentiation of sex-chromosomes (Ar-Rushdi, 1963).   At present it is difficult to support the first hypothesis since there is evidence (Fig.3) that chiasmata might be formed in earlier prophase-I before the presence of the distal parallel pairing which seems to be due simply to a particularly wide synaptic connection between homologues (Fig.4) (Colombera et al., 1975). Differentiated sex-chromosomes can be present in cyclopoids

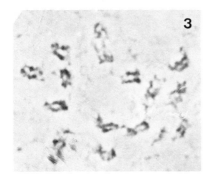
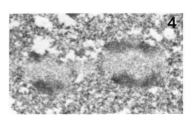

Figure 3. Diffuse stage chromosomes in oocyte of Tisbe reluctans (Colombera and Lazzaretto-Colombera, 1972).

Figure 4. Lateral view of metaphase-I chromosomes in oocyte of Tisbe holothuriae observed by E. M. (Colombera et al., 1975).

and in calanoids; both sexes may be heterogametic with the two forms 2n + XY or 2n + XO (Tables 1 and 3). Sex-chromosomes can be distinguished in species with both high and low chromosome numbers and in genera which possess species without differentiated sex-chromosomes. It must be emphasised that in some cases the determination of sex-chromosomes seems to be very doubtful.

Another peculiarity of copepod karyology is represented by the loss of chromatin during mitotic division in cleaving eggs (Beermann, 1959) but at present it is not possible to evaluate the significance of this character in chromosome evolution.

Harding (1950) observed that in copepods the primitive members have the highest chromosome numbers. In fact Calanoida, the most primitive order, possess the highest diploid numbers of 32 and 34, whereas in Harpacticoida and in Cyclopoida the highest value is 2n = 24. Moreover in Calanoida (Table 1) these higher values are found in genera which have been considered as primitive (Giesbrecht, 1892). It is of some interest to point out that the evolution from marine to fresh water species seems to have occurred very long ago since presumptive primitive chromosome numbers are present in both forms.

In the order Harpacticoida (Table 2) on the basis of chromosome analysis of the genus Tisbe and from comparisons with other data it might be deduced that 24 is the primitive

TABLE 1

Chromosome numbers of Copepods: Calanoida

CALANOIDA

Species	2 n	Sex-chrom.	Habitat	Reference
Calanus finmarchicus	34	XY ?	m	Harding, 1963
Calanus gracilis	32	–	m	Kornhauser, 1915
Calanus helgolandicus	34	XY ?	m	Harding, 1963
Calanus robustior	34	–	m	Tsytsugina, 1974
Eucheta marina	34	–	m	Kornhauser, 1915
Centropages typicus	27	♂ XO	m	Heberer, 1932
Centropages furcatus	6	–	est	Goswami and Goswami, 1974
Limnocalanus macrurus	32	–	fw	Holmquist, 1970
Limnocalanus johanseni	32	–	fw	Holmquist, 1970
Heterocope saliens	32	–	fw	Matschek, 1909
Heterocope weismanni	32	–	fw	Matschek, 1909
Diaptomus coeruleus	28	–	fw	Krimmel, 1910
Diaptomus castor	34	–	fw	Heberer, 1924
Diaptomus salinus	30–34	–	fw	Heberer, 1924
Diaptomus denticornis	34	–	fw	Matschek, 1909
Diaptomus laciniatus	32	–	fw	Matschek, 1909
Diaptomus gracilis	32	–	fw	Rückert, 1894
Temora turbinata	6	–	est	Goswami and Goswami, 1974
Pleuromamma abdominalis	32–34	–	m	Heberer, 1932
Pleuromamma gracilis	32–34	–	m	Heberer, 1932
Anomalocera patersonii	32	–	m	Heberer, 1924
Labidocera pavo	20	–	est	Goswami and Goswami, 1974
Labidocera pectinata	20	–	est	Goswami and Goswami, 1974

Table 1 (continued)

Labidocera acuta	20	—	est	Goswami and Goswami, 1974
Labidocera kroeyeri	20	—	est	Goswami and Goswami, 1974
Tortanus forcipatus	12	♂ XY	est	Goswami and Goswami, 1974
Tortanus barbatus	12	♂ XY	est	Goswami and Goswami, 1974
Tortanus gracilis	12	♂ XY	est	Goswami and Goswami, 1974
Acartiella keralensis	10	♂ XY	est	Goswami and Goswami, 1974
Acartiella gravelyi	10	♂ XY	est	Goswami and Goswami, 1974
Acartia spinicauda	11–12	♂ XO	est	Goswami and Goswami, 1974
Acartia negligens	11–12	♂ XO	est	Goswami and Goswami, 1974
Acartia centrura	11–12	♂ XO	est	Goswami and Goswami, 1974
Acartia plumosa	11–12	♂ XO	est	Goswami and Goswami, 1974

m   = marine waters
br.w. = brackish waters
f.w.   = fresh waters
est   = estuaries
(p)   = parasitic form
(sp)   = semiparasitic form

TABLE 2

Chromosome number of Copepods: Harpacticoida, Caligoida, Lerneopodoida

HARPACTICOIDA

Species	2 n	Sex-chrom.	Habitat	Reference
Tigriopus japonicus	24	—	m	Ar-Rushdi, 1963
Tigriopus brevicornis	24	—	m	Ar-Rushdi, 1963
Tigriopus californicus	24	—	m	Ar-Rushdi, 1963
Laophonte setosa	10	—	br w	Goswami and Goswami, 1974
Canthocamptus staphilinus	24	—	fw	Haecker, 1890
Canthocamptus trispinosus	22	—	fw	Krüger, 1911
Nitocra hibernica	16	—	fw	Krüger, 1911
Tisbe furcata	24	—	m	Lazzaretto-Colombera, 1976
Tisbe clodiensis	24	—	m	Lazzaretto-Colombera, 1976
Tisbe dobzhanskii	24	—	m	Lazzaretto-Colombera, 1976
Tisbe reticulata	24	—	m	Lazzaretto-Colombera, 1976
Tisbe aragoi	22	—	m	Lazzaretto-Colombera, et al., in prep.
Tisbe marmorata	24	—	m	Lazzaretto-Colombera, et al., in prep.
Tisbe holothuriae	24	—	m	Lazzaretto-Colombera, 1976
Tisbe battagliai	24	—	m	Lazzaretto-Colombera, 1976
Tisbe pontina	24	—	m	Lazzaretto-Colombera, 1976
Tisbe biminiensis	16	—	m	Lazzaretto-Colombera, 1976
Tisbe pori	24	—	m	Lazzaretto-Colombera, in prep.
Tisbe gracilis	22	—	m	Lazzaretto-Colombera, in prep.
Tisbe reluctans	24	—	m	Lazzaretto-Colombera, 1976
Tisbe persimilis	24	—	m	Lazzaretto-Colombera, 1976

Table 2 (continued)

Tisbe bulbisetosa	22	–	m	Lazzaretto-Colombera, in prep.

CALIGOIDA

Laemargus maricatus	16	–	(p)m	McClendon, 1906
Pandarus sinuatus	16	–	(p)m	McClendon, 1906
Orthagoriscola muricata	16	–	(p)m	McClendon, 1910

LERNEOPODOIDA

Lernaea cyprinacea	16	–	(p)fw	Fratello and Sabatini, 1972

TABLE 3

Chromosome number of copepods: Cyclopoida

CYCLOPOIDA

Species	2 n	Sex-chrom.	Habitat	Reference
Cyclops affinis	14	♀ XO ?	fw	Matschek, 1910
Cyclops albidus	14	♀ —	fw	Braun, 1909
Cyclops bicolor	10	—	bw	Goswami and Goswami, 1972
Cyclops bicuspidatus	18	—	fw	Braun, 1909
Cyclops diaphanus	12	—	fw	Braun, 1909
Cyclops dybowskii	18	—	fw	Braun, 1909
Cyclops fuscus	14	—	fw	Braun, 1909
Cyclops fuscus distinctus	14	—	fw	Stella, 1931
Cyclops gracilis	6	—	fw	Braun, 1909
Cyclops insignis	22	—	fw	Braun, 1909
Cyclops leuckarti	14	—	fw	Braun, 1909
Cyclops phaleratus	13	♀ XO	fw	Braun, 1909
Cyclops prasinus	11	♀ XO	fw	Braun, 1909
Cyclops modestus	8	—	fw	Chambers, 1912
Cyclops serrulatus	14	♀ XX	fw	Matschek, 1910
Cyclops strenuus	22	—	fw	Matschek, 1910
Cyclops vernalis	10	♀ XO	fw	Matschek, 1910
Cyclops signatus	8	—	fw	Haecker, 1890
Cyclops viridis brevicornis	24	—	fw	Haecker, 1890
Cyclops viridis americanus	10	—	fw	Chambers, 1912
Cyclops viridis brevispinosus	4	—	fw	Chambers, 1912

Table 3 (continued)

Cyclops viridis parc.	6	—		fw	Chambers, 1912
Cyclops viridis	12	—		fw	Heberer, 1927
Acantocyclops robustus	6	—		fw	Rüsch, 1960
Acantocyclops vernalis	10	♀ XY		fw	Rüsch, 1960
Megacyclops viridis	12	♀ XY		fw	Rüsch, 1960
Macrocyclops fuscus	14	—		fw	Rüsch, 1960
Eucyclops serrulatus	14	♀ XY		fw	Rüsch, 1960
Ectocyclops strenzkey	12	♀ XO		fw	Beermann, 1954
Oithona rigida	22	—		m	Goswami and Goswami, 1972
Copilia denticornis	16	—		m	Heberer, 1932
Sapphirina ovatolanceolata	16	—	(sp)	m	Heberer, 1932
Hersilia apodiformis	24	♂ XY ?	(sp)	m	Kornhauser, 1915
Myticola intestinalis	22	♀ XY	(sp)	m	Orlando, 1973
Lichomolgus forficula	20	—	(sp)	m	Kornhauser, 1915

TABLE 4

Chromosome numbers of Echinoderms: Echinoidea

Species	prior to 1914	prior to 1948	after 1947	Reference
Class Echinoidea				
Family Cidaridae				
Cidaris cidaris			44	Colombera et al., 1977b
Family Pourtalesidae				
Pourtalesia jeffreysi			32	Colombera et al., 1976
Family Centrechinidae				
Arbacia lixula	40			Baltzer, 1910
Arbacia pustulosa	18			Wilson, 1895
Arbacia pustulosa			44	Colombera, 1974
Arbacia punctulata		40		Tennent, 1912
Arbacia punctulata		38		Matsui, 1924
Arbacia punctulata		38-40		Harvey, 1940
Arbacia punctulata			44	Auclair, 1964
Arbacia punctulata			44	German, 1964
Echinus acutus	38			Doncaster and Gray, 1911
Echinus esculentus	18			Delage, 1901
Echinus esculentus	38			Doncaster and Gray, 1911
Echinus miliaris	20			Morgan, 1895
Echinus miliaris	34			Doncaster and Gray, 1911
Psammechinus microtuberculatus	18			Wilson, 1895
Psammechinus microtuberculatus	38			Stevens, 1902
Psammechinus microtuberculatus	18			Godlewski, 1905
Psammechinus microtuberculatus			42	Colombera, 1974
Paracentrotus lividus	18			Wilson, 1895

Table 4 (continued)

Species			Reference
Paracentrotus lividus	18		Delage, 1961
Paracentrotus lividus	18	42	Colombera et al., 1977a
Paracentrotus purpuratus			Hindle, 1911
Paracentrotus intermedius		50	Makino and Niiyama, 1947
Sphaerechinus granularis	18		Wilson, 1895
Sphaerechinus granularis		42	Colombera, 1974
Sphaerechinus parma		44–46	Auclair, 1965
Hemicentrotus pulcherrimus		42	Nishikawa, 1961
Anthocidaris crassispina		42	Nishikawa, 1961
Toxopneustes variegatus	38		Wilson, 1901
Toxopneustes variegatus	36		Heffner, 1910
Toxopneustes variegatus	38		Tennent, 1912
Hipponoe esculenta	32		Pinney, 1911
Family Clipeasteridae			
Clypeaster rosaceus		44–46	Gardiner, 1927
Family Schizasteridae			
Moira atrops	46		Pinney, 1911
Family Brissidae			
Brissus unicolor		42	Colombera, 1974
Echinocardium cordatum	18		Wilson, 1895
Echinocardium mediterraneum		42	Ubisch, 1923

diploid number (Lazzaretto-Colombera, 1976). It follows that, as far as we know, evolution of this order has been characterized by a preferential reduction of chromosome number from 2n = 24 to 22, 16 and 10. It is remarkable that "sibling species" and other very closely related species of the genus Tisbe can possess different chromosome numbers (Lazzaretto-Colombera et al., in prep; Lazzaretto-Colombera. in prep.), although there is considerable general uniformity of chromosome number in this genus.

As already observed for Calanoida and Harpacticoida, the genus Cyclops, order Cyclopoida (Table 3), has undergone an evolution toward reduction of chromosome number along different pathways (Harding, 1950).

Chromosome morphology has not been employed in phylogenetic speculation on copepods because karyotypes have been rarely described in the literature and from the analysis of meiotic bivalents it seems that the chromosomes of copepods are almost all metacentric or submetacentric. This has been confirmed for the genus Tisbe by analysis of mitotic chromosomes of cleaving eggs (Colombera and Lazzaretto-Colombera, 1972; Lazzaretto-Colombera et al., 1976).

Echinodermata. At present we have at our disposal karyological data for only 50 species out of about 6,000 species of extant echinoderms. Even worse, information is scanty and most of the oldest reports seem to be unreliable (Tables 4 and 5). Mitotic chromosomes have been described as acrocentric in Echinocardium mediterraneum (Ubisch, 1923), Asterias amurensis and Aphelasterias nipponica (Makino and Niiyama, 1947) or mainly acrocentric with a few metacentric in Toxopneustes variegatus (Heffner, 1910), Henricia nipponica (Makino and Niiyama, 1947) but the latest research on echinoderm chromosomes does not confirm these findings (Colombera, 1974; Colombera et al., 1976; Colombera et al., 1977a). We, therefore, at present suppose that differentiated sex-chromosomes are not present in the phylus Echinodermata. A "pre-anaphase stretch" has been observed in some echinoids (Colombera and Venier, 1976) and somatic pairing is probably usual in cleaving eggs (Colombera, 1974).

The dimension of echinoderm chromosomes are relatively small and DNA measurements (Hinegardner, 1974; Colombera et al., 1977a) indicate that asteroids, echinoids and crinoids have low amounts whereas ophiuroids and holothuroids average amounts of chromosomal DNA which are two to three times higher. It is of some interest to note that the available evidence. as well as paleontological data (Cuenot, 1948) indicate that

TABLE 5

Chromosome number of echinoderms: Asteroidae, Ophiuroidea, Holothuroidea, Crinoidea

Species	prior to 1914	prior to 1948	after 1947	Reference
Class Asteroidea				
Family Asteridae				
Asterias amurensis		30		Makino and Niiyama, 1947
Asterias glacialis	18			Wilson, 1895
Asterias glacialis	18			Delage, 1901
Marthasterias glacialis			44	Colombera et al., 1976b
Asterias vulgaris	36			Tennent, 1907
Asterias forbesi	18			Wilson, 1895
Asterias forbesi	36			Tennent and Hogue, 1906
Aphaelasterias nipponica		48		Makino and Niiyama, 1947
Family Echinasteridae				
Echinaster sepositus	18			Wilson, 1895
Echinaster sepositus			44	Delobel, 1971
Echinaster sepositus			44	Colombera, 1974
Henricia nipponica		54		Makino and Niiyama, 1947
Cribella sanguinolenta	36			Jordan, 1910
Family Asterinidae				
Asterina pectiniformis		40		Makino and Niiyama, 1947
Asterina gibbosa			44	Colombera et al., 1975
Family Ophidiasteridae				
Ophidiaster ophidiaster			44	Colombera, 1974

Table 5 (continued)

Hacelia attenuata	44	Colombera, 1974
Family Astropectinidae		
Astropecten bispinosus	44	Colombera, 1974
Class Ophiuroidea		
Family Ophiomixidae		
Ophiomixa pentagona	18	Wilson, 1895
Ophiomixa pentagona	42	Colombera, 1974
Family Ophiotricidae		
Ophiothrix fragilis	18	Wilson, 1895
Ophiothrix fragilis	44	Colombera, 1974
Family Ophiodermatidae		
Ophioderma longicauda	18	Wilson, 1895
Ophioderma longicauda	42	Colombera, 1974
Family Ophiocomidae		
Ophiocomina nigra	44	Colombera, 1974
Class Holothuroidea		
Family Holothuridae		
Stichopus regalis	18	Wilson, 1895
Holothuria polii	18	Wilson, 1895
Holothuria tubulosa	44	Colombera, 1974
Family Leptosynaptidae		
Leptosynapta gallienei	44	Colombera et al., 1976
Family Elpididae		
Elpidia glacialis	42	Colombera et al., 1976
Family Cucumariidae		
Cucumaria cucumis	18	Wilson, 1895
Cucumaria planci	44	Colombera et al., 1977a

Table 5 (continued)

Class Crinoidea		
Family Antedonidae		
Antedon rosacea	18	Wilson, 1895
Antedon rosacea	42	Colombera et al., 1977a

the classes such as Crinoidea, Echinoidea and Asteroidea, which suffered a general crisis during Permian and Triassic times, possess a smaller quantity of DNA, whereas ophiuroids and holothoroids, with more DNA at their disposal, were better able to withstand the Permian crisis.

The course of male meiosis in Echinodermata has been analyzed only in Asterina gibbosa and here it seems characterized by chiasmatic bivalents and a "lampbrush-like" phase.

The description of spermatogenetic bivalents for other species led us to deduce that spermatogenesis is chiasmatic in echinoderms. The morphology of male bivalents, limited to 17 species, allows the grouping of echinoderms into three sections: (1) asteroids (Fig.5) and echinoids (Fig.6); (2) holothuroids (Fig.7) and ophiuroids (Fig.8); (3) crinoids (Fig.9), in contrast with the indications obtained employing other parameters such as paleontological (Moore, 1966; Cuenot,

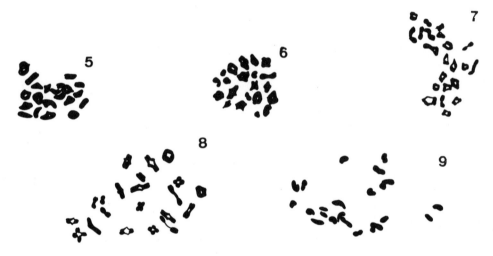

Figure 5. Spermatocyte bivalents of Asterina gibbosa (Colombera et al., 1975).

Figure 6. Spermatocyte bivalents of Paracentrotus lividus (Colombera et al., 1977).

Figure 7. Spermatocyte bivalents of Holothuria tubulosa (Colombera, 1974).

Figure 8. Spermatocyte bivalents of Ophiomixa pentagona (Colombera, 1974)

Figure 9. Spermatogenetic bivalents of Antedon rosacea (Colombera et al., 1977).

1948), biochemical (Yasumoto et al., 1966; Singh et al., 1967; Gupta and Scheurer, 1968) or embryological ones (Hyman, 1955), etc.

If only reliable data are considered it appears that in 20 families, 26 genera and in all the five classes of extant echinoderms, only the haploid numbers of 21 and/or 22 are found, with the sole exception of an aberrant echinoid, Pourtalesia jeffreysi, which possess a haploid number of 16 (Colombera et al., 1976). The scattered available data suggests that the haploid numbers of 21 and 22 are largely preponderant in the phylum Echinodermata.

As it seems highly improbable that these values 21 and 22 have arisen independently during echinoderm evolution, we assume that these chromosome numbers have been very stable in the phylum Echinodermata and considering the origin of the five classes (Fig.10), it follows that this stability might have been fixed in early Cambrian or Precambrian times. The presence of a haploid number of 16 in the deep sea Pourtalesia jeffreysi, belonging to the order Holasteroidea, and datable from Cretaceous times, opens the question as to how this chromosomal stability can be maintained and overcome.

Larvacea. Owing to the discrepancies between old and recent data (Colombera and Fenaux, 1973) we consider as reliable only the results summarized in Table 6.

In spite of the scantiness of the available data we can summarise some karyological characteristics of larvaceans:

(1) the extremely low chromosome number in the genus Fritillaria;

(2) the absence of differentiated sex-chromosomes;

(3) the presence of chiasmatic bivalents;

(4) the occurrence of metacentric and/or submetacentric chromosomes which seem to be typical of, if not exclusive to, the genera Fritillaria and Oikopleura.

Fritillaria is more recent than Oikopleura and here we have another example of evolution accompanied by reduction of chromosome numbers.

The peculiarity of the chromosomes of larvaceans in the ambit of Tunicates supports the view, already suggested on the basis of other characteristics (Van Name, 1921; Garstang, 1921, 1928; Fenaux, 1971), that larvaceans are by no means ancestral but very specialized forms.

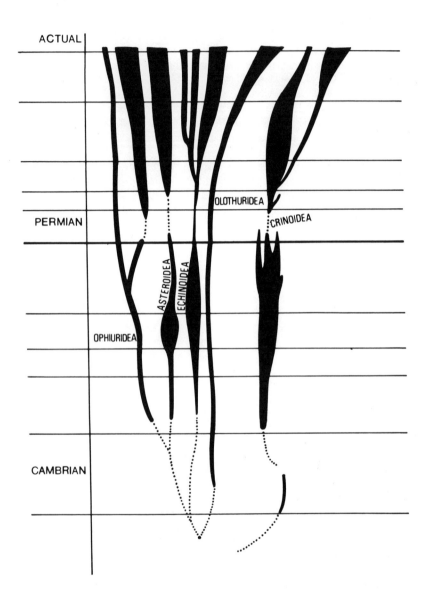

Figure 10. Evolutionary tree of echinoderms (Cuenot, 1948).

TABLE 6

Chromosome numbers of Larvacea

Species	Haploid chromosome numbers	sex-chromosomes	chiasmatic bivalents	acro-chromosomes	meta-telo-chromosomes	Reference
Oikopleura dioida	8	–	?	–	+	–	Colombera & Fenaux, 1973
Oikopleura albicans	8	–	+	–	+	–	Colombera & Fenaux, 1973
Oikopleura fusiformis	8	–	+	–	+	–	Colombera & Fenaux, 1973
Oikopleura longicauda	8	–	?	?	?	?	Colombera & Fenaux, 1973
Fritillaria pellucida	4	–	+	–	+	–	Colombera & Fenaux, 1973

Thaliacea. A recent study of thaliacean chromosomes (Colombera, 1974b) seems to deny any validity to previous researches so that the only reliable data are those summarized in Table 7.

The considerable uniformity in chromosome numbers is not accompanied by uniformity in chromosome dimension and form; for instance Pegea confederata has very small chromosomes whereas in Pyrosoma atlanticum they are considerably larger. Meiotic chromosomes are often very peculiar but the data are not sufficient for a comparison at this level.

Thaliaceans possess lower chromosome numbers than their ancestors, if we accept the hypothesis that they derive from ascidians (Tokioka, 1971; Van Name, 1921; Berrill, 1950; Barrington, 1965; Lohmann, 1933; Ihle, 1935) and in particular from some Polycitoridae-like forms (Kasas, 1972).

Ascidiacea. The results of our cytotaxonomic analyses can be summarized as follows:

(1) Ciona and Diazona, with the same haploid number (Colombera, 1974a) and very similar spermatogenetic bivalents (Colombera et al., in prep.b) consistently support the opinion of Arnback (1934) who places the above mentioned genera in the same family.

(2) Chromosome haploid numbers range from 6 up to 24; these numbers are distributed mainly around the values 20, 15, and 9: the highest values are in the suborder Aplousobranchiata, the lowest in the suborder Phlebobranchiata (Tables, 8, 9, 10).

(3) At the genus level chromosome numbers are very closely related and consistent similarities can be found at the family level too (Tables 8, 9, 10).

(4) Evolution is usually accompanied by decreases in chromosome numbers. In particular, primitive families have higher chromosome numbers than related specialized families. Within the suborder Phlebobranchiata we found $n = 14$ in Cionidae and Diazonidae, $n = 17$ in Perophoridae, all primitive families, whereas the specialized Ascidiidae have $n = 9-8$ chromosomes and Corellidae, the most evolved along this trend, have only $n = 6$. In the suborder Stolidobranchiata the haploid number 16 is found in the three families Styelidae, Pyuridae and Molgulidae but this value is found in the family Pyuridae only in the genus Boltenia, considered primitive by virtue of the sensory equipment of the larvae. In the most specialized genera Pyura and Microcosmus, the haploid value of 11 is found with the single intermediate value of $n = 13$ in Pyura micro-

TABLE 7

Chromosome numbers of thaliaceans

Species	Chiasmatic bivalents	n	acro-	telo-	meta-	differentiated sex-chromosomes	Reference
Ord. Pyrosomida							
Pyrosoma atlanticum	+	11	+	+	+	−	Colombera, 1974b.
Ord. Salpida							
Salpa maxima	+	13	+	?	+	−	Colombera, 1974b
Salpa fusiformis	+	12	+	?	+	−	Colombera, 1974b
Pegea confederata	?	12	+	+	+	−	Colombera, 1974b
Thalia democratica	+	12	+	?	+	−	Colombera, 1974b
Ord. Doliolida							
Doliolum denticolatum	+	11	+	+	+	−	Colombera, 1974b

TABLE 8

Chromosome number of ascidians:

Aplousobranchiata

Species	n	Reference
Order ENTEROGONA		
Suborder Aplousobranchiata		
Fam. Clavelinidae		
Subf. Polycitorinae		
Eudistoma magnum	22	Colombera et al., in prep. e
Polycitor adriaticus	24	Colombera, 1974a
Cystodites dellechiaiei	20	Colombera, 1974a
Subf. Clavelininae		
Clavelina lepadiformis	9	Colombera, 1971b
Clavelina nana	10	Colombera et al., in prep. e
Subf. Holozoinae		
Distaplia rosea	11	Colombera et al., in prep. e
Fam. Polyclinidae		
Subf. Polyclininae		
Polyclinum aurantium	13	Colombera et al., in prep. e
Sydnium turbinata	20	Colombera, 1974a
Sydnium fuscum	18	Colombera et al., in prep. e
Aplidium proliferum	20	Colombera et al., in prep. e
Aplidium densum	20	Colombera et al., in prep. e
Aplidium aereolatum	18	Colombera et al., in prep. e
Morchellium argus	20	Colombera, 1974a
Fam. Didemnidae		
Didemnum candidum	15	Colombera, 1974a
Trididemnum tenerum	15	Colombera, 1974a
Diplosoma listerianun	15	Colombera & Sala, 1972a
Lyssoclinum wegelei	23	Colombera et al., in prep. e
Polysincraton lacazei	26	Colombera et al., in prep. e

TABLE 9

Chromosome number in ascidians:
Phlebobranchiata

Species	n	Reference
Order ENTEROGONA		
Suborder Phlebobranchiata		
Fam. Cionidae		
Ciona intestinalis	14	Taylor, 1967
Ciona rulei	14	Colombera et al., in prep. f
Fam. Diazonidae		
Diazona violacea	14	Colombera, 1974a
Rhopalea neapolitana	14	Colombera et al., in prep. f
Fam. Perophoridae		
Perophora listeri	17	Colombera et al., in prep. f
Ecteinascidia turbinata	17	Colombera et al., in prep. f
Fam. Ascidiidae		
Ascidia mentula	9	Colombera & Tenca, 1975
Ascidia paratropa	9	Minganti, 1967
Ascidia callosa	9	Minganti, 1967
Ascidia virginea	9	Colombera, 1974a
Ascidia malaca	8	Minganti, 1967
Ascidia conchilega	8	Colombera, 1974a
Ascidiella aspersa	9	Colombera, 1971a
Ascidiella scabra	9	Colombera, 1968
Ascidiella patula	9	Floderus, 1896
Phallusia mammillata	8	Hill, 1895
Phallusia fumigata	8	Colombera, 1974a
Fam. Corellidae		
Corella parallelogramma	6	Colombera, 1974a
Corella willmeriana	6	Minganti, 1967

TABLE 10

Chromosome number in ascidians:
Stolidobranchiata

Species	n	Reference

Order PLEUROGONA

Suborder Stolidobranchiata

Fam. Styelidae

Subf. Styelinae

Styela gibsii	16	Minganti, 1967
Styela plicata	16	Minouchi, 1936
Styela coriacea	16	Colombera, 1974a
Styela clava	16	Colombera, 1974a
Styela partita	16	Colombera & Vitturi, 1976
Polycarpa gracilis	16	Colombera et al., in prep. c
Polycarpa pomaria	16	Colombera, 1974a
Dendrodoa grossularia	22	Colombera, 1974a

Subf. Botryllinae

Botryllus schlosseri	16	Colombera, 1963
Botrylloides leachi	16	Colombera, 1967
Polyandrocarpa sp.	16	Colombera et al., in prep. c
Distomus variolosus	24	Colombera et al., in prep. c
Stolonica socialis	24	Colombera et al., in prep. c

Fam. Pyuridae

Boltenia villosa	16	Minganti, 1967
Halocynthia papillosa	16	Colombera et al., in prep. d
Pyura microcosmus	13	Colombera et al., in prep. d
Pyura haustor	11	Minganti, 1967
Pyura squamulosa	11	Colombera, 1974a
Pyura dura	11	Colombera et al., in prep. d
Microcosmus sulcatus	11	Colombera, 1974a
Microcosmus sabatieri	11	Colombera, 1974a
Microcosmus polymorphus	11	Colombera et al., in prep. d

Fam. Molgulidae

Molgula manhattensis	16	Colombera, 1974a

cosmus.

The only doubtful case seems to occur in the family Clavelinidae. In fact Clavelininae are considered primitive by virtue of their relatively unspecialized type of budding a and larva (Millar, 1966) but Berrill (1950) thought Polycitorinae to be more primitive and Holozoinae intermediate between the two. We would like to point out that the relative primitiveness of Clavelina can be disputed on the basis of the observation that this "colonial" species is extremely large compared with all other Aplousobranchiata considered here.

(5) Evolution can sometimes be accompanied by increase of chromosome number and DNA amount: the genera Dendrodoa, Stolonica and Distomus in the family Styelidae (Colombera et al., in prep. c), the genus Polysincraton within Didemnidae (Colombera et al., in prep. e). In the family Ascidiidae the situation has not been cleared up because the position of Ascidiella aspersa is disputed (Colombera et al., in prep. a).

(6) Chromosome morphology is less constant than chromosome number. For instance in the family Styelidae (Colombera et al., in prep. c), there are numerous genera with n = 16 but Styela has only telocentric chromosomes, Polycarpa has metacentric or submetacentric chromosomes, while Botryllus has metacentric, submetacentric, acrocentric and subacrocentric chromosomes.

(7) The dimensions of chromosomes are relatively more uniform in specialized forms. Examples are given in the family Styelidae (Colombera, 1969; Colombera and Vitturi, 1976; Colombera et al., in prep. c) and in the suborder Phlebobranchiata (Colombera et al., in prep. f).

(8) There are consistent differences between oocyte and spermatocyte bivalents. So far 5 types of oocyte bivalents have been distinguished: the Clavelina lepadiformis type (Fig. 11) (Colombera, 1971b) within the Phlebobranchiata, the Ciona- (Fig.12) and Corella-type (Fig.13) (Colombera and Vitturi, 1977) within the Phlebobranchiata (Colombera et al., in prep. f), and the Botryllus- (Fig.14) (Colombera, 1969) and Styela-type (Fig.15) (Colombera and Vitturi, 1976) within the Aplousobranchiata.

(9) Somatic pairing and preanaphase stretch seem to be typical of mitotic chromosomes in cleaving eggs (Colombera, 1972b, 1973a, 1973c).

(10) Differentiated sex-chromosomes are absent (Colombera, 1972b, 1974a).

Figure 11.   Oocyte bivalents of <u>Clavelina lepadiformis</u> (Colombera, 1971b).

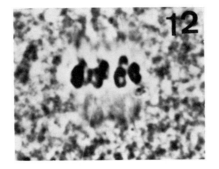

Figure 12.   Oocyte bivalents of <u>Ciona intestinalis</u> (Colombera et al., in prep. b).

Figure 13.   Oocyte bivalents of <u>Corella parallelogramma</u> (Colombera and Vitturi, in press).

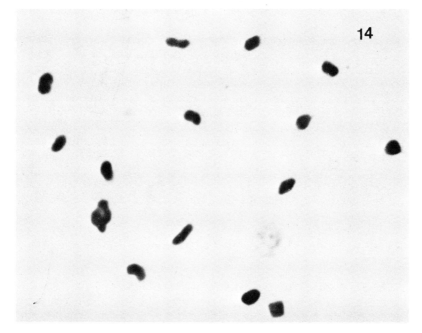

Figure 14. Oocyte bivalents of <u>Botryllus schlosseri</u> (Colombera, 1969).

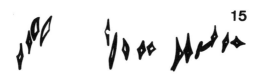

Figure 15. Oocyte bivalents of <u>Styela plicata</u> (Colombera and Vitturi, 1976).

Tactics and Strategies of Chromosome Evolution

A comparison of the karyotypes of echinoderms, copepods and tunicates indicates that these groups are characterized by different patterns of chromosome evolution: (1) very great stability of chromosome numbers together with consistent variability of chromosome form and dimension in echinoderms, (2) considerable uniformity of chromosome morphologies and relatively low chromosome numbers in copepods and (3) high variability of both chromosome form and numbers in tunicates.

However, in these groups the course of evolution seems to have been consistently accompanied by decrease of chromosome numbers and DNA amounts and an increase in chromosome symmetry and uniformity. Moreover, when it is possible to define a species as highly specialized, its karyotype is always characterized by relatively low chromosome number, high chromosome symmetry and uniformity of chromosomal dimensions.

In order to explain these analogies between karyotypes of highly specialized species we advance the hypothesis that evolution of karyotype is determined by general random factors (such as chromosome deletions, inversions, fission and fusion, etc.) selected by contingent genetic conditions (which are expected to be as numerous as species are) and controlled by two evolutionary strategies. These two strategies are opposite in their effects and can be summarized as follows: (a) when selective pressure rewards an increase of genetic potentiality and variability, karyotypes will evolve by increasing chromosome numbers and DNA amounts; (b) when selective pressure rewards an increase of genetic correlation, i.e. increased epistatic interactions, karyotypes will evolve by reduction of chromosome numbers and DNA amounts.

In the case of selection of type (a), low chromosome symmetry and chromosomes of very different sizes are to be expected. For instance, centric fission increases chromosome number and decreases chromosome symmetry; the probability of gene duplication in a chromosomal arm is obviously proportional to the number of genes in that arm, thus increasing chromosome asymmetry.

On the other hand, in the case of selection of type (b) karyotypes will tend to be characterized by chromosomes of high symmetry and similar length. In fact, symmetry increases because: (1) the _a priori_ probability of a loss of chromatin is obviously proportional to the arm lengths of chromosomes; (2) centric fusion reduces chromosome numbers and asymmetry; (3) when chromosome numbers tend to very low values, in order to maintain the indispensable minimum of DNA, the DNA must be mainly packed in chromosomes of maximum size and therefore metacentric.

If neither (a) nor (b) selection is acting, we would expect that the "general random factors" would determine karyotype with statistically more probable patterns. Of course, tactical factors, which can be considered as initial conditions, can limit karyotype evolution.

If this hypothesis is valid, as it seems to be for the marine invertebrates examined here, it must have a general validity and should therefore be tested in other animal phyla.

General Remarks

The considerable but not definitive stability of chromosome numbers in the phylum Echinodermata opens the question as to whether and how these partially "frozen karyotypes" can be correlated with an hypothetical "evolutionary weariness" of echinoderms. Although the situation is far less dramatic for copepods, again we are faced with a chromosome evolution which seems to be considerably hampered, since it seems to proceed only by reduction of chromosome numbers.

An analogous situation of chromosomal inertia seems to occur in larvaceans and thaliaceans, although the scanty data at hand might suggest wrong conclusions. Ascidians, on the other hand, seem to be the only marine invertebrates here investigated with an enormous evolutionary potential. In fact, numerous different trends of chromosomal evolution have been distinguished in different suborders and families with a wide variation of chromosome patterns and numbers. Moreover meiotic chromosomes are so different from one suborder to the other that ascidians might be arranged in a larger taxonomic range than that ordered on the basis of gross morphological characters.

On the whole it seems that in the groups examined here, evolution has been mainly accompanied by a reduction of chromosome numbers and in increase of chromosome symmetry, whereas in only a few cases has a different evolutionary trend been observed.

The correlation between specialized species and karyotypes with low chromosome numbers has been interpreted in general terms by considering the hypothesis that karyotype evolution is influenced by the genetic significance of chromosome numbers and DNA amounts.

Summary

Echinoderms are characterized by an astonishing uniformity of chromosome number, with the exception of <u>Pourtalesia jeffreysi</u>, an aberrant echinoid of recent origin. This uniformity of chromosome number is accompanied by a considerable variety of chromosome patterns and dimensions.

Tunicates are consistently differentiated as far as chromosome numbers, forms and dimensions are concerned: the class Copelata is typified by the low chromosome numbers found in the genus <u>Fritillaria</u>; the class Thaliacea is interesting for its relatively great uniformity of chromosome number; the class Ascidiacea, which has been extensively studied, is characterized by a considerable variety of cytoevolutionary trends.

Echinodermata and Tunicata are akin in lacking differentiated sex-chromosomes and in showing somatic pairing in cleaving eggs.

Copepoda are peculiar in the great symmetry of their chromosomes and in showing a distal parallel pairing of homologues during prophase- and metaphase-I in oocytes. Very low chromosome numbers are found in both calanoids and cyclopoids but calanoids have higher chromosome numbers, and cyclopoids lower, with harpacticoids in an intermediate position.

It is interesting to note that, although chromosome evolution has followed different trends in the three groups here investigated, highly specialized species are consistently characterized by analogous karyotypes.

An hypothesis has been advanced in order to explain this unexpected correlation.

REFERENCES

(A) Copepoda

Aacker, V. 1892. Die Eibildung bei Cyclops und Canthocamptus. Zool. Jahrb. 5, 211-248.

Ar-Rushdi, A. H. 1963. The cytology of achiasmatic meiosis in the female Tigriopus (Copepoda). Chromosoma 13, 526-539.

Beermann, S. 1959. Chromatin-diminution bei Copepoden. Chromosoma 10, 504-514.

Beermann. W. 1954. Weibliche Heterogamatie bei Copepoden. Chromosoma 6, 381-396.

Bocquet, C. 1951. Recherches sur Tisbe reticulata. Arch. de Zool. exp. gen. 87, 335-416.

Braun, H. 1909. Die spezifischen Chromosomenzahlen der einheimischen Arten der Gattung Cyclops. Arch. f. Zellforsch 3, 449-482.

Chambers, R. 1912. Quoted in Makino, S. 1951.

Colombera, D. and Lazzaretto-Colombera, I. 1972. The karyology of Tisbe reluctans (Copepoda, Harpacticoida). Caryologia 25, 525-529.

Colombera, D., Lazzaretto-Colombera, I and Ongaro, L. 1975. Ultrastructure of distant parallel homologues in oocyte of Tisbe holothuriae (Copepoda). Experientia 31, 225-226.

Fratello, B. and Sabatini, M. A. 1972. Cariologia e sistematica di Lernaea cyprinacea L. (Crustacea, Copepoda). Rend. Acc. Lincei 53, 209-213.

Giesbrecht, W. 1892. Quoted in Lang. K. 1948.

Goswami, U. and Goswami, S. C. 1972. Quoted in Genetics Abstracts 4 (9), 51, 1972.

Goswami, U. and Goswami, S. C. 1974. Quoted in Genetics Abstracts 7 (6), 158, 1975.

Haecker, V. 1890. Uver die Reifungsvorgange bei Cyclops. Zool. Anz. 13, (346), 551-558.

Harding, J. P. 1950. Cytology, genetics and classification. Nature 166, 769-771.

Harding, J. P. 1963. The chromosomes of Calanus finmarchicus and C. helgolandicus. Crustaceana 6, 81-88.

Heberer, G. 1924. Die spermatogenese der Copepoden. I. Z. wiss. Zool. 123, 555-646.

Heberer, G. 1927. Quoted in Makino, S. 1951.

Heberer, G. 1932. Die spermatogenese der Copepoden. II Z. wiss. Zool. 142, 191-253.

Heberer, G. 1938. Analyse der Genimi von Heterocope weismanni IMHOF (= H. borealis Fischer). Biol. Zbl. 58, 343-356.

Holmquist, C. 1970. The genus Limnocalanus (Crustacea, Copepoda). Zeitschr. fur Zool. Systematik u. Evolutionforschung 8, 273-296.

Kornhauser, S. J. 1915. A cytological study of the semiparasitic Copepod Hersilia apodiformis (PHIL), with some general considerations of Copepod chromosomes. Arch. f. Zellforsch 13, 399-445.

Krimmel. O. 1910. Die Chromosomenverhaltnisse in generativen und somatischen Zellen bei Diaptomus coeruleus nebst Bemerkungen uber die Entwicklung der Geschlachtorgane. Zool. Anz. 35, 778-793.

Kruger, P. 1911. Beitrage zur Kenntnis der Oogenese bei Harpacticiden, nebst biologischen Beobachtungen. Arch. Zellforsch. 6, 165-189.

Lang, K. 1948. Monographie der Harpacticiden. Ohlssen, H. Lund .

Lazzaretto-Colombera, I. 1976. Chromosome number in ten species of the genus Tisbe (Copepoda, Harpacticoida). Mar. Biol. 38, 159-162.

Lazzaretto-Colombera, I. (in prep.) Number of chromosomes in four species of the gracilis group of the genus Tisbe (Copepoda).

Lazzaretto-Colombera, I. Colombera, D. and Pitacco, G. 1976. La cariologia della specie Tisbe biminiensis (Copepoda,

Harpacticoida). Rend. Acc. Lincei 60, 303-308.

Lazzaretto-Colombera, I. et al., (in prep.) I cromosomi di 3 specie gemelle del genere Tisbe: T. aragoi, T. marmorata e T. reticulata.

Makino, S. 1951. An Atlas of the Chromosome Numbers in Animals. Ames Iowa.

Matscheck, H. 1909. Zur kenntnis der Eireifung und Eiablage bei Copepoden. Zool. Anz. 34, 42-54.

Matscheck, H. 1910. Uber Eireifung und Eiablage bei Copepoden. Arch. Zellforsch. 5, 36-119.

McClendon, J.F. 1906. quoted in Makino S. 1915.

McClendon, J.F. 1910. quoted in Makino S. 1951.

Orlando, E. 1973. Heterochromosomes in Mytilicola intestinalis Steuer (Copepoda). Genetica 44, 244-248.

Rückert, J. 1894. Quoted in Makino, S. 1951.

Rüsch, M. E. 1960. Untersuchungen uber geschlechtsbestimmungsmechanismen bei Copepoden. Chromosoma 11, 419-432.

Stella, E. 1931. Die Cytologie der Geschlechtszellen einiger Cyclopiden mit Bezug auf ihre systematische Stellung. Intern. Rev. Hydrobiol. 26, 111-142.

Tsytsugina, V. G. 1974. Karyological investigation of Calanus (Neocalanus) robustior Giesbrecht (Calanidae). Zool. Zh. 53, 1566-1569.

(B) Echinodermata

Auclair, W. 1964. The chromosome number of Arbacia punctulata. Biol. Bull. 127, 359.

Auclair, W. 1965. The chromosome of sea urchins, especially Arbacia punctulata: a method for studying unsectioned eggs at first cleavage. Biol. Bull. 128, 169-176.

Baltzer, F. 1910. Über die Beziehung zwischen dem Chromatin und der Entwicklung und Vererbungsrichtung bei Echinodermenbastarden. Arch. Zellforsch. 5, 497-621.

Colombera, D. 1974. Chromosome evolution in the phylum Echinodermata. Zeitschr. für Zool. Systematik u. Evolutionforschung 12, 299-308.

Colombera, D., Venier, G. and Morra, A. 1975. Meiotic chromosomes of males of Asterina gibbosa. J. Heredity 66, 245-246.

Colombera, D. and Venier, G. 1976. I cromosomi degli echinodermi. Caryologia 29, 35-40.

Colombera, D., Sibuet, M. and Venier, G. 1976. Constance du nombre chromosomique chez les Echinodermes. Cahiers de Biologie Marine 17, 433-438.

Colombera D., Venier, G. and Vitturi, R. 1977a. Chromosomes and DNA in the evolution of Echinodermata. Biol. Zbl. 96, 43-49.

Colombera D., Vitturi, R. and Zanirato. 1977b. Chromosome number of Cidaris cidaris. Acta Zoologica 58, 185-186.

Cuenot, L. 1948. Anatomie, ethologie et systematique des echinodermes. In, Traite de Zoologie. Vol XI. Grasse, P. P. (ed). Masson et Cie., Paris.

Delage, Y. 1901. Studes experimentales sur la maturation citoplasmatique et sur la parthenogenes artificielle chez les echinodermes. Arch. Zool. Exp. Gen. 9, 285-326.

Delobel, N. 1971. Determination de nombre chromosomique chez une asteride: Echinaster sepositus. Caryologia 24, 247-250.

Doncaster, L. and Gray, J. 1911. Cytological observation on cross fertilized echinoderm eggs. Proceedings Cambr. Phil. Soc. 16, 414-417.

Gardiner, M. S. 1927. The chromosomes of Clypeaster rosaceus. J. Morph. 43, 547-554.

German, J. 1964. The chromosomal complements of blastomeres in Arbacia punctulata. Biol. Bull. 127, 370-371.

Godlewski, J. E. 1905. Untersuchungen uber die Bastardierung der Echiniden und Crinoiden-familie. Arch. Entwicklungsmech. Organ. 20, 579-643.

Gupta, K. C. and Scheurer, P. J. 1968. Echinoderm sterols. Tetrahedron 24, 5431-5437.

Harvey, E. B. 1940. A comparison of the development of nucleate and non-nucleate eggs of Arbacia punctulata. Biol. Bull. 79, 166-187.

Heffner, B. 1910. A study of the chromosomes of Toxopneustes variegatus which show individual peculiarities of form. Biol. Bull. 19, 195-203.

Hill, B. A. 1895. Notes on the fecundation of the egg of Sphaerechinus granularis and on the maturation and fertilization of the egg of Phallusia mammillata. Quart. J. Micr. Sci. 38, 315-330.

Hindle, E. 1911. A cytological study of artificial parthenogenesis in Strongylocentrotus purpuratus. Arch. Entwicklungsmech. Organ. 31, 145-163.

Hinegardner, R. 1974. Cellular DNA content of the Echinodermata. Comp. Biochem. Physiol. B. 49, 219-226.

Hyman, L. H. 1955. The Invertebrates: Echinodermata. McGraw-Hill, New York.

Jordan, H. E. 1910. The relation of mucleoli to chromosomes in the eggs of Cribrella sanguinolenta. Arch. f. Zellforsch. 5, 394-405.

Makino, S. and Niiyama, H. 1947. A study of chromosomes in echinoderms. J. Fac. Sci. Hokkaido Univ. Serv. VI. Zool. 9, 225-232.

Matsui, K. 1924. Studies on the hybridization among ehinoderms, with special reference to the behaviour of the chromosomes. J. Coll. Agr. Tokyo Imp. Univ. 7, 211-236.

Moore, R. C. 1966. Treatise on Invertebrate Paleontology, Echinodermata, Part U, Echinodermata 3. Vols. 1 and 2, Geological Society of America and University of Kansas Press.

Morgan, T. H. 1895. The fertilization of non nucleate fragments of echinoderm eggs. Arch. Entwicklungsmech. Organ. 2, 268-280.

Nishikawa, S. 1961. Notes on the chromosomes of the echinoderms Hemicentrotus pulcherrimus and Antocidaris crassispina. Zool. Mag. (Tokyo) 70, 425-428.

Pinney, E. 1911. A study of the chromosomes of Hipponoe esculenta and Moira atrops. Biol. Bull. 21, 168-186.

Singh, H., Moore, R. E. and Scheurer, P. J. 1967. The distribution of quinone pigments in echinoderms. Experientia 23, 624-626.

Stevens, N. M. 1902. Experimental studies on eggs of Echinus microtuberculatus. Arch. Entwicklungsmech. Organ. 15, 421-428.

Tennent, H. D. 1907. Further studies on the parthenogenetic development of the starfish egg. Biol. Bull. 13, 309-316.

Tennent, H. D. 1912. The behaviour of the chromosomes in cross fertilized echinoid eggs. J. Morph. 23, 17-30.

Tennent, H. D. and Hogue, M. J. 1906. Studies on the development of the starfish eggs. J. Exp. Zool. 3, 517-541.

Ubisch, M. von. 1923. Ergebnisse einiger Bastardierungsversuche an Spatangiden mit Echiniden. Arch. f. Zellforsch. 17, 261-288.

Wilson, E. B. 1895. Archoplasm, centrosome and chromatin in the sea urchin egg. J. Morph. 11, 443-478.

Wilson, E. B. 1901. A cytological study of artificial parthenogenesis in sea urchin eggs. Arch. Entwicklungsmech. Organ. 12, 529-596.

Yasumoto, T., Tanaka, M. and Hashimoto, Y. 1966. Distribution of saponin in echinoderms. Bull. Jap. Soc. scient. Fish. 32, 673-676.

(C) Tunicata

Arnback, A. C. L. 1934. Northern and arctic invertebrates in the collection of the Swedish State Museum. XII. Tunicata. Kungl. Svenske Vet. Akad., Hadl. 13, 1-91.

Barrington, E. J. W. 1965. The biology of Hemichordata and Protochordata. Oliver and Boyd, Edinburgh. pp. 176.

Berrill, N. J. 1950. The Tunicata, with an account of the British species. The Ray Society pp. 354.

Colombera, D. 1963. I cromosomi di Botryllus schlosseri. La Ricerca Scientifica 33, 443-448.

Colombera, D. 1967. I cromosomi di Botryllus schlosseri e Botrylloides leachi. Boll. Zool. 34, 105.

Colombera, D. 1968. Il numero dei cromosomi in alcuni ascidiacei. Boll. Zool. 35, 446.

Colombera, D. 1969. The karyology of the colonial ascidian Botryllus schlosseri (Pallas). Caryologia 22, 339-350.

Colombera, D. 1971a. Number and morphology of chromosomes in three geographical populations of Ascidiella aspersa (Ascidiacea). Marine Biology 11, 149-151.

Colombera, D. 1971b. The karyology of Clavelina lepadiformis Mueller (Ascidiacea). Caryologia 24, 65-70.

Colombera, D. 1972b. The chromosomes of ascidians in relation to taxonomy. Fifth European Marine Biology Symposium, Piccin Ed. 287-294.

Colombera, D. 1973a. Somatic pairing in Botryllus schlosseri Pallas (Ascidiacea). Caryologia 26, 27-34.

Colombera, D. 1973c. Pre anaphase separation of daughter kinetochores in ascidian embryos. Caryologia 26, 35-42.

Colombera, D. 1973b. The chromosomes of lower chordates. In, Cytotaxonomy and vertebrate evolution. Chiarelli, A. B. and Capanna, E. (eds). Acad. Press, London.

Colombera, D. 1974a. Chromosome number within the class ascidiacea. Marine Biology 26, 63-68.

Colombera, D. 1974b. Chromosome of thaliaceans. Caryologia 27, 331-338.

Colombera, D. and Sala, M. 1972a. The male chromosomes of Diplosoma listerianum (Ascidiacea). Caryologia 25, 409-415.

Colombera, D. and Fenaux, R. 1973d. Chromosome form and number of Larvaceans. Boll. Zool. 40, 347-353.

Colombera, D. and Tenca, M. 1975. Chromosome of Ascidia mentula. Cytologia 40.

Colombera, D. and Vitturi, R. 1976. The chromosomes in the genus Styela. Biol. Zbl. 95, 735-740.

Colombera, D. and Vitturi, R. 1978. I cromosomi meiotici di Corella parallelogramma. Caryologia in press.

Colombera, D. et al. (a) Analisi cariologia di sei specie della famiglia Ascididae. in prep.a.

Colombera, D. et al. (b) Affinita cariologica tra le specie Ciona intestinalis e Diazone violacea. in prep.b.

Colombera, D. et al. (c) Cromosomi ed evoluzione nella famiglia Styelidae. in prep. c.

Colombera, D. et al. (d) Les chromosomes de 6 especes de la famille Pyuridae. in prep. d.

Colombera, D. et al. (e) Introduction a la caryologie du sous-ordre Aplousobranchiata. in prep. e.

Colombera, D. et al. (f) Les chromosomes chez le subordre Phlebobranchiata. in prep. f.

Fenaux, R. 1971 La couche oikoplastique de l'Appendiculaire Oikopleura albicans (Leuckart) (Tunicata). Z. Morph. 69, 184-200.

Floderus, M. 1896. Ueber die Bildung der Follikelhullen bei der Ascidien. Z. wiss. Zool. 61, 163-260.

Garstang, W. 1921. The theory of recapitulation: a critical re-statement of the biogenetic law. J. Linn. Soc. Zool. 35, 81-392.

Garstang, W. 1928. The morphology of the tunicata and its bearing on the phylogeny of the Cordata. Quart. J. Microsc. Sci. 72, 51-187.

Hill, B. A. 1895. Notes on the fecundation of the egg of Sphaerechinus granularis and on the maturation and fertilization of the egg of Phallusia mammillata. Quart. J. Micr. Sci. 38, 315-330.

Ihle, J. E. W. 1935. Desmomyaria-Kukental und Krumbach. Handb. d. Zool. 5, 401-532.

Kazas-Ivanova, O. M. 1972. Asexual reproduction in Tunicata, its origin and evolution. Leningrad Society of Naturalists

78, 267-311.

Lohmann, H. 1933. Tunicata, allgemeine Einleitung in die Naturgeschichte der Tunicata. Kukenthal und Krumbach. Handb. d. Zool. 5, 3-14.

Millar, R. H. 1966. Evolution in Ascidians. In, Some Contemporary Studies in Marine Science. Barnes, H. (ed). pp. 519-534.

Minganti, A. 1967. Ricerche su ibridi di ascidie. Boll. Zool. 34, 142-143.

Minouchi, O. 1936. Notiz uber die Chromosomen von Tethyum plicatum Les. (Ascidia). Z. Zellforsch. Mikrosk. Anat. 23, 790.

Taylor, K. M. 1967. The chromosomes of some lower chordates. Chromosoma 21, 181-188.

Tokioka, T. 1971. Phylogenetic speculation of the Tunicata. Publ. Seto Mar. Biol. Lab. 19, 43-63.

Van Name, W. G. 1921. Budding in compound Ascidians and other invertebrates and its bearing on the question of the early ancestry of the vertebrates. Bull. Am. Mus. Hist. 44, 278-282.

SECTION 4

SEX DETERMINATION, BREEDING SYSTEMS AND ISOLATING MECHANISMS

VARIABILITY OF THE MODES OF SEX DETERMINATION IN LITTORAL AMPHIPODS

H.-P. Bulnheim

Biologische Anstalt, Helgoland

Federal Republic of Germany

Introduction

In recent years, various aspects of sexuality in crustaceans, particularly in isopods and amphipods, have been the subject of detailed investigation. Several studies have led to a variety of insights into the mechanisms of hormonal regulation of sex differentiation as well as into the role of genetic and non-genetic factors which control the determination and genetics of sex. Although some basic principles have been established with respect to the processes leading to male or female development (cf. Charniaux-Cotton, 1972), very different modes of sex determination have been found to occur in the species hitherto studied (cf. Bocquet, 1967). They are represented by genetic systems including heterogamety in males or females, multiple sex chromosomes and polygenic sex determination, as well as by external factors and parasitization, any of which may exert a more or less pronounced influence on sex ratio.

The present paper concentrates upon the modes of sex determination that have been analyzed in some amphipod species inhabiting the littoral zone of estuarine and coastal environments. The studies made thus far refer only to members of the families Gammaridae and Talitridae comprising the genera Gammarus and Orchestia. They are based on extensive rearing and crossbreeding experiments conducted under controlled laboratory conditions as cytological investigations have not allowed the formulation of general conclusions concerning the mechanism of sex determination in amphipods.

The most common haploid chromosome number is 26 in Gammarus species and 25 in Orchestia species (e.g. Poisson and Le Calvez, 1948; Le Calvez and Certain, 1951; Orian and Callan, 1957). Except for Anisogammarus annandalei, which exhibits male digamety (Niiyama, 1950), the existence of heteromorphic sex chromosomes has not been established. This finding does not imply a general absence of homo-heterogamety mechanisms, since sex digamety might not necessarily be expressed by morphologically distinguishable heterochromosomes.

In this report a survey of the role of the various factors controlling sex ratio in amphipods is given. New results obtained from studies on several Gammarus duebeni populations are included and are compared with recent findings on sex determination in Orchestia species.

Sex Determination in Gammaridae

Fresh-water, brackish-water and marine representatives of Gammaridae, including Gammarus pulex subterraneus (Schneider), G. duebeni Liljeborg, G. chevreuxi Sexton, G. zaddachi Sexton, G. locusta (L.) have been used for investigations of sex determination (Kinne, 1952, 1953; Anders, 1957; Traut, 1962; Bulnheim, 1967a, b, 1969, 1972, 1975a). A great deal of work has been concerned with G. duebeni duebeni This euryhaline brackish-water form shows considerable deviation from the 1:1 sex ratio. Traut (1962) demonstrated the occurrence of monogeny, the production of unisexual offspring by certain pairs, manifested either as thelygeny (all-female progeny) or arrhenogeny (all-male progeny), as well as amphogeny which is expressed by progeny of both sexes in different percentages. Following Traut's conclusions, sex determination, except for thelygenic strains, results from an interplay of a polyfactorial system of sex genes and modifying environmental factors. Thelygeny, however, was assumed to be maternally inherited, since this sex-ratio condition, persisting from generation to generation, was shown to be transferred only by females and not to be affected by the genotype of the male. Traut, therefore, after having been able to exclude other explanations, suggested cytoplasmic inheritance.

Investigations designed to ascertain which environmental factors influence sex ratio in G. d. duebeni provided experimental proof for the importance of photoperiod on sex expression in amphogenic strains. Depending on the day-length to which offspring are subjected, a variable shift of sex ratio may occur. Generally, ♀♀ prevail in short-day photoperiods (LD 8:16), whereas in long-day photoperiods (LD 16:8) there is a preponderance of ♂♂ (Fig.1a).

Consecutive sets of experiments have disclosed a photosensitivity of sex determination during the sexually indifferent period after hatching (Bulnheim, 1967b, 1969). By dividing broods after birth into equal halves exposed to different light-dark cycles, highly significant differences in sex ratio could be established in progenies of most parental animals obtained from the estuary of the River Elbe in the Federal Republic of Germany. Considerable variation can be observed with respect to the shift of the sex proportions induced by different photoperiods. This may range from all-femaleness under short-day photoperiods and all-maleness under long-day photoperiods to an almost constant level of sex ratio despite exposure to different daylengths. However, cases of little or no photoperiodic response have been observed only rarely in offspring from pairs of this particular population.

This marked variation in expression of an environmental influence on sex ratio undoubtedly reflects genetic variation. Accepting the theory of polygenic sex determination (Kosswig, 1964), there may be a balance between the freely combinable male and female sex-deciding genes localized in various autosomes. Consequently, such an equilibrium allows environmental factors to play a more or less effective role in sex determination.

Progenies of samples from other populations have also been tested with regard to their photoperiodic responses. _G. d. duebeni_ specimens from the Baltic Sea Fjord of Schlei did not show any differences from the population of the Elbe estuary. By contrast, a much less pronounced photoperiodic response was seen when offspring of specimens from Newfoundland (Canada) were examined after dividing broods of various pairs into equal halves and subjecting them to short-day and long-day condition, respectively (Fig.1b). Although significant differences exist within the pooled data, the sex ratio of the descendants produced by several pairs could not be modified by exposure to LD 8:16 and LD 16:8. The differences noted between the two populations studied suggest that genetic factors are responsible. This is supported by the findings reported by Steele and Steele (1969), who could not find marked deviations from the 1:1 sex ratio in natural populations of a Newfoundland study area.

Investigations to elucidate the role of photoperiod in sex determination have been extended to two other euryhaline _Gammarus_ species, _G. zaddachi_ and _G. locusta_. By applying

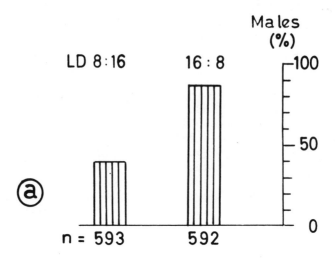

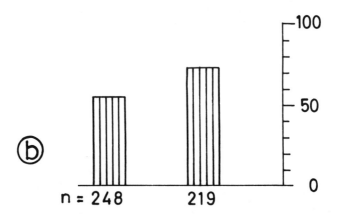

Figure 1. Influence of short-day and long-day photoperiod (8 and 16 hours of light per day; 50 - 100 lux) on sex ratio of progeny reared at a salinity of 10‰ and 15°C; n = number of sexed individuals. a: <u>Gammarus duebeni duebeni</u> (Elbe estuary); pooled data obtained from 12 pairs expressed as percentage of males. Statistics: X^2_1 = 298.52, $P < 0.001$. b: <u>Gammarus duebeni duebeni</u> (St. Philip's, Newfoundland); pooled data obtained from 10 pairs. Statistics: X^2_1 = 16.64, $P < 0.001$.

the same experimental procedure described above (see also Bulnheim, 1972), a shift of sex ratio induced by different daylengths has also been recorded in G. zaddachi obtained from the Elbe estuary. Resembling G. d. duebeni but in a less pronounced way, ♂♂ outnumber ♀♀ following exposure to long-day photoperiod (Fig.2a).

On the other hand, a photoperiodic response was not seen in G. locusta obtained from the island of Helgoland (North Sea). Progeny produced by a number of pairs revealed an almost balanced sex ratio under both photoperiodic regimes (Fig.2b). In this species, there is some indication of a homo-heterogamety switch mechanism; however, this assumption could not be verified by demonstrating the occurrence of heteromorphic sex chromosomes or sex-linked inheritance.

Besides the monogeny phenomena which may be induced by different daylengths, thelygeny occurs in G. d. duebeni and this cannot be modified by photoperiod. As mentioned above, such thelygenic ♀♀ selected from the Elbe estuary population produce progeny consisting exclusively or predominantly of female descendants. Surprisingly, the origin and perpetuation of thelygeny was demonstrated to be a consequence of parasitization by certain microsporidians transferred via the eggs from generation to generation (Bulnheim, 1967a, 1969). Since all or most eggs contain the parasites, the emerging progeny develop completely, or predominantly, into females. Male descendants, which occasionally arise in such thelygenic strains, develop from uninfected eggs. On the other hand, intersexes, which may also appear among the offspring produced by thelygenic ♀♀, are usually infected.

Investigation of the microsporidians showed that they comprise two distinct species, Octosporea effeminans and Thelohania heredieteria (Bulnheim and Vavra, 1968; Bulnheim, 1971). Either of them, independently of each other, may exert a feminizing influence when they pass into the young. However, except for their influence on sex ratio, they do not affect any important life functions of their host including fecundity. Both microsporidians parasitize the ovaries and T. heredieteria also invades the muscles.

Rearing experiments performed with single individuals parasitized by O. effeminans or T. heredieteria, respectively, revealed some differences in the extent of shift of sex ratio in favour of females. Further differences were established by comparing sex ratios in the offspring of infected individuals derived from various populations.

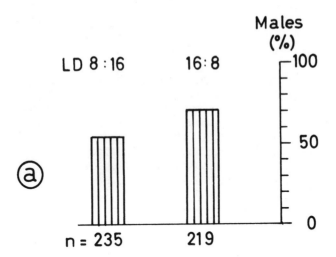

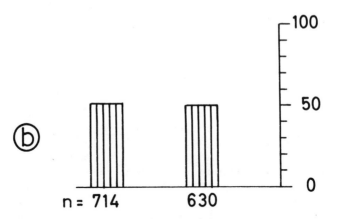

Figure 2. Influence of short-day and long-day photoperiod on sex ratio of progeny. a: <u>Gammarus zaddachi</u> (Elbe estuary); pooled data obtained from 12 pairs (S = 10‰, 15°C). Statistics: χ^2_1 = 12.84, P < 0.001. b: <u>Gammarus locusta</u> (Helgoland, North Sea); pooled data obtained from 8 pairs (S = 30‰, 15°C). Statistics: χ^2_1 = 0.26, P > 0.5.

Figure 3 shows the results of breeding experiments conducted with descendants of a thelygenic female infested by O. effeminans. This thelygenic line was maintained through 8 generations by continuously outcrossing female descendants with males from non-infected strains. In contrast to the all-female offspring observed from most pairs, the progeny of pair M 36 was found to consist almost completely of males. As revealed by microscopic studies of the eggs deposited, the mother responsible for this sex ratio in the progeny was, by chance, not infected. Thus, sex differentiation of the young was being governed by both genetic and photoperiodic conditions. Since all broods were subjected to long-day photoperiod, males prevailed. A further deviation in the sex ratio of the progeny was observed in the 4th filial generation. About one third of the young produced by pair M 46 developed into males. There was evidence that some of the eggs released were not infected; most or all of the eggs free of microsporidians differentiated into males, whereas, of course, females developed from all parasitized eggs. This lower prevalence of infection, however, was not maintained in the following generation; as demonstrated by breeding from five infected descendants (M 54 - M 58) of this pair, all-femaleness eventually reappeared.

In total, 13 Octosporea-infected females collected in the field have been examined under laboratory conditions with respect to the sex ratio of the offspring. All of them gave birth to thelygenic broods.

In addition, females infected by T. herediteria have been used for breeding experiments. Progeny of most infected females also revealed complete or almost complete absence of males which was maintained in successive generations. Figure 4a shows the sex ratios obtained from broods of female descendants which, reared through 5 filial generations, had been outcrossed with various males. In this strain only a few of the infected ♀♀ gave rise to a few males and/or intersexes; such occasional males proved to be free of microsporidians.

From 15 parasitized females from the Elbe estuary tested for sex ratio, two individuals were selected which produced mixed progeny composed of infected females, uninfected females and uninfected males. Since sex differentiation of infected young is not influenced by photoperiod, the proportion of uninfected males to uninfected females is dependent upon the genetic background and the photoperiod conditions under which the broods are raised. The infected daughters of these imperfectly thelygenic ♀♀ also proved to be bisexual-type ♀♀, as demonstrated by breeding continued over 4 generations (Fig.4b).

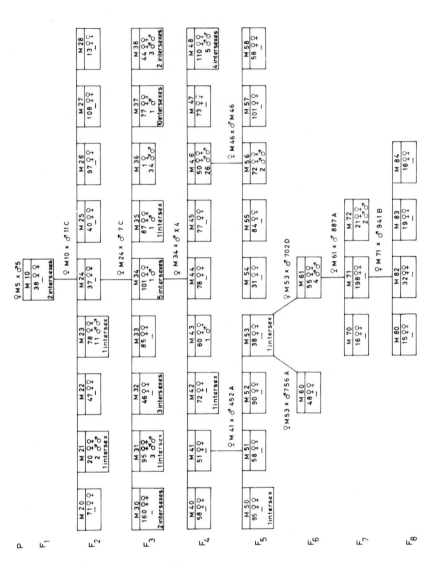

Figure 3. Sex ratio of descendants derived from a *Gammarus duebeni duebeni* ♀ (Elbe estuary) parasitized by *Octosporea effeminans* (S = 10‰, 15°C, LD 16:8).

MODES OF SEX DETERMINATION 537

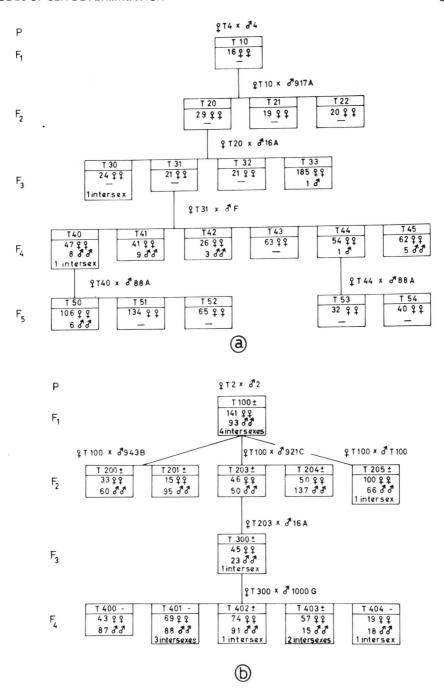

Figure 4. Sex ratio of descendants derived from two ♀♀ (Elbe estuary) parasitized by <u>Thelohania herediteria</u> (S = 10‰, 15°C, LD 16:8). a: perfect thelygeny, b: imperfect thelygeny; + indicates partial infection, - no infection of the eggs produced.

The reasons for the incomplete parasitization of the oocytes in some Thelohania-infected ♀♀ and their descendants are not clear. It is suggested that this phenomenon may reflect a genetically controlled parasite-host relationship. Since the sex-ratio condition for mixed progeny is maintained in successive generations despite matings with males of different genotypes it seems that the genetic background of the host is not responsible for this result. Environmental factors can also be excluded as being responsible for this imperfect thelygeny. Evidently, certain strains of microsporidians may exist which are responsible for either perfect or imperfect thelygeny. As a consequence of their transovarial transmission, these lines persist in the female descendants of their crustacean host.

Thelygeny caused by the feminizing influence of microsporidians has been shown not to be irreversibly imposed on infected females of the Elbe population studied. Also, environmental factors may play a role as interacting components by exerting an indirect influence on sex determination of parasitized females. O. effeminans - but not T. heredeteria - has been demonstrated to be sensitive to salinity. Following transfer from salinity 10‰ or 20‰ to 30‰, Octosporea-infected females deposit uninfected eggs which develop normally into females and males according to the genetic and photoperiodic conditions (Bulnheim, 1969). Thus, as a consequence of the loss of the parasites, a normalization of the mode of sex determination can be established. In addition, low temperatures may also have an adverse effect on the microsporidians. If the infected females are exposed to temperatures below $4^{\circ}C$ before oviposition, the microsporidians occasionally do not invade all eggs before these are released. Mixed progeny may, therefore, arise (Bulnheim, 1978).

In the course of this study several other G. d. duebeni populations have also been examined in order to find out whether or not parasitization by microsporidians occurs. In most populations studied, either O. effeminans or T. heredeteria or both protozoans could be detected. Some breeding experiments have also been performed.

A relatively high prevalence of O. effeminans (ca.- 50% of the females examined) has been found in a population from the Fjord of Schlei (Western Baltic Sea). Most females gave rise to all-female progeny; some, however, produced imperfectly thelygenic broods owing to partial infections of the oocytes. Whether or not mixed progeny may also arise in successive generations of imperfectly thelygenic females has not yet been tested.

In a rock-pool population from the Baltic Sea island of Bornholm, infections by O. effeminans as well as by T. herediteria have been found. Rearing experiments demonstrated that perfectly thelygenic broods resulted from Thelohania-infected females. On the other hand, bisexual offspring could be obtained from Octosporea-parasitized females, although female descendants prevailed. Surprisingly, in all broods tested, either all or most males were also infected. In such males the gonads contained both schizonts and spores of the parasite. Parasitized males did not exhibit any alteration of sex phenotype. This is in contrast to findings obtained from experimentally infected males of strains derived from the Elbe estuary. Following artifical infection by O. effeminans, parasitized males from this population acquire an intersexual appearance by developing oostegites (Bulnheim, 1977).

In addition, populations of Gammarus duebeni celticus have been included in surveys of the occurrence of microsporidian parasites and their influence on sex ratio. G. d. celticus which has been given subspecific rank by Stock and Pinkster (1970) is confined to waters with low ion concentrations and is distributed in certain localities of Ireland, the Isle of Man, Brittany and other areas.

Two different populations of this subspecies, obtained from Brasparts (Finistère, France) and the Isle of Man (Port Erin, Poyllvaaish) have been studied. The French population examined revealed a high prevalence of both microsporidians, these occurring in both females and males. Dual parasitization was also observed in both sexes. However, parasitized males did not display any alteration of external and internal sex characters (Bulnheim, 1978).

These findings were confirmed by investigations of populations from the Isle of Man. G. d. celticus from this area only show parasitization by T. herediteria. Again, this microsporidian has been found in females as well as in males.

The reasons for this resistance to the feminizing influence of the microsporidians are not yet clear. Possibly, the absence of the feminizing effect normally exerted by the parasites may be caused by a more stable sex genotype of the host. This suggestion derives from observations made with other euryhaline gammarids which are also parasitized by microsporidians but do not exhibit aberrant sex ratios. G. locusta and G. salinus may be parasitized by Thelohania species which differ from T. herediteria in some minor morphological characters (Bulnheim, 1975b). In both amphipods, both sexes are infected.

Sex Determination in Talitridae

In the following, some aspects of the genetics of sex determination in <u>Orchestia</u> species, as revealed by Ginsburger-Vogel (1973 - 1975), will be outlined. Hitherto, three species have been studied: <u>Orchestia gammarellus</u> (Pallas), <u>O. montagui</u> (Audouin), and <u>O. cavimana</u> Heller. Monogeny phenomena have been found to occur in <u>O. gammarellus</u> and <u>O. montagui</u>; in both species thelygeny is associated with the appearance of intersexuality.

Two categories of <u>O. gammarellus</u> populations can be ditinguished as revealed by differences in sex ratios (Ginsburger-Vogel, 1973): (1) Populations of estuaries exhibiting reduced percentages of males (25 - 30%) and relatively high numbers of male intersexes (ca. 15% of males); the latter are equipped with oostegites and reduced gnathopods. (2) Populations of pebble beaches which display sex ratios near to unity and low numbers of intersexes (1 - 2%).

In the progeny of animals obtained from estuarine populations (as studied at Penzé, Brittany, and other localities) the occurrence of amphogeny as well as of perfect and imperfect arrhenogeny and thelygeny has been established. As in the all-female condition observed in <u>Gammarus duebeni duebeni</u>, thelygenic females of <u>O. gammarellus</u> play an important role in the expression of thelygeny and intersexuality. This sex-ratio condition can be maintained through successive generations by crosses between females of such lines and amphogenic males. Since male intersexes and occasional normal males transmit neither thelygeny nor intersexuality when mated to amphogenic females, maternal inheritance, evidently mediated by a cytoplasmic factor, is suggested. The expression of thelygeny linked to intersexuality was demonstrated to be temperature-dependent. Crosses at 17°C between thelygenic females and amphogenic males yielded high rates of females and intersexes, whereas the same parental animals, following transfer to 22°C, gave rise to bisexual offspring or even to imperfectly or perfectly arrhenogenic broods (Ginsburger-Vogel, 1974, 1975c).

These and other observations on sex ratios obtained from various crosses led to a new interpretation of the mode of sex determination in <u>O. gammarellus</u>. Whereas in his first reports Ginsburger-Vogel (1967, 1970) tended to assume polygenic sex determination, he later (1972, 1975a) provided evidence for the existence of sex digamety in <u>O. gammarellus</u>

and O. cavimana by crosses between amphogenic ♀♀ and neo-males; the latter are genetic females transformed into functional males. Masculinization of juvenile females can be achieved before initiation of sex differentiation by implantation of androgenic glands, the source of the male sex hormone in Malacostraca. Such neo-males crossed with amphogenic females give rise only to female descendants. This result indicates female homogamety (XX) and male heterogamety (XY).

The transformation of thelygenic ♀♀ into arrhenogenic ♀♀, by high temperature, is attributed to an inversion of the sex phenotype of genetically determined males. The male genotype of such females is masked at 17°C, but expressed at 22°C. As evidenced by the sex ratio of progeny following matings to amphogenic ♂♂ at 22°C, the existence of XX-♀♀ as well as of XY-♀♀ and YY-♀♀ has been suggested (Fig.5). At the higher temperature XX-♀♀ produce mixed progeny, whereas XY-♀♀ give birth to imperfectly arrhenogenic broods (75% ♂♂) and YY-♀♀ to perfectly arrhenogenic broods (100% ♂♂). Intersexes born

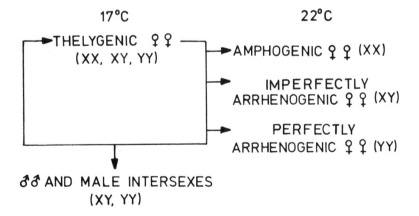

Figure 5. Interpretation of different categories of Orchestia gammarellus ♀♀ observed at 17° and 22°C. (Modified after Ginsburger-Vogel, 1974, 1975c).

from thelygenic females are regarded as incompletely feminized males (XY, YY). However, in this interpretation it is unclear which factors determine whether or not XY and YY individuals develop into females or male intersexes.

The mode of action of the cytoplasmic factor which is suggested as being involved in the maternal inheritance of thelygeny, and its inactivation by elevated temperature, is still unclear. Recently, a parasitic organism has been discovered to be associated with thelygenic O. gammarellus ♀♀; it is similar to a haplosporidian (Ginsburger-Vogel et al., 1976). Possibly it is responsible for the aberrant sex-ratio condition established.

In O. montagui thelygeny linked to intersexuality has also been found; it is explained by the action of a still unknown epigenetic feminizing factor. Evidence has been provided that it is not sensitive to raised temperatures (Ginsburger-Vogel, 1975b).

The mode of sex determination detected in O. gammarellus populations of pebble beaches (studied at Pointe de Bloscon near Roscoff, Brittany) is in some respects different from that established for populations of estuaries (Ginsburger-Vogel, 1975a). The sex ratio does not differ markedly from 1:1. Breeding experiments revealed a low frequency of arrhenogenic and thelygenic pairs. The monogeny phenomena observed appear to be different from those established for estuarine populations. Three categories of males have been found: they are either of the arrhenogenic, thelygenic or amphogenic type. On the other hand, females are all amphogenic; arrhenogenic and thelygenic males are responsible for the transmission of monogeny to their male offspring. Arrhenogeny is transferred to half of the males in each generation, whereas thelygeny is transmitted to a higher ratio of males, amounting to about 2/3.

The important role of males for the maintenance of monogeny was also verified by crosses with neo-males and females of different origin. Results of crosses between neo-males and females of amphogenic strains support the assumption of female homogamety, since progeny were shown to be all, or almost totally, female. Neo-males from thelygenic strains mated to females from amphogenic strains produced imperfectly thelygenic broods. The appearance of males in such broods is interpreted as being due to a still unknown masculinizing epigenetic factor which provokes the inversion of the sex phenotype of some **genetic females**. However, the action of autosomal factors interfering with the sex chromosomes cannot be excluded. This must be taken into consideration in the case of colour genes which affect sex determination in Gammarus pulex subterraneus (Anders, 1957). In this freshwater amphipod a series of three allelic colour genes (R^2, r, r^+) exert a more or less strong male determining influence on offspring. It is possible that a polyfactorial genetic mechanism, which may modify or overrule the action of the sex chromosomes, also operates in O. gammarellus.

Conclusions

The few amphipod species hitherto studied reveal rather complex modes of sex determination. Significant inter- and intraspecific differences became evident by comparing closely related species, including geographical populations. In this respect, some parallels can be established with isopods, in which several representatives have been investigated with reference to determination and inheritance of sex (cf. Legrand and Legrand-Hamelin, 1975).

In amphipods, proof for the existence of sex digamety (although not verified cytologically), as well as of sex plurigamety (Bacci, 1965), sensu polygenic sex determination has been obtained. Depending on the balance between the system of multiple masculinizing and feminizing genes, extrinsic factors, such as photoperiod, may exert a more or less pronounced influence on sex determination as well. Moreover, there may be an indirect effect by external factors, such as temperature and salinity. These environmental variables may interact with the epigenetic feminizing factors responsible for the occurrence of thelygeny. The agents inducing sex phenotype inversion can be clearly recognized as transovarially transferred microsporidians in G. duebeni duebeni; in Orchestia further research on their nature is necessary.

Thelygeny caused by micro-organisms has also been recorded from Armadillidium vulgare (Martin et al., 1973). Bacteria-like organisms which are transmitted via the eggs to the offspring have been found to be associated with thelygenic females and intersexes of this isopod. Monophologically similar micro-organisms have been reported to infect the nervous system of male intersexes in Ligia oceanica and to be responsible for the changed sex phenotype (Martin et al,, 1974). In addition, Juchault and Legrand (1975) suggest that a masculinizing infectious epigenetic factor of still unknown nature is responsible for intersexuality in Porcellio dilatatus dilatatus.

These recently detected parasitic or symbiotic associations should stimulate further analyses of the monogeny phenomena known to occur in some other pericaridean crustaceans. It should be noted, however, that apart from the presence of the micro-organisms, in their respective hosts, there was no indication of the existence of any plasmatic constituents that might be acting as plasmagenes. This indicates that extra-chromosomal inheritance, often assumed to occur in some of the amphipod and isopod species considered, has been mimicked by the transovarial transfer of micro-organisms. Their mode of action on the process of sexual differentiation deserves further detailed investigations.

Summary

The case of rearing and crossbreeding amphipod species in the laboratory has allowed study of the modes of sex determination in representatives of species inhabiting marine, brackish-water and fresh-water environments, comprising members of the families Gammaridae and Talitridae. This paper concentrates on studies of three euryhaline Gammarus species, in particular Gammarus duebeni duebeni. Genetic as well as various non-genetic factors acting in a hierarchically graduated system are shown to influence sex determination in this amphipod. The occurrence of thelygeny, caused by microsporidian infections and transmitted through the maternal line, is described. Comparisons between various populations of G. d. duebeni and G. d. celticus show some variability of their sex-determining mechanisms. The results obtained are compared with the monogeny phenomena (thelygeny, arrhenogeny) observed in Orchestia species by Ginsburger-Vogel (1973, 1974, 1975). In O. gammarellus and O. montagui thelygeny is linked with intersexuality. Though primarily regulated by an epigenetic feminizing factor, the expression of thelygeny is also environmentally influenced in O. gammarellus. In contrast to the polygenic sex determination seen in G. d. duebeni, sex digamety, as expressed by female homogamety and male heterogamety, is assumed to exist in Orchestia. Differences in the role of various sex-determining factors in closely related species, including geographic populations, are considered.

Acknowledgements

I am indebted to Drs. D. H. Steele (St. John's, Nfld., Canada) and J. Dieleman (Amsterdam, The Netherlands) for assistance in sampling some of the test animals, to

M. Mühlenkamp for help with the extensive breeding experiments and to J. Marschall for preparation of the figures.

REFERENCES

Anders, F. 1957. Uber die geschlechtsbeeinflussende Wirkung von Farballelen bei Gammarus pulex ssp. subterraneus (Schneider). Z. Vererbungsl. 88, 291-332.

Bacci, G. 1965. Sex Determination. Pergamon Press, Oxford,

Bocquet, C. 1967. Structure génétique du sexe chez les Crustacés. Ann. Biol. 6, 225-239.

Bulnheim, H. -P. 1967a. Mikrosporidieninfektion und Geschlechtsbestimmung bei Gammarus duebeni. Zool. Anz., Suppl. 30 (Verh. Dtsch. Zool. Ges. 1966), 432-442.

Bulnheim, H. -P. 1967b. Uber den Einfluβ der Photoperiode auf die Geschlechtsrealisation bei Gammarus duebeni. Helgoländer wiss. Meeresunters. 16, 69-83.

Bulnheim, H. -P. 1969. Zur Analyse geschlechtsbestimmender Faktoren bei Gammarus duebeni (Crustacea, Amphipoda). Zool. Anz., Suppl. 32 (Verh. Dtsch. Zool. Ges. 1968), 244-260.

Bulnheim, H. -P. 1971. Entwicklung, Übertragung und Parasit-Wirt-Beziehungen von Thelohania herediteria sp.n. (Protozoa, Microsporidia). Z. Parasitkde 35, 241-262.

Bulnheim, H. -P. 1972. On sex-determining factors in some euryhaline Gammarus species. In, Proceedings of the Fifth European Marine Biology Symposium. Battaglia, B. (ed). Piccin, Padova, pp. 115-130.

Bulnheim, H. -P. 1975a. Intersexuality in Gammaridae and its conditions. Pubbl. Staz. Zool. Napoli 39, Suppl. 399-416.

Bulnheim, H. -P. 1975b. Microsporidian infections of amphipods with special reference to host-parasite relationships: A review. Mar. Fish. Rev. 37 (5-6), 39-45.

Bulnheim, H. -P. 1977. Geschlechtsumstimmung bei Gammarus duebeni (Crustacea, Amphipoda) unter dem Einfluβ hormonaler und parasitärer Faktoren. Biol. Zbl. 96, 61-78.

Bulnheim, H. -P. 1978. Interaction between genetic, external and parasitic factors in sex determination of the crustacean amphipod Gammarus duebeni. Helgoländer wiss. Meeresunters. 31, 1-33

Bulnheim, H. -P. and Vávra, J. 1968. Infection by the microsporidian Octosporea effeminans sp.n., and its sex determining influence in the amphipod Gammarus duebeni. J. Parasitol. 54, 241-248.

Charniaux-Cotton, H. 1972. Recherches récentes sur la différenciation sexuelle et l'activité genitale chez divers Crustacés supérieurs. In, Hormones et Differenciation Sexuelle chez les Invertébrés. Vol 3. Wolff, E. (ed). Gordon & Breach, Paris. pp. 127-178.

Ginsburger-Vogel, T. 1967. Détermination du sexe et monogénie chez Orchestia gammarella (Pallas). C.R. Soc. Biol. (Paris) 161, 23-29.

Ginsburger-Vogel, T. 1970. L'inversion expérimentale du sexe chez les femelles d'Orchestia gammarella (Crustacé Amphipode) et l'étude du déterminisme génétique du sexe. Ann. biol. 9, 441-446.

Ginsburger-Vogel, T. 1972. Démonstration expérimentale de l'homogamétie femelle chez Orchestia cavimana (Heller) (Crustacés Amphipodes Talitridae). C.R. Acad. Sci., Paris 275 (Sér. D). 1811-1813.

Ginsburger-Vogel, T. 1973. Détermination génétique du sexe, monogénie et intersexualité chez Orchestia gammarella Pallas (Crustacé Amphipode Talitridae). I. Phénomènes de monogénie dans la population de Penzé. Arch. Zool. exp. gén. 114, 397-438.

Ginsburger-Vogel, T. 1974. Détermination génétique du sexe, monogénie et intersexualité chez Orchestia gammarella Pallas (Crustacé Amphipode Talitridae). II. Etude des relations entre la monogénie et l'intersexualité. Influence de la température. Arch. Zool. exp. gén. 115, 93-127.

Ginsburger-Vogel, T. 1975a. Détermination génétique du sexe, monogénie et intersexualité chez Orchestia gammarella Pallas (Crustacé Amphipode Talitridae). III. Etude de phénomènes de monogénie indépendants de l'intersexualité. Arch. Zool. exp. gén. 116, 615-647.

Ginsburger-Vogel, T. 1975b. Les phénomènes de monogénie liés a l'intersexualité et leur thermosensibilite; etude comparée chez Orchestia gammarella Pallas et Orchestia montagui (Audouin) (Crustaces Amphipodes Talitridae). Pubbl. Staz. Zool. Napoli 39, Suppl., 417-442.

Ginsburger-Vogel, T. 1975c. Temperature-sensitive intersexuality and its determinism in Orchestia gammarella Pallas. In, Intersexuality in the Animal Kingdom. Reinboth R. (ed). Springer, Berlin, pp.106-120.

Ginsburger-Vogel, T., Desportes I. et Zerbib, C. 1976. Présence chez l'Amphipode Orchestia gammarellus (Pallas) d'un Protiste parasite; ses affinités avec Marteilia refringens agent de l'épizootie de l'huître plate. C. R. Acad. Sci., Paris 283 (Sér. D). 939-942.

Juchault, P. et Legrand, J.-J. 1975. Modalités de la transmission héréditaire du facteur épigénétique responsable de l'intersexualité masculinisante chez le crustacé oniscoide Porcellio dilatatus dilatatus Brandt. Bull. Soc. zool. Fr. 100, 467-476.

Kinne, O. 1952. Zur Biologie und Physiologie von Gammarus duebeni Lillj., III. Zahlenverhältnis der Geschlechter und Geschlechtsbestimmung. Kieler Meeresforsch. 9, 126-133.

Kinne, O. 1953. Zur Biologie und Physiologie von Gammarus duebeni Lillj., VII. Über die Temperaturabhangigkeit der Geschlechtsbestimmung. Biol. Zbl. 72, 260-270.

Kosswig, C. 1964. Polygenic sex determination. Experientia 20, 190-199.

Le Calvez, J. et Certain, P. 1951. Gammarus chevreuxi Sext. et la caryologie des Gammariens. Arch. Zool. exp. gén. 88, 131-141.

Legrand, J.-J. et Legrand-Hamelin, E. 1975. Déterminisme de l'intersexualité et de la monogenie chez les Crustacés Isopodes. Pubbl. Staz. Zool. Napoli 39, Suppl., 443-461.

Martin, G., Juchault, P. et Legrand, J.-J. 1973. Mise en évidence d'un micro-organism intracytoplasmique symbiote de l'oniscoide Armadillidium vulgare Latr. dont la présence accompagne l'intersexualité ou la féminisation totale des mâles génétiques de la lignée thélygène. C. R. Acad. Sci., Paris 276 (Sér. D), 2313-2316.

Martin, G., Maissiat, R., Juchault, P. and Legrand, J. -J. 1974. Mise en évidence d'un micro-organisme intracytoplasmique symbiotique chez les intersexués (mâles à oostégites) du Crustace Ligia oceanica L. (Isopode Oniscoide). C.R. Acad. Aci., Paris 278 (Sér.D), 3375-3378.

Niiyama, H. 1950. The X-Y mechanism of the sex chromosome in the male of Anisogammarus annandalei (Tattersall) (Crustacea: Amphipoda). Ann. Zool. Jap. 23, 58-62.

Orian, A. J. E. and Callan, H. G. 1957. Chromosome numbers of gammarids. J. mar. biol. Ass. U.K. 36, 129-142.

Poisson, R. et Le Calvez, J. 1948. La garniture chromosomique de quelques Crustacés Amphipodes. C. R. Acad. Sci., Paris 227, 228-230.

Steele, D. H. and Steele, V. J. 1969. The biology of Gammarus (Crustacea, Amphipoda) in the northwestern Atlantic. I. Gammarus duebeni Lillj. Can. J. Zool. 47, 235-244.

Stock, J. H. and Pinkster, S. 1970. Irish and French fresh water populations of Gammarus duebeni subspecifically different from rackish water populations. Nature 228, 874-875.

Traut, W. 1962. Zur Geschlechtsbestimmung bei Gammarus duebeni Lillj. und verwandten Arten (Crustacea, Amphipoda). Z. wiss. Zool. 167, 1-72.

GENETICS OF SEX DETERMINATION IN OPHRYOTROCHA (ANNELIDA POLYCHAETA)

G. Bacci

Istituto di Zoologia,

Università di Torino, Italy

Environmental and Genetic Control of Sex Expression

Ophryotrocha puerilis, which is the earliest described species of the genus Ophryotrocha, is a protandrous hermaphrodite Polychaete. It was regarded (with Bonellia viridis) as the model of the so-called phenotypical determination of sex after Hartmann (1936-1943) showed that all female phase individuals revert to the male phase when subjected to treatments such as starvation or abnormal concentrations of different ions. Of particular importance is the so-called "pair culture effect" consisting in the reversal to the male phase of the less ripe female phase individual when two female phase individuals are put together in a small container. This effect has a clear adaptive significance as it allows the production of progeny from two individuals which were both originally in the female phase.

The uniform response of all individuals to such environmental treatment led to the postulate that the sex genotype of O. puerilis can be indicated by the general formula FFMM, where male and female factors are perfectly balanced (F=M) (Hartmann, 1956). Such formulation was regarded as valid for all hermaphroditic species and for all known examples of so-called phenotypical sex determination as opposed to genetical sex determination.

The clear cut division between phenotypical and genotypical mechanisms of sex determination was accepted with much favour in the biological literature because of its simplicity.

Nevertheless it brought to a standstill practically all
genetical research on the sex determination of hermaphrodites
as it assumed the complete uniformity of the sex genotypes and
interpreted variability of sex phenotypes in terms of environmental influences only.

Research by Coe (1948) on the Prosobranch Gastropod
Crepidula plana showed that a few male and female individuals
are present in predominantly hermaphroditic populations and
indicated that such sexual phenotypes are genetically determined. Also Baltzer (1937) at the end of his work on
Bonellia viridis, an Echiurid worm which shows an extreme
degree of male-female dimorphism, concluded that a few larvae
are genetically determined either as males or as females
while the majority of them become males or females according
to environmental influences. A new approach to the problem
of sex determination in hermaphroditic populations and to
research on the genetical and environmental control of sex
expressions originated from statistical investigations on the
Mediterranean limpet Patella coerulea (Bacci, 1949). Continuous variation between male and female individuals through
a majority of protandrous hermaphroditic individuals was
demonstrated in a Naples population of limpets and this was
interpreted as an example of polygenic sex determination.
Such interpretation was regarded as valid for a number of
hermaphroditic species.

Ophryotrocha puerilis proved to be the best material in
which to test the hypothesis. It has, however, some disadvantages, as its entire reproductive cycle takes more than
two months at 18°C and natural populations of O. puerilis
show limited variability of sex expression.

The time when the change from the male to the female
phase takes place is determined by the number of chaetigerous
segments shown by each individual when the first free oocytes
appear in the coelomatic cavities.

The data on which Figure 1 is based are derived from
483 individuals, that are the progeny of 4 pairs of O. puerilis
which were collected in Naples in the spring and summer months
of 1976. They change to the female phase at a mean length of
18·50 chaetigerous segments.

Individuals that changed sex at a length of 22 chaetigerous segments obviously experienced a male phase considerably
longer than the male phase of the individuals that underwent
sex inversion at a length of 15 segments. No pure males or
females, however, have been observed so far in the progeny of

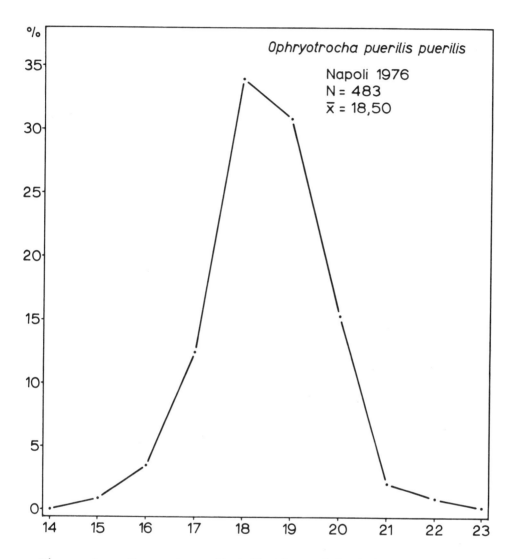

Figure 1. Percentage distribution of individuals of the Mediterranean subspecies which pass from the male to the female phase at different numbers of segments.

individuals which came directly from natural populations.

Brother-sister matings can be achieved between individuals that change early to the female phase and between those that change late because, when two female phase individuals are put together in a small container, one of the partners reverts to the male phase and fertilization can take place. Selection for early change to the female phase or for the prolongation of the male phase could thus be carried on and the mean lengths at which sex inversion takes place were significantly shifted at each generation (Fig.2). Lines selected for early manifestation of the female phase are called thelygenous and those selected for its retardation (or for the duration of the male phase) are called arrhenogenous lines.

Individuals where either the male or the female phase was abolished began to appear at generations 5 or 6, according to the strains, and they are designated as pure female and pure males lines respectively (Fig.3). Isolated oocytes have been observed in pure females even at the length of 4 chaetigerous segments and pure males, when examined cytologically, show, only in a few instances, abortive oogenesis. Crosses between such male and female individuals were highly fertile and produced, as expected, progenies which were composed of hermaphroditic individuals only and which were heterotic.

The hypothesis of polygenic sex determination was thus proved and Figure 4 provides a very simplified model for interpreting sex variation among hermaphroditic populations.

It is assumed that the different sex phenotypes are obtained through the interaction between four pairs of independent, additive factors, indicated by $M_1, 2, 3, 4$ for the male and by $m_1, 2, 3, 4$ for the female factors. Four M and four m factors in the genotype are supposed to give phenotypes that show a balanced expression of the male and of the female phases, more M (or male) factors give hermaphrodites with longer male phases, more m (or female) factors give longer female phases, until the balance is shifted to maleness or to femaleness in such a way that pure males or females are obtained. Pure M or m genotypes must be very rare and are supposed to be inviable, or at least, infertile for reasons that will be discussed later. Figure 5 shows how the model can be employed to interpret various instances of sex variability not only among hermaphroditic but also among gonochoric species like Bonellia whose members react in different ways with the same environmental factors according to their different sex genotypes. Pure males and females are absent from the

Gen.	P	N	x̄	Number of chaetigerous segments																								
				14	15	16	17	18	19	20	21	22	23	24	25	26	27	28	29	30	31	32	33–40
0		148	18,79			5	15	43	45	26	12	1	1										
1	17–17	145	17,98		2	12	36	53	22	18	2												
	20–20	122	20,56					8	13	35	40	22	3	1									
2	17–17	103	16,80	3	10	23	38	26	3														
	23–23	127	22,00						4	10	29	37	37	6	4								
3	15–15	170	16,44	2	29	54	63	21	1														
	23–23	72	22,76							1	13	18	17	16	7									
4	14–14	133	16,70	2	11	40	54	23	3														
	25–26	98	22,97						4	5	14	20	14	19	13	5	3	1						
5	28–31	6	24,00									2	–	1	2	–	–	–	–	–	–	and 5 pure ♂♂
	34–35	10	27,40									1	2	2	1	–	–	–	–	1	–	–	1	and 2 pure ♂♂
6	30–30	10	30,70							1		1	1		–	–	–	–	–	1	–	1	1	1	and 1 pure ♂
	36–37	4	26,75									1	1		2	–	–	–	–	–	–	–	–		and 6 pure ♂♂

Figure 2. Selection for the prolongation of the male and the female phase in arrhenogenous and thelygenous lines respectively. P indicates the number of chaetigerous segments when the parents of each generation changed sex. The arrhenogenous line has produced pure males but selection was stopped at the fourth generation of the thelygenous line (after Bacci, 1975).

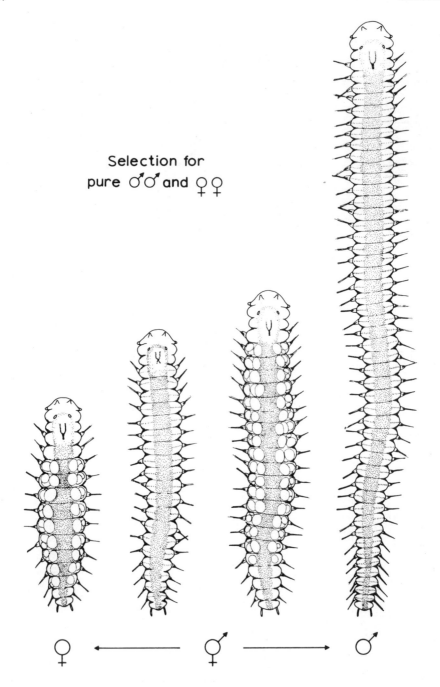

Figure 3. Pure females (left) and males (right) have been obtained through selection from purely protandrous hermaphroditic strains (after Bacci, 1965).

GENETICS OF SEX DETERMINATION IN *Ophryotrocha* 555

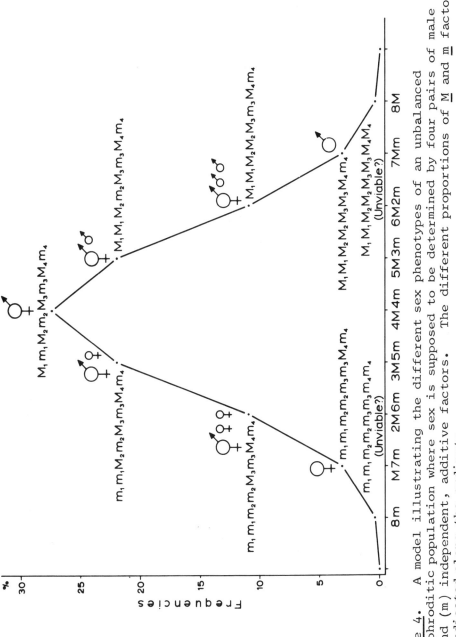

Figure 4. A model illustrating the different sex phenotypes of an unbalanced hermaphroditic population where sex is supposed to be determined by four pairs of male (M) and (m) independent, additive factors. The different proportions of M and m factors are indicated along the ordinate.

GENOTYPES	% FREQ.	PHENOTYPES IN:			
		OPHRYOTROCHA PUERILIS	PATELLA COERULEA	CREPIDULA PLANA	BONELLIA VIRIDIS
8 M —	0.4		Probably unfertile		
7 M 1 m	3.1	Pure ♂ (in the laboratory only)	Pure ♂ (in the nature)	Pure ♂ (associated)	Pure (genetic) ♂ (Spätmännchen)
6 M 2 m	10.9		♂ with long ♂ phase		Individuals which are easily masculinized
5 M 3 m	21.9				
4 M 4 m	27.3		♂ with equal lenghts of ♂ and ♀ phases		Individuals which are equally easily masculinized or feminized
3 M 5 m	21.9		♂ with long ♀ phase		Individuals which are easily feminized
2 M 6 m	10.9				
1 M 7 m	3.1	Pure ♀ (in the laboratory only)	Pure ♀ (in the nature)	Pure ♀ (isolated)	Pure (genetic) ♀ (larvae which will not adhere to the adult female)
— 8 m	0.4		Probably unfertile		

Figure 5. The variability of sex expression is interpreted both in unbalanced hermaphroditic strains and in Bonellia on the basis of the polyfactorial theory of sex determination (after Bacci, 1965, modified).

natural populations of O. puerilis but they exist in populations of Patella coerulea, which is regarded as a typical unbalanced hermaphroditic species. The hypothesis of the existence of multiple sex genotypes in Bonellia (Bacci, 1965) has been supported by recent observations (Leutert, 1975).

The number of effective factors involved in the determination of sex in O. puerilis has not been properly ascertained and unpublished data by Sella indicate 17 pairs which, however, we regard as not very reliable yet and requiring further tests.

Estimates of the heritability of length of sex phase made by Sella (1970) on three thelygenous and three arrhenogenous lines gave values of approximately 0.35. It must also be pointed out that a drastic fall in the fitness takes place during the selection experiments after the 5th or 6th generation, just when the highest or lowest levels of the mean inversion lengths are reached.

Such a fall in fitness is certainly related to the phenotypical variability of the strains becoming greater and greater at each generation.

It is thus clear why no male or female individuals have ever been observed among the progeny of O. puerilis specimens obtained from the sea. It appears that the conditions prevailing in the natural environment exert a kind of stabilizing selection thus maintaining the mean inversion length of the Mediterranean population (O. puerilis puerilis) at a mean of 18,5 chaetigerous segments and the mean inversion length of the Atlantic populations (O. puerilis siberti) at a mean level of about 20 segments (Bacci and La Greca, 1953).

Interactions between Sex Factors

Cytological investigations on the gonadal differentiation of pure male and female individuals show the existence of a homeostatic mechanism, which is inherent in the maturation of the gametes in the arrhenogenous and thelygenous lines (Bacci and Bortesi, 1967). In fact prolonged selection in favour of either sex phase in O. puerilis gradually postpones and prevents the development of the sex elements against which selection has been carried on.

It has been remarked, however, that those same sex cells, for an increase of which selection has been carried on, differentiate and reach maturity in small numbers and irregularly. Such an effect becomes apparent at the 3rd and 4th generation both in the arrhenogenous and thelygenous lines. If, for instance, selection has been carried on for the prolongation of the male phase not only is degeneration of oocytes and nurse cells observed but the maturation of spermatogonia, spermatocytes etc. is also slowed down.

The most conspicuous effects are observed at the 5th or 6th generations when uncontrolled proliferation of germinal cells takes place in most individuals. After abortive beginnings of sex differentiation, some of them dedifferentiate, reproduce actively by mitosis and eventually fill up the coelomatic cavities of some segments. Spermatogenesis or oogenesis in arrhenogenous or thelygenous lines respectively continues in the remaining segments.

The classic work by Goldschmidt on Lymantria dispar and by Bridges on Drosophila melanogaster, which are both gonochoric species, showed that the alteration of the sex balance produces intersexes when the differences between the male and female valences drop below the epistatic minimum or when the X/A ratios fall between 0.5 and 1 respectively. In the hermaphroditic O. puerilis, on the contrary, pure or predominantly male or female individuals are obtained by shifting the balance in favour of either sex factor and therefore by increasing the differences in the numbers of the M (male) and m (female) factors or, one might say, between the male and female valences in the different lines or individual genotypes.

It can be concluded, therefore, that the damage to the production and differentiation of the sex cells for which selection has been carried on is a consequence of the breakdown of the balance between male and female sex factors, which is maintained in the natural populations of O. puerilis.

A further and more important conclusion is obtained from these observations, namely that a certain number of factors for the other sex are necessary in order to obtain the normal expression of the genes of the complementary sex. In fact the normal expression of sex is fully re-established when hermaphroditic individuals (which are heterozygous for male and female factors) are obtained through matings between pure homozygous male and female Ophryotrocha.

In Figure 6 we see furthermore that normal development of male and female gonads is obtained also when pure males and females (resulting from prolonged divergent selection) are put together and interaction between prevalently male and female genotypes is achieved, possibly through the production

GENETICS OF SEX DETERMINATION IN *Ophryotrocha*

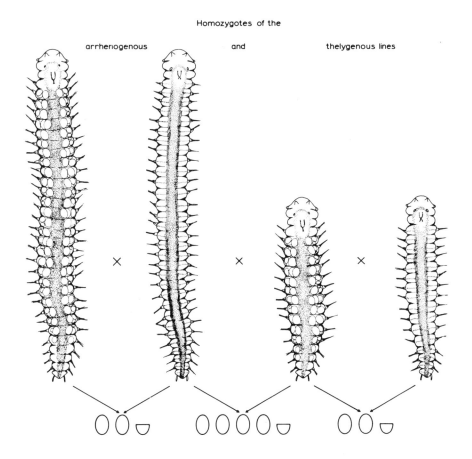

<u>Figure 6</u>. The reproductive rate (no. of eggs per day) is doubled when nearly homozygous individuals from arrhenogenous and thelygenous strains are mated. (Bacci & Bortesi, 1967.)

of pheromones produced by both the male and the female. In
this instance prevalently female genotypes are activated by
prevalently male genotypes and vice versa. A sort of
"compound supergenotype", which is perfectly balanced from
the point of view of sex factors, is thus obtained through
phenotypical interactions between the partners and as a
result the reproductive rate of the pair is doubled in comparison with the reproductive rates shown by the homozygous
pairs of the same generation, which are sexually unbalanced.

The assumption that the presence of factors for both
sexes is necessary for the activation of either the male or
female factors or both now provides an interpretation for the
general occurrence of genetic sex bipotency, a principle based
on the work of Bridges on _Drosophila_ intersexes and supported
by a number of other researches both in the field of genetics
and in the field of embryology.

Sexuality in eucaryotes appears more and more to be the
expression of a complex system of genetic and environmental
factors in which the interaction between male and female
factors plays a major role.

It is now evident that in _O. puerilis_ (as well as in
Crepidula, in _Bonellia_ and in a number of other species) the
genetic factors of one sex activate the factors of the other
sex by means of hormone-like substances or pheromones which
exert their action either by contact (as in _Ophryotrocha_) or
by diffusion in water (as in _Bonellia_). Such interactions
at the phenotypical level help to explain the mechanisms that
render necessary the joint presence of genetic factors of
both sexes in each male and female phenotype. In fact highly
heterozygous genotypes for both male and female factors (e.g.
4$\underline{M}$4$\underline{m}$, 5$\underline{M}$3$\underline{m}$ or 3$\underline{M}$5$\underline{m}$ genotypes) show the same high degree of
fertility as the pairs of nearly homozygous genotypes for
either male or female factors (e.g. 7$\underline{M}$1$\underline{m}$ or 1$\underline{M}$7$\underline{m}$ genotypes).
This means that interactions between $\underline{M}$ and $\underline{m}$ factors must take
place in the heterozygous individuals in similar ways either
by means of repressing or derepressing mechanisms. As a
matter of fact the "compound supergenotypes" referred to
earlier formed by the pairs composed of near homozygotes
for the male and for the female sex (7$\underline{M1m}$ + 1$\underline{M7m}$ = 8$\underline{M8m}$)
and in the pairs formed by heterozygotic hermaphrodites
(4M4m + 4M4m = 8M8m) have practically the same composition,
and are perfectly balanced, as shown by employing the simple
notation of diagram 4.

At this point one can advance the hypothesis of the

existence of circuits between operons of different sex polarity. Even if such genetic mechanisms have not been conclusively demonstrated in eucaryotic organisms, models of this type may help in understanding interactions between complementary sex factors and hormones.

The first assumption is that a small amount of hormone of one sex is necessary in order to derepress the genes of the opposite sex. In such instances regulator genes of the two kind of operons must be regarded as repressor genes and h hormones therefore function as derepressors which are capable of inactivating the repressors coded by each regulator gene.

Figure 7a is a representation of a pair of "male" and "female" operons where the product of the regulator gene (indicated with R^R) is repressed by the products (hormones or pheromones) of the structural genes of the complementary sex, which therefore function as derepressors. Two operons only are indicated in the simplified scheme but the existence of several operons for sex must be assumed in Ophryotrocha according to experimental results. M_1......M_n and m .....m_n indicate structural genes in male and female operons respectively.

The scheme fits with our knowledge of sex expression in Ophryotrocha. The altered differentiation and reduced production of male sexual elements in the arrhenogenous lines subjected to strong selection against female (m) sex factors, can in fact be ascribed to reduced production by the female structural genes. Only a partial derepression of the operons for the male sex can be obtained as a consequence and therefore only a limited proportion of germinal cells differentiate properly in the male direction. A similar model can explain sex differentiation in the thelygenous lines. When, on the other hand, predominantly male or female genotypes, form a pair - and a balanced "compound genotype" is originated - heterologous pheromones dissolved in the medium reinstate the balance and activate the structural genes to the normal levels. (Figure 6.)

In some instances, however, activation of sex factors is obtained through an inhibitory mechanism and the repression of the ovarian development by the androgenic gland in Crustaceans is the clearest example of such a mechanism.

It can be assumed therefore that some sex operons are regulated through a circuit (Fig.7b) where the genes of one sex repress the genes of the other sex by means of their

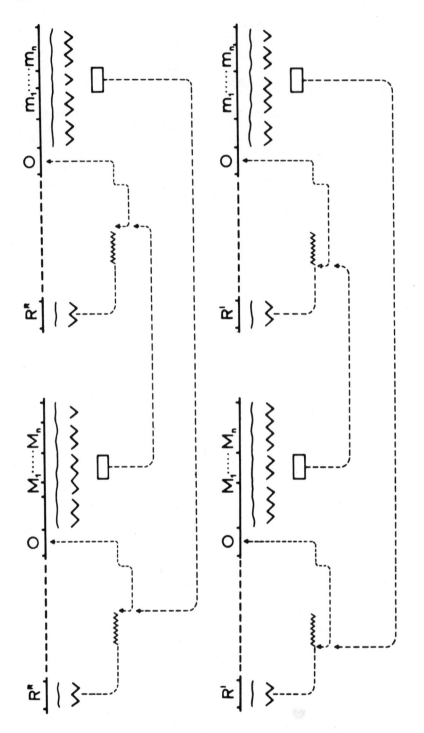

Figure 7. Circuits between hypothetical sex operons: 7a the repressor is inhibited and 7b the repressor is induced.

products and inactivate the substance produced by the respective regulator gene, which is indicated by R^I. The derepression of each sex is thus obtained through the inactivation of the structural genes of the opposite sex. The above mentioned mechanism of action of the androgenic gland referred to above can be interpreted in this way.

A combination of the two types of interaction can be suggested for the pair culture effect in O. puerilis: the partner that has the stronger female genotype inhibits the expression of the female factors of the partner with stronger male genotype, which in turn stimulates the female factors in the former (Bacci, 1952).

The balance between sex factors and the influence of environmental factors decide which type of interaction (activation or repression) will take place and at what stage of the individual life cycle. In this way the signals and instructions issued by the environment are differently modulated according to the structure of the different genotypes available in the populations (Bacci, 1975).

Crosses are now being carried on in order to obtain thelygenous and arrhenogenous lines of Ophryotrocha puerilis without inbreeding in order to check whether an altered balance between M and m factors or the reduced fitness (due to inbreeding depression) is responsible for the altered differentiation of the sex cells.

Sexual Races and Species of Ophryotrocha

A review of the existing species of Ophryotrocha has recently been published by Åkesson (1975): seven of them show different kinds of hermaphroditism, four species are gonochoric, two are not sufficiently investigated from the point of view of sex conditions and one of them, O. labronica, is represented by both gonochoric and hermaphroditic populations.

Details of O. labronica were first published in 1962 (La Greca and Bacci) and the first observations on sex phase were made on a protandrous hermaphroditic population changing sex at a mean length of about 13 segments. Parenti (1960) also proved that the population collected in the Leghorn Aquarium could reproduce by self-fertilisation and therefore O. labronica was regarded as rather unsuitable for genetic research. However, current research shows that rare instances of self-fertilisation also occur in O. puerilis.

Later observations by Åkesson (1970-1975), led to completely different conclusions: a number of populations from the Mediterranean and from the Pacific coast of the USA are strictly gonochoric and show sex ratios ranging from 28% (Malaga) to 49.7% (Palma de Mallorca) male individuals. Sex ratios also vary significantly in consistent samples taken in different years from the same localities. Samples recently taken from Venice (1976) proved also to be gonochoric.

Åkesson, however, also found a few hermaphrodites in a strain collected in southern Portugal and in a recent paper (1976) Pfannenstiel has described a highly variable population from Naples where a high percentage of the individuals change from the male to the female phase and the others are either pure males or females. In such population there are therefore those males and females that in O. puerilis can only be obtained as a result of selection in the laboratory. Female phase individuals cannot however be fertilized either by pure males or by male phase hermaphrodites.

O. labronica is thus represented by purely gonochoric populations, by partly hermaphroditic populations (or unbalanced hermaphrodites, like the limpets) and by purely hermaphroditic populations which are also capable of self-fertilization. O. labronica can be regarded therefore as a valuable tool for the study of sex evolution and old and recent results require a complete reappraisal.

Environmental factors such as variations in salinity, temperature, water quality, K^+ in excess and different foods do not affect sex ratios in gonochoric populations, as shown by Åkesson (1972), but recent investigations carried out in Turin shows that both sex ratio and the time when the first oocytes appear can be significantly shifted by interactions which take place between partners in pairs formed by juveniles. Åkesson was able to shift sex ratios in favour of the male sex but not in favour of the female sex by selection experiments and also found that inbreeding increases the percentages of males (from 32 to 49.6%), a result that is in line with Battaglia's observations (1958, 1961) on the Copepod Tisbe reticulata.

It appears therefore that in gonochoric populations of O. labronica the presence of dominant factors for the female sex and of recessive, non-allelic factors for the male sex masks the possible presence of the multiple, additive factors for maleness and for femaleness demonstrated in O. puerilis and possibly also present in O. labronica.

Crosses between geographically distant gonochoric populations of O. labronica produced no progeny in a few instances. In other cases reciprocal crosses gave quite different results: males of the Naples II strain crossed with either Leghorn, Rovinj or Los Angeles females gave from 96 to 100% male progeny but the reciprocal crosses produced males in proportions approaching 50%. Significant shifts of the sex ratios were also observed in other crosses.

The result of 100% males obtained with crosses of males of the Naples II strain with females from other strains deserves special attention. 100% male hybrids are clearly not obtained as a consequence of prolonged inbreeding but, on the contrary, they are highly heterozygous. Therefore the absence of females must be imputed to an imbalance of the sex genes in the hybrid progenies due to the high valence of the male factors introduced by the Naples II strain. Such strains must be regarded therefore as belonging to a strong sexual race in the sense of Goldschmidt. Instead of intersexes of different degrees, as in Lymantria, purely male phenotypes are obtained from such crosses, as shown in Figure 8, where the valences of the + (male) and - (female) factors are expressed by mean values and vary in the different genotypes and gametes.

O. labronica is thus composed of a number of different sexual races even within the group of purely gonochoric populations. Sexual races were described also in O. hartmanni (Parenti, 1962) and in a sense the atlantic and mediterranean races of O. puerilis can be regarded as sex races because their times of sex inversion are significantly different.

As both hermaphroditic and gonochoric populations, and both weak and strong races of Ophryotrocha can be collected in the same localities it seems advisable to regard them as sex races because sexual characters are the outstanding differentiating features.

More or less advanced stages of reproductive isolation are correlated with such fragmentation of O. labronica into a number of sex races. The occurrence of self-fertilisation (although we do not know how common this might be) certainly increases the possibility that some populations will become more and more isolated.

It is also worth mentioning that, according to Åkesson, protracted inbreeding does not reduce the reproductive rate of some labronica strains, and this explains why self-fertilisation could take place in the original Leghorn strain

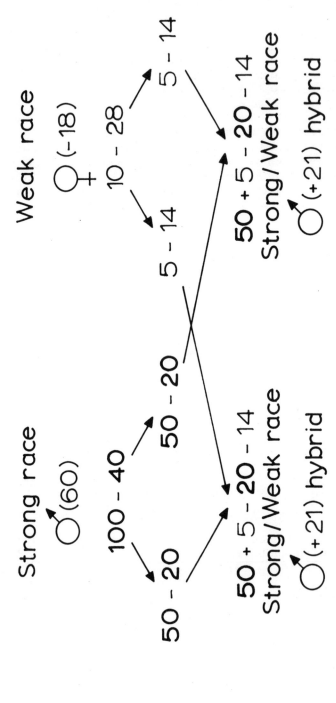

Figure 8. An interpretation of the appearance of totally male progenies from crosses between strong race males and weak race females. Numbers at the sides of each sex phenotype indicate the epistatic minima.

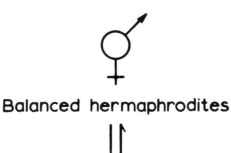
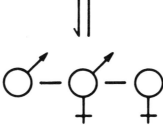
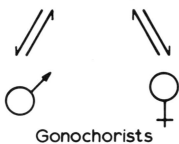

Figure 9. A scheme of the sex conditions found in different populations of O. labronica and of their possible evolutionary pathways.

without impairing the fitness of the population. It also explains why no pure male and female individuals have been observed in hermaphroditic populations of labronica and are instead eliminated by homeostatic mechanisms from O. puerilis populations in nature.

The extreme variety of sexual conditions in the genus Ophryotrocha appears to have originated basically from polyfactorial mechanisms, which allow very effective controls of sex determination and which can adjust themselves very rapidly to varied selective factors. Some species of Ophryotrocha can be regarded as evolutionary "culs de sac" from the point of view of the sex expression but O. labronica (which may well be considered as an aggregate of sibling species) represents a remarkable tool for the study of sex evolution. The umbalanced hermaphroditic populations, with their extreme variability of sex expressions, appear to be at the centre of evolution of sex in labronica (Fig.9). Stabilizing selection (of the type operating on O. puerilis) can originate balanced hermaphroditic populations, that is hermaphrodites with reduced sex variability. On the other hand divergent selection can give rise to gonochoric populations with the help of some chromosomal mechanism favouring more or less close linkages between either male or female factors on different chromosomes. The existence of strong and weak races indicates that the process of building up gonochoric populations is still going on in O. labronica. Interbreeding between different populations and strong selective forces in heavily polluted habitates certainly influence the sexual structure of most populations of O. labronica and cause transitions from one condition to another.

Summary

Ophryotrocha puerilis is a protandrous hermaphroditic species which shows the so-called 'pair culture effect' of reversal to the male phase of one animal when two female phase individuals of the species are kept together in the same culture vessel. Selection for the prolongation or the reduction of the male phase is effective and lines characterised by pure males or pure females have been obtained.

The joint influence of multiple genetic sex factors and of environmental factors thus accounts for the variability of sex expression in O. puerilis.

A shift of the balance of sex determining factors in favour of either male or female factors eventually not only adversely affects the differentiation of the sex cells against which selection has been carried on but it results in a reduction in the production of germ cells appropriate to the sex favoured by selection. It is assumed therefore that a minimum number of the factors of one sex are necessary in order to activate the factors of the complementary sex. Such an interpretation provides an explanation of the well known principle of genetic sex biopotency.

Different kinds of interaction between complementary sex operons (through either induction or repression) are envisaged.

About twenty species of Ophryotrocha are known, some of them are hermaphroditic, some are gonochoric. O. labronica comprises both gonochoric and hermaphroditic populations and reproductive isolation of varying degrees favours the segregation of sexual races. The first O. labronica population described is protandrously hermaphroditic and shows self-fertilisation, two populations are unbalanced hermaphroditic, and many populations are gonochoric. Both weak and strong sexual races are represented among gonochoric populations.

REFERENCES

Åkesson, B. 1970. Sexual conditions in a population of the Polychaete Ophryotrocha labronica La Greca & Bacci from Naples. Ophelia 7, 110-119.

Åkesson, B. 1972a. Incipient reproductive isolation between geographic populations of Ophryocha labronica (Polychaeta, Dorvilleidae). Zool. Scr. 1, 207-210.

Åkesson, B. 1972b. Sex determination in Ophryotrocha labronica (Polychaeta, Dorvilleidae). In, Fifth European Marine Biology Symposium. Battaglia, B. (ed). Piccin, Padova.

Åkesson, B. 1975. Reproduction in the genus Ophryotrocha. Pubbl. Staz. Zool. Napoli 39 Suppl. 377-398.

Bacci, G. 1949. Osservazioni statistiche sulla determinazione dei sessi in Patella coerulea. Pubbl. Staz. Zool. Napoli 22, 26-39.

Bacci, G. 1952. Diverso comportamento sessuale delle Ophryotrocha puerilis di Napoli e di Plymouth. Boll. Soc. It. Biol. Sper. 28, 1293.

Bacci, G. 1965. Sex Determination. Oxford, Pergamon Press.

Bacci, G. 1975. Genetic and environmental controls of sex determination in marine animals. Pubbl. Staz. Zool. Napoli 39 Suppl. 366-376.

Bacci, G. and Bortesi, O. 1967. The restablishment of sex balance in Ophryotrocha puerilis through interactions between individuals from arrhenogenous and thelygenous strains. Experientia 23, 448.

Bacci, G. and La Greca, M. 1953. Genetic and morphological evidence for subspecific differences between Naples and Plymouth populations of Ophryotrocha puerilis. Nature 171, 1115.

Baltzer, F. 1937. Entwicklungsmechanische Untersuchungen an Bonellia viridis. III. Ueber die Entwicklung und Bestimmung des Geschlechtes und die Anwendbarkeit des Goldschmidtschen Zeitgesetzes der Intersexualitat bei Bonellia viridis. Pubbl. Staz. Zool. Napoli 16, 89-159.

Battaglia, B. 1958. Balanced polymorphism in Tisbe reticulata, a marine Copepod. Evolution 12, 358-364.

Battaglia, B. 1961. Rapporti fra geni per la pigmentazione e la sessualitá in Tisbe reticulata. Atti Ass. genet. ital. 6, 439-447.

Battaglia, B. and Volkmann-Rocco, B. 1969. Gradienti di isolamento reprodivo in popolazioni geografiche del Copepode Tisbe clodiensis. Atti Inst. veneto sci. (Cl. Sci. Mat. Nat.) 127, 371-381.

Coe, W.R. 1948. Variations in the expression of sexuality in the normally protandric gastropod Crepidula plana. J. exp. Zool. 208, 155-170.

Hartmann, M. 1956. Die Sexualität, 2nd ed. Fischer. Stuttgart.

Hartmann, M. and Huth, W. 1936. Untersuchungen u. Geschlechtsbestimmung und Geschlechtsumwandlung von Ophryotrocha puerilis.Zool. Jahrb. (Phys.) 56, 389-439.

Hartmann, M. and Lewinski, G. 1938. Untersuchungen ... II Versuche uber die Wirkung von Kalium, Magnesium u. Kupfer. Zool. Jahrb. (Phys.) 58, 551-574.

Hartmann, M. and Lewinski, G. 1940. Untersuchungen ...
   III. Die stoffliche Natur der vermannlichenden Wirkung
   "starker" Weibchen (Eistoffe). Zool. Jahrb. (Phys.)
   60, 1-12.

La Greca, M. and Bacci. G. 1962. Una nuova specie di
   Ophryotrocha delle coste tirreniche. Boll. Zool. 29,
   13-23.

Leutert, R. 1975. Sex Determination in Bonellia. In,
   Intersexuality in the Animal kingdom. Reinboth (ed).
   Springer Verlag, Berlin.

Parenti, U. 1960. Self-fertilization in Ophryotrocha
   labronica. Experientia 16, 413-414.

Parenti, U. 1962. Variabilitá sessuale di una nuova
   sottospecie di Ophryotrocha hartmanni del Mediterraneo.
   Rend. Acc. Naz. Lincei 33, 78-83.

Pfannenstiel, H.D. 1976. Ist der Polychaet Ophryotrocha
   labronica ein proterandrischer Hermaphrodit? Mar.
   Biol. 38, 169-178.

Sella, G. 1970. Osservazioni sulla selezione per la
   modificazione delle fasi sessuali in Ophryotrocha
   puerilis puerilis. Pubbl. Staz. Zool. Napoli 37,
   630-640.

A NEW OPHRYOTROCHA SPECIES OF THE LABRONICA GROUP (POLYCHAETA, DORVILLEIDAE) REVEALED IN CROSSBREEDING EXPERIMENTS

B. Åkesson

Department of Zoology

University of Gottenburg, Sweden

Introduction

In her report on marine annelids of North Carolina Hartman (1945) recorded the first member of the dorvilleid genus Ophryotrocha from the eastern American coast. She obtained O. puerilis from shallow water around Pivers Island in Beaufort, North Carolina.

From Hartman's description it is obvious that she confused the species. The characteristic cutting edge of the mandibles (Hartman, Plate 5, Fig.7) is found not in O. puerilis, but in all known members of the labronica group; a group of sibling species which differs from most other Ophryotrocha species in chromosome number, morphology and reproductive pattern (Akesson, 1975). The information given by Hartman about size and number of segments conforms more closely with members of the labronica group than with O. puerilis. The first species of this group, O. labronica, was described in 1962 by La Greca and Bacci.

Four members of the sibling labronica group have been cultivated for some years in Gothenburg. They are O. labronica, O. robusta and O. macrovifera from the Mediterranean Sea and O. notoglandulata from Japan (Akesson, 1975). These four species are completely intersterile in all possible crossing combinations. They differ in minor morphological details and in reproductive patterns.

Material and Methods.

During a visit in 1974, to Duke University Marine Laboratory in Beaufort, I found some hundred small Ophryotrocha specimens in shallow water around Pivers Island. The fouling community under rafts constituted the best habitat for collecting. The polychaetes proved to be even more abundant along the docks of Morehead City about eight kilometers from Beaufort.

In 1975 I found a similar Ophryotrocha in Bermuda as a contaminant in the aquaria at the Bermuda Biological Station for Research.

Like most Ophryotrocha species the material from Beaufort, Morehead City and Bermuda proved to be easy to culture. The three strains were brought to Gothenburg for identification, and the polychaetes were treated according to the methods described by Åkesson (1970, 1975).

During the preliminary work with the new material it became evident that the Morehead City and Bermuda strains were mixed populations of two species. In crosses with previously known species one of the two species was identified as O. macrovifera.

Ophryotrocha macrovifera differs from other members of the sibling group by having a larger egg diameter and larval release at a more advanced stage (Åkesson, 1975). I have collected the same species in southern Portugal and in Cyprus. In both these regions it is sympatric with O. labronica. O. macrovifera is also recorded from Tampa Bay, Florida, by Dr. Joseph Simon who kindly provided a strain for my comparative studies.

All these geographically distant populations of O. macrovifera are wholly interfertile. No genetic incompatibility has been revealed in the crosses.

The remaining specimens from Morehead City, Bermuda and Beaufort all proved to be from one single species which is reproductively isolated from other members of the labronica group.

Crossbreeding Experiments with the New Species

The new species is intersterile in crosses with O. notoglandulata, O. macrovifera and O. robusta; neither males nor females of the new species accept mating partners from any of these species. Both sexes of O. labronica are accepted in mating. These crosses are sterile when the female belongs to the American species, but are partially fertile in the reciprocal cross. Twelve strains of O. labronica were utilized in these experiments.

The genetic incompatibility in crosses between O. labronica and the new species is expressed in F1 hybrids as high mortality during early development and as release of inviable larvae from egg masses. Breakdown of reproductive potential in F2 hybrids and in back crosses has been observed. The probability of gene exchange in nature seems to be insignificant and the American species can therefore, be distinguished as a separate species, Ophryotrocha costlowi. It has been named after the Director of Duke University Marine Laboratory, Dr. John D. Costlow Jr., who facilitated my work at the laboratory in all possible ways.

Interspecific Crosses and F1 Hybrids

In intraspecific matings of both O. labronica and O. costlowi more than 95 per cent of the eggs developed into viable larvae. In the hybrid crosses no larvae were released from about one third of the egg masses. Most embryos died before the stage when jaws become visible. On average 30-35 per cent of the eggs in all hybrid egg masses developed into larvae which were released. Presumably these figures do not give a correct picture of the viability of individual embryos. Viable embryos may be killed within the egg mass by decomposing neighbours.

The decrease of viability varied among the released hybrid larvae. Some never started to feed and died within a few days. The mortality of those feeding was recorded from two days after larval release to the time when the fastest growing animals attained sexual maturity, indicated by the first appearance of egg masses in the bowls in which they were cultured. In experiments performed at 20-21°O the mortality of F1 hybrids during the growth and maturation period was significantly higher than in O. labronica and O. costlowi (Table 1). The rate of mortality in the Naples I strain of O. labronica has previously been determined as 10.0 ± 4.9 per cent at a temperature of 20°C (Åkesson, 1976). The newer estimate in Table 1 is in good agreement with this.

Perhaps an even better expression of reduced hybrid viability is obtained from a comparison of the mean number

TABLE 1

Mortality rates in Ophryotrocha labronica (Naples I strain), O. costlowi (Beaufort strain) and F1 hybrids. Each bowl was started with 2 day old feeding larvae, (50 per bowl). Each bowl was scored when the first egg mass appeared.

	No. of bowls	% mortality Mean ± S.E.	t	P
O. labronica	7	12.0 ± 2.6		
			3.244	0.001<P<0.01
O. labr. x O. costl.	7	23.7 ± 2.5		
			2.744	0.01 <P<0.05
O. costlowi	7	14.3 ± 2.3		

of adults per egg mass in intra- and interspecific matings (Table 2). The females were always taken from the Naples I strain of O. labronica. In both intra- and interspecific crosses these females were cultured from larvae released, following the standard procedure, and were distributed at random just prior to the first mating. At that time the oocytes of the first brood already have attained their maximum size and the average number of eggs per egg mass was therefore, assumed to be the same.

Table 2 shows that the mean number of adult F1 hybrids obtained in crosses with all three strains of O. costlowi was less than half the number in the Naples I strain of O. labronica. The considerable variation in the number of F1 hybrids reflects the great variation in reproductive success between individual matings.

The sex ratio of the Bermuda F1 hybrids was drastically altered compared with the ratios of the parent strains. Only female hybrids were produced in about 75 per cent of the matings. The sex ratio of the remaining cultures was about equal to that of the strains of the male parent (26.6 ± 5.5 per cent males). The sex ratios of the other two hybrids were also close to those of the male parents.

TABLE 2

Sex ratio and number of progeny in strains of Ophryotrocha labronica and O. costlowi and in F1 hybrids. (Calculated from the females' first brood.)

Parents (♂ × ♂)	% males			Adults produced per egg mass		
	(n)	$\bar{X} \pm$ S.E.	(n)	$\bar{X} \pm$ S.E.	Range	
O. labronica, Naples I	43	31.6 ± 1.1	43	118.7 ± 6.1	29 - 210	
O. costlowi, Beaufort	12	50.9 ± 1.6	12	77.4 ± 6.2	43 - 113	
O. costlowi, Bermuda	11	28.7 ± 4.6	11	109.4 ± 14.1	47 - 198	
O. costlowi, Morehead	12	59.4 ± 2.1	12	69.3 ± 4.1	53 - 95	
O. labr. N. × O. costl. Beauf.	19	54.6 ± 3.4	19	21.1 ± 5.9	0 - 74	
O. labr. N. × O. costl. Berm.	37	7.4 ± 2.3	31	40.7 ± 7.9	0 - 163	
O. labr. N. × O. costl. Moreh.	24	60.6 ± 1.8	12	50.3 ± 10.1	14 - 88	

In a growth rate experiment the Beaufort F1 hybrids were compared with the two parent strains. After synchronous spawning 350 larvae of each kind were distributed among culture bowls of 80 ml in volume with 50 larvae per bowl. They were reared to the adult stage following the standard procedure. When the first egg masses appeared in a bowl the animals were counted and the number of segments was recorded for each individual (Fig.1). The material was the same as that used in the mortality experiment (Table 1). As it was difficult

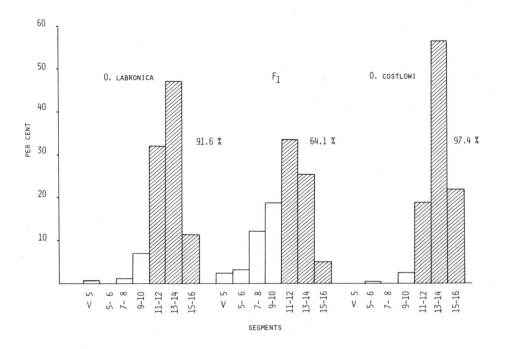

Figure 1. Size distribution of <u>Ophryotrocha labronica</u>, <u>O. costlowi</u> and their F1 hybrid 22 days after larval release from the egg masses. Hatched portions and percentage figures indicate size classes where more than 50 per cent of the females have free oocytes in the coelom. Based on the same material as in Table 1.

to obtain 50 larvae from a single hybrid egg mass, the larvae represent mixed populations obtained from many egg masses.

The hybrids spawned as early as did animals of the parent strains at an average of about 20 days from the start of the experiment. The mean segment numbers were: $\underline{O.\ costlowi}$ 13.4, $\underline{O.\ labronica}$ 12.7 and the F1 hybrid 11.0. All three figures are significantly different from each other ($P < 0.001$). The parent strains had a low percentage of small animals. In the F1 hybrids about 36 per cent had less than 11 segments. The subdivision of the histogram (Fig.1) between 10 and 11 segments has been chosen because at 11 segments more than 50 per cent of the females had free oocytes in the coelom.

Similar retarded growth rates due to an increased number of inviable larvae have also been observed in crosses between distant strains of $\underline{O.\ labronica}$ (Åkesson, 1972a). On the other hand, hybrids of $\underline{O.\ puerilis}$ are heterotic (Åkesson, 1977a).

Hybrid Breakdown

In routine cultures 100-150 adults of $\underline{O.\ labronica}$ or $\underline{O.\ costlowi}$ can be reared in bowls of 80 ml volume. At this density they thrive well enough to produce numerous egg masses and generally about 95 per cent of the deposited eggs develop into viable larvae.

The F1 hybrids are fertile, but the F2 embryos have reduced viability. When F1 hybrids were kept at the same density or even under less crowded conditions, they produced as many egg masses as the parent strains, but no embryos developed to the stage at which larvae are released. In most embryos the development was confined to early cleavage stages. In a few egg masses some embryos developed to a stage when sclerotized jaw pieces became visible as a dark spot (the dark jaw stage of Table 3).

In 8 bowls, each with 25 adult F1 hybrids (15 females and 10 males) from a cross $\underline{O.\ labronica}$, Naples I strain x $\underline{O.\ costlowi}$, Beaufort strain, more than 100 egg masses were deposited in a 4-week period, but not a single larva survived. A replicate experiment with the Bermuda F1 hybrids yielded the same result.

The conditions in 80 ml bowls with 25 animals per bowl are much more favourable than those in the natural habitat. There is no reason, therefore, to presume that the reproduc-

TABLE 3

Hybrid breakdown in F2 and in backcrosses to the parent strains, Ophryotrocha labronica, Naples I, and O. costlowi, Beaufort. All numbers refer to egg masses.

	F1 x F1	F1 x O. labr.	F1 x O. costl.	O. labr. x F1	O. costl. x F1
Total number of egg masses	44	26	34	10	12
Development confined to early cleavage	35	15	20	0	0
Some embryos develop to the "dark jaw" stage	1	6	8	0	0
Some larvae are released but die before the 6-setiger stage	5	1	6	0	0
Some released larvae grow to sexual maturity	3	4	0	10	12

tive success of F1 hybrids would be better in nature.

In order to provide an environment as close to optimum as possible the F1 hybrids were crossed and backcrossed in a series of experiments with only one female and one or two males in each bowl. The results of these experiments are summarized in Table 3. Of the 44 F1 hybrid matings, which yielded approximately 5,000 eggs, only 3 egg masses produced sexually mature F2 hybrids, in total 12 individuals, 5 females and 7 males. When these F2 hybrids were crossed all embryos died at an early cleavage stage.

Backcrosses of F1 females to males of the parent strains yielded similar results. Reciprocal crosses were more successful. Sexually mature progeny developed from all matings. However poor viability was reflected in the low mean number of adults per egg mass. The mean numbers of adults per egg mass are summarized for the parent strains, F1 and F2 hybrids and the backcrosses in Table 4.

Ethological Isolation

As crosses of O. costlowi female x O. labronica male are sterile, "female choice" experiments could be performed where a female of O. costlowi has the choice between a conspecific male and an O. labronica male. If the eggs develop, they are the result of homogamic matings. The very rare cases, when both males mate simultaneously with the female, can be neglected. Should this happen, the egg mass could be identified by the presence of sections with all embryos dead and other sections where all eggs develop. Results of female choice experiments between the Beaufort strain of O. costlowi and 12 strains of O. labronica have been reported by Akesson (1977a). Stalker's isolation index, expressed as homogamic matings minus heterogamic matings, divided by the total number of matings, varied from 1 to -0.33. An almost complete isolation could be demonstrated in crosses with three strains of O. labronica from the eastern Mediterranean Sea. From previous crosses it is known that these eastern strains also exhibit a high degree of ethological isolation at the intraspecies level (Åkesson, 1975).

In 8 out of 12 combinations there were no significant differences from random mating. One of these combinations involved the Naples I strain of O. labronica. As this strain has been used in the other crosses it was also chosen for a series of experiments involving selection for increased ethological isolation.

TABLE 4

Mean number of adults per egg mass in parent strains, in crosses and backcrosses between Ophryotrocha labronica, Naples I strain, and O. costlowi, Beaufort strain.

Strain or cross (♀ x ♂)	No. of egg masses	Adults per egg mass $\bar{X} \pm$ S.E.
O. labronica	43	118.7 $\pm$ 6.1
O. costlowi	12	77.4 $\pm$ 6.2
O. labr. x O. costl.	19	21.1 $\pm$ 5.9
O. costl. x O. labr.	24	0.0
F1 x F1	44	0.3 $\pm$ 0.2
F1 x O. labr.	26	0.4 $\pm$ 0.2
F1 x O. costl.	34	0.0
O. labr. x F1	10	16.7 $\pm$ 4.2
O. costl. x F1	12	8.3 $\pm$ 2.4

In generation I a synchronous spawning of O. costlowi from the unselected Beaufort strain yielded a population of larvae which was reared to a stage close to sexual maturity. A female choice experiment was set up with one female and two males in each of 24 bowls. The O. labronica males were of equal age and of equal size to those of O. costlowi. If a male became sick or looked inferior to the other one, both males were replaced.

When two or three egg masses had been produced in each bowl, the progeny of two females with three consecutive homogamic matings were selected for the next generation. In this and succeeding generations of selection the O. costlowi male and female of individual bowls were siblings.

The second and third generations were treated in a similar manner to the first one. With the one difference that the material was subdivided into four subgroups each of which represented descendants of one O. costlowi egg mass.

The ratio of homogamic/heterogamic matings in generation

II was significantly different from random mating, but in generation III a return to random mating is seen. So far the selection had not been successful (Table 5).

The results of the first three generations also make it clear that within the subgroups the ratio of homogamic/heterogamic matings over a period of one month or more was not significantly different from the ratio calculated on the females' first mating ($0.1 < P < 0.5$). In the next two generations, therefore, the experiment was interrupted when all females had produced their first brood and the selection was made from comparisons between the subgroups.

An extra series was added in generation V. The bowls contained two equally mature females of O. costlowi and one male of each species. If both females are ready to mate simultaneously and if there is a tendency to discriminate against the alien male, one female will mate with the conspecific male. The second female has the choice of waiting for the same male or mating with the alien male.

Due to technical reasons the selection was interrupted after generation V. The progeny was inbred by sib-mating for four generations in four subgroups. One more female choice experiment was performed in generation X, one series with one female per bowl and a second one with two females per bowl. The generations of inbreeding did not affect the level of ethological isolation.

The selection series are summarized in Table 5 together with the control series. The first of these has been reported previously (Åkesson, 1977a). Generation I of the unselected stock served as the second control series. The third control series was run parallel to generation IV and the fourth one was run together with generation X. The null hypothesis of the chi-square test which postulates no change of ethological isolation due to selection must clearly be rejected.

In the previously reported female choice experiment with 12 strains of O. labronica (Åkesson, 1977a) there was a certain preference for alien males in the series with the Leghorn strain. Thirty egg masses were produced as the result of 10 homogamic and 20 heterogamic matings. The result was not significantly different from random mating ($\chi^2 = 3.33$, $0.05 < P < 0.10$). In a corresponding series with selected O. costlowi individuals from generation X, all females mated with conspecific males. Obviously the successful selection in favour of ethological isolation between O. costlowi and

TABLE 5

Selection in favour of increased reproductive isolation in the Beaufort strain of *Ophryotrocha costlowi*. In each replicate one or two females of *O. costlowi* had a choice between a conspecific male and a male of *O. labronica*, Naples I strain. The isolation index is calculated as Homogamic − Heterogamic /Total number of matings.

Generations	No. of replicates	No. of ♀♀/bowl	Matings Homogamic (+)	Heterogamic (−)	Isolation index	χ^2
I (not selected)	24	1	28+	26−	0.04	
II	24	1	43+	23−	0.30	7.501**
III	24	1	27+	28−	−0.02	0.014
IV	24	1	17+	7−	0.42	4.878*
V:1	24	1	24+	0−	1.00	25.685***
V:2	12	2	20+	3−	0.74	13.761***
VI − IX Inbreeding between siblings in 4 separate substrains						
X:1	24	1	44+	0−	1.00	47.089***
X:2	12	2	26+	1−	0.93	24.901***
Control 1	12	1	10+	14−	−0.17	
Control 2	24	1	28+	26−	0.04	
Control 3	12	1	11+	13−	−0.08	
Control 4	12	1	8+	8−	0.00	
All controls	60	1	57−	61−	−0.03	

* = $P < 0.05$ and > 0.01, ** = $P < 0.01$ and > 0.001, *** = $P < 0.001$

the Naples I strain of O. labronica was not confined to this particular combination.

Discussion

The results of the crosses make it obvious that O. costlowi is more closely related to O. labronica than to any other member of the labronica group. The members of this group are very similar in morphology and reproductive patterns. They also seem to prefer the same kind of polluted or semi-polluted habitat, such as harbours, lagoons and estuaries. This does not mean, however, that these sibling species are the results of recent speciation. On the contrary, studies on parasite-host relationships indicate that at least some species have evolved independently for a long time (Akesson, 1975, 1977b).

The reciprocal differences in crosses between O. costlowi and O. labronica has its counterpart in interpopulation crosses in O. labronica and in intersubspecies crosses in O. puerilis (Akesson, 1972a, 1975, 1977a). Similar intraspecific reciprocal differences have also been reported in Tigriopus fulvus by Bozic (1960) and in Tisbe clodiensis by Battaglia and Volkmann-Rocco (1969, 1973); Volkmann (1977). They may be explained as due to a kind of cytoplasmic inheritance and are possibly connected with the presence of extranuclear DNA in the oocytes (Emanuelsson, 1969, 1971). In selection experiments Akesson (1972b) demonstrated a maternal hereditary factor in sex determination of O. labronica.

When comparing growth rates in the interspecies cross O. labronica x O. costlowi (Fig.1) and the intersubspecies cross O. puerilis puerilis x O. puerilis siberti (Akesson, 1977a), the frequency of size classes indicated heterosis in O. puerilis but decreased viability in F1 hybrids of the labronica/costlowi cross. Part of this difference, however, may be due to the fact that inviable F1 embryos die at an earlier stage of development in O. puerilis than do hybrids of O. labronica x O. costlowi. In O. puerilis there seem to be two outcomes, either larvae die at an early cleavage stage or they are released as normal, even heterotic larvae. In the labronica/costlowi hybrids the inviability is expressed at more levels. Zygotes may die at early cleavage stages, at the "dark jaw" stage or be released as morphologically abnormal larvae which never start feeding. Even among the feeding individuals some are inviable, which is expressed as high mortality and a retarded growth rate (Table 1, Fig.1).

The mean segment numbers of the two kinds of hybrids and their parents are compared in Table 6. Members of the puerilis group were reared for 30 days from larval release, members of the labronica/costlowi group for 22 days. The different length of the experimental periods reflects the time difference from larval release to sexual maturity in the two groups. In O. puerilis the hybrids are significantly larger than any of the parents. In the other group O. costlowi grows faster than O. labronica and both are superior to the hybrids.

The sex ratios of the three O. costlowi strains were significantly different from each other ($0.001 < P < 0.01$). In hybrids with male parents from Beaufort or Morehead the sex ratio was close to that of the male parent strain. O. costlowi from Bermuda had the lowest natural male ratio. In the hybrids it was even lower (Table 2). 28 out of 37 hybrid matings with Bermuda males produced all female progenies.

Such a drastic decrease of the male ratio has not been observed in interpopulation crosses of O. labronica. On the contrary, the male ratio is significantly increased in these crosses (Åkesson, 1972a, 1975).

The successful and rapid selection in favour of an increased ethological isolation indicates that few genes are involved. The two species are allopatric and in no need of a barrier to prevent gene flow. The unselected Beaufort stock culture is significantly isolated only from three strains of O. labronica (Åkesson, 1977a). They are all strains from the eastern Mediterranean Sea with strong ethological isolation also in intraspecies crosses (Akesson, 1975).

If these strains are excluded, the remaining nine combinations have a distribution of 139 homogamic and 133 heterogamic matings which agrees well with random mating expectation. Obviously there is no ethological barrier preventing mating between O. costlowi from Beaufort and most strains of O. labronica. But the genetic material from which ethological barriers can be built is already present. If the distribution areas of the two species come into contact, gene exchange would be prevented by a rapidly developing ethological isolation.

Summary

An Ophryotrocha from North Carolina and Bermuda is described as a new member of the sibling labronica group. The

TABLE 6

Comparisons between mean segment numbers of parent strains and hybrids in (A) *Ophryotrocha puerilis puerilis* 30 days after larval release, and (B) *O. costlowi* and *O. labronica* after 22 days.

Parents (♀ × ♂)	n	$\bar{X}$ + s.e.	t	P
(A) O. puerilis siberti, Plymouth	288	13.08 ± 0.21		
O. puerilis puerilis, Leghorn	340	13.50 ± 0.16	1.630	ns
F1 siberti × puerilis	332	14.39 ± 0.21	3.426	P < 0.001
F1 puerilis × siberti	325	14.68 ± 0.20	1.030	ns
(B) O. costlowi, Beaufort	300	13.41 ± 0.09		
O. labronica, Naples I	308	12.73 ± 0.10	4.995	P < 0.001
F1 O. labronica × O. costlowi	267	10.99 ± 0.16	9.426	P < 0.001

new species, Ophryotrocha costlowi, proved to be intersterile in all crossing combinations with four other members of the group, except in crosses between females of O. labronica and males of O. costlowi. Low hybrid viability was expressed as high mortality and a retarded growth rate. In one combination the hybrid sex ratio was changed in favour of females. The hybrid breakdown was almost complete in the F2 generation and in backcrosses to males of the parent strains. Backcrosses with females of the parent strains yielded 10-15 per cent adult progeny as compared to the fecundity of the parents.

Matings seem to be at random between O. costlowi and most strains of O. labronica. After ten generations of selection for homogamic mating, ethological isolation in a line of O. costlowi was almost complete.

The results are discussed and compared with results from intersubspecies crosses in O. puerilis.

REFERENCES

Åkesson, B. 1970. Ophryotrocha labronica as test animal for the study of marine pollution. Helgolander wiss. Meeresunters 20, 293-303.

Åkesson, B. 1972a. Incipient reproductive isolation between geographic populations of Ophryotrocha labronica (Polychaeta, Dorvilleidae). Zool. Scr. 1, 207-210.

Åkesson, B. 1972b. Sex determination in Ophryotrocha labronica (Polychaeta, Dorvilleidae). In, Fifth European Marine Biology. Battaglia, B. (ed). 163-172. Piccin Editore, Padova and London.

Åkesson, B. 1975. Reproduction in the genus Ophryotrocha (Polychaeta, Dorvilleidae). Pubbl. Staz. Zool. Napoli 39, Suppl., 377-398.

Åkesson, B. 1976. Temperature and life cycle in Ophryotrocha labronica (Polychaeta, Dorvilleidae). Ophelia 15(1), 37-47.

Åkesson, B. 1977a. Crossbreeding and geographic races, experiments with the polychaete genus Ophryotrocha. Mikrofauna Meeresboden 61, 7-14.

Åkesson, B. 1977b. Parasite-host relationships and phylogenetic systematics. The taxonomic position of dinophilids. Mikrofauna Meeresboden 61, 15-24.

Battaglia, B. and Volkmann-Rocco, B. 1969. Gradienti di isolamento reproduttivo in popolazioni geografiche del Copepode Tisbe clodiensis. Atti. Ist. veneto Sci. (Cl. Sci. Mat. Nat.) 127, 371-381.

Battaglia, B. and Volkmann-Rocco, B. 1973. Geographic and reproductive isolation in the marine harpacticoid copepod Tisbe. Mar. Biol. 19, 156-160.

Bozic, B. 1960. La genre Tigriopus Norman (Cop. Harpact.) et ses formes europeennes: recherches morphologiques et experimentales. Arch. Zool. exp. gen. 98, 176-269.

Emanuelsson, H. 1969. Electronmicroscopical observations on yolk and yolk formation in Ophryotrocha labronica La Greca and Bacci, Z. Zellforsch. 95, 19-36.

Emanuelsson, H. 1971. Metabolism and distribution of yolk

DNA in embryos of Ophryotrocha labronica La Greca and Bacci. Z. Zellforsch. 113, 450-460.

Hartman, O. 1945. The marine annelids of North Carolina Duke. Univ. Marine Stat. Bull. No. 2, 1-54.

La Greca, M. and Bacci, G. 1962. Una nuova specie di Ophryotrocha delle coste tirreniche. Bull. Zool. 29, 13-23.

Volkmann, B. 1979. Geographic and reproductive isolation in the genus Tisbe (Copepoda, Harpacticoida). Mikrofauna Meeresboden 61, 313-314.

BREEDING SYSTEMS, SPECIES RELATIONSHIPS AND EVOLUTIONARY TRENDS IN SOME MARINE SPECIES OF EUPLOTIDAE (HYPOTRICHIDA CILIATA)

R. Nobili, P. Luporini and F. Dini

Istituto di Zoologia

Università di Pisa, Italy

Introduction

Ciliates are common inhabitants of the sea, as about one third of the known species of the subphylum live in this environment. Apart from tintinnids and a few other forms, the majority of marine ciliates usually live as micro-benthos associations, especially of the littoral zone. Many have a cosmopolitan distribution and may thrive equally well in different biotopes. Free-living ciliates exhibit, in fact, a remarkably wide range of tolerance with regard to environmental conditions (Noland and Gojdics, 1967; Fauré-Fremiet, 1967; Dragesco, 1974) with food as the most important factor influencing their distribution.

To fully appreciate the role played by ciliates in a benthic community one should know something about the typical characteristics of the species populations such as organization, structure, genetic make-up, life cycle, growth and maintenance. Inferences can then be drawn as to the evolutionary strategies adopted by the species in meeting environmental demands. It is with this background that we report here in some detail what is known about the breeding system, the life cycle and the genetic make-up of some marine species of the family Euplotidae.

Within the genus <u>Euplotes</u>, 36 out of 51 well described species (Curds, 1975) are marine dwelling organisms all characterized by a flattened ventral surface, in which the mouth opens, encircled by ciliary membranelles used mainly as devices for food capturing. Agglutinated cilia (cirri) in a fixed number are also located on the ventral surface and are used by the animals to crawl about on the bottom substrata. They can also swim, but not for long because they are well adapted to live in the interstitial environment on account of their rather flattened body; they are commonly found on sandy substrata of varying granulometry and among algae.

The species we have studied i.e. E. vannus, E. crassus, E. minuta, E. cristatus all show the single-vannus type argyrome according to the terminology of Curds (1975). Two more species, E. mutabilis and E. balticus, are included in the vannus group and will be touched upon later.

The specimens used in this work were collected in different geographical areas (i.e. several regions of the Mediterranean Sea, North Sea and the Indian Ocean along the Somalian coast) from the sandy bottom of the eulittoral zone at depths varying from 0 to 40 m and stocks were maintained under laboratory conditions by feeding them on the green flagellate, Dunaliella salina.

A - Reproduction and Breeding System

These Euplotes species are characterized by typical cortical features (Fig.1). They reproduce asexually by transverse division, and this is their mode of increase in number. They reproduce sexually too, either by conjugation or autogamy, the latter process being limited to some populations of certain species as will be shown later. Conjugation ordinarily occurs between individuals belonging to complementary mating types. The term "mating type" refers to physiological differences between individuals that mark them off into mating classes, as first shown by Sonneborn (1937) in Paramecium aurelia. The mating type of an animal is inherited by its asexually produced descendants. Thus a clone is usually treated as a unit in the study of the breeding system or for other genetic purposes. We may assume that when competent animals of complementary types meet each other under favourable conditions, they conjugate. During conjugation the two partially fused animals undergo cross fertilization, then they separate and start dividing again, yielding exconjugant clones. Thus sex and reproduction are separated in Euplotes as in all ciliates, and this phenomenon might prove to be a useful tool for comparing individual flexibility versus genotypic plasticity within a clone.

As sex is the process for incorporating and transmitting

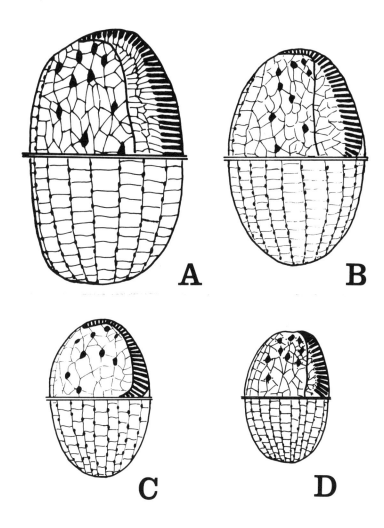

Figure 1. Cortical patterns (argyrome) revealed by silver impregnation of: A = E. vannus; B = E. crassus; C = E. cristatus; D = E. minuta. In each drawing the anterior part corresponds to the ventral surface with cirri and adoral membranelles of the mouth region; , the posterior part corresponds to the dorsal surface with rows of cilia (kineties) and a single regular mesh between them. (E. vannus is the largest species in agreement with Heckmann's paper of 1963).

genetic variation, sexual patterns and the breeding system
are very important in the evolutionary and adaptive strategies
of species. In this connection, it should be noted that
all the species considered here are organized in populations
within which a multiple mating type system exists. This
feature can easily be assessed by collecting as many specimens
as possible from a single station, then growing clones (stocks)
separately from each specimen and mixing the clones two by
two. The clones are assigned to the same mating type whenever
they do not react with each other. In Table 1, are given
data on 26 stocks of E. crassus collected on the same day from
an area of about 1 m^2 of a sandy beach on the Island of
Montecristo (Tuscan archipelago) analysed according to the
above procedure, and grouped into ten different mating types.
From the same sample many stocks of E. minuta were also recovered; they all fell into 13 mating types (Bracchi and
Luporini, unpublished).

The question as to how many mating types are present
within a population then appears to be relevant. As the
species show a cosmopolitan distribution, the answer to the
question requires the collection of stocks from different
localities and the testing of their mating types. Three
stations, distributed along the Tuscan coast-line between
the mouth of the river Arno to Piombino, were sampled regularly for one year; 167 stocks of E. crassus were gathered
and cultured in the laboratory. They fell into 20 different
mating types (Table 2). In a more restricted zone of the
same geographical area, Magagnini and Santangelo (1977)
examined seven stations, all included within 10^4 m^2 for two
years and found 11 mating types of E. crassus. Similar
results have been found time and again from samples of other
collecting areas not visited so regularly.

Because of the painstaking and very laborious procedure
required for maintaining and assessing a very large number of
stocks, not all of the necessary tests were performed between
the available mating types of the species. Nevertheless,
when randomly chosen stocks from distant or remote sources
are tested, the overall picture does not change. The number
of mating types shared by stocks from distant origin remains
very low both in E. minuta and in E. crassus (Table 3). In
the latter species, 29 stocks (three from the Eastern
Mediterranean, 12 from the Central Mediterranean Sea and 14
from the Somalian coast) proved to be of different and complementary types when recently analyzed by Dini and Luporini
(in preparation). All this evidence supports the hypothesis
that the breeding system of the above species comprises a
large number of mating types.

That the actual number may be even higher than that shown by the data is reasonably expected for two reasons. The first is that the mating types are genetically determined, as we shall see later, and secondly the bias associated with the way stocks are sampled from nature tends to produce an underestimate of the mating type number. Usually one or more days elapse from the time of collecting samples to the time the animals are isolated as single animals in the laboratory. This allows binary fission to occur, which increases the probability of isolating cells of the same clone, particularly if the mating types are present in different frequencies in the sample from nature. For this reason the sampling error tends to underrate the number of mating types in each sample and accordingly in the whole population. In addition, a small number of samples from a given area cannot be representative of the population dwelling in it. In spite of constraints and of the limited data pertaining to different geographical regions of the species distribution, one is justified in assuming that the mating types of E. crassus and E. minuta may number hundreds. This assumption is probably valid, too, for E. vannus (Heckmann, 1963; Génermont et al., 1976) and E. cristatus (Wichterman, 1967) which are also endowed with a multiple mating type system.

This multi-mating type system is genetically determined by a simple multi-allelic series at the mt locus. Heckmann (1963, 1964) thoroughly worked out the genetic determination of the mating type system in E. vannus and E. crassus. The same situation was found in E. minuta (Nobili, 1966a) and probably holds true for E. cristatus (Wichterman, 1967). Each gene of the multi-allelic series codes for a different mating type and a "peck-order" type of dominance exists among the alleles as shown in Table 4.

With a multi-allelic system the compatibility of clonal sib matings relative to non-clonal sib matings is reduced as the number of the alleles increases. Moreover, any new allele that arises will have the advantage of being compatible with all the others. Theoretically, this should result in a strong selection for a large number of alleles to produce new-universal acceptability when each clone is of a different mating type.

Power (1976) has discussed at length the selective forces in the evolution of mating types, coming to the conclusion that the maximum number of types in a population is probably limited by counter-selection operating at different levels. The data here presented indicate a finite number of mating types within each population but a high number within

TABLE 1

The breeding structure of 26 stocks of *E. crassus* collected on the same day in about 1m² of a sandy sample from Cala Maestra, Montecristo Island, Tuscan archipelago.

	2	4	5	9	10	12	15	16	17	20	27	1	14	24	11	19	26	6	21	13	23	7	8	18	22	25
2		−	−	−	−	−	−	−	−	−	−	+	+	+	+	+	+	+	+	+	+	+	+	+	+	+
4			−	−	−	−	−	−	−	−	−	+	+	+	+	+	+	+	+	+	+	+	+	+	+	+
5				−	−	−	−	−	−	−	−	+	+	+	+	+	+	+	+	+	+	+	+	+	+	+
9					−	−	−	−	−	−	−	+	+	+	+	+	+	+	+	+	+	+	+	+	+	+
10						−	−	−	−	−	−	+	+	+	+	+	+	+	+	+	+	+	+	+	+	+
12							−	−	−	−	−	+	+	+	+	+	+	+	+	+	+	+	+	+	+	+
15								−	−	−	−	+	+	+	+	+	+	+	+	+	+	+	+	+	+	+
16									−	−	−	+	+	+	+	+	+	+	+	+	+	+	+	+	+	+
17										−	−	+	+	+	+	+	+	+	+	+	+	+	+	+	+	+
20											−	+	+	+	+	+	+	+	+	+	+	+	+	+	+	+
27												+	+	+	+	+	+	+	+	+	+	+	+	+	+	+
1													−	−	+	+	+	+	+	+	+	+	+	+	+	+
14														−	+	+	+	+	+	+	+	+	+	+	+	+
24															+	+	+	+	+	+	+	+	+	+	+	+
11																+	+	+	+	+	+	+	+	+	−	−
19																	+	+	+	+	+	+	+	+	−	−

Table 1 (continued)

	2	4	5	9	10	12	15	16	17	20	27	1	14	24	11	19	26	6	21	13	23	7	8	18	22	25
26																	−									
6																	+	−								
21																	+	−	−							
13																	+	+	+	−						
23																	+	+	+	−	−					
7																	+	+	+	+	+	−				
8																	+	+	+	+	+	+	−			
18																	+	+	+	+	+	+	+	−		
22																	+	+	+	+	+	+	+	+	−	
25																	+	+	+	+	+	+	+	+	+	−

TABLE 2

Number of mating types of E. crassus found in three sandy places along the Tuscan coast about 30 km apart from each other in samples collected over a period of one year.

Stations	No. stocks	No. mating types	Mating types in common
A	65	7	2
B	48	5	1
C	54	8	1
	167	20	2

TABLE 3

Mating types of some natural stocks collected in different geographical areas. The number of mating types in common between two stations is given in parenthesis.

	E. minuta				E. crassus	
No. stocks	No. mating types	Locality	No. stocks	No. mating types	Locality
?	4	Villefranche	?	5	Sylt Island
27	6(1)	Marina Pisa	18	9	Marina Pisa
18	5(2)	Livorno	12	6(2)	Livorno
31	13	Montecristo Island	26	10	Montecristo Island
13	7	Gulf of Naples	30	11	Gulf of Naples
3	1	Venice	4	2	Venice
Total 92	33		90	41	

TABLE 4

Mating type determination likely valid for all E. vannus group of species. Mating types are indicated by Roman numerals; the corresponding genotypes are assigned on the basis of the peck order dominance given at the bottom of the table.

Mating types	Genotypes			
VII	mt^7/mt^7	mt^7/mt^6	....	mt^7/mt^1
VI	mt^6/mt^6	mt^6/mt^5	....	mt^6/mt^1
V	mt^5/mt^5	mt^5/mt^4	....	mt^5/mt^1
IV	mt^4/mt^4	mt^4/mt^3	....	mt^4/mt^1

$$mt^7 > mt^6 > mt^5 > mt^4 > \ldots\ldots\ldots mt^1$$

a species increasing as the number of areas sampled increases.

This outbreeding feature of the species, viz. the presence of many mating types, is reinforced by the species life cycle and other characteristics as well, which are not random but consistently point to outbreeding. Specifically, longer immaturity periods are associated with even longer maturity periods. In all four species, a period of sexual immaturity ensues at the initiation of the life cycle after conjugation during which asexual reproduction occurs and the animals so produced are unable to conjugate even when provided with mature potential mates under very favourable condition for crossing. The longer the immaturity period the higher the probability that animals will spread and meet potential mates of different genetic extraction (Sonneborn, 1957) 1957). The length of immaturity may average from 40 to over 100 fissions (this means a period of at least a month) which itself appears to be genetically determined probably on a multifactorial basis.(Luporini, 1968, 1970, and unpublished results). Once sexual maturity is reached, it is maintained for several hundred fissions in all the species. This long period is apt to favour the encounter of animals of complementary types before senescence phenomena set in.

Conjugating Euplotes do not always give rise to two exconjugant isogenic clones. This depends on the peculiar nuclear events taking place at conjugation and on other causes not well understood at present (Nobili and Luporini, 1967; Luporini and Nobili, 1967b) which might operate at the chromosomal level during meiotic processes.

All the species of Euplotes have a nuclear apparatus consisting of a large C-shaped macronucleus and a small roundish micronucleus. On conjugation, the micronucleus of each conjugant divides mitotically and the two resulting nuclei pass through meiosis, giving rise to eight haploid nuclei. While six of these degenerate, the remaining two divide once more. Out of the four products only two become the functional pronuclei. One pronucleus of each conjugant then migrates into the partner, uniting with the latter's stationary pronucleus to form a diploid fertilization nucleus. The normal nuclear apparatus is then reconstituted by the differentiation of a new macronucleus and a new micronucleus from the products of two successive mitosis of the fertilization nucleus (Fig.2).

All E. vannus-like species conform to the above scheme of nuclear behaviour, with the exception of the form studied by Luecken (1973) in which the third pregamic division is lacking. Now, if there is random survival of any two of the eight nuclei produced through meiosis and of any two of the four pronuclei, the ratio of isogenic: heterogenic exconjugant clones may readily be predicted according to the genotype of the mates (Sonneborn and Behme in Nobili and Luporini, 1967). In E. crassus and E. vannus random survival of pronuclei was actually found by Heckmann (1963, 1964). However, in E. minuta isogenic clones prevail over heterogenic ones which means that karyogamy occurs more frequently between genetically identical pronuclei (Luporini and Nobili, 1967a; Nobili and Luporini, 1967). Luporini and Dini (1977) found that crosses between Somalian stocks of E. crassus yielded significantly more heterogenic clones than expected on a random basis of karyogamy.

Should the variable ratio of isogenic: heterogenic exconjugant clones be relevant to the adaptability to a particular environment, then the difference reported above between species and populations within a species are to be expected. Indeed, when two exconjugant clones of a pair are of a different mating type they can conjugate again between themselves. While this diminishes the outbreeding potentialities associated with the multiple mating type system, it preserves the relative fitness in that habitat already enjoyed by the parent genotypes. Because of the patchiness of the environ-

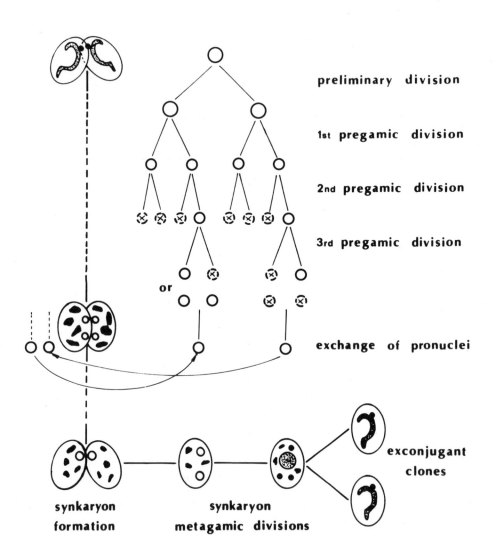

Figure 2. Schematic representation of nuclear events during conjugation of Euplotes.

ment in space and time, the tactics adopted by different populations need not be the same within the general strategy of= outbreeding. The scope of this strategy to ensure outcrossing is realized by the presence of several, or many, mating types living in each area. However, by varying the duration of sexual immaturity and the ratio of isogenic: heterogenic segregation of mating types in each kind of cross a certain amount of inbreeding can occur within a population.

The factors considered by Power (1976) such as (a) selection against genetically incompatible pairing, (b) competition and (c) sexual selection among mating types, by limiting their number may favour inbreeding too. Other important variables are dispersal potentialities and population density. As to the former, the species analyzed do not encyst and their dispersal can be accomplished only by the passive transport of vegetative cells, which have the advantage of being immigrants sexually compatible with the members with which they merge, but which probably cannot travel long distances by this means. As to the population density, it largely depends on resources at each particular habitat. When conditions are favourable and resources abundant, asexual reproduction will be at an immediate advantage and the multiplicative capacity of the animals will be fully expressed. However, if members of a single mating type are the only animals present they could not survive for a long time because of the decrease in their growth rate and final death which comes with clonal aging (Nobili et al., unpublished results). In clones of E. crassus, heterozygous at the mt locus, the recessive allele can be expressed and selfing (conjugation within a clone between cells of a different mating type) takes place (Heckmann, 1967), reinitiating a new life cycle.

Thus, even in a very predictable environment, at least two mating types must be present to ensure population survival. But the best chance for conjugation is offered to the animals as they reach a high density after fast multiplication when the food available has been exploited and several or many mating types are present simultaneously.

In a changing environment inbreeders may survive by rapidly expressing new mutations. In this way they can genetically track the environment, adapting and deadapting at each of the frequent and successive sexual events (Nyberg, 1974) as the environmental changes require. Outbreeders may attain the same result by the recombination of alleles provided that many sexually mature mating types are present when changes occur. Indeed, the finding of Luporini and Dini (1977) of a high degree of heterozygotes at the mt locus in the Somalian population, indicative of a very high number of mating types, is in favour of the recombination hypothesis.

Allowing for all these factors, we take the view that outbreeding prevails over inbreeding in natural populations of these species which can finely tune their sexual potentiality according to the conditions they have to meet in nature.

B - Phyletic Relationships of the Species

In revising the morphology and systematics of the genus Euplotes, first Touffrau (1960) and more recently Curds (1975) arrived at the conclusion that the six species mentioned in the introduction show a remarkable resemblance in all the features of taxonomical importance, which set them apart from the other known species of the genus. In particular the geometrical pattern of the dorsal argyrome (Fig.1), the meshwork structure revealed by the silver impregnation technique on the dorsal surface, is identical in all six species. What varies is the number of dorsal ciliary rows and the cell size which are both used to identify the species. The four species: E. vannus, E. crassus, E. minuta and E. cristatus were also shown to share common physiological characteristics such as a multiple mating type breeding system, a very similar life cycle and a genetically identical mating type determination (cf. Section A). All of these similarities point to the existence of a phyletic relationship among these four species and most likely with E. mutabilis and E. balticus too.

Indeed, when animals of a pure mating type ready to mate were mixed with comparable animals of either one of the other four species a loose cell aggregation was observed (Nobili, 1964, 1965, 1966b and unpublished results). Cell agglutination is considered a specific preconjugant cell interaction in mixtures of complementary types of a species. Agglutination was followed by hybrid pair unions which formed in a more or less unpredictable way and in varying number in the interspecies mixtures. Whatever their number in any mixture of the four species, such unions unfailingly proved to be sterile.

Recently Génermont et al. (1976) reported that crossfertilization occurs freely between E. mutabilis and E. crassus species with a normal "hybrid" survival. On this basis they no longer regarded E. mutabilis as a valid biological species and lumped it with E. crassus. They also say that some hybrids can survive when E. crassus is crossed with E. vannus although no data are reported. Our data, summarized in Table 5, indicate that all hybrid pairs, were either unable to separate and died out, or they separated, but no exconjugant clone survived the first few fissions after

TABLE 5

Outcome of Hybrid Conjugation

	E. minuta	E. cristatus	E. crassus	E. vannus
E. minuta	–	pairs separate, exconjugant death	pairs never separate, meiosis arrest	pairs never separate, meiosis arrest
E. cristatus		–	not done	not done
E. crassus			–	pairs separate, exconjugant death
E. vannus				–

conjugation. In the former type (minuta mixed with crassus and vannus) nuclear changes were arrested at the first meiotic division, in the latter type (crassus with vannus and cristatus with minuta) nuclear changes went on normally. Whether pronuclei were exchanged was not established. In our opinion, there appears to be no gene introgression between the stocks of species we analyzed.

Still in this connection, it is important to note that pairs can be formed between cells belonging to the same mating type of one species (Table 6) in almost all mixtures of any two species (Nobili, 1967, and unpublished results). Such pairs, as the two cells involved do not change mating type, may be called homotypic pairs. When homotypic pairing takes place in mixtures of E. vannus with E. crassus or E. minuta with E. cristatus, it is quite hard to distinguish hybrid pairs from homotypic ones because the size distributions of the two species overlap and they can easily be confused. Morphological or genetic marker have to be used to tell the two types

TABLE 6

Induction of homotypic pairing in mixtures of two species. + indicates homotypic pair formation at least in one of the mixed species. All stocks used belong to a different mating type. Among the many interspecies mixtures, only those involving three stocks per species are reported.

Species	stock	crassus 1 4 7	vannus A H V	minuta 3, 22, 31	cristatus a, o, f
crassus	1		+ − +	+ + +	+ − −
crassus	4		+ + +	+ + −	+ + +
crassus	7		+ + +	+ + +	− + +
vannus	A	+ + +		+ − +	− + +
vannus	H	− + +		− + +	+ + −
vannus	V	+ + +		+ + +	+ + +
minuta	3	+ + +	+ − +		+ + +
minuta	22	+ + +	− + +		+ + +
minuta	31	+ − +	+ + +		+ + +
cristatus	a	+ + −	− + +	+ + +	
cristatus	b	− + +	+ + +	+ + +	
cristatus	f	− + +	+ − +	+ + +	

of pairs apart since homotypic pairs do survive and usually give rise to vigorous exconjugant clones.

Homotypic pairing is as relevant as hybrid pairing in the context with which we are dealing for at least two reasons: (1) the conditions required for the occurrence of both phenomena are satisfied by mixing sexually mature and ready to mate animals; (2) immature or non-reactive animals or animals belonging to species other than the four used are unable to evoke agglutination and any form of pairing. Thus the conditions promoting mating in the interspecies mixtures are identical to those at work in the conjugation of complementary types of a species. Moreover, both hybrid and homotypic cell unions are restricted to mixtures of the <u>vannus</u> group of species (<u>E. mutabilis</u> should be included with <u>E. crassus</u>, thus only <u>E. balticus</u> is left out). It is reasonable to conclude that the mating signals of the four species still have some elements in common, which can be recognized by each one of them. But this amounts to saying that a very close relatedness exists among the species, which may be considered as being at an intermediate level of evolutionary divergence and relationship, compared with other morphologically different species of the genus. How large their genetic divergence is, remains to be explored. Unfortunately nothing is known about the chemical nature of the mating signals.

When two complementary types of a species are put together, 1-2 hours elapse (waiting period) before conjugant pairs are formed in the mixture. The waiting period is usually longer in interspecies mixtures. Heckmann and Siegel (1964) postulated that complementary cells exchange specific information during the waiting period so that they can unite in pairs. Dini and Miyake (1977), studying the kinetics of cell interaction during the waiting period with exposure to a protein synthesis inhibitor, found that the cells interact with a factor that is sensitive to the inhibitor and decays during about 4 h of treatment. They further inferred that this factor promotes the production of another factor which is insensitive to the inhibitor and is also involved in the interaction process. Protein synthesis goes on during sexual cell interaction. One could assume that the specific mating signal, whatever it may be, induces the production of a factor indispensable to modifying cells, making them apt to unite in pairs (Miyake, 1977). Homotypic or hybrid pairing could be explained as an aspecific response to specific and recognized mating signals of another species.

The breakdown of incompatibility between species and within a mating type should play different and minor roles in nature. The hybrids are sterile, while homotypic pairing occurs when conspecific complementary cells are not available. The interspecific sexual interaction might, however, help in arousing the "aptitude" to forming pairs in animals that live side by side in nature, as shown in Section A.

The present, widely overlapping distribution of the extant species of the <u>vannus</u> group raises the problem of how and when the speciation process started. We can only speculate about its remoteness and the possible routes taken because the necessary data are not yet available. A fruitful approach could be the isolation and chemical characterization of the mating substance(s) including the relative modification of their structure in the different species, associated with a better characterization of the ecological needs of the species.

C - Autogamic Reproduction and its Evolutionary role within the Species

In the preceding sections we have described some mechanisms by which inbreeding could occur in these species otherwise committed to outbreeding for so many of their characteristics. The extreme form of inbreeding is autogamy which consists in the fusion or karyogamy of two pronuclei formed in the one unpaired cell. Most ciliates produce two haploid pronuclei functionally differentiated in the migrating and stationary pronuclei during conjugation. In early cell pairing, nuclear changes conducive to meiosis and pronuclei formation are induced. A mutation which starts the cascade phenomena of sex at nuclear level without the need for mating would suffice to induce autogamic reproduction, since all the metabolic machinery for its occurrence would already be at disposal of the cell. Such a mutation, though rare enough not to be a burden for outbreeding populations, can however offer short-term advantages and therefore be exploited rapidly by the species. It is, then, not surprising that in the two cosmopolitan species E. minuta and E. crassus there are populations which can reproduce by autogamy (Siegel and Heckmann, 1966; Nobili, 1966a,b; Luporini, 1970; Luporini, 1974; Luporini and Dini, 1977).

Autogamous stocks are restricted in number and interspersed between non-autogamous ones as can be seen by comparing the distribution data reported in Table 7 with that of Tables 2 - 3. All stocks of E. minuta from the Somalian coast were grouped in six mating types, none of them autogamous (Luporini, unpublished results). The sparse distribution over a given area occupied by the species is reminiscent of the distribution of self-compatible sagebrush, Stephanomeria, found by Gottlieb (1976) among incompatible plants. As in plants, the shift from incompatible to self-compatible systems (in our case autogamy) does not completely disrupt the system, since autogamous stocks can still conjugate with complementary autogamous and non-autogamous types. The mortality rate, however, of true "hybrids" between autogamous per non-autogamous stocks of each species is high, or very high, in both back-crosses and F2 and F3 autogamic generations, while control

TABLE 7

Number of stocks and mating types of autogamous populations (subspecies) of E. minuta and E. crassus and their geographic distribution.

	E. minuta		E. crassus	
Locality	no. stocks	no mating types	no. stocks	no. mating types
Tyrrhenian Sea	12	3	1	1
North Sea	1	1	-	-
Indian Ocean (Somalian Coast)	-	-	11	3

mortality ranges from 5 to 15%. The data on mortality rate are reported in Table 8. From the table one can see that hybrid mortality in E. crassus is lower than in E. minuta and that hybrids from some autogamous stocks perform better than others (see Luporini, 1974). On the whole, these results strongly support the view that some sort of post-zygotic incompatibility exists between the two types of populations, namely the autogamous and the non-autogamous ones.

Other isolating mechanisms act at the pre-zygotic level between the two kinds of stocks. Mating intensity, i.e. the number of pairs formed within 6 h of mixing ready-to-mate autogamous with non-autogamous animals, is usually much weaker in both species than in control mixtures of autogamous with autogamous and non-autogamous with non-autogamous animals (Luporini and Nobili, 1967a; Luporini and Dini, 1977). All conditions being favourable, autogamy or conjugation are triggered by food depletion in animals predominantly in the G_1 cell cycle phase (Luporini and Dini, 1975, and unpublished results). But while conjugation between non-autogamous stocks usually starts within 2 h after mixing, conjugation between autogamous and non-autogamous cells **requires** usually 4 h or more of coexistence. Moreover, autogamous cells **acquire** the competence to undergo autogamy, either before or at the same time they become competent to conjugate, independently of the sexual

TABLE 8

Clone mortality rate (in %) at successive sexual generation of "hybrids" between autogamous and non-autogamous stocks of E. crassus and E. minuta. Data for an autogamous stock of E. crassus found to be homozygous at several loci are given in parenthesis.

	1st generation by conjugation (R_1)	2nd generation by backcross	2nd generation by autogamy	3rd generation by backcross (R_2)	3rd generation by autogamy
E. crassus					
No. of crosses examined	12 (5)	10 (5)	10 (5)	8 (3)	7
range	33–98 (3–29)	13–60 (12–38)	27–83 (35–53)	15–54 (20–24)	60–96
mean and S.E.	67.7±6.2 (15.6±5.7)	29.7±4.2 (24.2±4.6)	47.2±5.1 (45±2.8)	36.0±4.0 (21.3±1.3)	78.7±5.1
E. minuta					
No. of crosses examined	11	6	5	2	1
range	10–35	84–96	90–98	72–94	98.1
mean and S.E.	19.7±2.5	91.6±1.9	95.2±1.3	83.4±11.1	98.1

processes whereby they are formed, (Luporini, 1970; Luporini and Dini, 1977; Luporini and Nobili, 1967a). This pattern of life cycle of autogamous stocks appears not to be random but consistent with the other features favouring autogamic reproduction or crossing within autogamous stocks rather than between autogamous and non-autogamous stocks.

However, that the incompatibility of the two kinds of stocks has not passed the point of no return even in nature, is proved by the presence of heterozygotes at the mt and a loci (a = autogamy trait) in autogamous natural stocks of E. crassus (Luporini and Dini, 1977). The same stocks do show a very high mortality rate after autogamy; they probably originated in the field through crosses with non-autogamous stocks.

The latter fact, associated with the lower mortality rate of hybrids between the two kinds of populations of E. crassus, indicates a weaker isolating mechanism operating in this species than in the E. minuta so far studied. Several reasons might account for this, such as a relatively more recent origin of the autogamous stocks of E. crassus in the Somalian area. In these stocks autogamy is controlled by a single gene at the a locus dominant over the wild type (Luporini and Dini, 1977). The same kind of inheritance is very probably present in autogamous stocks of E. minuta (Siegel and Heckmann, 1966). The mutation at the a locus may have started the development of sexual isolation by the disruption of the breeding system. Afterwards, the autogamous and non-autogamous stocks may have come to be separated by multiple gene differences determining ethological and cytogenetic incompatibility. The amount of divergence between them could then be a function of the time elapsed from the mutational events up to the present.

If autogamic reproduction depends on a fortuitous mutation at the a locus, this event could have occurred many times in the past history of the species and might still occur. Thus, the limited spread of this sexual process among populations of the species cannot be ascribed solely to its relatively recent origin. The advantage of monoparental reproduction must be strongly countered by disadvantages such as the loss of the potentiality to share the gene pool variation of the whole population.

When autogamy can alternate with conjugation in an adjustable fashion, as in the case of the freshwater species of the Paramecium aurelia complex (Sonneborn, 1975) the advantages of the two modes of reproduction can be better exploited. A certain degree of crossing goes on in the autogamous populations of the Euplotes species. In addition, there is a tendency to maintain a heterozygous state even through successive autogamies (Nobili and Luporini, 1967; Luporini and Dini, 1977). From all of this it follows that autogamous stocks have a genetic spectrum of adaptive capabilities, which allows them to proliferate in some limited parts of the areal distribution of the species. On this basis, because the autogamous stocks of both species are partially reproductively isolated from the progenitor populations, they warrant subspecific status whatever the probability of their long-term persistence and the subsequent evolutionary paths.

While refraining from naming the subspecies, we can conclude that the biology of these species offers several interesting peculiarities deserving further investigations and of interest in relation to problems of speciation. An analysis of the genetic distance between and within the four closely related species, on the line of the work on Tetrahymena by Borden et al. (1977), could be very rewarding on account of their cosmopolitan distribution, great ecological tolerance and the multiple mating type system which contrast with the strictly dioecious species usually examined. Moreover, the incipient divergence of autogamous stocks offers an excellent opportunity to investigate how evolutionary forces are working in these animals in an apparently stable environment.

Summary

The breeding structure of hypotrichous ciliates appears to have evolved according to different strategies. The multiple mating type system is by far the most common in the family Euplotidae. Each marine species so far analyzed consists of a large number of mating types between which conjugation is possible. In some cosmopolitan marine species of Euplotes, this system is controlled by a multi-allelic series at the mt locus. Each allele codes for a different mating type and complete (ranked) dominance prevails among them. The outbreeding characteristics of the system is reinforced by the species life cycle. A period of sexual immaturity, lasting many fissions, ensues after conjugation. It is followed by a long period of sexual maturity that precedes senescence. This kind of life cycle increases the outbreeding efficiency based on the multiple allelic system which greatly reduces per se sib matings relative to non-sib matings.

Cross conjugations, although abortive, occur between some species of Euplotes which are closely related to each other in most morphological features. Similarly, intraclonal conjugation may be induced by cell to cell contacts with individuals of a strictly related species. These phenomena indicate an intimate relatedness among these Euplotes species.

Two subspecies, one of E. minuta and one of E. crassus may have recently arisen by saltational speciation. They are both made up of self-fertilizing individuals (autogamic), are geographically restricted, and probably originated from the widespread, out-crossing species through mutation. The different breeding system, the highly reduced fertility of the hybrids and some premating ethological isolation are the main mechanisms favouring reproductive isolation between the subspecies and the two mother species living sympatrically.

REFERENCES

Borden, D., Miller, E. T., Whitt, G. S. and Nanney, D. L. 1977. Electrophoretic analysis of evolutionary relationships in Tetrahymena. Evolution 31, 91-102.

Curds, C. R. 1975. A guide to the species of the genus Euplotes (hypotrichida, Ciliatea). Bull. Brit. Mus. Zool. 28, 1-61.

Dini, F. and Miyake, A. 1977. Precongugant cell interaction in Euplotes crassus. J. Protozool. 24 (Suppl.), 32A.

Dragesco, J. 1974. Ecologie des protistes marins. In, Actualites Protozoologiques. Puytorac De, P. and Grain, J (eds). Univ. de Clermont-Ferrand 1, 219-228.

Fauré-Fremiet, E. 1967. Chemical aspects of ecology. In, Chemical Zoology. Florkin, M., Scheer, B. T. and Kidder, G. W. (eds). 1, 21-54. Acad. Press.

Génermont, J., Machelon, V. and Truffau, M. 1976. Données experimentales relative au probleme de l'espéce dans la genre Euplotes (Ciliés Hypotriches). Protistologica 12, 239-248.

Gottlieb, L. D. 1976. Biochemical consequences of speciation in plants. In, Molecular Evolution. Ayala, F. J. (ed). Sinauer Associates, Inc., Sunderland, Massachusetts 123-140.

Heckmann, K. 1963. Paarungssystem un Genabhangige Paarungstyp-differenzierung bei dem Hypotrichen Ciliaten Euplotes vannus O. F. Muller. Arch. Protistenk. 106, 393-421.

Heckmann, K. 1964. Experimentelle Untersuchungen an Euplotes crassus. I. Paarungssystem, konjugation und Determination der Paarungstypen. Z. Vererblehr. 95, 114-124.

Heckmann, K. 1967. Age-dependant intraclonal conjugation in Euplotes crassus. J. Exp. Zool. 165, 269-278.

Heckmann, K. and Siegel, R. W. 1964. Evidence for the induction of mating type substances by cell to cell contacts. Exp. Cell Res. 36, 688-691.

Luecken, W. W. 1973. A marine Euplotes (Ciliophora, Hypotrichida) with reduced number of prezygotic micronuclear divisions. J. Protozool. 20, 143-145.

Luporini, P. 1968. Ciclo biologico in ceppi autogamici di Euplotes minuta Yocom (Ciliophora, Hypotrichida). Boll. Zool. 35, 433.

Luporini, P. 1970. Life cycle of autogamous strains of Euplotes minuta. J. Protozool. 17, 324-328.

Luporini, P. 1974. Killer and autogamous strains of Euplotes crassus (Dujardin) from the Somalian coast. Monitore zool. ital. (N.S.), 11 (Suppl.), 129-132.

Luporini, P. and Dini, F. 1975. Relationships between cell cycle and conjugation in 3 hypotrichs. J. Protozool 22, 541-544.

Luporini, P. and Dini, F. 1977. The breeding system and the genetic relationship between autogamous and non-autogamous sympatric populations of Euplotes crassus (Dujardin) (Ciliata, Hypotrichida). Monitore zool. ital. (N.S.), 11, 119-154.

Luporini, P. and Nobili, R. 1967a. New mating types and the problem of one or more syngens in Euplotes minuta Yocom (Ciliata, Hypotrichida). Atti Ass. Genet. It. 12, 345-360.

Luporini, P. and Nobili, R. 1967b. Heterozygosity and autogamy in strain A-31 of Euplotes minuta (Ciliata, Hypotrichida). J. Protozool. 14 (Suppl.), 36-37.

Magagnini, G. and Santangelo, G. 1977. Temporal distribution of mating types in a natural population of Euplotes crassus (Dujardin) (Ciliata, Hypotrichida). Monitore zool. ital. (N.S.), 11, 223-230.

Miyake, A. 1967. Cell communication, cell union and initiation of meiosis in ciliate conjugation. In, Current Topics in Develop. Biol. Moscona, A. A. and Monroy, A. (eds). Acad. Press 12, 37-78.

Nobili, R. 1964. On conjugation between Euplotes vannus O. F. Muller and Euplotes minuta Yocom. Caryologia 17, 393-397.

Nobili, R. 1965. Coniugazione ibrida tra specie di Euplotes (Ciliata, Hypotrichida). Boll. Zool. 31, 1339-1348.

Nobili, R. 1966a. Mating types and mating type inheritance

in Euplotes minuta Yocom (Ciliata, Hypotrichida). J. Protozool.13, 38-41.

Nobili, R. 1966b. La riproduzione sessuale nei Ciliati. Boll. Zool. 32, 93-131.

Nobili, R. 1967. Induzione di autoconiugazione in mescolanze interspecifiche di Euplotes (Ciliata, Hypotrichida). Arch. Zool. It. 52, 51-64.

Nobili, R. and Luporini, P. 1967. Maintenance of heterozygosity at the mt locus after autogamy in Euplotes minuta (Ciliata, Hypotrichida). Genet. Res. 10, 35-43.

Noland, L. E. and Gojdics, M. 1967. Ecology of free-living Protozoa. In, Research in Protozoology. Chen. T. T. (ed). 2, 215-266 Pergamon Press.

Nyberg, D. 1974., Breeding system and resistance to environmental stress in Ciliates. Evolution 28, 367-380.

Power, H. W. 1976. On forces of selection in the evolution of mating types. Am. Nat. 110, 937-944.

Siegel, R. W. and Heckmann, K. 1966. Inheritance of autogamy and the killer trait in Euplotes minuta. J. Protozool. 13, 34-38.

Sonneborn, T. M. 1937. Sex, sex inheritance and sex determination in Paramecium aurelia. Proc. Nat. Acad. Sci. Wash. 23, 378-385.

Sonneborn, T. M. 1957. Breeding, reproductive methods and species problems in Protozoa. In, The Species Problem. Mayr, E. (ed). Amer. Ass. Adv. Sci. Wash. 155-324.

Sonneborn, T. M. 1975. The Paramecium aurelai complex of fourteen sibling species. Trans. Amer. Microsc. Soc. 94, 155-178.

Touffrau, M. 1960. Revision du genre Euplotes, fondée sur la comparison des structures superficielles. Hydrobiologie 15, 1-77.

Wichterman, R. 1967. Mating types, breeding system, conjugation and nuclear phenomena in the marine ciliate Euplotes cristatus Kahl from the Gulf of Naples. J. Protozool. 14, 49-58.

A STUDY OF REPRODUCTIVE ISOLATION WITHIN THE SUPER-SPECIES TISBE CLODIENSIS (COPEPODA, HARPACTICOIDA)

B. Volkmann[1], B. Battaglia[1,2] and V Varotto[1,2]

[1]Ist. di Biologia del Mare, Venezia, and

[2]Ist. di Biologia Animale, Università di Padova, Italy

A series of investigations carried out in recent years on the harpacticoid copepod Tisbe has shown that relatively little interspecific morphological differentiation has occurred, which makes the taxonomy of this genus particularly complex (Volkmann-Rocco, 1971).

The detection of reproductive barriers by means of crossbreeding experiments clearly indicates that certain populations previously considered, because of their morphological similarity, as belonging to the same species actually belong to different species. In fact, increasing numbers of sibling species groups and closely related species within the genus are now being recognized (Volkmann-Rocco and Battaglia, 1972; Volkmann-Rocco, 1973a and b).

On the other hand, it has generally been found that different strains of Tisbe are able to mate and produce fertile offspring only when they belong to the same species, regardless of their geographic origin (Battaglia and Volkmann-Rocco, 1973; Volkmann, unpublished data).

Laboratory crosses between several Mediterranean populations of the recently discovered species T. clodiensis gave unexpected results (Battaglia and Volkmann-Rocco, 1969). When the first crosses were made, it came as a surprise that while some were fertile, in others no hybrids were produced. This effect appeared to depend upon the geographic origin of the female or male and it was found that reproductive success within this species ranges from full interfertility to practically complete intersterility.

In order to provide a satisfactory answer to the problem, cross-breeding experiments were extended to a larger number of strains, obtained by sampling Atlantic populations from Europe and from the East Coast of North America.

The purpose of the present study is to examine the degree of reproductive isolation between Mediterranean and Atlantic populations. Although further analysis will be necessary to unravel the isolating mechanisms involved in the process of active speciation present in T. clodiensis, an attempt is made here to understand the interesting, thus far unique, situation observed within the genus Tisbe.

Materials and Methods

The first series of experiments were carried out in 1968. In 1972, an opportunity was available to collect new strains of T. clodiensis and to adapt them to laboratory conditions. Systematic intercrosses were started in February 1973 in the Laboratoire de Génétique Evolutive, CNRS, Gif-sur-Yvette, France and were continued in November 1974 in Padua.

The material consisted of 6 geographical populations (Figure 1) collected as follows: Venice (V), summer 1967; Beaufort (BE), N.C., USA, summer 1971; Ponza (P), Italy, autumn 1969; Roscoff (R), France, spring 1973; Arcachon (A) and Banyuls-sur-Mer (B), France, summer 1972. These populations were raised under standard laboratory conditions as described by Battaglia (1970). For collecting methods see also Fava and Volkmann (1975). All cross-breeding experiments were carried out at a constant temperature of 18°C and a salinity of about 34-36 o/oo.

20 to 30 ovigerous females were taken randomly from the two laboratory cultures destined to be crossed, and these were isolated in single dishes. As soon as the nauplii hatched the females were removed. When the young reached the 4th or 5th copepodite stage, they were mixed and transferred to 5 dishes containing 100 cc of sea-water. In this way healthier females with bigger egg-sacs were obtained than when females are taken directly from mass culture. Moreover, the method allows the choice of progenies of the first or second egg-sacs which are more numerous and usually in better condition than the descendants of the last egg-sacs; when selecting a female from a culture, it is impossible to ascertain which egg-sac she is carrying.

REPRODUCTIVE ISOLATION WITHIN T. clodiensis 619

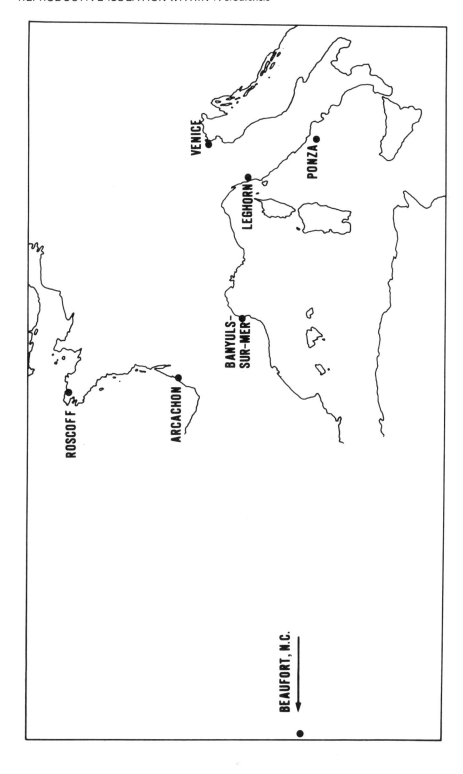

Figure 1. Geographic location of populations of Tisbe clodiensis.

The ovigerous females obtained in this way were isolated and their offspring used to set up pairs: one female and one male copepodite per dish. In T. clodiensis at 18°C the 5th copepodite stage is reached 8-9 days after hatching of the nauplii. For each type of cross at least 60 pairs of interpopulation crosses were set up, 30 of each reciprocal type. At the same time an approximately equal number of intra-population crosses was made, which served as controls, in order to compare characters such as speed of development, fecundity, survival, and sex-ratio. This last character was taken into account because a shift from the normal sex-ratio 1:1 in favour of males, is believed also to indicate genetic incompatibility (Battaglia, 1957).

For most population intercrosses from which adult offspring were obtained, first back crosses were also made using methods similar to those described above.

Unfortunately, the Banyuls strain was lost at the end of 1973, and, thus, there are only preliminary results of the cross Arcachon x Banyuls, while the cross Banyuls x Roscoff could not be made. The results of the cross Arcachon x Ponza are not complete; they are given in Table 1, but not discussed in the text.

TABLE 1

Crossability between strains of different geographic origin of T. clodiensis. + = perfect compatibility; (+) = non-reciprocal incompatibility; O = no hybrids; N† = hybrids dead at nauplius stage; C† = hybrids dead at copepodite stage.

♀	♂	V	B	P	L-A	R	A	BE
Venezia			(+)	(+)	(+)	(+)	(+)	C†
Banyuls		O		+	+		(+)	O
Ponza		N†	+		+	+	O	+
Leghorn-Anzio		O	+	+				
Roscoff		O		+			(+)	+
Arcachon		N†	N†	N†		N†		N†
Beaufort, NC		O	N†	+		+	(+)	

T. clodiensis exhibits a polychromatism, consisting in the presence or absence of violet pigment on some of the segments of the second, third and fourth pair of legs in the female. This is controlled by a simple Mendelian factor, the presence of the pigment being dominant. Although all populations collected so far exhibit this polymorphism, dominant and recessive strains homozygous for the genes concerned have been obtained only from the Venice population. In the present experiments the dominant homozygote of the Venice strain was utilized.

Results

Crossability and Fertility Tests. All strains were fully fertile inter se; hatching percentages of nauplii in intrapopulation crosses were generally about 90%. The minimum generation time for T. clodiensis is about 12 days depending on the number of copepods per dish (Volkmann-Rocco and Battaglia, 1972). The higher the crowding the slower the development time and the smaller the egg-sacs.

Whenever the dominant genotype (Venice) was used for interpopulation crosses (V x A; V x B; etc.), female hybrids showed the typical pigmentation indicating that, in spite of the reproductive barrier between the populations, the character "violet legs" is always dominant.

Geographic populations are practically indistinguishable morphologically. Biometric differentiation may exist, but thus far an appropriate analysis has not been made.

In interpopulation crosses characteristic differences were observed. As appears from Tables 1 and 2, some of the interstrain crosses showed full compatibility in both directions (P x B, P x R; BE x P, BE x R), even in populations as far distant as Beaufort and Roscoff, and Beaufort and Ponza. There was, however, a distinct reduction in the number of offspring of the F_1 and first backcrosses in the transatlantic crosses.

On the other hand, in some of the interstrain crosses incompatibility in both directions was observed. While the cross P x R gave healthy offspring, the cross P x A was completely unproductive. There was, however, a difference between the ♀P x ♂A and its reciprocal cross. In the latter, 60% of the females produced egg-sacs but the nauplii hatching normally died at the 2nd or 3rd naupliar stage. The reciprocal cross gave no viable offspring.

TABLE 2

Crosses showing perfect crossability: P x B, P x R, B x R, B x P. a = females with undeveloped egg-sacs; b = females with adult offspring; c = average number of individuals per family; d = sex-ratio (% ♀♀). ND = normal development; n = number of pairs.

	Mating ♀ ♂	n	a	b	c	d
Controls	P x P	20	–	20 (ND)	35	34%
	B x B	20	–	18 (ND)	26	31%
P1 Crosses	P x B	30	3	22 (ND)	42	33.2%
	B x P	30	1	27 (ND)	55.5	48.3%
F1 Crosses	BP x BP	22	–	22 (ND)	50	36%
	PB x PB	15	–	15 (ND)	36	28%
Controls	P x P	48	–	46 (ND)	61.5	28.2%
	R x R	38	–	33 (ND)	61	26.3%
P1 Crosses	P x R	58	2	50 (ND)	77	32.5%
	R x P	58	1	54 (ND)	76.6	37%
F1 Crosses	PR x RP	15	–	15 (ND)	58.9	35%
	RP x PR	15	–	14 (ND)	74.7	22%
1st Backcrosses	RP x P	15	–	15 (ND)	81.6	34%
	PR x P	15	1	14 (ND)	66.1	44%
P1 Crosses	BE x R	32	–	32 (ND)	50.4	40.2%
	R x BE	34	1	33 (ND)	60.3	40.7%
F1 crosses	BER x RBE	19	–	19 (ND)	38.1	33.3%
	RBE x BER	15	–	15 (ND)	30.8	40.0%
1st Backcrosses	RBE x BE	19	–	19 (ND)	31.9	26.4%
	BER x BE	20	–	20 (ND)	35.3	34.3%
P1 Crosses	BE x P	31	–	31 (ND)	49.1	51.9%
	P x BE	32	–	32 (ND)	47.1	53.5%
F1 Crosses	BEP x PBE	20	–	20 (ND)	33.1	42.1%
	PBE x BEP	21	–	21 (ND)	31.4	40.0%
1st Backcrosses	BEP x P	17	–	17 (ND)	29.5	40.0%
	PBE x P	19	–	19 (ND)	31.3	32.8%

Similar results were obtained in the transatlantic crosses Beaufort x Banyuls, although the outcome was even more drastic (Battaglia and Volkmann-Rocco, 1973). Only very few nauplii hatched in the ♀BE x ♂B cross, all of which died within 4 days. 72% of the females from Venice produced viable offspring with the Beaufort ♂♂. Many died as nauplii, but in 42% of the families small copepodites appeared (2nd - 3rd stage) which died after 6 - 24 days. The reciprocal cross was completely sterile.

The most interesting results emerged from interstrain crosses showing reciprocal differences in incompatibility characterized by fertile hybrids (mostly males) in one direction, and either no offspring or inviable offspring in the reciprocal cross (Tables 1, 3, 4). Great heterogeneity in the outcome of different crosses was, however, observed, due mainly to different hatchability and numbers of offspring. The rate of development of the hybrids was considerably retarded in all crosses showing "relative intraspecific incompatibility". This cross incompatibility cannot be ascribed to an intrinsic lack of fertility of one of the sexes, since intrapopulation controls are normal.

Because of the almost complete shift of the sex-ratio towards males (except in the V x A cross) only a few F1 crosses could be carried out, while backcrosses were generally made with the hybrid males. Whenever backcrosses succeeded, development of the offspring was normal, but they were often less numerous than in the parental crosses, and showed a strong shift in sex ratio in the direction of males.

Hybrid females were smaller and not as healthy as females from either parental strain. In the few crosses where F1 females were involved, their fertility was clearly reduced (e.g., VR x VR and VR x R). As seen in Table 4, the ♀V x ♂B cross gave very few adults and in order to make the backcrosses, 4 - 5 egg-sacs had to be isolated. Descendants of the first backcrosses (to V parent) were perfectly fertile _inter se_, all females producing adult offspring. The second generation backcrosses with the V parent resulted in adult progenies in 90 percent of the crosses. With the B parent (V B) V x B and B x (V B)V adult offspring were produced in only a few families. In the third generation backcrosses the following results were obtained:

(V B) V) B x B: from 10 ♀♀ two gave adult offspring;
(V B) V) V x V: from 10 ♀♀ one gave adult offspring;
B x (V B) V) V: from 13 ♀♀ one gave adult offspring;
V x (V B) V) B: all 5 ♀♀ gave normal adult offspring.

TABLE 3

Crosses with non-reciprocal incompatibility: Beaufort x Arcachon, Roscoff x Arcachon, Venezia x Arcachon. a = females with undeveloped egg-sacs; b = offspring dead at nauplii stage (days); c = offspring dead at copepodite stage (days); d = females with adult offspring; e = average number of individuals per family; f = sex-ratio (% ♀♂); n = number of pairs; ND = normal development.

	Mating ♀ ♂	n	a	b	c	d	e	f
Controls	BE x BE	58	1	—	—	49(ND)	42.4	40%
	A x A	60	6	—	—	47(ND)	58.8	44%
P1 Crosses	BE x A	57	7	1	—	32(12-35d)	18	0.1%
	A x BE	60	14	27(16-25d)	—	—	—	—
F1 Crosses	BEA x BEA	3	3	—	—	—	—	—
1st Backcrosses	BE x BEA	22	—	—	—	12(ND)	27	6.8%
	A x BEA	20	4	—	—	12(ND)	35.2	24%
Controls	R x R	51	7	—	—	44(ND)	42	47.5%
	A x A	42	3	—	—	38(ND)	49	38.7%
P1 Crosses	R x A	49	12	—	—	14(18-35d)	26.3	0.5%
	A x R	48	13	30(13-22d)	—	—	—	—
F1 Crosses	RA x RA	9	1	1	—	1(12d)	1♂	—

Table 3 (continued)

	Mating ♀ × ♂	n	a	b	c	d	e	f
1st Backcrosses	R × RA	32	3	–	–	23(ND)	48.7	12%
	A × RA	32	5	–	–	23(ND)	26	26%
Controls	V × V	55	–	–	–	53(ND)	43.8	41%
	A × A	60	6	–	–	47(ND)	50.8	44.1%
P1 Crosses	V × A	58	2	1(9d)	–	48(16–35d)	37.7	42.4%
	A × V	59	7	39(16–27d)	1(20d)	–	–	–
F1 Crosses	VA × VA	31	1	–	–	29(nd)	4.8	38.6%
1st Backcrosses	V × VA	15	–	–	–	14(ND)	32.3	35%
	A × VA	15	–	–	–	14(ND)	43.7	19.8%
	VA × V	15	5	–	–	10(ND)	37.5	47.7%
	VA × A	29	5	20(2–7d)	–	–	–	–

TABLE 4
Crosses with non-reciprocal incompatibility: Venezia x Banyuls, Venezia x Ponza, Venezia x Roscoff. Abbreviations as in Table 3.

	Mating ♀ ♂	n	a	b	b	c	d	e	f
Controls	V x V	21	–	–	–	18(ND)	no data	
	B x B	24	–	–	–	20(ND)		
P1 Crosses	V x B	43	–	–	–	6(21-29d)	very few	–
	B x V	30	–	1(16-20d)	–	–	–	–
F1 Crosses	VB x VB	25	–	1(-18d)	–	–	–	–
1st Backcrosses	V x VB	16	–	–	–	2(ND)	27.5	37%
	B x VB	4	–	–	–	?	–	–
	VB x V	19	–	1	–	4(ND)	17	53%
	VB x B	16	–	–	–	–	–	–
Controls	V x V	20	1	–	–	19(ND)	no data	
	P x P	20	–	–	–	18(ND)		
P1 Crosses	V x P	36	6	1(21d)	12(21-35d)	10(25-39d)	8.4	10%
	P x V	30	6	9(8-20d)	–	1(33d)	14	0.7%
F1 Crosses	VP x VP	12	–	–	–	–	–	–
1st Backcrosses	V x VP	10	3	–	–	6(ND)	no data	
	P x VP	10	2	–	–	1(ND)	19	2.6%

Table 4 (continued)

	Mating ♀ × ♂	n	a	b	c	d	e	f
Controls	V × V	29	—	—	—	29(ND)	35.5	30%
	R × R	32	2	—	—	29(ND)	33	27.3%
P1 Crosses	V × R	62	6	3(8–17d)	6(16–29d)	20(24–39d)	16	10.3%
	R × V	50	17	3(5–16d)	—	—	—	—
F1 Crosses	VR × VR	27	4	7(2–5d)	1(20d)	2(ND)	1♀ + ♂	—
1st Backcrosses	V × VR	30	—	—	—	24(ND)	35	21%
	R × VR	28	1	—	—	25(ND)	33.7	4.6%
	VR × R	10	1	—	1(14d)	—	—	—

From some of the unilaterally viable crosses small cultures were set up using ovigerous females. We started with the second egg-sac since the first egg-sacs were employed for the other experiments. In the cases of ♀B x ♂R and ♀R x ♂A cultures could not be established. The females gave adult offspring, but these apparently failed to reproduce. However, in the crosses ♀V x ♂R and ♀V x ♂A the reproductive barrier was overcome and the cultures were productive for months.

Ethological Isolation. Since males and females of $\underline{T.}$ $\underline{clodiensis}$ often copulate for several hours or even as long as a day it is possible to detect mating of a certain percentage of pairs with only one check per day. Even in those crosses that yielded no viable offspring, mating took place normally.

In order to investigate the possible existence of an ethological barrier, a "male choice" experiment was carried out with the Venice and Arcachon strains. In all crosses involving sibling species, only very few cases of ♂♂ and ♀♀ of different species mating were found. Females and males from Arcachon and Venice were aged separately for about two days as described in Volkmann-Rocco (1972) in order to prevent mating. Two females, one from Arcachon and the other from Venice, were placed in a dish. In thirty replicates, a single male from Arcachon was added to the dish and another thirty replicates were set up with Venice males. In the case of the Venice strain, we utilized the violet form (see above) which enabled the females to be easily distinguished.

Most of the matings occurred within the first 20 minutes. Within 40 minutes, all pairs were mating. Of the 30 Arcachon ♂♂, 16 mated with females of their own strain while in the reciprocal experiment, of 30 Venice males, 18 preferred females of their origin. Nearly all females extruded an egg-sac on the following day.

There seems to be no discrimination between partners of different origin and thus the Venice and Arcachon strains are not isolated ethologically. Similar results may be expected for the other strains, even those which are geographically closer (e.g., Arcachon and Roscoff). During the crossbreeding experiments involving these two strains (A and R), a high percentage of mating pairs were observed, but many of these females did not extrude an egg-sac or their eggs did not develop.

Discussion

Further collecting and experimental work will be needed to elucidate fully the distributional relationships and genetic structure of the geographic populations of T. clodiensis. These results, however, in conjunction with earlier data (Battaglia and Volkmann-Rocco, 1969) make some aspects of the situation clearer.

The tables show that in the compatible interstrain crosses all results are well within the normal range of variation as seen in the intrastrain crosses, and sometimes they may even be superior.

Crosses with unilateral production of adult offspring were quite variable insofar as the compatible crosses produced different percentages of adult progenies (e.g., V x A 83%, and V x B 1.4% of the crosses). There were also differences in the reciprocal crosses, since in the crosses A x BE, A x V, A x R nauplii hatch normally and then die, while very few eggs develop to the naupliar stage in the R x V, B x V and P x V crosses.

Nevertheless, the data demonstrate that there are basically three levels of crossing relationships: adult offspring obtained in both directions, adult offspring obtained in one direction only, no offspring or inviable larvae obtained in both directions.

This inability of certain populations to interbreed and to give adult offspring (P x A, BE x B, BE x V) might suggest that the geographic strains belong to different species. However, it seems better to consider them as belonging to a single species, mainly because of the existence of bridging populations which produce fertile hybrids with those populations which are reproductively isolated from each other (as V x A, B x P, BE x P). It is, however, evident that T. clodiensis has a deeply cleft gene pool, and that we have here a borderline case, where the process of species splitting has in some populations reached the critical stage of transition from race to species. We therefore suggest that the different strains be considered as incipient species or semispecies, and T. clodiensis, a superspecies.

From Table 1 it appears that all crosses involving the Venice or the Arcachon strain show reproductive isolation, that is non-reciprocal or even bilateral incompatibility. The Venice strain showed the strongest incompatibility with other strains when used as males while in the Arcachon strain reproductive isolation was greater when females were utilized. A surprising result was the complete sterility of the Beaufort x Banyuls cross, while both the Beaufort x Ponza and Ponza x Banyuls crosses were fertile.

All the other populations seems to interbreed well with each other, and the interstrain crosses are slightly superior to the intrastrain crosses, in terms of survival, fecundity and sex-ratio. The results seem to indicate that the populations inhabiting lagoons, especially those from Venice and Arcachon, have undergone genetic differentiation which has isolated them partially or completely from the other populations. To a certain extent these two lagoons possess similar features. Due to fresh waters inflow, both are brackish, polyhaline environments exhibiting wider fluctuations of physical and chemical characters than the adjacent seas.

All attempts to collect T. clodiensis south of the Po River have so far failed; it has not been found in either Porto Corsini or Pesaro. It is conceivable that the Po acts as a strong barrier to gene flow, since T. clodiensis occurs outside the Lagoon of Venice at the Lido.

The same may be suspected for the Arcachon population, where the rivers Garonne and Dordogne may create a barrier, but we do not know whether the species occurs immediately north of the Gironde Bay; though farther north at Concarneau it could not be collected.

For the strain from the Beaufort Inlet, which may also be considered as an estuarine environment, reproductive isolation is much less pronounced. This may be partly due to the fact that there seems to exist an uninterrupted gene flow between the Beaufort population and the open sea. At 27 miles off Cape Lookout (34°19.4 N - 76°56.2 W) T. clodiensis was found at a depth of 85 feet.

On the basis of the present experiments, we may conclude that in T. clodiensis ecological factors are far more important than distance per se in the process of isolation and genetic differentiation.

A comparison with Drosophila paulistorum, a sibling species of the willistoni group, seems legitimate. It is considered a single Mendelian population, but represents at the same time a cluster of species in statu nascendi, a superspecies consisting of six semi-species (Dobzhansky and Spassky, 1959; Dobzhansky et al., 1964) which exhibit various degrees of reproductive isolation from each other. They are mostly allopatric but in some places their distributions overlap; in this case two or more of the semispecies may coexist without apparent interbreeding. It was found that two genetically different isolating mechanisms are involved in this speciation process: a polygenically controlled sexual isolation (ethological), and a form of hybrid sterility acting through a genically induced modification of the egg-cytoplasm (Ehrman, 1961, 1962). The sexual isolation is almost certainly the more effective of the two mechanisms in preventing gene exchange between the incipient species in the natural habitat (Malogolowkin et al., 1965).

Such a mechanism can certainly not account for the reproductive isolation of T. clodiensis. As shown above there is no evidence for ethological or mechanical isolation. This may be partially due to the fact that the clodiensis populations analyzed so far are all allopatric, and such an isolating mechanism would be efficient only where different races meet. On the other hand, experiments with T. clodiensis and its sibling species T. dobzhanskii have shown that sexual isolation is complete between these species, whether they are allopatric or sympatric. (It is not known so far if this is due to behavioural differences.)

As to the second mechanism, hybrid sterility does not seem to play a decisive role. The hybrid males are perfectly fertile. They can be backcrossed to females of the parental races, and the backcross progenies develop normally and consist of fertile daughters and sons. There seems to exist only one exception involving the V x B strains which show the strongest non-reciprocal incompatibility. The barrier, however, is generally overcome in the second generation backcrosses.

Since females are rare in the interstrain crosses (generally 0.5% - 2% ♀♀ with the exception of the V x A cross), very few crosses involving F1 females could be made. From these it does appear, however, that their fertility is reduced, although the success of most of the crosses seemed to depend also on the type of cross. This is evident in the F1 and backcrosses of the ♀V x ♂A combination. Hybrid females gave adult offspring with V males, but nauplii with A males. The F1 crosses were 95% fertile, but yielded only a few adults, due to high mortality of nauplii and copepodites. Thus, it was found that at least some of the hybrid females of the non-reciprocal interstrain crosses were fertile. However, the numbers of offspring except in the VA x V cross, which was normally fertile, were low.

It becomes evident that although we have a situation superficially similar to Drosophila paulistorum, that is the breaking up of a superspecies in partly reproductively isolated strains, the mechanisms which maintain this isolation must be different.

One of the most striking and best known cases of unilateral incompatibility is found in mosquitoes, especially in Culex pipiens (Laven, 1959). The phenomenon has recently also been described from the midge Clunio marinus (Neumann, 1971). The situation is similar to that in T. clodiensis, although in the Culex crosses showing unilateral incompatibility there is usually full fertility in one direction (average hatch of 87% with a sex-ratio close to 1:1). This is not the case for T. clodiensis, except for the V x A cross. The outcome of all the other unilaterally successful crosses is more negative; isolation is already stronger.

In an analytically brilliant experiment of long duration (over 50 generations of backcrosses) Laven (1957) found that the determinants for the different crossing types must be of a cytoplasmic nature. Thus, speciation in Culex and in mosquitoes in general seems to depend upon two different, but complementary systems: cytoplasmic differentiation as a means of isolation and prevention of gene flow, and then the ordinary pressures of variation and/or selection.

Clunio marinus shows the same non-reciprocal cross sterility as Culex and backcrosses seem to indicate that the genetic control of the character is similar to that found in Culex (Neumann, 1971).

"Relative intraspecific incompatibility" has also been described in the polychaete Ophryotrocha (Akesson, 1972, 1975). It is expressed as a change of sex-ratio, decreased mating propensity and preferential mating. The ethological barrier is considered as the main isolating mechanism.

As seen from the experiments described above, the reproductive barrier in T. clodiensis is mostly overcome in the first backcrosses.

This might indicate that the character "non-crossability" is controlled directly by at least one nuclear gene, but it seems more likely to be determined by a multiple system of incompatibility genes, which in some crosses are responsible for hybrid mortality at the naupliar stage and in others for retarded development, drastic shift of the sex-ratio and varying degrees of survival to adulthood. The strongest isolating factor of the unilateral crosses is doubtlessly the nearly complete absence of hybrid females.

The questions that arise now, are: How could such a non-reciprocal sterility develop? Why are there populations as distant as Roscoff and Ponza, that interbreed perfectly, while others, as close as Arcachon and Roscoff are partly inter-sterile? A possible explanation could be that the incompatibility of certain strains developed as a by-product of adaptation of allopatric races to environments found in their respective region of distribution. It can be assumed that the ecological factors played a determining role in diversifying the various populations, and that such differentiation has taken place especially in the lagoon populations of Venice and Arcachon.

The conclusion that in Tisbe clodiensis, we are dealing with a cluster of semispecies or species in statu nascendi as in Drosophila paulistorum, is further supported by the preliminary results of a biochemical comparison. Affinity indices have been worked out for groups of morphologically distinct species, sibling species, and geographic races, by means of acrylamide-gel electrophoresis for 15 enzyme systems (Bisol, 1976).

These indices were calculated on the basis of the similarity in number, mobility and activity of electrophoretic bands. The index of biochemical affinity between the various populations of Tisbe clodiensis lies between those generally regarded as relating to geographic races and those characterising sibling species, the figures showing a clear cut differentiation.

Summary

A number of crosses involving 3 Mediterranean, 2 Eastern Atlantic and 1 Western Atlantic population of the species Tisbe clodiensis were carried out. In interpopulation crosses characteristic differences were found showing basically three kinds of results: (a) adult offspring obtained in both directions; (b) adult offspring in one direction; (c) no offspring or dying larvae in both directions. In the second case development of the hybrids was greatly retarded; also the number of progeny per female was lower when compared to the controls, and there was a nearly complete shift of the sex-ratio towards males. Non-reciprocal incompatibility, or even incompatibility in both directions was evident in all crosses involving the Venice or the Arcachon populations;

the Venice strain showed the strongest isolation from others when the male was used, while in the Arcachon strain the reproductive barrier was greater utilizing the female. The fact that this incompatibility was generally overcome in the first generation backcrosses would suggest that the character "non-crossability" is controlled directly by at least one mendelian factor. It is concluded that in T. clodiensis we are dealing with a cluster of semispecies, and that ecological factors are far more important than distance per se in the process of isolation and genetic differentiation.

Acknowledgments

We are grateful to the late Professor Th. Dobzhansky for critically reading a first version of the manuscript. Part of the work was carried out by one of us (B.V.) at the Laboratoire de Génétique Evolutive et de Biométrie, Gif-sur-Yvette, France and was made possible by support from the National Research Councils of France and Italy.

REFERENCES

Akesson, B. 1972. Incipient reproductive isolation between geographic populations of Ophryotrocha labronica. Zool. Scripta 1, 207-210.

Akesson, B. 1975. Reproduction in the genus Ophryotrocha (Polychaeta Dorvilleidae). Pubbl. Staz. Zool. N poli 39 Suppl.,377-398.

Battaglia, B. 1957. Ecological differentiation and incipient intraspecific isolation in marine Copepods. Année. biol. 33, 259-268.

Battaglia, B. 1970. Cultivation of marine copepods for genetics and evolutionary research. Helgoländer Wiss. Meeresunters.20, 421-438.

Battaglia, B. and Volkmann-Rocco, B. 1969. Gradienti di isolamento riproduttivo in popolazioni geografiche del copepode Tisbe clodiensis. Atti Ist. veneto Sci. 127, 371-381.

Battaglia, B. and Volkmann-Rocco, B. 1973. Geographic and reproductive isolation in the marine harpacticoid copepod Tisbe. Mar. Biol. 19, 156-160.

Bisol, P. M. 1976. Polimorfismi enzimatici ed affinità tassonomiche in Tisbe (Copepoda, Harpacticoida). Atti Accad. Naz. Lincei Rc. 60, 864-870.

Dobzhansky, Th., Ehrman, L., Pavlosky, O. and Spassky, B. 1964. The superspecies Drosophila paulistorum. Proc. Nat. Acad. Sci. 51, 3-9.

Dobzhansky, Th. and Spassky, B. 1959. Drosophila paulistorum, a cluster of species in statu nascendi. Proc. Nat. Acad. Sci. 45, 419-428.

Fava, G. and Volkmann, B. 1975. Tisbe (Copepoda, Harpacticoida) species from the Lagoon of Venice. I. Seasonal fluctuations and ecology. Mar. Biol. 30, 151-165.

Ehrman, L. 1961. The genetics of sexual isolation in Drosophila paulistorum. Genetics 46, 1025-1038.

Ehrman, L. 1962. Hybrid sterility as an isolating mechanism in the genus Drosophila. Quart. Rev. Biol. 37, 279-302.

Laven, H. 1957. Vererbung durch Kerngene und das Problem der ausserkaryotischen Vererbung bei Culex pipiens. II. Ausserkaryotische Vererbung. Z. indukt. Abstamm. u. Vererb. -L. 88, 478-516.

Laven, H. 1959. Speciation by cytoplasmic isolation in the Culex pipiens complex. Cold Spr. Harb. symp. Quant. Biol. 24, 166-173.

Malogolowkin-Cohen, CH., Solima-Simmons, A. and Levene, H. 1965. A study of sexual isolation between certain strains of Drosophila paulistorum. Evolution 19, 95-103.

Neumann, D. 1971. Eine nicht-reziproke Kreuzungssterilität zwischen ökologischen Rassen der Mücke Clunio marinus. Oecologia (Berl.) 8, 1-20.

Volkmann-Rocco, B. 1971. Some critical remarks on the taxonomy of Tisbe (Copepoda, Harpacticoida). Crustaceana 21, 127-132.

Volkmann-Rocco, B. 1972. The effect of delayed fertilization in some species of the genus Tisbe (Copepoda, Harpacticoida). Biol. Bull. 142, 520-529.

Volkmann-Rocco, B. 1973a. Tisbe biminiensis (Copepoda, Harpacticoida) a new species of the gracilis group. Archo Oceanogr. Limnol. 18, 71-90.

Volkmann-Rocco, B. 1973b. Etude de quatre espèces jumelles du groupe Tisbe reticulata Bocquet (Copepoda, Harpacticoida). Arch. Zool. exp. gén. 114, 317-348.

Volkmann-Rocco, B.and Battaglia, B. 1972. A new case of sibling species in the genus Tisbe (Copepoda, Harpacticoida). Proc. 5th Eur. mar. Biol. Symp. Piccin, Padova, 67-80.

GENETICS OF ETHOLOGICAL ISOLATING MECHANISMS IN THE SPECIES

COMPLEX JAERA ALBIFRONS (CRUSTACEA, ISOPODA).

M. Solignac

Lab. de Biologie et Génétique Évolutives, C.N.R.S.

Gif-sur-Yvette, France.

Introduction

The Jaera albifrons complex includes five related species found in cold and temperate coastal regions of the North Atlantic. The species are limited to the intertidal zone and live under stones or among algae.

Females of the five species are morphologically indistinguishable, while differences in the male pereiopods allow specific identification (Bocquet, 1953). In each species, the sexual variants (secondary sexual characters) are closely related to the stimulus-signals emitted by the male during the precopulation phase (Solignac, 1975).

Three species, Jaera (albifrons) albifrons, J. (a.) ischiosetosa and J. (a.) praehirsuta have a widespread distribution on both sides of the Atlantic. The two other species, J. (a.) posthirsuta and J. (a.) forsmani are limited to the east coast of North America and to western Europe respectively.

It is fairly easy to obtain interspecific hybrids in the laboratory when no choice of mate is allowed. The ten possible hybrid combinations and their reciprocals can all be obtained but with varying degrees of success. In nature, however, where mixed populations are not uncommon, the frequency of hybrids generally does not exceed 1%.

In the absence of temporal (Jones and Naylor, 1971),

mechanical and gametic isolation, the sexual barrier guarantees specific integrity in mixed populations. Hybrids of the first generation are usually perfectly viable, except in some crosses (and then only in one direction) where females (the heterogametic sex) are produced in reduced numbers. The F_1 hybrids produce numerous and apparently normal gametes, but offspring of hybrids crossed among themselves or with parents are significantly fewer in number (Solignac, 1976).

The present work is limited to a study of sexual isolation. Some aspects of the genetic determination and of the variation between pairs of species of this isolating mechanism are analysed. In the sexual behaviour of this isopod, two steps are likely to be important (Solignac, 1975):

1. Discrimination by males during pair formation: males take the initiative and the study was made by the male choice technique.
2. Female reactivity to heterospecific overtures: studies were made on isolated heterospecific couples.

Sexual Choice by Males

The male choice technique has been widely used for the study of the degree of isolation between species of <u>Drosophila</u> and of the incipient isolation between geographic strains (reviewed by Anderson and Ehrman, 1969). Results have generally been based upon the percentage of homogamic and heterogamic insemination. In <u>Jaera albifrons</u>, ethological isolation between species is very strong, both in nature and in the laboratory. As observation of heterospecific impregnation in <u>Jaera</u> would be relatively infrequent, this study was limited to pair formation. This is, of course, only one component of isolation, but the level of discrimination shown by males is important in avoiding the formation of heterospecific couples with consequent wastage of time by partners in a pairing without issue.

<u>Analysis of sexual choice</u>. Some preliminary experiments were performed before the technique was standardized in order to establish the optimal conditions for repeatability, and to establish the limits of reliability of the data. Two populations belonging to different species, <u>ischiosetosa</u> and <u>albifrons</u>, both originating from Kiel (Germany) were studied in some detail.

Equal numbers of females of the two species (20 of each) were introduced into a sea water tank. Females, being morphologically indistinguishable, were identified by genetic markers. A choice was then given to a sample (generally about 40) of the males of the first species: the test was considered as positive when a male mounted a female in the "head-to-tail" position characteristic of the initial precopulation phase in the Jaera albifrons complex. The two partners were then separated, the male was removed and the female put back in the tank. The nature of the female chosen was noted. After 20', most of the males had made a choice and had been substituted by males of the second species which were tested in the same manner.

These experiments were repeated several times with the same animals, intervals of one week separating each test. Results were generally homogeneous and summed.

Following this protocol, a series of nine experiments was performed on these two populations using different generations, over a period of more than four years. The results are shown in Table 1. They are homogeneous for albifrons and slightly heterogeneous for ischiosetosa. Nevertheless, considering the diversity of tested animals, particularly in body size and in the genetic markers carried, the results may be taken as reasonably consistent. Another presentation is given in Figure 6: the variability observed for this pair of populations is small compared to those obtained by changing the nature of populations.

The data indicate two rather extreme cases: albifrons males appear to be indifferent in their choice of female, while ischiosetosa males choose their own females far more often than alien ones, showing only an average rate of error of twenty percent.

The nine experiments described above were performed with animals isolated from birth and replaced in individual dishes after each test. Other males extracted from mass cultures, and kept with their own females between tests were also studied. The ninth experiment in Table 1 was the control (isolated males) for this experiment. Males extracted from mass culture were less active than isolated males (Table 2). In both species, males from mass culture showed more pronounced homogamic tendencies than isolated males.

From these results, it can be concluded that the degree

TABLE 1

Choice by ischiosetosa and albifrons males between ischiosetosa (♀I) and albifrons (♀A) females and percentages of albifrons females chosen. Each experiment represents the sum of several tests.

DATE	ISCHIOSETOSA MALES			ALBIFRONS MALES		
	♀I	♀A	%♀A	♀A	♀I	%♀A
IV 72	81	10	11	53	50	51
VIII 72	24	4	14	18	24	43
II 73	59	11	16	40	41	49
V 73	150	34	18	65	78	45
V 74	133	27	17	197	168	54
X 75	71	17	19	50	42	54
II 76	196	39	14	80	69	54
IX 76	216	63	23	392	267	59
X 76	414	137	25	390	314	55
MEAN			17.5			51.8
TOTAL	1344	342	20.3	1285	1053	55.0

Homogeneity

$\chi^2_8 = 17.99$ $\chi^2_8 = 14.98$

$0.01 < P < 0.05$ $P > 0.05$

TABLE 2

Sexual activity of and discrimination by ischiosetosa and albifrons males. For both species, isolated and non-isolated males were tested.

		% of males having chosen:		ischiosetosa females chosen	albifrons females chosen	% ♀ A
		20♀	60♀			
albifrons males	isolated	89		314	390	55.4
	from mass culture	42	63	202	304	60.8
ischiosetosa males	isolated	92		414	137	24.8
	from mass culture	18	32	170	21	11.0

of sexual choice of males, for a given pair of populations, is reproducible in defined conditions and is hereditarily transmissible. Learning cannot be excluded, but the greater discrimination of males with previous contact with females is perhaps a secondary result of their lower sexual activity.

A modified technique was used in an attempt to search for variation in the sexual choice of males. Results from experiments of this sort could reveal behavioural heterogeneity of males, some of them with homogamic tendencies, others being indifferent or preferentially heterogamic.

Males of the two species, isolated from birth, were allowed five successive choices at intervals of one week. After five experiments, males were distributed in six classes based upon having chosen one kind of female (in this case ischiosetosa) 0 to 5 times. In Figure 1, the experimental distributions are compared to the theoretical binomial ones calculated with the same respective mean of choice. The experimental distributions for the two species, for the five successive tests are not significantly different from expected.

Although the technique is far from sensitive these results suggest the absence of any obvious intra-specific genetic variability for this behaviour. All males seem to have the same probability of error in making their choices.

Using this experimental procedure, strictly homogamic or heterogamic albifrons males were selected. After the four generations of selection so far completed no consistent results have been produced in promoting choice of their own or alien females. The number of generations of selection is low, but these early results are not in conflict with the inference drawn from the shape of the distributions in Figure 1.

Genetic determination of sexual choice. As the variation in sexual choice is weak in parents, the study of the genetic determinism of these choices must be approached by the shift of interspecific hybrids. Two independent experiments were performed.

In the first experiment, pure albifrons and ischiosetosa males and F_1 hybrid males were given a choice between hybrid and/or pure females (Table 3). The results were as follows:

ischiosetosa males always rejected albifrons females and apparently confused hybrid females with their own;

GENETICS OF ETHOLOGICAL ISOLATING MECHANISMS

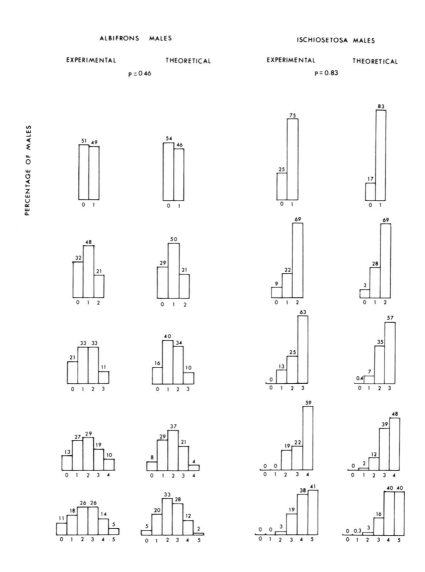

Figure 1. Experimental and theoretical binomial distribution in five successive mate choice experiments with <u>albifrons</u> and <u>ischiosetosa</u> males. The theoretical distribution is established using the value of p obtained as a mean of the five experiments.

TABLE 3

Choices by <u>ischiosetosa</u>, <u>albifrons</u> and F_1 hybrid males between hybrid and/or pure-bred females.

♂ \ ♀	I N	I %	A N	A %	IA N	IA %	AI N	AI %
I	133	83	168	46	212	55	93	53
A	27	17 *	197	54	173	45 *	82	47
I	137	55	95	45	169	41	77	44
AI	114	45	118	55	244	59 *	98	56
AI+IA	78	89	74	45	167	54	81	58
A	10	11 *	89	55	144	46	59	42
Σ%/3 I		46		30		32		32
Σ%/3 A		9		36		30		30
Σ%/3 Hyb.		45		33		38		38

* $P < 0.05$

- <u>albifrons</u> males, as in the preceding experiments, showed a low discriminatory capacity.

- hybrid males obtained from the two reciprocal parental crosses showed the same behaviour with, generally, a slight preference for the hybrid females.

A second experiment was performed with pure males and various types of hybrid males, but with a choice only between parental female types. 18 kinds of males were tested:

- pure <u>albifrons</u> and <u>ischiosetosa</u> males;

- reciprocal hybrids of the first generation;

- second generation hybrids from crosses ♀ <u>AI</u> x ♂ <u>AI</u>;

— male products of the three first generations of recurrent or non-recurrent back crosses (BC).

The back cross males were obtained by crossing hybrid females AI to the two types of pure males (BS1 $\underline{I}$ and BC1 $\underline{A}$); female offspring of these two back crosses were then crossed either to ischiosetosa or to albifrons males, giving four kinds of BC2: $\underline{II}$, $\underline{IA}$, $\underline{AI}$ and $\underline{AA}$, considering only the nature of males to which F1 and BC1 females were back crossed. Of the 8 kinds of the third back cross, only seven were tested, the number of $\underline{IIA}$ males being too small to test.

Males can be characterized by an average genetic constitution, in terms of the percentage of genes of one species they carry (in this case albifrons).

The correlation between the percentage of albifrons females chosen and the genetic constitution of the males was very strong ($r = +0.92$) (Fig.2). The experiment involved a total of 4,376 choices.

It can therefore be concluded that there is genetic determination of male sexual choice. A monofactorial determination cannot be totally excluded because each point corresponds to a mean based upon numerous males (45 as an average for each category). It seems, however, reasonable to attribute a polyfactorial basis to this character.

Variability in sexual choice of males. Male choice experiments were carried out with a number of populations belonging to the same species or to different species. Genetic markers were again used to recognise the two kinds of females. Intrapopulational references were also needed to test the effects of these markers.

Choice indices of males 1 and 2 were calculated. These two categories were males of different colour morphs within the same population or were individuals from different populations within the same species or were individuals of different species. Male 1 choices were classified as homogamic (or positively assortative) if females of type 1 were chosen, heterogamic (or negatively assortative) if females of type 2 were chosen. The choice index is the difference between the number of homogamic and heterogamic choices, divided by the sum of the two. The index ranges from -1 to $+1$, positive values denoting homogamy, negative values heterogamy, and a zero value, random choice. This simple index is derived from the isolation index of Stalker (1942) calculated on the

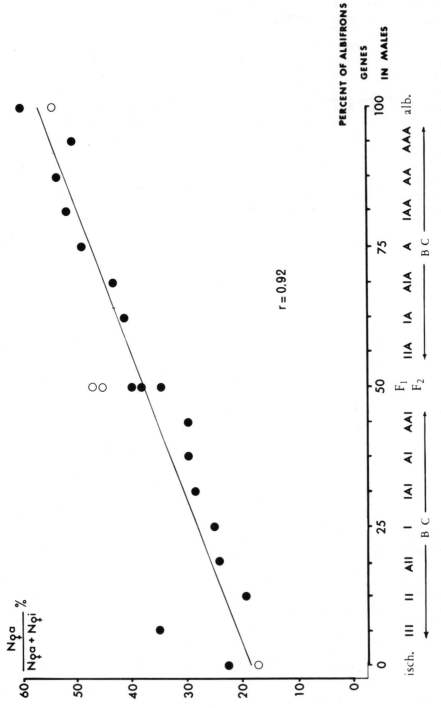

Figure 2. Choice between *ischiosetosa* and *albifrons* females shown by pure-bred males, F_1 and F_2 hybrid males and males produced by the three first back crosses. Open circles represent results obtained with the experiments reported in Table 3.

percentage of inseminated females. The reciprocal index is calculated in the same way for males of type 2.

The results are presented graphically (Fig.3): the index of male 1 is plotted along the abscissa, male 2 being plotted on the ordinate. The two kinds of males are permutable and there is a symmetry along the NE-SW axis. The male with the highest index was always designated as male 2 (all points are on the same side of the axis of symmetry). The joint choice index (half of the sum of the two reciprocal indices) can be read on the NW-SE diagonals.

The distance of points from the axis of symmetry gives the attractivity index of females (half of the difference between the two reciprocal indices, derived from Bateman's index, 1948). According to our convention, the index is positive for female 2, negative with the same absolute value for female 1. Absolute values are on the same scale as the joint isolation index.

The positions of points on the plan define immediately the quality of the choice.

(a) Intrapopulational choice: the two sexes were marked by genes of the natural polychromatism of these isopods. Variations affect the colour of the pigment, the pattern of chromatic cells distribution, the development of chromogranules and the degeneration of the pigmentry system (Bocquet, 1953; Cléret, 1970). Alleles controlling the phenotypes were maintained, not in separate stocks, but in a polymorphic state in mass culture. The so-called wild type phenotype (Cléret, 1970) <u>nigrum granulatum uniforme</u> is characterized by black chromatophores uniformly distributed.

The deviations are relatively moderate (Fig.4) compared with random choice: most of the points occupy a very small area, near the origins. Nevertheless, the nonrandomness is statistically significant ($P = <0.05$) in 7 out of 28 chi-squares (calculated for males 1 and 2 in 14 experiments): in some instances selective choice exists. One kind of male (male 2 according to our convention) chose the category of female marked by the same allele as he possessed (female 2) (homogamy). The second kind of male also preferred the type 2 females (heterogamy). The choice shown by males was independent of the genes they carried, but one kind of female was more attractive to the two kinds of males.

In the genus <u>Drosophila</u>, experiments using mutant and wild-type stocks have demonstrated discrimination by females,

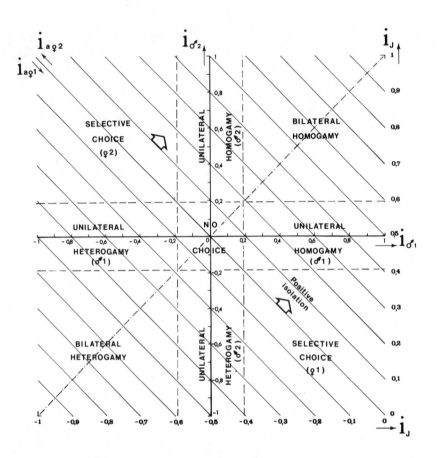

Figure 3. Graphic representation of male choice. Reciprocal indices of choice ($i\ \delta_1$ and $i\ \delta_2$) are plotted along the co-ordinates. Joint isolation index and female attractivity can be read along diagonals. Broken lines indicate limits of significance of χ^2, compared with random choice, for an arbitrary number of 100 choices. The occurence of points in various sectors indicates the quality of choices.

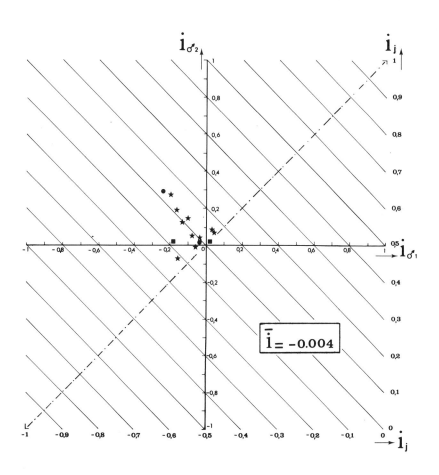

INTRAPOPULATION

Figure 4. Distribution of points observed with individuals marked by genes for colour polymorphism within populations (circles: albifrons, stars: ischiosetosa, squares: praehirsuta).

and the absence of male discriminating ability (Bateman, 1948). In his review, Spiess (1970) concluded: "genotypes of males may be selectively different and females are the selective agents in the process". The conclusions in <u>Jaera</u> are different but the technique used does not show whether any dissimilarity exists in the courtship propensity of males.

Some other points of interest, still under investigation, appeared during this experiment.

The most notable is that selective choices (combining both significant or non-significant deviations) are in favour of wild type females in 9 cases out of 10 (four other experiments were performed with non-wild type animals). In the last one (<u>praehirsuta</u> of Woods Hole), <u>Aurantiacum</u> females were preferred. It is interesting to note that this phenotype has a high frequency in this strain.

Genes affecting the colour polymorphism of <u>Jaera</u> have pleiotropic effects on behaviour. Colour alleles do not appear to influence the extent of isolation (the average index of choice is -0.004) but may, in some cases, influence selective choice.

(b) Interpopulation choices: the results of these tests are illustrated in Figure 5, and are seen to be very different from the intrapopulation tests. Some pairs of populations show an absence of choice in a few cases, but frequently one type of male and often the two types show homogamic tendencies. Considering couple formation, the existence of an incipient isolation between geographic strains within a species can be concluded. The average index is + 0.200. Further analysis of data indicates a lack of correlation between the geographical distance separating populations and the degree of isolation.

(c) Interspecific choices: the average value is just a little higher (+ 0.279) and the distribution of points in Figure 6 is very similar to that of Figure 5. For some pairs of populations, isolation is rather high: the number of heterospecific couples formed is reduced by the discriminatory faculties of males. In nature, isolation by male choice is probably higher since previous contacts render males more homogamic as we have seen.

Comparison of the three diagrams (Fig. 4, 5 and 6) allows further conclusions. Single genes influencing colour change the attractivity of females, but do not generate isolation within populations. Incipient isolation between strains and

GENETICS OF ETHOLOGICAL ISOLATING MECHANISMS

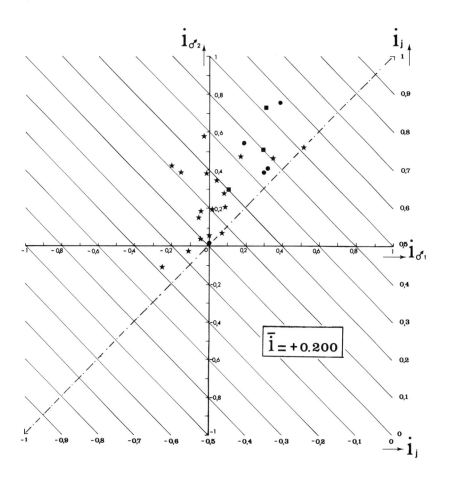

INTERPOPULATION

Figure 5. Distribution of points for mate choice between populations within species (same symbols as Fig.4).

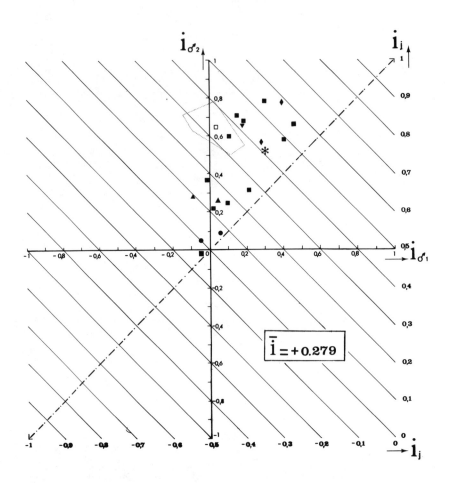

INTERSPECIFIC

Figure 6. Distribution of points for mate choice between different species. Each symbol corresponds to a given pair of species. The dotted polygon is the envelop of points obtained from data of Table 1 for <u>ischiosetosa</u> and <u>albifrons</u> populations from Kiel.

GENETICS OF ETHOLOGICAL ISOLATING MECHANISMS

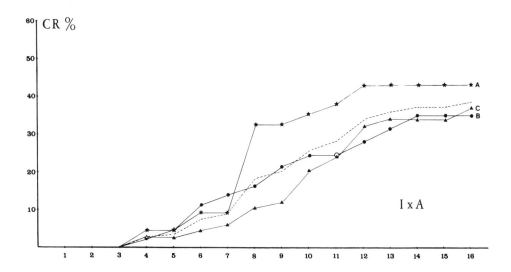

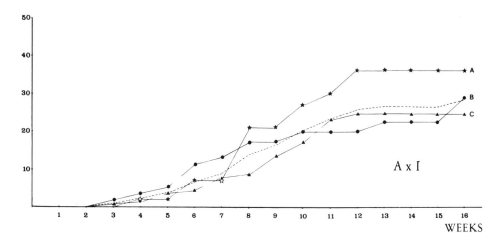

Figure 7. Crossing rate (CR) expressed as a cumulative percentage as a function of time (in weeks) in reciprocal crosses between ischiosetosa and albifrons (the first letter indicates the species of the female).

species (considering male choice) cannot result from the simple accumulation of such point mutations.

The homogamy observable in interspecific choices is close to that obtained between strains of the same species. These incipient isolations must be considered more as a by-product of the geographical divergence of populations rather than as a consequence of a previous differentiation preceding or accompanying speciation.

Crosses without Choice and Female Reactivity

Using the same two populations of <u>albifrons</u> and <u>ischiosetosa</u> from Kiel, hybridizations were performed with heterospecific isolated couples. The success of crosses was estimated by the appearance of hybrids. If couples are infertile it is judged that impregnation has not taken place. Offspring appear 2 to 5 weeks after insemination, depending on the time of copulation in the female sexual cycle.

<u>Analysis and repetitivity</u>. The curves of Figure 7 depict the cumulative percentage of fertile couples over time. This kind of presentation is preferred because once a female has been inseminated, even by a heterospecific male, she continues to produce broods. These data were corrected to take into account mortality (Solignac, 1976).

Hybridization progressed only slowly during the course of weeks and even months. Three independent experiments were performed, in each direction of crossing, with the same result. The two reciprocal crosses give approximately the same results, but this may not always be the case (see below). Reciprocal hybridization rates can be considered as characteristic of a given pair of populations.

The probability of finding new fertile **couples** among those which were previously unmated seems to be more or less constant over the weeks: the curve is damped as the number of virgin females diminishes. This suggests the involvement of a random process in the partial success of hybridization, rather than a high receptivity of certain females to heterospecific sexual overtures. This was also supported by complementary experiments. <u>Albifrons</u> females were crossed with wild-type <u>ischiosetosa</u> males. As soon as a female produced a brood, the male was changed for another <u>ischiosetosa</u> male bearing a dominant marker. First and second heterospecific fertilizations

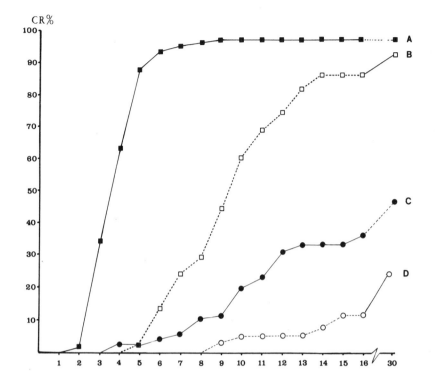

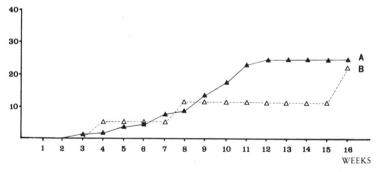

Figure 8. Crossing rate (cumulative percentage) as a function of time (in weeks) in various kinds of intra- and interspecific crosses: upper curves: ischiosetosa females: A: virgin ischiosetosa females crossed to their own males B: already inseminated ischiosetosa females crossed to their own males (second conspecific insemination) C: virgin ischiosetosa females mated to albifrons males: lower curves: albifrons females A: hybridization of virgin albifrons females with ischiosetosa males B: second interspecific insemination of albifrons females by ischiosetosa males.

are represented in Figure 8b. Such previously impregnated females presented to a second male were only rarely fertilized a second time.

Data on intraspecific controls are seen in the curves A and B of Figure 8a depicting progressions of the first and second fertilizations in ischiosetosa. Curves C and D correspond to the crosses I x A using virgin females and females previously impregnated by their own males respectively.

The partial success of hybridization (frequently between 30 and 50% after 16 weeks of contact) is not due to the existence of true variation in the readiness to hybridise in males or females, but to a similar reproductive threshold for all pairs, overcome only by chance.

Genetic determinism of hybridization. Pure albifrons and ischiosetosa females, and females with various degrees of hybridity (obtained in the same manner as the males described above) were crossed to both kinds of pure males. Hybridization rate was estimated by the percentage of fertile couples after 16 weeks of contact, corrected for mortality. Figure 9 shows the variation of success of crossing as a function of female genotype. For both kinds of males, successful matings were observed (more than 90%) when the female possessed at least 50% of genes in common with the male. Futch (1973) obtained similar results with Drosophila ananassae and D. pallidosa.

The parameter discussed here (hybridization rate) gives a slightly distorted impression however. A more detailed analysis indicated that the speed of crosses increased the more the two partners have genes in common. The impression of a kind of dominance given in Figure 9 then disappeared.

Each couple was observed weekly to follow the progress of crosses. The occurrence of partners in precopula was noted. For each kind of cross, the frequency of observed precopula was established. The observation of a "head-to-tail" position does not necessarily mean that copulation occurred, but it is an index of the sexual drive of the male, with a given category of female. The frequency of precopula for both pure male types is indicated in Figure 10 as a function of female genotype.

With albifrons males, this activity is clearly independent of female genotype. In experiments on male choice, these males appeared to be nearly indiscriminant between the two

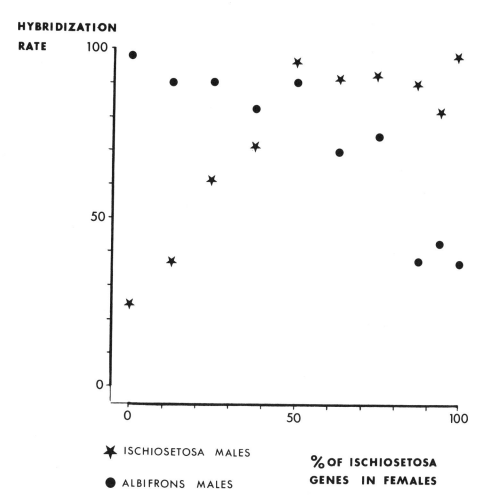

Figure 9. Hybridization rate of ischiosetosa and albifrons males with pure or hybrid females. Hybridization rate (percentage of successful crosses after 16 weeks, corrected for mortality) is expressed as a function of female genotype.

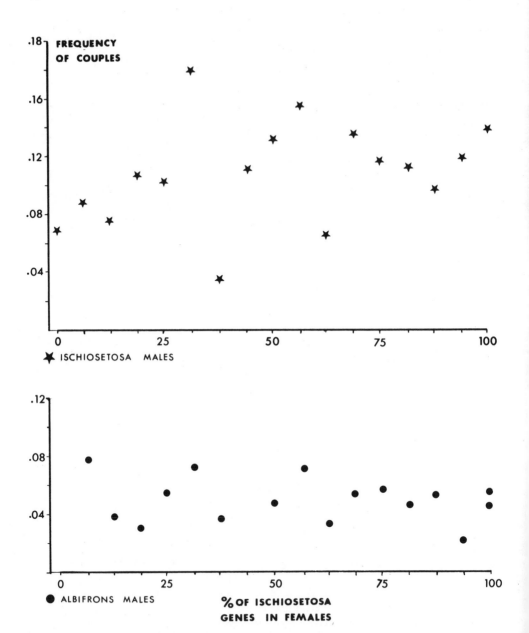

Figure 10. Frequency of precopula phase observed during weekly examination of crossings for both ischiosetosa and albifrons males as a function of the genotype of the females to which they were crossed.

kinds of pure females. In this case, the result of crossing is evidently due to female reactivity to male stimuli.

With ischiosetosa males, the distribution of points is not so clear, but a similar conclusion can reasonably be drawn.

This kind of comparison must be limited to one type of male with different types of females. Indeed, the length of precopulation is very variable between populations or from one species to another.

It seems logical to suppose that the success of these crosses, in the absence of choice, depends largely on female reactivity. Such a reaction is genetically determined but it is not possible to conclude much as to the nature of the genetical determinants. Indeed, most of the results can be explained by assuming the existence of two alleles, each of them characterizing one species. Nevertheless, as in the male choice experiments, the hypothesis of a polyfactorial basis of female reactivity is highly probable.

Variations of hybridization rate. When varying the pairs of species or even the populations within the species, considerable variation in results was obtained. Often, the final hybridization rate was between 30 and 50%. But female posthirsuta originating from Newport (United States), crossed with albifrons or ischiosetosa males as rapidly as they did with their own. On the contrary, praehirsuta males, from Roscoff (France) produce a hybridization rate always close to zero with ischiosetosa and albifrons females.

Some curves are given in Figure 11. It can be seen that the results obtained from the two reciprocal crosses are sometimes strikingly different. For the moment, no general rule can be drawn: the data are few due to the long time needed for such experiments.

Conclusions

An attempt will be made to correlate the results obtained by the two types of experiments. When choice of mate is available, the role of the two sexes is rather clear:

- due to male discrimination in certain pairs of populations, most couples are of the homogamic type;

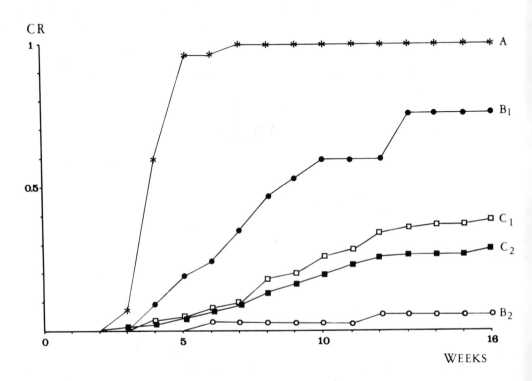

Figure 11. Crossing rate (CR, cumulative percentage) in various pairs of species: A: ♀ posthirsuta (Newport) x ♂ ischiosetosa (Ringkøbing). B$_1$: ♀ praehirsuta (Roscoff) x ♂ ischiosetosa (Kiel); B$_2$: reciprocal cross. C$_1$: ♀ ischiosetosa (Kiel) x ♂ albifrons (Kiel); C : reciprocal cross.

- females are rapidly receptive to males of their own kind and show a low receptivity to alien males;

- fertilized females are less receptive than virgin ones, especially to heterospecific courtship.

When no mate - choice is allowed, the situation is not so clear and three kinds of rather conflicting conclusions can be drawn:

- in many experiments on hybridization, the rate of fertilization depends on the discriminating faculties of the male. The more discriminating the male during choice, the lower the success of hybridization with that male. Consequently, the role of the male seems important.

- nevertheless, the frequency of precopula in isolated heterospecific couples does not confirm the above conclusion. The attempts of males to copulate seem to be characteristic of the kind of male and independent of the genotype of the female. The reactivity of the female seems most important in crosses, depending on the female genotype.

- this last impression is also supported by inter-population crosses. Males are sometimes very discriminating in their choices, as much as in interspecific choices. Nevertheless, in the absence of choice they fertilize females from alien populations very rapidly.

Consequently, the hybridization rate seems to be the consequence of both frequency of attempts made by the male (male activity) and of female reactivity. Nevertheless, in the absence of choice, the female receptivity is probably the most important factor.

More difficult, but also more important is the problem in nature. The frequency of hybrids can be established in males without difficulty, F1 hybrids and often products of first back cross being morphologically identifiable. The hybrid rate is generally between 1% and 0.1%. The actual rate of hybridization in nature can also be established. Hybrids in *Jaera albifrons* develop quite well under laboratory conditions, as well in fact as the parental forms. Offspring raised in the laboratory from inseminated females collected in the field permit the estimation of the actual hybridization rate. The frequency of hybrids in the offspring of such females is never higher than the hybrid frequency observed in natural conditions among samples of males collected at the same time as the females. In other words, the real frequency of hybridization is close to the observed frequency of hybrids.

In the laboratory, in experimentally mixed populations, species remain isolated with the same intensity as in nature (Solignac, 1976). Experimental analysis seems therefore to give an accurate picture of what occurs in nature. This is, however, not a general conclusion valid for all organisms. For instance, Drosophila pseudoobscura and D. persimilis, are more isolated in the field than in the laboratory for reasons which are not clear (see for example Dobzhansky, 1951).

Summary

The complex Jaera albifrons includes five related species of the cold and temperate coasts of the North Atlantic. In the laboratory, species can be hybridized if there is no choice of mates, but in nature, they remain reproductively isolated in the zones of contact or overlap. A genetic study of ethological isolating mechanisms has been carried out. In Jaera albifrons, couples are established following male initiative. In male choice experiments, using males and females from the same population marked by different alleles of loci controlling the natural colour polymorphisms, sexual choice is either at random or is strictly selective i.e. one kind of female is preferred by both types of males. On the contrary, interspecific choice, and at a lower level interpopulation preferences within a species, are of the homogamic type. The choice made by parental males and various kinds of hybrid males (F1, F2 products of recurrent or non-recurrent back crosses) show a genetic determination of sexual preferences, probably of a multifactorial nature.

Female refusal to alien male stimuli is the main bar to hybridization. In the absence of mating choice, the success of crosses between females or various degrees of hybrity and parental males depends on the degree of genetic resemblance between the two parents. Full success of the cross is assured if the female possesses at least 50% of genes in common with the male. Both under natural conditions and in laboratory populations, the frequency of hybrids does not exceed 1%. The frequency of hybrids, estimated from males, is a good index of the frequency of hybridization and hence the ethological differences appear to be established as the main isolating mechanism in nature.

REFERENCES

Anderson, W. W. and Ehrman, L. 1969. Mating choice in crosses between geographic populations of Drosophila pseudoobscura. Am. Midl. Nat. 81, 47-53.

Bateman, A. J. 1948. Intra-sexual selection in Drosophila. Heredity 2, 349-368.

Bateman, A. J. 1949. Analysis of data on sexual isolation. Evolution 3, 174-177.

Bocquet, C. 1953. Recherches sur le polymorphisme naturel des Jaera marina (Fabr.) (Isopodes Asellotes). Essai de systematique évolutive. Arch. Zool. exp. gén. 90, 187-450.

Cléret, J. J. 1970. La pigmentation des Jaera albifrons Leach. V. Equilibre du système pigmentaire et gènes responsables du polychromatisme. Arch. Zool. exp. gén. 111, 447-493.

Dobzhansky, Th. 1951. Experiments on sexual isolation in Drosophila. X. Reproductive isolation between Drosophila pseudoobscura and Drosophila persimilis under natural and under laboratory conditions. Proc. Natl. Acad. Sci. U.S.A. 37, 792-796.

Futch, D. G. 1973. On the ethological differentiation of Drosophila ananassae and Drosophila pallidosa in Samoa. Evolution 27, 456-467.

Jones, M. B. and Naylor, E. 1971. Breeding and bionomics of the British members of the Jaera albifrons Leach group of species (Isopoda: Asellota). J. Zool., London 165, 183-199.

Solignac, M. 1975. Isolement reproductif entre les membres de la super-espèce d'Isopode littoral Jaera albifrons Leach. Publ. Staz. Zool. Napoli 39 Suppl., 696-716.

Solignac, M. 1976. Demographic aspects of interspecific hybridization. A study of the Jaera albifrons species complex (Crustacea, Isopoda, Asellota). Oecologia 26, 33-52.

Spiess, E. B. 1970. Mating propensity and its genetic basis

in _Drosophila_. In, _Essays in Evolution and Genetics in Honor of Theodosius Dobzhansky_. Hecht, M. K. and Steere, W. C. (eds). pp.315-379. Appleton-Century-Crofts, New York.

Stalker, H. D. 1942. Sexual isolation studies in the species complex _Drosophila virilis_. Genetics 27, 238-257.

SECTION 5

SOME APPLIED ASPECTS

EFFECTS OF ORGANOPHOSPHATE PESTICIDES ON FISH ESTERASES:

A CASE FOR AN ISOZYME APPROACH

M. Krajnović-Ozretić[1] and W. de Ligny[2]

[1]Institute "Rudjer Bošković", Rovinj, Yugoslavia

[2]Neth. Inst. for Fish. Invs. IJmuiden, The Netherlands

Introduction

The increasing use of organophosphate pesticides in agricultural pest control has raised considerable concern regarding their potentially hazardous effect on aquatic environments which they enter either indirectly through accidental contamination after agricultural application, or directly for the purpose of aquatic insect control. In the past, organophosphate pesticides were thought to degrade rapidly and were not considered a chronic hazard to the environment. Later, however, it was shown that under certain conditions some of these compounds retain anticholinesterase activity for a year or more while in aqueous solution (Weiss and Gakstatter, 1965).

The generally accepted mode of action of organophosphate pesticides on vertebrates is through disruption of nerve impulse transmission in the central and peripheral nervous system by inhibition of acetylcholinesterase (AChE), the enzyme that regulates the amount of the neurotransmitter acetylcholine (O'Brien, 1960, 1967; Aldridge, 1971). The inhibition of AChE is cumulative and persistent because of the relatively irreversiable phosphorylation of the enzymes.

Several authors (Weiss, 1958, 1959, 1961; Williams and Sova, 1966; Holland et al., 1967; Gibson et al., 1969; Coppage, 1972) have used inhibition of AChE in fish as a biological monitor of water pollution by organophosphate pesticides, but the interpretation of the relationship of inhibition and death is controversial. Death of fish after exposure to organophosphate pesticides in the laboratory has been reported to occur at brain AChE activities in a range from 5.4 to 92 per cent of the normal activity (Weiss, 1958, 1961), while fish have survived at activities as low as 10 to 20 per cent of the normal activity (Gibson et al., 1969; Weiss, 1961).

However, comparative studies of four species of fish by Coppage and Matthews (1974, 1975), showed that death occurred at relatively constant levels of brain AChE inhibition. The report by Gibson et al.(1969) suggests explanations for the disparity of the results reported in the literature, such as variability of AChE activity with time and between population and variation in the mode of storage in the brain.

An additional source of variation was reported by Hogan (1970), who showed that an increase in ambient temperature results in an increase of AChE specific activity. However, Baslow and Nigrelli (1964) found that in killifish brain AChE activity varied inversely with the temperature of acclimation and Ludtke and Ohnesorge (1966) also reported no positive correlation in the tench.

The work of Baldwin (1971) showed that the effect of the temperature of assay on the substrate affinity of AChE from various species of fish is related to the temperature range in the habitat of the species. Variation in the relative proportion of different molecular forms of brain AChE in goldfish depending on the ambient temperature (Guillon and Massoulie, 1976) may further complicate, and add to, the variation of AChE activity measurements.

In general fish can tolerate markedly depressed brain AChE activity better than higher vertebrates (Benke et al., 1974a). This suggests that the relative rate of formation and accumulation of acetylcholine and the relative susceptibility to the lethal effect of excess of acetylcholine is different in various species.

In a study of the rate of AChE recovery in fish very little, if any, early recovery occurred and return to the control level is so low as to suggest that it may be due to synthesis of new enzyme. Recovery may take a month or more. Slow recovery of AChE in fish may make them more susceptible to cumulative injury by organophosphates (Benke and Murphy, 1974b).

The final toxic impact of the organophosphate pesticides

on a species also depends on the enzyme systems involved in the metabolism of the compounds. These enzymes are the mixed-function oxidases (MFO) of the liver, the glutathione conjugating enzymes and the enzymes able to hydrolyse organophosphate esters. The activity of these enzyme systems results either in detoxification or activation of many organophosphate pesticides. Brodie and Maickel (1962) reported that fish lack significant MFO activity and that many lipid soluble compounds pose "no problem to fish since they rapidly diffuse through lipoidal membranes of the gills". Contrary to this report Ludke et al., (1972) found that certain fresh water fishes do possess the enzymes necessary for the metabolism of lipid soluble xenobiotics. The reactions catalysed by these enzymes are analogous to those demonstrated in mammals and insects; parathion is converted to paraoxon, but the rate at which this occurs in fish is lower than in the rabbit (Potter and O'Brien, 1964) and mice (Benke et al., 1974). The latter report indicates that fish livers may also be capable of glutathione dependent detoxification of certain organophosphate compounds, but again at a considerably lower rate than mouse liver.

The third mechanism by which organophosphates are converted and detoxified is enzymatic hydrolysis either of the phosphate ester or the carboxy ester linkage by arylesterases (Zech and Zurcher, 1974; Mendoza et al., 1976) and some carboxylesterases (Motoyama and Dauterman, 1974). The occurrence of enzymes of this kind in fish will be discussed in more detail later in this paper. It should, however, be mentioned here that the capacity of these esterases to degrade organophosphates differentiates them not only from cholinesterases, discussed previously, but also from the majority of the carboxyl-or aliesterases which, like the cholinesterases, are strongly inhibited by organophosphate esters.

The controversial results of the cholinesterase inhibition studies, the simultaneous occurrence of organophosphate sensitive aliesterases and the possible implications of the presence of organophosphate degrading esterases in fish has led us to consider the potential value of electrophoretic analysis and inhibition assays of fish esterases as an additional tool in monitoring for the presence of organophosphates in the water. The introduction of zone electrophoresis in starch (Smithies, 1955) or polyacrylamide gels (Ornstein, 1964) and the application of histochemical reagents to the gels (Hunter and Markert, 1957) has revealed the presence of numerous molecular forms of carboxylesterases, and cholinesterases. Application of inhibitors during the development of zones of enzyme activity in the gels has made the demonstration of inhibition easier and more specific. So far, however, very little effort has been spent on the assessment of differential inhibition of multiple molecular forms of esterases by various organophosphates.

Implications of isozymes for inhibition studies

In using inhibition of (acetyl) cholinesterases and carboxylesterases as indicators of the exposure of aquatic animals to organophosphates in the environment it is important to know if one or more esterases are involved in the activity measurements (Hogan and Knowles, 1968a; Hogan, 1971). This is also relevant even if a more specific substrate for AChE such as acetyl (thio) choline is used. This was shown by Tripathi and O'Brien (1973, 1975), who found four AChE isozymes in housefly brain, with different inhibition characteristics, towards various organophosphate compounds. From these reports it seems likely that one isozyme (V) is the crucial target for organophosphate and that inhibition of this isozyme causes death, in spite of the fact that it represents only 25 per cent of total AChE activity. Tsakas and Krimbas (1970) studying the olive fruit fly found a difference in the acetylcholinesterase genotypes that survived or were killed by dimethoate. Dimethoate preferentially killed insects that were homozygous and heterozygous for a silent allele of the esterase A locus. Knowles, Arurkar and Hogan (1968) detected multiple zones of brain AChE and carboxylesterases in bluegill and catfish with different sensitivity to Paraoxon and DFP.

Davis and Agranoff (1968) demonstrated differential isozyme recovery following inhibition of rat retinal AChE by DFP. The faster recovery of one of the isozymes was related to its more rapid metabolic turnover. Vijayan and Brownson (1975) demonstrated that rat brain AChE may exist as three isozymes with different susceptibility to parathion, and that the major isozyme (isozyme 3) exhibits the greatest sensitivity to the inhibitor in all areas of the brain.

This multiplicity of AChE and carboxylesterase types is of importance for the application of esterase inhibition measurements as indicators of the presence of organophosphates in the following respects:

1. Measurements of "overall" inhibition of cholinesterases or carboxylesterases may give results which do not show a consistent relation with the detrimental effect of the organophosphates present because of underlying differences

in the sensitivity of multiple forms of the enzymes (isozymes).

2. If "mixed tissue" such as gills is used for measurements, different ratios of tissue specific isozymes such as AChE from blood red cells and gill tissue with different inhibition characteristics may give variable results.

3. Measurements of brain AChE inhibition as an overall indicator of contamination with organophosphate compounds may be unreliable if the different individuals and populations being compared carry genetically determined variant enzymes (allozymes) with different inhibition characteristics.

4. Differences between isozymes of AChE or carboxylesterases in sensitivity of inhibition by specific organophosphates may be applied to provide a more selective indicator of exposure to certain compounds. This appears particularly promising in view of the contrasting results of inhibition studies with various organophosphate esters in electrophoretic analyses as reported by Hogan and Knowles (1968a) for brain tissue, and by Metcalf et al. (1972) and Krajnović-Ozretić and Žikić (1975) for other tissues.

5. The possibility of genetic polymorphism in particular of the AChE of the brain and the gills also justifies an isozyme approach, as differences in sensitivity between genetic variants may imply differences in resistance to toxic effects of organophosphate compounds, which in the long term may result in shifts in gene frequencies.

Importance of isozymes for organophosphate breakdown

Various hydrolytic enzymes including arylesterases and some carboxylesterases play a role in the detoxification system of the body by their ability to metabolize organophosphate esters. The activity of these esterases may influence the toxicity of organophosphate compounds and also must be considered in the evaluation of monitoring studies. The presence of such enzymes in the organism used as an indicator of the presence of organophosphates (by inhibition studies) will impair the value of the monitoring system as the compounds may be broken down in vivo, prior to assay. They also may interfere with the use of inhibitors in electrophoretic assays as if present in the gel, this may break down added inhibitors during the usual preincubation period.

Hogan and Knowles (1968b) found enzymes in the liver of bluegill and catfish capable of degrading DFP and dichlorvos. Breakdown of paraoxon has been observed in the serum of various mammals and been ascribed to arylesterases (Zech and Zurcher, 1974). Although arylesterases have not been found in the serum of fish, their presence in the brain, liver and muscle of the white crappie was reported by Metcalf et al. (1972), and in the liver, muscle and intestine of Fundulus by Holmes and Whitt (1970). DFP resistant esterases, which were not classified as arylesterase were also found in liver tissue of the Adriatic sardine (Krajnović-Ozretić and Žikić, 1975). Dettbarn and Hoskin (1975) found a very low toxicity of DFP toward squids which appears to be reflection of the high level of DFP-ase in the squid nervous system.

A number of paraoxon hydrolysing enzymes was reported for rat liver and four enzymes were identified by Kojima and O'Brien, (1968). In the human brain Sakai and Matsumura (1971) found 6 esterases of the A-type and one DFP-ase hydrolysing toxic phosphates and carbamates. Electrophoretic studies indicate that each insecticidal substrate can be hydrolysed by several esterasee and that some insecticides (parathion and diazinon) are degraded by a similar group of enzymes while in other cases specific enzymes (e.g. DFP-ase) are responsible for the breakdown of specific substrates.

The variability in the occurrence and activity of paraoxon hydrolysing enzymes in man as reported by Zech and Zurcher (1974) shows a distribution suggestive of control by a two allele system. Some carboxylesterases of the house fly have been shown to possess phosphatase activity, and these enzymes show an increase in an organophosphate resistant strain (Ahmad, 1974).

To analyse the presence of **organophosphate** degrading enzymes in fish electrophoretic techniques may be applicable if other analytical methods are not available (e.g. the manometric methods for assay of breakdown of DFP and dichlorvos used by Hogan and Knowles, 1968b). Strips containing esterase zones may be excised from duplicate gel layers and incubated with the organophosphate compounds. The incubation mixtures then may be tested for remaining inhibitory activity towards other esterases sensitive to these inhibitors. This assay also may be carried out using gel strips after previous characterization of the sensitivity of various esterase bands for the inhibitors. To study the breakdown of pesticides, esterase gels can also be cut in pieces, extracted and incubated with radioactive labelled insecticides and the resulting radio-activity measured by liquid scintillation

(Sakai and Matsumura, 1971). Paraoxon degrading activity may be localized in certain isozymes by use of gel strips, incubated with this compound, and subsequent colorimetric analysis of the liberated nitrophenol (Zech and Zurcher, 1974).

Differences in resistance to organophosphate esters between species and between individuals may result either from differences in the ability to degrade these compounds or from differences in the sensitivity of the AChE and carboxylesterases to their inhibitory effect. Genetically determined differences of organophosphate degrading enzymes therefore also may contribute to a long term monitoring method.

Summary

The literature on the effect of organophosphate pesticides on fish esterases is reviewed. It is pointed out that inhibition of fish acetylcholinesterase is not always a consistent measure of water pollution by OP esters, or of the adverse effect on the fish itself. It is suggested that this inconsistency may be due to the presence of multiple forms of cholinesterases differing in their sensitivity to the inhibitors or to the presence of esterases capable of degrading OP esters. The literature describing such multiplicity of esterases in various species is cited. The importance of electrophoretic analysis for the differentiation of various forms of esterases is emphasized, and it is suggested that such analysis may enable more specific monitoring for certain OP esters and allow detection of their detrimental impact on fish at the population level.

Acknowledgement

This work was supported by a grant from the Netherlands Organization for Pure Scientific Research (Z.W.O.).

REFERENCES

Ahmad, S. 1974. An electrophoretic study of the negative correlation of certain carboxylesterases to insectide resistance in Musca domestica L. Experientia 30, 331-333.

Aldridge, W. N. 1971. The nature of the reaction of organophosphorus compounds and carbamates with esterases. Bull. WHO 44, 25-30.

Baldwin, J. 1971. Adaptation of enzymes to temperature: Acetylcholinesterases in the central nervous system of fishes. Comp. Biochem. Physiol. 40B, 181-187.

Baslow, M. H. and Nigrelli, R. F. 1964. The effect of thermal acclimation on brain cholinesterase activity of the killifish, Fundulus heteroclitus. Zoologica 49, 41-51.

Benke, G. M., Cheever, K. L., Mirer, F. E. and Murphy, S. D. 1974a. Comparative toxicity, anticholinesterase action and metabolism of methyl parathion and parathion in sunfish and mice. Toxicol. Appl. Pharmacol. 28, 97-109.

Benke, G. M. and Murphy, S. D. 1974b. Anticholinesterase action of methyl parathion, parathion and azinophosmethyl in mice and fish: Onset and recovery of inhibition. Bull. Environ. Contam. Toxicol. 12, 117-122.

Brodie, B. B. and Maickel, R. P. 1962. Comparative biochemistry of drug metabolism. Proc. 1st Pharmacol. Meeting 6, 299-324.

Coppage, D. L. 1972. Organophosphate pesticides: specific level of brain AChE inhibition related to death in sheepshead minnows. Trans. Amer. Fish. Soc. 101, 534-536.

Coppage, D. L. and Matthews, E. 1974. Short-term effects of organophosphate pesticides on cholinesterases of estuarine fishes and pink shrimp. Bull. Environ. Contam. Toxicol. 11, 483-488.

Coppage, D. L. and Matthews, E. 1975. Brain-acetylcholinesterase inhibition in a marine teleost during lethal and sublethal exposures to 1, 2-Dibromo-2, 2-dichloroethyl dimethyl phosphate (Naled) in seawater. Toxicol. Appl. Pharmacol. 31, 120-133.

Davis, G. A. and Agranoff, B. W. 1968. Metabolic behaviour

of isozymes of acetylcholinesterase. Nature 220, 277-280.

Dettbarn, W. D. and Hoskin, G. S. 1975. Toxicity of DFP and related compounds to squids in relation to cholinesterase inhibition and detoxifing enzyme levels. Bull. Environ. Contam. Toxicol. 13 (2), 133-140.

Gibson, J. R., Ludke, J. L. and Ferguson, D. E. 1969. Sources of error in the use of fish brain acetylcholinesterase activity as a monitor for pollution. Bull. Environ. Contam. Toxicol. 4 (1), 17-23.

Guillon, G. and Massoulie, J. 1976. Multiplicité des formes moléculaires de l'acétylcholinestérase et acclimation thermique chez le Carassius auratus. Biochimie 58, 465-471.

Hogan, J. W. 1970. Water temperature as a source of variation in specific activity of brain acetylcholinesterase of bluegills. Bull. Environ. Contam. Toxicol. 5, 347-352.

Hogan, J. W. 1971. Brain acetylcholinesterase from cutthroat trout. Trans. Amer. Fish. Soc. 100, 672-675.

Hogan, J. W. and Knowles, C. O. 1968a. Some enzymatic properties of brain acetylcholinesterase from bluegill and chanel catfish. J. Fish. Res. Bd. Can. 25, 615-623.

Hogan, J. W. and Knowles, C. O. 1968b. Degradation of organophosphates by fish liver phosphatases. J. Fish. Res. Bd. Can. 25, 1571-1579.

Holland, H. T., Coppage, D. L. and Butler, P. A. 1967. Use of fish brain acetylcholinesterase to monitor pollution by organophosphorus pesticides. Bull. Environ. Contamin. Toxicol. 2, 156-162.

Holmes, R. S. and Whitte, G. S. 1970. Developmental genetics of the esterase isozymes of *Fundulus heteroclitus*. Biochem. Genet. 4, 471-480.

Hunter, R. L. and Markert, C. L. 1957. Histochemical demonstration of enzymes separated by zone electrophoresis in starch gels. Science 125, 1294.

Knowles, C. O., Arurkar, S. K. and Hogan. J. W. 1968. Electrophoretic separation of fish brain esterases. J. Fish.

Res. Bd. Can. 25, 1517-1519.

Kojima, K. and O'Brien, R. D. 1968. Paraoxon hydrolyzing enzymes in rat liver. J. Agri. Food. Chem. 16, 574.

Krajnović-Ozretić, M. and Žikić, R. 1975. Esterase polymorphism in the Adriatic sardine (Sardina pilchardus Walb.) 1. Electrophoretic and biochemical properties of the serum and tissue esterases. Anim. Blood Grps. biochem. Genet. 6, 201-213.

Ludtke, A. H. and Ohnesorge, F. K. 1966. Characterization of cholinesterases in various tissues of the tench (Tinca vulgaris) and of the rabbit. Z. Vergleich. Physiol. 52, 260-275.

Ludke, J. L., Gibson, J. R. and Lusk, C. I. 1972. Mixed function oxidase activity in freshwater fishes: Aldrin epoxidation and parathion activation. Toxicol. Appl. Pharmacol. 21, 89-97.

Mendoza, C. E., Shields, J. B. and Augustinsson, K. B. 1976. Arylesterases from various mammalian sera in relation to cholinesterase, carboxylesterases and their activity towards some pesticides. Comp. Biochem. Physiol. 55C, 23-26.

Metcalf, R. A., Whitt, G. S., Childers, W. F. and Metcalf, R. L. 1972. A comparative analysis of the tissue esterases of the white crappie (Pomoxis annularis Rafinesque) and black crappie (Pomoxis nigromaculatus Lesueur) by electrophoresis and selective inhibitors. Comp. Biochem. Physiol. 41B, 27-38.

Motoyama, N. and Dauterman, W. C. 1974. The role of nonoxidative metabolism in organophosphorus resistance. J. Agr. Food Chem. 22, 350-356.

O'Brien, R. D. 1960. Toxic Phosphorus Esters, Chemistry, Metabolism and Biological Effects. Academic Press, New York.

O'Brien, R. D. 1967. Insecticides, Action and Metabolism, Academic Press. New York.

Ornstein, L. 1964. Disc electrophoresis. I. Background and theory. Ann. New York Acad. Sci. 121, 321-349.

Potter, J. L. and O'Brien, R. D. 1964. Parathion activation by livers of aquatic and terrestrial vertebrates. Science 144, 55-56.

Sakai, K. and Matsumura, F. 1971. Degradation of certain organophosphate and carbamate incesticides by human brain esterases. Toxicol. Appl. Pharmacol. 19, 660-666.

Smithies, O. 1955. Zone electrophoresis in starch gels: group variations in the serum proteins of normal adults. Biochem. J. 61, 629-641.

Tripathi, R. K. and O'Brien, R. D. 1973. Effect of organophosphates in vivo upon acetylcholinesterase isozymes from housefly head and thorax. Pest. Biochem. Physiol. 2, 418-424.

Tripathi, R. K. and O'Brien, R. D. 1975. The significance of multiple molecular forms of acetylcholinesterase in the sensitivity of houseflies to organophosphorus poisoning. In, Isozymes II: Physiological Function. Markert, C. L. (ed). Academic Press, New York.

Tsakas, S. and Krimbas, C. B. 1970. The genetics of Dacus oleae. IV. Relation between adult esterase genotype and survival to organophosphate insecticides. Evolution 24, 807-815.

Vijayan, V. K. and Brownson, R. H. 1975. Polyacrylamide gel electrophoresis of rat brain acetylcholinesterase: isoenzyme changes following parathion poisoning. J. Neurochem. 24, 105-110.

Weiss, C. M. 1958. The determination of cholinesterase in the brain tissue of three species of fresh water fish and its inactivation in vivo. Ecology 39, 194-199.

Weiss, C. M. 1959. Response of fish to sub-lethal exposure of organic phosphorus insecticides. Sew. Ind. Wastes 31, 580-593.

Weiss, C. M. 1961. Physiological effect of organic phosphorus insecticides on several species of fish. Trans. Amer. Fish. Soc. 90, 143-152.

Weiss, C. M. and Gakstatter, J. H. 1965. The decay of anticholinesterase activity of organic phosphorus insecticides on storage in waters of different pH. Proc. Second Int.

Water Poll. Res. Conf. Tokyo, 83-95.

Williams, A. K. and Sova, C. R. 1966. Acetylcholinesterase levels in brains of fishes from polluted waters. Bull. Environ. Contam. Toxicol. 1, 198-204.

Zech, R. and Zürcher, K. 1974. Organophosphate splitting serum enzymes in different mammals. Comp. Biochem. Physiol. 48B, 427-433.

QUANTITATIVE GENETIC VARIATION IN FISH - ITS SIGNIFICANCE FOR SALMONID CULTURE

M. Holm and G. Naevdal

Institute of Marine Research

Directorate of Fisheries, Bergen, Norway

Introduction

The extent of genetic variation in fish seems to have been largely unexplored until around 1950. In the early literature one often encounters statements that an observed character is "possibly heritable", but it is evident that when variation was observed, it was usually not known whether such variation was controlled by environmental or by genetic factors. As far as we know, the first unquestionable evidence of influence by genetic factors in fish resulted from hybridization experiments made by Gordon (1931) concerning variations in melanophores. Other early investigations of heredity in fish were made by Svärdsson (1950, 1952) who studied the numbers of gill rakers in white fish (Coregonus lavaretus).

Following the investigations of Sick (1961) on the two-allele system of cod (Gadus morhua) and whiting (Gadus melanogrammus) hemoglobins, qualitative genetic characters such as blood types and protein polymorphisms have been extensively studied in fish since the early 1960's. Polymorphism can be studied from population data and the aim of the studies on polymorphic characters in fish has been mostly to investigate the population structure of species in nature.

Studies of quantitative genetic variation in fish have been started in connection with the growing interest in fish farming and sea-ranching (release of fish to improve the fisheries). To study quantitative characters, controlled breeding experiments over at least two and preferably more generations are necessary. Biometrical analysis of collected data is essential in order to enable estimates of the magnitude of the different sources of variation and in order to assess whether the variation arises from genetic or from environmental reasons.

The main objective for such experiments is to estimate the degree of control of additive genetic factors on observed variation i.e. to estimate the heritability (h^2) the magnitude of which would indicate if selective breeding would be efficient for genetic gain in the desirable character. One could also aim at estimating the extent of control by non-additive genetic factors which again would indicate how useful hybridization of pure lines would be as the main tool for genetic gain.

Other topics in the study of quantitative genetics have been investigations of the effect of inbreeding, genetic and environmental correlations, population crosses and the hybridisation of species.

In the present paper a short account of earlier experiments in quantitative genetic variation in fish is given. Some original data concerning Atlantic salmon and rainbow trout are presented and the results are discussed in relation to their significance for cultivation of salmonids.

Review

In spite of the increasing interest in aquaculture the study of quantitative genetics in fish still lies far behind that of farmed warmblooded animals. Most of the reported experiments are on the common carp (<u>Cyprinus carpio</u>).

A lot of selection work has been done by fish breeders and they have provided "indirect" evidence for the existence of genetic variability in fish. The carps have been cultured for centuries in the far east as well as in Europe and by selecting the largest fish for generation after generation several distinctive strains which seem to be improved breeds have evolved. The most important breeding work on carp has been done by Israeli, Soviet and West German breeders (cf. Moav 1976; Kirpichnikov, 1971, 1973; Golovinskaya, 1972 and Schäperclaus, 1961).

Fish breeders have been working extensively with salmonids since the 1920's. Hayford and Embody (1930) obtained increased growth rate, increased average number of eggs/female and higher resistance to disease after 10 years of selection with the eastern brook trout (Salvelinus fontinalis).

Lewis (1944) succeeded in producing a brood stock of rainbow trout that spawned in October instead of in spring which is the normal spawning time. Donaldson and Olson (1957) and Donaldson (1968) created a famous strain of rainbow trout by selecting the largest, strongest and earliest maturing individuals from several strains which spawned at the desired season. After 25 years of selection the fish grew to 1.5lb in one year, a weight the initial strains reached in 4 years. The age of maturation and the time of spawning were also altered considerably. Donaldson and Menasveta (1961) and Donaldson (1968) also worked with the chinook salmon (Oncorhynchus tshawytscha) and obtained faster growth rate, higher fecundity, higher tolerance to high temperatures and resistance to disease.

Successful breeding of rainbow trout is reported by Savostyanova (1969) from the USSR, where growth rates and age of maturity have been altered. Millenbach (1973) advanced the mean spawning time of rainbow trout by 2 months in 14 years.

Although successful, all these experiments suffer, from a scientific point of view, from the drawback that they lack control lines of unselected material. Therefore it is difficult to tell whether the changes produced, especially in growth rate, arose from real genetic improvement or whether they arose from improved rearing techniques.

Some large scale, controlled experiments have been carried out by Israeli scientists who have worked extensively on the common carp. Experiments were carried out with inbreds, crossbreds and full-sib families and with breeds from Europe and the Far East (Wohlfarth et al., 1964; Moav and Wohlfarth, 1966, 1973; Moav et al., 1975b). The experiments revealed that even inbred lines of the domesticated European carp are very heterozygous. The genetic variance component was estimated to be 30-40% of the total phenotypic variance and was found to be mostly non-additive.

A two-way selection experiment for growth in common carp (Moav, 1976) showed that mass selection for high growth rate for five generations produced no response while selection for slow growth gave a heritability estimate of around 0.30.

Stegman (1958) and Wohlfarth et al., (1971) have investigated on heritability of body height/length ratio of carp. It is concluded that in spite of a heritability factor of 40%, selection for this trait is dubious since the genetic correlation of growth rate to this and to other body conformation characters is found to be near zero. (Ankorion, 1966.)

The possibility of reducing intramuscular bones by selective breeding has attracted some attention. However, the results of the investigations are contradictory. Von Sengbusch (1963) found wide phenotypic variation in carp and Kossman (1972) found a significant genetic component for intramuscular bones. Lieder (1961) found small phenotypic variance for this trait and Moav et al., (1975a) found only small phenotypic and practically no genetic variation for intramuscular bones.

During the last ten years several investigations of heritability in salmonids have been made. Many of the experiments deal with growth characteristics, but other traits such as mortality, egg number and volume, smoltification, tolerance to disease and to low pH have also been investigated.

Calaprice (1969) reported an h^2 of 0.25 for "size" of three species of pacific salmonids. Aulstad et al., (1972) investigated variation in growth as measured by body length and weight of three year classes of rainbow trout. Based on full sib groups, intra class correlations ranging from 0.09 to 0.17 were found. Based on paternal and maternal half-sib groups heritability factors of h^2_S = 0.17-0.32 and h^2_D = 0.03-0.07 for weight at 150 and 280 days of age respectively were estimated. Non-additive genetic factors were found to be of little importance for these traits.

Heritability estimates made by Kincaid (1972) for 150 day weight of hatchery rainbow trout varied between 0.26 and 0.29, Gall (1975) estimated a h^2 of 0.21 for postspawning body weight, but he states that apparently the genetic variability for this trait is of a non-additive nature. Chevassus (1976) estimated the h^2 for length and weight of rainbow trout to be in the range of 0.045-0.375 $\pm \approx 0.10$ based on covariance between full sibs.

In Atlantic salmon heritability for weight at recapture was estimated by Ryman (1971) to be 0.22. However, in Ryman's opinion this is an overestimate and he states that only a minor part of variation is of genetic origin. Lindroth (1972) found a heritability factor of 0.6 for weight of Atlantic salmon parr.

Other traits that have been investigated in salmonid fish are survival of eggs and alevins. Ayles (1974) found a low heritability for survival of uneyed eggs of splake hybrids, ($h^2 = 0.06 \pm 0.07$ and 0.09 ± 0.11). The h^2 for survival of alevins was 0.41 ± 0.18 which Ayles refers to as attributable to genetic differences in resistance to blue sack disease. Kanis et al., (1976) conclude in their analysis of egg and alevin mortality that there are significant differences between different strains of Atlantic salmon, brown trout and rainbow trout. The heritability ranges were 0.03-0.3, based on the sire component and 0.09-0.6 based on the dam component. The authors suggest that the high dam components are due to maternal effects as well as environmental interactions and that in general low heritabilities could be expected since the traits in question being fitness traits, are controlled mainly by non-additive genetic factors.

Resistance to vibrio disease has been investigated by Gjedrem and Aulstad (1974) and they found significant differences in resistance between river strains of Atlantic salmon parr. The calculated estimates of heritability within strains were 0.12 and 0.07 respectively for the dam and the sire component. The estimates were statistically significant from zero and the authors argue that resistance to vibrio disease can be altered by selection.

Refstie et al., (1977) report highly significant variation between river strains of Atlantic Salmon in percent parr smoltifying in one year. The smolt percent was found to be strongly correlated with the average weight of the families. Low heritability estimates for sire component, averaging 0.06, to gether with significant sire x dam interaction suggest a considerable non-additive genetic variance controlling smoltification at one year of age.

Gall (1975) investigated several **reproductive** traits in hatchery reared rainbow trout and estimated the heritability based on full sib comparison to be about 0.20 for volume of eggs spawned, egg number, egg size and eggs per 100 g body weight for 2 year old females. Dominance and epistatic variance was evident for all traits except egg number. Genetic, environmental and phenotypic correlations were calculated and it was suggested that selection for egg volume might be the most effective way of improving reproductive performance.

Material and Methods

Starting in 1971, fertilized eggs of Atlantic salmon were collected each year from several Norwegian rivers (Fig.1) and from two commercial fish farms. In 1971 eggs from one Swedish and two Canadian rivers were included. In general a set of two females and two males from each river were used to make four F_1 progeny groups in a 2 x 2 factorial design. From 1975 onwards the emphasis has been on producing the F_2 generation and only a few F_1 groups were made. Family selection was used to produce the F_2 - crossings within and between strains as well as inbred groups.

The eggs were incubated and hatched at a field research station where heated water from a hydroelectric power station keeps the hatchery water several degrees above natural water temperature.

The groups were kept separate until they were about 6 months old. The fish were then marked by amputation of a fin and put together in large tanks. After smoltification, at 1+ or at 2+ age, the fish were transferred to large floating pens in full strength seawater at a commercial fish farm 200 km north of the field research station. Fish which did not reach smolt stage at 1+ age were transferred to floating cages in brackish water (22-25 ppt salinity) in the fjord outside the hatchery and kept there until they were smolts and ready to be transferred to sea water at about 2+ age. Before stocking the groups into larger pens the fish were either tagged individually or cold branded with a group symbol. Efforts were made to maintain the environment as similar as possible for all the groups in order to avoid systematic error.

Records were kept of mortality, smoltification and maturation and the fish were weighed and measured every six months. Conventional biometrical methods (cf. Sokal and Rohlf, 1969) were used for statistical treatment of the data and the heritability factors were estimated following the methods of Bogyo and Becker (1965) and Becker (1967).

In 1972 and 1973 the egg and milt material for the F_1 generations of rainbow trout came from two commercial fish farms. The ancestors of the parent fish were imported from Denmark and the owners of the farms had for some generations carried out mass selection for high growth rate and later maturity, but the effect of this selection is unknown due to lack of controls.

QUANTITATIVE GENETIC VARIATION IN FISH

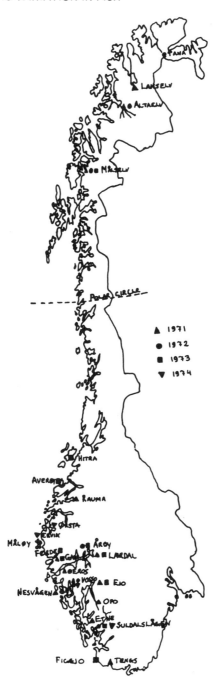

Figure 1. The Norwegian rivers from which eggs of Atlantic salmon were collected in 1971, 72, 73 and 74.

For the rainbow trout the 2 x 2 factorial design described above was used for making the progeny groups. The groups were given similar treatment to the salmon, but the rainbow trout were, in general, transferred to brackish water when 8 months old and to full strength seawater when 14 months old. At two and a half years of age, 20-30 fish of each of the best performing families are selected as parents for the F_2 generation. From 1975 onwards inbred groups as well as non-related crosses within, and between, strains have been made.

A more detailed description of the experimental design is given by Naevdal *et al*., (1975a,b, 1976).

Results

The survival of the eggs was in general fairly high for most of the strains of Atlantic salmon in all year classes although considerable variation was noted. However, this variation is difficult to evaluate since in some instances it might have arisen from trouble with the water supply. Differences in transport and handling might also have caused differences in viability. In the 1976 year class there seems to be a tendency towards higher viability of the eggs of the wild strains than of the eggs of the F_2 generation. The inbred groups also seem to have a lower viability than their crossbred half-sib groups.

A wide variation in growth rate was found between salmon populations from different rivers. In some cases there was also considerable variation between groups of the same river origin, but in general the intrariver variation was small.

To see whether growth rate in early life may be used as an index of growth ability at a later age, correlation and regression coefficients between mean lengths for the same groups at different ages were calculated for the 1972 year class. The coefficients are presented in Table 1, which shows that correlation is high between measurements taken at short intervals of time, but correlation between the earliest and latest measurements is near zero. Growth (measured as group mean lengths) at early life therefore seems to be of negligible value for predicting the growth at later stages and selection for high growth rate cannot be based on presmolt growth rate.

When regarding the total experimental material of a year class as one "population", variable, and usually high,

TABLE 1

Correlation coefficients (above the diagonal) for growth at different ages of sib groups of Atlantic salmon of the 1972 year class. Regression coefficients are shown below diagonal.

Age Months	6	12	18	24	30	36	42
6	–	0.76	0.27	0.004	0.01	0.05	-0.03
12	1.27	–	0.49	0.32	0.20	0.18	0.10
18	0.51	0.56	–	0.60	0.49	0.48	0.19
24	0.01	0.48	0.70	–	0.69	0.64	0.50
30	0.03	0.87	1.72	2.04	–	0.93	0.72
36	0.39	0.88	1.87	2.15	1.00	–	0.76
42	0.17	0.43	0.63	1.39	0.65	0.64	–

heritability factors for length variation were found. Based on paternal (S) and maternal (D) half sib groups, ranges of h^2_S = 0.15-1.00 and h^2_D = 0.03-1.00 were found. Some of these estimates are clearly overestimates indicating that environmental factors have influenced the results. Maternal effect (i.e. egg size) could also account for part of the overestimation, since the highest heritabilities in general were found for length at 6 and 12 months.

However, the high estimates to a great extent reflect the variation between river populations, and are not heritability factors in a true sense. The heritability factors for length of full sibs within half sib groups were in general lower (in the range 0.00-0.35) than the estimates based on half sib groups. Because these calculations were made within populations the estimates are true heritability factors, and although they may be influenced by non-additive genetic factors, the estimates suggest that genetic gain for growth rate may be expected by selection.

The proportion of fish transforming to smolts after one year varied greatly between the groups in the 1974 as well as in the 1976 year class, being in the range of 0.6 - 72%

and 0 - 30% respectively. The estimates of heritability factors for proportions of smolts at 1 year were in the range h^2_S = 0.3-0.4 and h^2_D = 0.1-0.3.

From experience in practical fish farming it is well known that there is a near connection between growth and smoltification. From Figure 2, showing two arbitrarily chosen groups of the 1974 year class it can be seen that, compared to the parr, the smolts must probably have had an accelerated growth pattern several months before the fish could be classified as smolts based on morphological characters. A two way dependence between growth and smoltification seems to occur - high presmolt growth gives a high proportion of smolts and starting smoltification causes an acceleration in growth rate, thus giving the bimodal curve seen in Figure 2.

The connection between growth rate and smoltification was studied with a simple correlation test where mean length of the parr was used as the independent variable reflecting the growth characteristics of the groups and the proportion of smolts transformed to inverse sine was taken as the dependent variable. Using the deviations of group means from the river localities the correlation coefficients were 0.85 for the 1974 year class and 0.44 for the 1975 year class.

We have drawn the conclusion that although smoltification is closely connected with presmolt growth rate, the growth characteristics alone cannot explain the total variation between the groups in the proportion of one year smolts. The estimates of heritability factors cited above, therefore seem to reflect both genetic factors controlling the smoltification directly and factors effecting the variation in growth rate.

The age when salmon mature varies greatly between natural river populations. Maturation is known to retard growth rate and cause increased mortality and therefore this trait is of great interest to salmon cultivation. In the 1972 year class some of the groups had a high percentage of males maturing in the first sea year. This was particularly so for the Canadian groups, where one of the rivers had as high as 28% maturation in the first sea year. In the second sea year the proportion of mature fish among the groups varied between 0-100%. A similar variation was also found in the 1973 year class. Intra-river variation was fairly small. The results were in general in good accordance with the life histories of the river populations from which the groups originated.

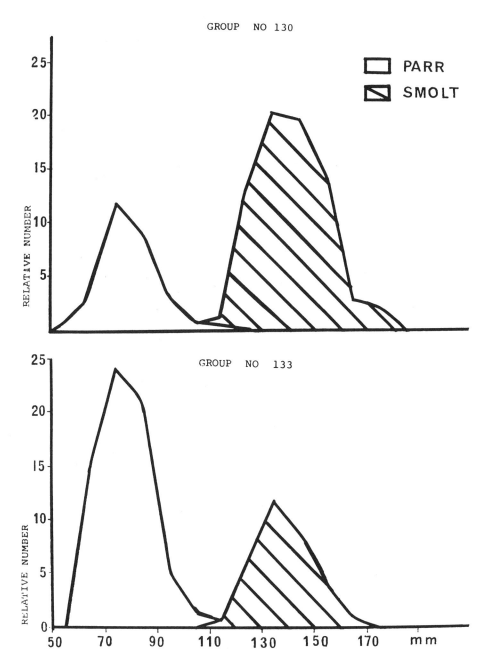

Figure 2. Length distribution of parr and smolt in two arbitrarily chosen groups of 1974 year class of Atlantic salmon.

If the total experimental material is regarded as one single population heritability estimates calculated from pairs of paternal and maternal half sibs are high, $h^2_s \simeq 1.0$ and $h^2_D \simeq 0.67$. However, these figures reflect mainly the variation of age at maturity between populations. Based on full sibs within pairs of half sibs the heritability factors were in the order of 0.05 - 0.10 indicating that heritability within population ("true heritability") is low. The data for the 1973 and 1974 year classes although not completely worked up, seem to confirm these results.

Growth rates of the rainbow trout vary between the groups, but the evaluation of the growth is complicated by the onset of the maturation. However, variation can also be seen when comparing immature fish. Data concerning the year class hatched in 1974 are presented in Figure 3.

It is evident (Fig.3a) that maturation retards growth rate since the growth of the mature fish at an age of 1+ during the winter (Dec. 1975 - April 1976) was small compared with immature fish. Maturation does not seem to be connected with individual growth because the mean lengths of mature and immature fish are very similar in most groups (Fig.3a). The same has also been observed for the 1973 year class of rainbow trout (Møller et al., 1976).

However, the immature fish seem to be smaller than the mature fish at the age of 2+ (Fig.3b). This is confirmed by the 1972 and 1973 year classes (Møller et al., 1976) and there seems to be a tendency for the females to be bigger than the males. At the age of 1+ almost all mature rainbows are males and when examining the individually tagged 1973 year class, it appeared that most of these males mature again at 2+ age. Since maturation has retarded growth rate at 1+ age it therefore could be expected that on an average the males would be smaller than the females at the age of 2+.

The age when the rainbow trout mature varies between the groups as well as between the year classes (Fig.4). Estimates of heritability factors based on half sib groups gave h^2 near zero while the estimates for 2+ age were high, $h^2 = 0.4$-0.7. For variations in length the estimates were $h^2 = 0.0$-0.3 at 1+ age. Growth rate as well as age at maturation therefore might be improved by selection. Because it is not possible to find any connection with growth or environmental factors, the low heritability estimates for variation of early maturity in males in the groups indicate that non-additive genetic factors exert

QUANTITATIVE GENETIC VARIATION IN FISH

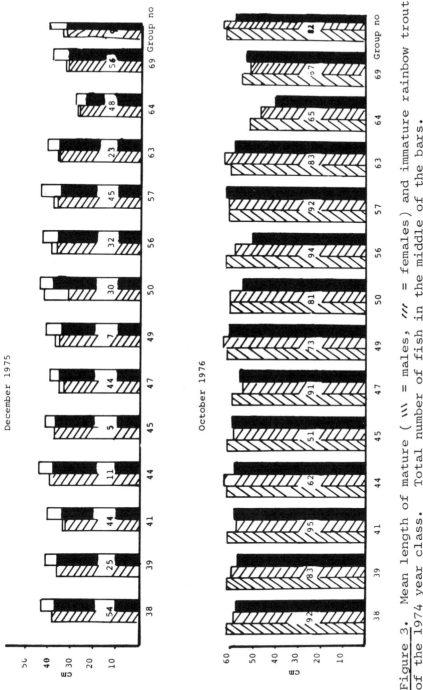

Figure 3. Mean length of mature (\\\ = males, /// = females) and immature rainbow trout of the 1974 year class. Total number of fish in the middle of the bars.
a (top): Mean length at 1+ years (Dec. 1975, hatched and black area) with the increase in mean length at 2 years (April, 1976, white area) on top.
b (below): Mean length of males, females and immatures at 2+ years (Dec. 1976).

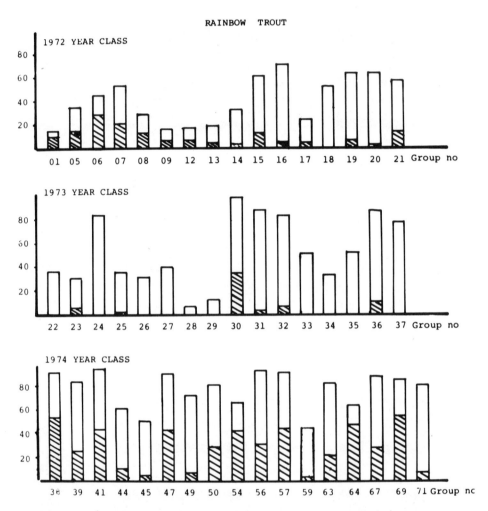

Figure 4. Percent mature fish at 1+ years (hatched area) and at 2+ years (white area) in groups of three year classes of rainbow trout.

some control on maturation. In addition the estimates of h^2 based on full sib groups were high and such estimates include dominance effects.

Discussion

From a fish farming point of view survival of eggs has not always been considered to be one of the most important characters, since the eggs have been fairly cheap and fertilized eggs have been abundant on the market (Kanis et al., 1976). When the prices rise or when the demand for eggs exceeds the supply survival becomes a commercially more important charater. For experimental purposes it is, however, most important to be able to maintain large groups for accurate estimates of the traits under investigation.

The heritability estimates for growth and age at maturation of salmon indicate that these traits are partly controlled by additive genetic factors. The estimates also seem high enough to expect genetic improvement by selection. Since the variation between populations is very large, selection of populations seems appropriate for changing the age at first maturity. For faster growth rate a combination of selection of families and selection of the largest individuals seems indicated. This conclusion is also supported by the results of Gjedrem (1976). Since there seems to be little correlation between growth at an early age and at a later age (Table 1) selection based on size at an early stage in life is not expected to give the best genetic gain. Selection should preferentially be based on size at normal time of slaughter i.e. after 2 or 2.5 sea years.

The significance for salmonid farming of obtaining late maturing fish can be illustrated by the following example: While the mean weight gain of the 1972 year class between two registrations was 2.5 kg for the immature salmon it was only 1.1 kg for the early matured females and 0.8 kg for the males.

Early smoltification is an important character for raising smolt for liberation into the sea or for fish farming purposes. By early smoltification the time the fish have to be reared in fresh water is reduced. Thereby the expenses of producing the fish can be cut down and the capacity of the rearing stations can be increased. The results indicate that smoltification has a genetic base and that it can be altered by selection. Refstie et al., (1977) conclude that that the best chance for improvement would be family selection and they also point out that to take advantage of the non-additive genetic variance for smoltification at 1 year a cross-breeding of strains would be appropriate.

The rather high heritability factors found for growth and age at first maturation of the rainbow trout, indicate that these characters can be improved by selection of phenotypes. This is in accordance with Aulstad et al., (1972) who found evidence of low non-additive genetic variance for growth of rainbow trout and argued that selection for either length or weight would give genetic improvement.

The findings of Gall (1975) indicate dominance and epistatic variance for several reproductive traits and for postspawning body weight. However, we have concluded that the best method of selection would be the selection of fast-growing individuals from families with greater age at first maturation.

Rainbow trout that are immature at the age of 2+ would mean the possibility of producing big "salmon like" fish suitable for smoking and for slaughtering throughout the year. The growth rate would not be retarded by the onset of maturation and an increase of e.g. 5 cm in mean length of the fish would make an increase in weight of approximately 500 g for each individual.

One thing has to be stressed: we do not yet know whether differences in growth rates arise from differences in food intake or from differences in food conversion efficiency. To solve this problem will be an important field of research in the future.

The results so far are only a first step towards the understanding of quantitative genetic variation in salmonids. Experiments on a parent-offspring design and a study of extreme variants are now being undertaken to test the results. By using salmonids instead of e.g. guppies we hope to obtain the double benefit of producing material fit for practical maricultural purposes and at the same time throw light on the nature of quantitative genetic variation in fish.

Summary

A short review of literature on quantitative genetic variation in fish is given. Data on growth smoltification and age at maturity of Atlantic salmon (Salmo salar) and rainbow trout (Salmo gairdneri) are presented. Heritability estimates for these characters are given. The results are discussed in relation to their significance for salmonid culture.

REFERENCES

Ankorion, J. 1966. Studies on the Heredity of a Number of Morphological Traits in Carp (in Hebrew). M.Sc. Thesis, Genetics Department, Hebrew University, Jerusalem.

Aulstad, D., Gjedrem, T. and Skjervold, H. 1972. Genetic and environmental sources of variation in length and weight of rainbow trout. J.Fish. Res. Bd. Can. 29, 237-241.

Ayles, G.B. 1974. Relative importance of additive genetic and maternal sources of variation in early survival of splake hybrids (Salvelinus fontinalis x S. namaycush). J. Fish. Res. Bd. Can. 31, 1499-1502.

Becker, W.A. 1967. Manual of Procedures in Quantitative Genetics. Washington State Univ. Press Wash.

Bogyo, T.P. and Becker, W.A. 1965. Estimates of heritability from transformed percentage sib data with unequal subclass numbers. Biometrics 21, 1001-1007.

Calaprice, J.R. 1969. Production and genetic factors in managed salmonid populations. In, Symposium on Salmon and Trout in Streams, 1968. H.R. Mac Millan Lectures in Fisheries. Northcote, T.G. (ed). Univ. British Columbia, Vancouver, B.C.

Chevassus, B. 1976. Variabilite et Heritabilite de Performances de Croissance chez la Truite Arc en Ciel (Salmo gairdneri Richardson). (Mimeo) 10 pp Laboratoire de Physiologie des Poissons. Joues en Josas.

Donaldson, L.R. 1968. Selective breeding of salmonid fishes. In, Marine Aquaculture. Selected papers from the Conference on Marine Aquaculture. McNeil, W.J. (ed). Oregon State University Press.

Donaldson, L.R. and Menasveta, D. 1961. Selective breeding of chinook salmon. Trans. Am. Fish. Soc. 90, 160-164.

Donaldson, L.R. and Olson, R.P. 1957. Development of rainbow trout brood stock by selective breeding. Trans. Am. Fish. Soc. 85, 93-101.

Gall, G. A. E. 1975. Genetics of reproduction in domesticated rainbow trout. J. Anim. Sci. 40, 19-28.

Gjedrem, T. 1976. Possibilities for genetic improvements in salmonids. J. Fish. Res. Bd. Can 33, 1094-1099.

Gjedrem, T. and Aulstad, D. 1974. Selection experiments with salmon. I. Differences in resistance to Vibrio disease of salmon parr (Salmo salar). Aquaculture 3, 51-59.

Golovinskaya, K.A. 1971. Breeding in fish culture. Rep. FAO/UNDP (TA), 2926, 292-300.

Gordon, M. 1931. The heredity basis for melanosis in hybrids of Mexican Killifishes. Proc. Nat. Acad. Sci. 17, 276-280.

Hayford, C.O. and Embody, G.L. 1930. Further progress in the selective breeding of brook trout at the New Jersey State Hatchery. Trans. Am. Fish. Soc. 60, 109-113.

Kanis, E., Refstie, T. and Gjedrem, T. 1976. A genetic analysis of egg, alevin and fry mortality in salmon (Salmo salar) sea trout (Salmo trutta) and rainbow trout (Salmo gairdneri) Aquaculture 8, 259-268.

Kincaid, H.L. 1972. A preliminary report of the genetic aspects of 150-day family weights in hatchery rainbow trout. Proc. 52nd. Ann. Conf. West Assoc. State Game Fish Comm. Portland, Oregon.

Kirpichnikov, V.S. 1971. Genetics of the common carp (Cyprinus carpio L.) and other edible fishes. Rep. FAO/UNDP (TA) 2926, 186-201.

Kirpichnikov, V.S. 1973. Genetic selection of common carp in the U.S.S.R. FAO Aquacult. Bull. 5, 5.

Kossman, H. 1972. Untersuchungen über die genetische Varianz der Zwischenmuskelgräten des Karpfens. Theor. Appl. Genet. 42, 130-135.

Lewis, R. C. 1944. Selective breeding of rainbow trout at Hot Creek Hatchery. Cal. Fish and Game 30, 95-97.

Lieder, U. 1961. Wie viel Gräten haben unsere Süsswasserfische? Dtsch. FischZtg. 8, 334-338.

Lindroth, A. 1972. Heritability estimates of growth in fish. Aquilo Ser. Zool. 13, 77-80.

Millenbach, C. 1973. Genetic selection of steelhead trout for management purposes. Int. Salmon Foundation. Special publ. ser. 4 (1), 253-257.

Moav, R. 1976. Genetic improvement in aquaculture industry. FIR: Ag/conf./R. 9, 1-23.

Moav, R. and Wohlfarth, G. 1966. Genetic improvement of yield in carp. FAO Fish. Rep. 44, 112-129.

Moav, R. and Wohlfarth, G. 1973. Carp breeding in Israel. In, Agricultural Genetics - Selected Topics. Wiley, New York.

Moav, R., Finkel, A. and Wohlfarth, G. 1975a. Variability of inter-muscular bones, vertebrae, ribs, dorsal fin rays and skeletal disorders in the common carp. Theor. Appl. Genet. 46, 33-43.

Moav, R., Hulata, G. and Wohlfarth, G. 1975b. Genetic differences between the Chinese and European races of the common carp. 1. Analysis of genotype - environment interactions for growth rate. Heredity 34, 323-340.

Møller, D., Naevdal, G., Holm, M. and Lerøy, R. 1976. Variation in growth rate and age at sexual maturity in rainbow trout. FAO Techn. Conf. on Aquaculture, May 1976/E 61, 1-7.

Naevdal, G., Holm, M., Lerøy, R. and Møller, D. 1975a. Variation in age at sexual maturity in rainbow trout. ICES, C.M. 1975/M 23, 1-7.

Naevdal, G., Holm, M., Møller, D and Østhus, O.D. 1975b. Experiments with selective breeding of Atlantic Salmon. ICES C.M. 1975/M 22, 1-10.

Naevdal, G., Holm, M., Møller, D. and Østhus, O.D. 1976. Variation in growth rate and age at sexual maturity in Atlantic salmon. ICES C.M. 1976/E 40, 1-10.

Refstie, T., Steine, T.A. and Gjedrem, T. 1977. Selection experiments with salmon. II. Proportion of Atlantic salmon smoltifying at 1 year of age. Aquaculture 10, 231-242.

Ryman, N. 1971. An analysis of growth capability in full sib families of Salmon (Salmo salar L.). Swedish Salmon Res. Inst. 5, 1-8.

Savostyanova, G.G. 1969. Comparative fishery characteristics of different groups of rainbow trout. In, Genetics Selection and Hybridization of Fish. pp 221-222. Acad. Sci. of the USSR Moscow.

Schaperclaus, W. 1961. Lehrbuch der Teichwirtschaft. Berlin.

Sick, K. 1961. Hemoglobin polymorphism in fishes. Nature 192, 894-896.

Sokal, R.R. and Rohlf, F.J. 1969. Biometry. Freeman, San Francisco.

Stegman, K. 1958. Die Karpfenzuchtung in Polen. Dtsch. FischZtg. 5, 179-184.

Svardsson, G. 1950. The coregonid problem. II. Morphology of two Coregonid species in different environments. Inst. Freshwater Res. Drottningholm 31, 151-162.

Svardsson, G. 1952. The coregonid problem IV. The significance of scales and gillrakers. Inst. Freshwater Res. Drottningholm 33, 204-232.

Von Sengbusch, R. 1963. Fische ohne Gräten. Züchter 33, 284-286.

Wohlfarth, W.G., Ankorion, J. and Moav, R. 1971. Genetic investigation and breeding methods of carp in Israel. Rep. FAO/UNDP(TA), (2926), 160-185.

Wohlfarth, G., Lahman, M. and Moav, R. 1964. Genetic improvement of carp. 5. Comparison of fall and winter growth of several carp progenies. Bamidgeh 19, 71-76.

THE EXTENDED SERIES OF Tf ALLELES IN ATLANTIC COD, GADUS MORHUA I

A. Jamieson and R. J. Turner

M.A.F.F. Fisheries Laboratories

Lowestoft, Suffolk, U.K.

Introduction

The Atlantic cod, Gadus morhua L., yields a catch of about 2.5 million tonnes per annum. The different fisheries are concentrated on spawning and feeding grounds on north Atlantic banks, continental shelf and coastal areas. Major oceanic stocks are fished about Newfoundland, Greenland, Iceland and the Norwegian Arctic. The distributions of the oceanic stocks of cod are influenced by the transport of pelagic eggs and larvae in ocean currents and the movements of marked specimens can be followed. Cod stocks showing less mobility are found inshore and on some more isolated banks. Although regional cod stocks have been described in much behavioural and phenotypic detail, genetic analysis of cod races started only recently. Most of the empirical subdivisions of cod stock taxonomy rely on observations of marked fish. These show that limited degrees of physical mixing of specimens between stock areas occurs. Some more recent scientific descriptions of cod populations attempt more critical analyses, questioning the genetic consummation of such mixing and also the status of sub-populations in historic or contemporary isolation. The techniques by which some modern evolutionists measure genetic distances are applicable also in fisheries biology, as wherever natural units of stock exist there is a need for these to be recognised for the purposes of practical management and conservation.

Geographic genetic variation in marine fish is usually

overlooked because the occasional long distance migrations of tagged fish imply continuity and the effects of genes are not immediately obvious. The development, functional morphology and behaviour of marine fishes are usually correlated with variable factors in their environment while the unseen genes have been disregarded but for a few exceptions. Heincke (1898) saw variation in meristic characters as indirect evidence for herring races, as did Schmidt (1930) for cod races. Punnett (1904) presented positive genetic evidence for merism in a viviparous shark, Etmopterus spinax (L.), (= Spinax niger L.). Similarly Schmidt (1919) used diallel crosses to estimate the heritability of meristic characters in trout, Salmo trutta L. Despite those early achievements genetic studies using marine fish lagged behind until zone electrophoresis in gel media exposed the products of structural genes. This technical innovation describes individual fish in great detail. Electrophoretic methods are capable of describing different races in species that are not amenable to experimentation.

Sick (1961) discovered a gene controlling the molecular structure of cod haemoglobin. The frequencies of the HbI^1 and HbI^2 alleles distinguish the races of cod in the Belt Sea and the Baltic sea (Sick, 1965a) and were applied in attempted racial analyses throughout the species range (Sick, 1965b; Frydenberg et al., 1965; Møller, 1968; Jamieson and Jonsson, 1971; Jamieson and Otterlind, 1971; Wilkins, 1971). Other polymorphic gene loci known in cod are transferrin (Møller, 1966a), esterase (Nyman, 1965), glucosephosphate isomerase (Dando, 1974), lactate dehydrogenase (Odense et al., 1966 and 1969) and phosphoglucomutase (Tills et al., 1971). Of those loci, transferrin has been used most widely for typing populations. Surveys of transferrins in cod populations started in Norway (Møller, 1966a and 1966b) were soon extended to Canada, Greenland, Iceland, Faroe and around the British Isles (Jamieson, 1967a and 1967b) and later in the Baltic Sea (Jamieson and Otterlind, 1971).

Success in using structural genes for racial studies rests on selecting gene loci that show promising variability, and spreading observations over material from a range of geographic sites. A chance observation, that the use of starch gel media to separate rivanol treated cod plasma almost doubled the resolution of cod transferrins, is the essence of the present contribution. As cod transferrins are known to show much variability in the north-west Atlantic (Jamieson, 1975) the improved transferrin typing method was applied to fresh material and to stored cod plasmas from that area.

In retrospect it was apparent that some of the newly resolved alleles had been overlooked and their products earlier dismissed as occasional satellite artefacts. Thus some of the earlier work inadvertently scored heterozygotes as single banded types before the rivanol method was established. Thus some apparent excesses of homozygotes in North American bank samples (Jamieson, 1975) can now be accounted for satisfactorily. The improved method of screening for variation at the Tf locus gives sharper statistical statements about races of cod and this paper gives data of this sort. Statistical tests are used to show the relevance of transferrin variation to the racial analyses of cod and genetic distances relate this practical work to evolutionary studies.

Material and Methods

The 3018 typed cod Gadus morhua, L. were caught and bled in the course of nine research cruises from 1966 to 1975. Figure 1 outlines a chart of the north-west Atlantic showing the positions and names of 18 grounds fished between Disko Island, Greenland and Georges Bank, Massachusetts. Table 1 gives the numbers of specimens sampled on different cruises. Almost half (1464) of the specimens were caught on 7 grounds along the Greenland side of the Davis Strait and the remaining 1554 cod were caught on 11 grounds off North America.

Live cod were caught in trawls and kept on deck in tanks of circulating seawater. Blood specimens were drawn from caudal veins into 2 ml evacuated glass tubes (Beckton, Dickinson and Company, Rutherford, New Jersey) containing 0.2 ml of 3.8% solution of Sodium citrate as anticoagulant and 0.04 mg of potassium sorbate as antimycotic agent.

Blood plasma specimens were separated by centrifugation at $+4^{\circ}C$ and were stored at $-25^{\circ}C$.

Transferrin typing methods have been applied to many organisms since Møller (1966a) reported serum transferrin polymorphism in cod. An improved test now separates more cod transferrins more clearly than the earlier tests. The changes in techniques are as follows:-

Blood plasma specimens were treated with rivanol before electrophoresis. This was the primary improvement because it gave clear separation and identification of transferrins by excluding other proteins. Rivanol (2-ethoxy-6, 9-diaminoacridine lactate) was used by Boettcher et al., (1958) to isolate metal combining proteins from blood plasma. It

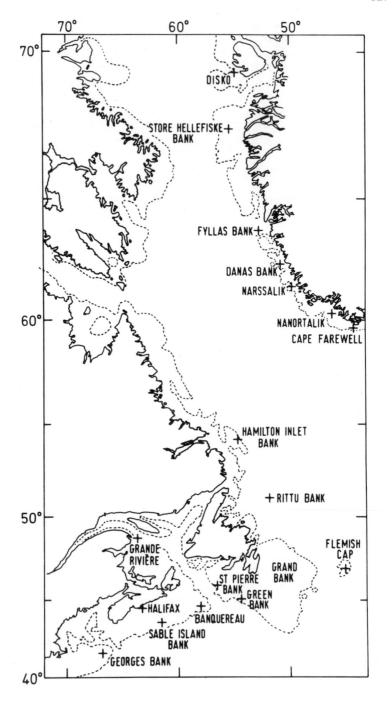

Figure 1. An outline chart shows the 18 fishing grounds sampled for cod.

TABLE 1

Blood sera specimens from 3018 Atlantic cod, Gadus morhua L., caught on Northwest Atlantic fishing banks were treated with rivanol and tested for transferrin variants. The numbers of typed specimens are tabulated against time and place of capture. Of the eighteeen banks sampled, eight yielded 2 or 3 sub-sample repeats.

	1966 Oct	1967 Jun	1967 Jul	1967 Nov	1971 Feb	1972 Mar	1973 Nov	1975 Jun	1975 Jul
Disko, Greenland	139								
Store Hellefiske, "	181								
Fyllas, "	201					118	70		
Danas, "						31	103		
Narssalik, "						56			
Nanortalik, "						71	203		
Cape Farewell, "	201					90			
Flemish Cap				13					188
Hamilton Inlet Bank, Labrador					108				188
Rittu Bank, Newfoundland				168	108				112
Grand Bank, "					108				31
Green Bank, "									
St. Pierre Bank, "								173	
Grande Rivière, Quebec			98						
Banquereau, Nova Scotia					109				56
Sable Island Bank, "									23
Halifax, "		99							
Georges Bank, Massachusetts					42				

has also been used to precipitate out non-transferrin proteins in brook trout (Hershberger, 1970) and hake (Pichôt, 1971; Mangaly, 1974).

Rivanol (0.6%) was dissolved in a buffer solution identical to that described below for the preparation of starch gel media. Heating without boiling dissolves the rivanol completely. The solution (at +4°C) was added to a small aliquot of plasma from each cod at a ratio of rivanol solution to plasma of 4:1 by volume. This mixture was centrifuged at +4°C. The supernatant was absorbed in 10 mm x 7 mm rectangles of filter paper (Whatman No. 17) which were then inserted along a slot cut into each slab of gel about 4 cm from the cathodic wick end. A potential of 4 volts per cm of gel, at right angles to the insert line, was maintained for 50 minutes. The paper inserts were removed and the run continued at 7 volts per cm until the borate boundary had moved 10 cm past the insert line. This borate was present in a discontinuous buffer system which replaced the earlier continuous system. The buffer in the electrode tanks contained:-

| Lithium hydroxide | LiOH | 0.06M (2.52 g/l) |
| Boric acid | H_3BO_3 | 0.03m (1.85 g/l) |

The buffer for the preparation of starch gel medium contained:-

| Tris (hydroxymethyl) aminomethane | 0.030m (3.6 g/l) |
| Citric acid | 0.005M (1.05 g/l) |

In starch gel preparation 1 ml of lithium-boric tank buffer was added to every 100 ml of tris-citric buffer. Hydrolysed starch (Connaught Laboratories Limited, Willowdale, Toronto) at 13% replaced the starch plus agar method used earlier. Starch mixed in buffer was boiled, de-gassed and poured hot into moulds 185 mm x 155 mm x 7 mm. Menisci were sliced off after the gels had set cold. Electrophoretic runs were horizontal at +4°C. Control cod sera of known genotype were interspersed along all gels. Slices of gels were stained in 1% Amido Black for 5 minutes. Excess stain was removed by an electrophoretic destainer (E-C Apparatus, St. Petersburg, Florida). Quick destaining allowed repeat tests to be made immediately whenever unusual control types were required.

Statistical methods compared the numbers of co-dominant genotypes and the numbers of all transferrin alleles in all the sub-samples taken on fishing grounds on different occasions. The tests tested the racial unity of the species by searching for examples of imbalance in the numbers of genotypes and searching for examples of allele frequency differences between sub-samples.

The transferrin allele frequency values were used to calculate values for genetic identity, I, and genetic distance, D, in pairwise comparisons of populations using the following coefficient by Nei (1971, 1972):-

$$I = \frac{\Sigma x_i y_i}{(\Sigma x_i^2 \Sigma y_i^2)^{\frac{1}{2}}}$$, where

x_i = gene frequency of the ith allele in population X
y_i =   "   "   "   "   "   "   "   "   Y,
and
$D = -\log_e I$

The coefficient of identity, I, is theoretically capable of assuming values ranging from zero to unity. In evolutionary studies comparing species it is usual to average distances over several tested loci. The present application is limited to comparing a single locus in local populations within one species, Gadus morhua.

Results

A large number of cod blood plasma specimens analysed by electrophoresis produced evidence for 13 co-dominant alleles at the transferrin locus, cod Tf. Rivanol treatment prior to electrophoresis effectively excluded other plasma proteins from the region of gel showing transferrin results. The treated and untreated controls in Figure 2 demonstrate this exclusion. Cod plasma samples which had been stored at -25°C for up to 10 years gave results as clear as those from fresh material. The separated transferrins appeared as single band patterns in homozygotes and double band patterns in heterozygotes. The relative positions of the 14 cod transferrin bands are displayed in Figure 3. Their positions indicate their rates of migration towards the anode. The increase in number of resolvable alleles required the old notation be revised. The 7 bands which are thought to approximate most closely to those seen by Møller and Naevdal (1966) are now called A_1 A_2 B_2 C_1 C_3 D_1 and E and the 6 alleles which straddle those are called B_1 B_3 B_4 C_2 C_4 and D_2. The letters of the earlier notation were retained, but the subscripts were added to revise and extend the notation.

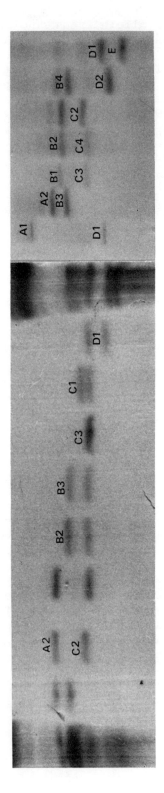

Figure 2. Cod transferrin bands separated out on starch gel electrophoresis. Each of the cod transferrins attributed to a series of 13 co-dominant alleles is represented in a photograph of assorted transferrin types in cod blood plasma specimens. Two of the 15 specimens were tested with, and without, Rivanol treatment. All of the remainder were treated. From left to right the samples are:-

untreated A_2B_2 treated A_2B_2 A_2C_2 A_2C_3 B_2C_3 C_3C_3 C_1C_3 C_3D_1 untreated C_3D_1
treated A_1D_1 A_2B_3 B_1C_3 B_2C_4 B_2C_2 B_4D_2 D_1E

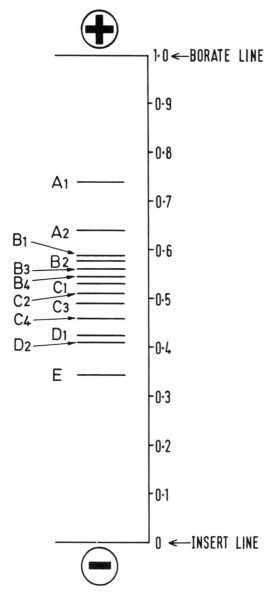

Figure 3. A series of cod transferrins were distinguished by their relative anodic mobilities during starch gel electrophoresis. Thirteen electromorphs were identified. They are plotted along a scale between the insert line at zero mobility and the borate buffer line at unit mobility. This unit distance averaged 10 cm in the actual gels. In reading routine tests, identifications were made by comparing unknowns with adjacent biological controls. Physical measurements relative to this scale were less critical than comparisons with the biological controls which were present in all tests.

A co-dominant series of n alleles produces n homozygotes and $0.5n(n-1)$ heterozygotes. Thus 7 alleles give 21 heterozygotes whereas 13 alleles give 78. The increase in the number of detectable cod transferrin types appeared as an increase in the number of heterozygotes showing examples of less frequent alleles. The 91 places in the triangle shown in Table 2 represent all of the possible genotypes at the Tf locus. The 38 blank places represent rare genotypes which did not appear, and were not expected to occur, in the overall total of 3018 cod. The numbers entered in each square are the observed number (upper) and the expected number (lower) per genotype specified by the alleles in the margins. The expected numbers assume conformity to Hardy-Weinberg expectation throughout the whole region sampled. The large numerical excess of homozygotes along the hypotenuse and the correspondingly large deficiency of heterozygotes in the rest of the triangular figure is strongly indicative of the presence of genetic isolates or races of cod in the north-west Atlantic area. ($\chi^2 = 58.42$, $p < 0.001$)

Statements of cod transferrin genotypes at west Greenland fishing grounds in Tables 3 and 4, and at North American grounds in Tables 5 and 6, show greater detail than the summary given in Table 2. Unlike the pooled data the separate sub-samples showed good genetic balance. Imbalance resulted from combining population samples with different allele frequencies, presumambly by combining races.

Contingency tests were applied to groups of transferrin allele frequencies from different sea areas to find out whether the sub-samples showed racial differences. The seven fishing grounds sampled at west Greenland showed significant heterogeneity ($\chi^2_{31} = 48.19$ $0.05 > p > 0.02$). The most prominent factor in this heterogeneity is a drop in the frequency of the allele B_2 south of Danas Bank. This gene frequency difference suggests the presence of two cod stocks at west Greenland, and fits well with the tagging evidence that many cod from south-west Greenland migrate to Iceland to spawn. Presumably they first arrive in Greenland as fry drifting from Iceland in the known currents. More northerly cod at West Greenland reproduce at West Greenland, (Tåning, 1937).

A contingency test on transferrins in cod on the North American Banks suggests that this material is divisible according to transferrin frequencies into one large cluster of samples from Hamilton Inlet Bank, Labrador to Halifax, Nova Scotia including Rittu, Grand, Green, St. Pierre, Banquereau and Grande Rivière, but excluding the cod typed on Flemish

TABLE 2

The 3018 individuals from fishing grounds in the north-west Atlantic variously grouped into 47 observed genotypes. The letters and subscripts along the top and right hand side indicate 13 co-dominant alleles. Observed numbers are placed above expected numbers (rounded off) in the appropriate genotype squares. Expected values of less than 0.5 are not included.

	A1	A2	B1	B2	B3	B4	C1	C2	C3	C4	D1	D2	E	
											1			A1
	28 13			57 103	1 2	1 1	14 30	4 6	224 177		35 42	1	5	A2
							1	1						B1
			255 209		8 8	5 5	167 122	29 24	606 714	1	183 168	1 3	27 21	B2
							3 2	1	16 14		3 3			B3
							4 1	3	4 8		1 2	1		B4
							34 18	16 7	130 209		47 49	1	10 6	C1
							1	16 41			13 10	3	2 1	C2
									718 611	1	259 288	4 5	22 35	C3
											1			C4
											41 34	2 1	15 8	D1
														D2
													1 1	E

TABLE 3

Cod at West Greenland between $62°$ and $70°N$ showed genetic balance.

Cod Tf Types	Disko Oct. 1966	Store Helle. Oct. 1966	Fyllas Bank Oct. 1966	Fyllas Bank March 1972	Fyllas Bank Nov. 1973	Danas Bank March 1972	Danas Bank Nov. 1973
A_1D_1			1				
A_2A_2	1 (1)	3 (2)	5 (2)	1 (1)	(1)		1 (2)
A_2B_2	5 (6)	6 (8)	9 (7)	8 (5)	3 (3)	(1)	2 (6)
A_2C_1	(1)	3 (1)	1 (1)	(1)			1 (1)
A_2C_2	(1)		2 (1)				
A_2C_3	15 (13)	15 (19)	18 (27)	13 (13)	9 (7)	5 (4)	19 (15)
A_2D_1	3 (2)	5 (3)	3 (3)	(2)	(1)	1	4 (2)
B_2B_2	10 (8)	11 (9)	5 (6)	5 (5)	3 (3)	2	8 (5)
B_2B_3				1			
B_2C_1	(2)	3 (3)	(1)	1 (2)			(1)
B_2C_2	2 (1)	1 (1)	2 (1)				
B_2C_3	34 (35)	39 (43)	36 (42)	24 (27)	19 (19)	3 (5)	23 (25)
B_2C_4			1				
B_2D_1	4 (5)	7 (8)	9 (4)	4 (5)	3 (2)		5 (4)
B_2E		1		1			
B_3C_2	1						
B_3C_3	(1)	1 (1)			(1)	1 (1)	
B_4C_2			1				
B_4C_3	1 (1)		1 (1)				
C_1C_3	4 (4)	4 (8)	4 (4)	6 (4)	1 (1)		4 (3)
C_1D_1	2 (1)	3 (1)	1		(1)		
C_1E	1	1					
C_2C_3	1 (3)	2 (2)	1 (4)			1 (1)	
C_2D_1	1		1				
C_2D_2	1						
C_3C_3	42 (39)	58 (51)	90 (77)	36 (37)	25 (26)	15 (15)	29 (30)
C_3D_1	9 (12)	16 (19)	9 (16)	16 (12)	7 (6)	3 (3)	7 (9)
C_3E	(1)	(1)	(1)	1 (1)			
D_1D_1	2 (1)	2 (2)	(1)	1 (1)			(1)
D_1E			1				

TABLE 4

Cod population samples were taken at 3 positions to the south of Frederikshaab (60 degrees north) along the southwest coast of Greenland. The observed numbers of cod in each transferrin genotype showed good agreement with expected numbers shown in brackets. Expected numbers assume that genetic balance occurred in each sub-sample.

Cod Tf Types	Narssalik March 1972	Nanortalik March 1972	Nanortalik November 1973	Farewell October 1966	Farewell March 1972
$A_2 A_2$	1 (1)	2 (1)	3 (2)	5 (4)	1 (2)
$A_2 B_2$	2 (2)	1 (2)	5 (7)	4 (8)	3 (4)
$A_2 C_1$		1		3 (1)	1
$A_2 C_3$	9 (9)	11 (12)	28 (27)	35 (35)	16 (15)
$A_2 D_1$		2 (2)	4 (4)	5 (4)	2 (2)
$B_2 B_2$	1 (1)	(1)	3 (5)	2 (4)	(2)
$B_2 B_3$				1	
$B_2 C_1$				1 (1)	
$B_2 C_3$	10 (10)	14 (11)	43 (39)	43 (35)	23 (18)
$B_2 D_1$	1 (1)	2 (2)	8 (6)	4 (4)	3 (2)
$B_3 C_3$			1 (1)	1 (2)	
$B_3 D_1$				1	
$B_4 C_3$			1 (1)		
$C_1 C_2$			1	1	
$C_1 C_3$	1 (1)	(1)	3 (3)	3 (5)	1 (1)
$C_2 C_3$	1 (1)		(1)	(1)	
$C_2 D_1$				1	
$C_3 C_3$	25 (26)	28 (29)	79 (81)	78 (78)	29 (33)
$C_3 D_1$	5 (4)	10 (9)	22 (24)	12 (16)	10 (9)
$C_3 D_2$					1 (1)
$D_1 D_1$		(1)	2 (2)	1 (1)	(1)

TABLE 5

A major cluster of cod transferrin types consistently covered eight of the banks sampled off North America.

Cod Tf Types	Hamil. Inlet Bank July 1975	Rittu Bank Feb. 1971	Rittu Bank July 1975	Grand Bank, Newf'land Nov. 1967	Grand Bank Feb. 1971	Grand Bank July 1975	Green Bank Feb. 1971	St Pre. Bank June 1975	Grande Riv. July 1967	Banquereau Feb. 1971	Banquereau July 1975	Hal. June 1967
$A_2 B_2$	1		0,1				1		1,1			
$A_2 C_1$		1		1								
$A_2 C_2$			1									
$A_2 D_1$									1			
$B_1 B_4$			1									
$B_1 C_1$								1				
$B_2 B_2$	15,18	12,14	18,22	34,33	17,13	5,4	17,13	19,23	12,14	22,18	7,8	20,17
$B_2 B_3$	1	0,1	0,1		1,1	0,1	0,1	1,3		1	0,1	1
$B_2 B_4$			2,2				0,2	2,1				1
$B_2 C_1$	17,13	14,15	21,14	33,26	6,9	5,3	14,15	21,26	10,11	8,10	6,5	6,8
$B_2 C_2$	2,2	1,2	3,2	2,2	2,3		1,1	1,2	3,1	4,4	2,1	1
$B_2 C_3$	26,26	24,21	19,19	27,33	17,22	7,7	14,22	43,39	34,23	17,25	14,14	22,22
$B_2 D_1$	13,12	12,10	17,15	16,19	11,11	1,2	10,6	14,12	3,10	12,11	4,4	9,14
$B_2 D_2$			0,1					1,1				
$B_2 E$	2,3	2,1	2,3	2,3	3,2	0,1	1,2	5,3		2,1	2,2	2,3
$B_3 C_1$								2,1		1		
$B_3 C_3$		2,1	1		1,1	2,1	2,1	3,2			1,1	
$B_3 D_1$								2,1				

Table 5 (continued)

B4 C1	1,1											
B4 C2								2,1				
B4 C3			0,1					2		1		
B4 D1			0,1				1,2	0,1				
C1 C1	0,2	3,4	1,2	8,5	1,2	0,1	5,4	5,5	4,2	2,1	1,1	3,1
C1 C2	2,1	3,1	0,1	1,1	2,1		1,1	1,1	7,9	3,1	3,4	5,5
C1 C3	9,9	9,11	2,6	4,13	12,8	2,3	11,12	17,17		6,7	3,4	1,3
C1 D1	3,4	7,5	3,5	5,8	4,4	0,1	3,3	3,5	3,4	3,3	0,1	0,1
C1 E	1,1	1,1	2,1	1,1	0,1	1	1,1	1,1			0,1	
C2 C3	0,1	1,1	0,1	0,1	2,2		0,1	1,1	0,1	0,3	0,1	
C2 D1	1,1	0,1	0,1	1,1	2,1	1		1		3,1		
C3 C3	10,10	9,8	5,4	16,8	10,10	3,3	15,9	15,16	5,9	16,9	8,6	5,7
C3 C4				1								
C3 C1	7,9	5,8	9,6	10,9	10,10	2,2	1,5	9,10	9,8	7,8	3,4	15,9
C3 D2	1							1,1				1
C3 E	4,2	0,1	1,1	0,1	2,2	1,1	4,2	2,2		1,1	1,2	1,2
C3 D1				1								
D1 D1	3,2	2,2	1,2	4,3	2,3	1	1,1	2,2	4,2	1,2	1,1	3,3
D1 D2			1						1			
D1 E	0,1		1,1	1,1	2,1		1,1	0,1			2,1	3,1
E E			1	1								

TABLE 6

The proportions of transferrin types in cod caught on Flemish Cap, Sable Island Bank and Georges Bank differ from those caught on adjacent fishing grounds. Expected numbers shown in brackets demonstrate good genetic equilibrium within each bank sample.

Cod Tf Types	Flemish Cap November 1967	Flemish Cap July 1975	Sable Island Bank July 1975	Georges Bank February 1971
$A_2 A_2$	1	4 (4)		
$A_2 B_2$	1 (1)	5 (5)		
$A_2 B_3$		1		
$A_2 B_4$		1		
$A_2 C_1$		2 (2)		
$A_2 C_2$		1 (2)		
$A_2 C_3$	1 (2)	30 (28)		
$A_2 D_1$	(1)	5 (8)		
$B_2 B_2$		1 (1)	1	5 (3)
$B_2 B_3$				1
$B_2 C_1$		1 (1)	(1)	(1)
$B_2 C_2$		2 (1)		(1)
$B_2 C_3$	2 (2)	20 (18)	2 (3)	8 (8)
$B_2 D_1$	1 (1)	3 (4)	1 (1)	5 (5)
$B_2 E$			1	
$B_4 D_2$		1		
$C_1 C_1$				1
$C_1 C_2$		1 (1)		
$C_1 C_3$	1 (1)	9 (8)	2 (2)	(1)
$C_1 D_1$		1 (2)	1	2 (1)
$C_1 E$			1	
$C_2 C_3$		6 (7)		(1)
$C_2 D_1$		1 (2)		(1)
$C_2 D_2$				2
$C_2 E$		2		
$C_3 C_3$	3 (3)	51 (54)	7 (6)	6 (5)
$C_3 D_1$	2 (2)	32 (28)	6 (5)	6 (6)
$C_3 E$	(1)	2 (3)	(1)	2 (1)
$D_1 D_1$		5 (4)	1 (1)	2 (2)
$D_1 E$	1	1 (1)		2 (1)

Cap, Sable Island Bank and Georges Bank. A contingency test on the data from this group of 8 Canadian banks shows no heterogeneity ($X^2_{36} = 38.732$ $0.50 > p > 0.30$). The grouping of material in T_{a}ble 5 shows the considerable differences in the frequencies of transferrin alleles in cod from 6 different areas in the north-west Atlantic. The chi squared values indicate the enormous regional differences in the Tf allele frequencies. All the allele frequencies are given in Tables 8 and 9 and illustrated in Figure 4.

Nei's coefficient of genetic distance was applied to the cod Tf locus data to produce pairwise comparisons among the 18 fishing grounds. The 153 measures of distance stated in Table 10 were used to construct the histogram in Figure 5. The greatest genetic distances are towards the right of the figure. Distances of between 0.29 to 0.49 were obtained by comparing Flemish Cap cod with 8 Canadian cod samples from the vicinity of Newfoundland and also by comparing this same group of Canadian cod samples with three samples from the southern part of the west Greenland fishery. The shortest genetic distances arise from comparisons of adjacent samples at west Greenland and among the 8 Canadian grounds. Flemish Cap is surprisingly close to the 7 west Greenland samples.

The Flemish Cap ecosystem has been studied by Templeman (1976). The Flemish Channel between Flemish Cap and the Grand Bank of Newfoundland is deeper than the 1000 metre isobath. Tagging evidence shows that marked cod are more disposed to move from Flemish Cap than towards it. Compared with cod on the main banks, those at Flemish Cap lack the nematode Terranova decipiens (Templeman et al., 1957) and the copepod Lernaeocera branchialis (L) (Templeman and Fleming, 1963). Flemish Cap cod spawn mainly in March which is earlier than adjacent cod populations.

The Sable Island Bank cod are distinctly different from the other Canadian samples but are closer to Flemish Cap cod. Flemish Cap and Sable Island are both offshore banks and both have hydrographic gyres which could retain drifting cod larvae.

The Fundian Channel separates the Georges Bank cod from the others and could contribute to their genetic isolation. The Laurentian Channel between St. Pierre Bank and Banquereau would be another potential barrier to cod movements yet the transferrin locus shows negligible genetic distance ($D = 0.01$) over the Laurentian Channel.

The divisions of the cod stocks reported above are subject

TABLE 7

The transferrin data for the cod samples from the north-west Atlantic demonstrated the presence of at least six races. At west Greenland there appears to be an indigenous northerly race; also a southerly race with Icelandic affinity. At North America the races of cod on Flemish Cap, Sable Island Bank and Georges Bank appear to be genetically isolated from the main group of cod in the vicinity of Newfoundland. A contingency test showed highly significant racial differences. The rare alleles A_1 B_1 C_4 and D_2 were not tabulated. Expected numbers are shown in brackets.

Putative races of cod	Series of co-dominant alleles at the Tf locus in Gadus morhua L.									χ^2	Prob.
	A_2	B_2	B_3	B_4	C_1	C_2	C_3	D_1	E		
D'ko to Danas	172 (110)	244 (443)	4 (8)	3 (5)	40 (130)	17 (25)	962 (759)	135 (179)	6 (22)	201.5	P<0.001
N'lik to Fwl.	156 (81)	180 (327)	4 (6)	1 (4)	16 (96)	4 (19)	782 (559)	98 (132)	0 (16)	328.4	P<0.001
F. Cap	57 (26)	37 (106)	1 (2)	2 (1)	15 (31)	13 (6)	213 (181)	57 (43)	6 (5)	108.7	P<0.001
S. Isl.	0 (3)	6 (12)	0	0	4 (4)	0 (1)	24 (21)	10 (5)	2 (1)	8.4	P<0.02
Ham. I to H'fx	7 (167)	1002 (677)	22 (13)	14 (8)	380 (198)	52 (39)	709 (1159)	320 (273)	64 (35)	703.8	P<0.01
Georges	0 (5)	24 (22)	1 (0)	0 (0)	4 (6)	2 (1)	28 (38)	19 (9)	4 (1)	14.6	P<0.01
χ^2	294.6	292.3	8.9	5.3	305.2	26.9	326.7	48.3	56.9		
Prob.	P<0.001	P<0.001	P<0.001	P<0.05	P<0.001	P<0.001	P<0.001	P<0.001	P<0.001		

TABLE 8

Atlantic cod, Gadus morhua, L., were sampled on seven fishing grounds at west Greenland from Disko Island to Cape Farewell. Their blood sera were tested for transferrin variants. The results indicated a co-dominant series of electromorphs, tabulated as allele frequencies.

Cod Tf locus alleles	DISKO	S. HELLEFISKE	FYLLAS	DANAS	NARSSALIK	NANORTALIK	FAREWELL
A1			0.001				
A2	0.090	0.097	0.100	0.127	0.116	0.113	0.139
B2	0.234	0.218	0.189	0.198	0.134	0.144	0.148
B3	0.004	0.003	0.001	0.004		0.002	0.005
B4	0.004		0.003			0.002	
C1	0.025	0.039	0.018	0.019	0.009	0.009	0.017
C2	0.022	0.008	0.009	0.004	0.009	0.002	0.003
C3	0.532	0.533	0.600	0.575	0.679	0.633	0.617
c†			0.001				
D1	0.083	0.097	0.073	0.075	0.054	0.095	0.067
D2	0.004						0.002
E	0.004	0.006	0.004				

TABLE 9

Atlantic cod, Gadus morhua L. were sampled on eleven North American banks from Labrador to Massachusetts. Tests for transferrin variants in their blood sera indicated a series of co-dominant electromorphs, tabulated as allele frequencies.

Cod Tf locus alleles	FLEMISH C	SABLE I	HAMILTON I	RITTU	GRAND	GREEN	ST PIERRE	G RIVIERE	BANQUEREAU	HALIFAX	GEORGES
A2	0.142		0.004	0.005	0.002	0.005		0.010			
B1				0.002			0.003				
B2	0.092	0.174	0.390	0.402	0.399	0.347	0.364	0.383	0.394	0.414	0.286
B3	0.003		0.004	0.007	0.007	0.009	0.023		0.009	0.005	0.012
B4	0.005			0.009		0.023	0.006	0.005	0.003	0.005	
C1	0.037	0.044	0.136	0.164	0.155	0.194	0.162	0.143	0.112	0.091	0.048
C2	0.032		0.021	0.020	0.021	0.019	0.012	0.015	0.039	0.005	0.024
C3	0.530	0.522	0.284	0.230	0.257	0.292	0.306	0.306	0.303	0.273	0.333
C4					0.003						
D1	0.142	0.217	0.127	0.139	0.132	0.079	0.095	0.133	0.115	0.172	0.226
D2	0.003		0.004	0.002			0.006	0.005		0.005	0.024
E	0.015	0.044	0.030	0.021	0.024	0.032	0.023		0.024	0.030	0.048

EXTENDED SERIES OF Tf ALLELES

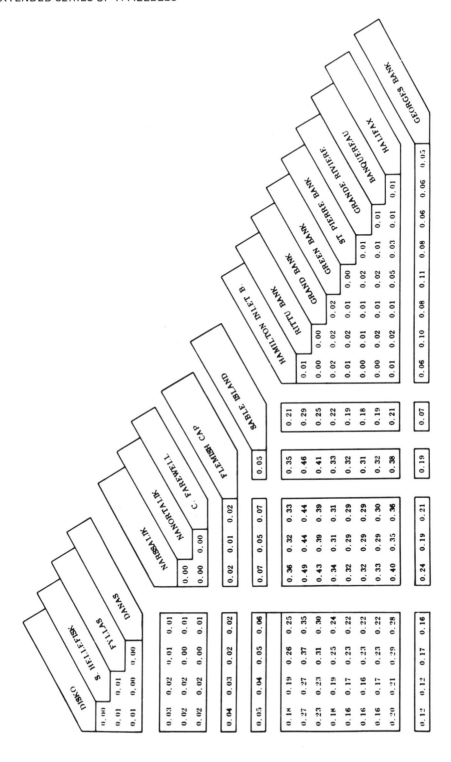

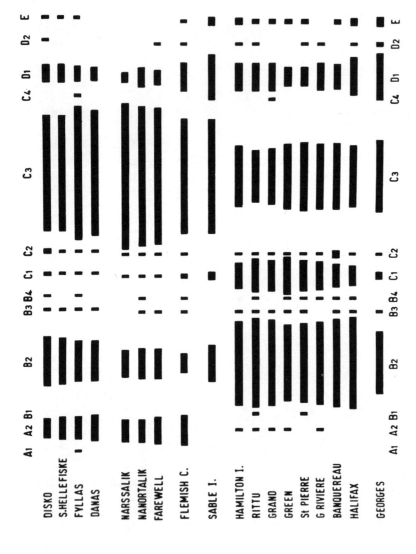

Figure 4. The proportions of transferrin alleles in cod samples from 18 fishing grounds in the north-west Atlantic are represented by the relative lengths of black lines. As all of the alleles are in series at one locus, the sum of the frequencies is unity for each ground. All of the frequencies are stated numerically in Tables 8 and 9.

EXTENDED SERIES OF Tf ALLELES

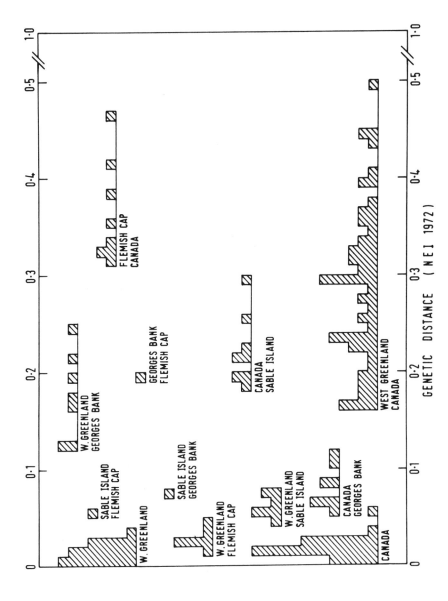

Figure 5. Histograms display 153 pairwise comparisons of cod populations on a common horizontal scale of genetic distance. Each histogram has its own zero baseline and a vertical scale of one unit square for each distance measurement.

to the limitations of (a) sampling and (b) our present knowledge of cod genotypes. More genetic information could reveal more races perhaps, but the findings to date accord practically completely with the divisions of cod reviewed by Templeman (1962).

General Discussion

The scientific management of commercial fisheries rests on a knowledge of specific and sub-specific divisions of fishes. If closely related species present taxonomic problems, a quick solution is to match a macerated extract of unidentified fillet against known controls in electrophoresis. Such a test is suitable for identifying species, but too crude for separating sub-populations within species. Resolution of individual proteins together with staining techniques show different versions of structural protein genes in individual fish. Sub-population infrastructures show assortments of alleles which can be tested for heterogeneity, genotypes which can be tested for genetic balance and different loci from which genetic distances may be computed.

Atlantic cod belong to a single species, Gadus morhua L., divided into several stocks for practical descriptive and commercial reasons. The accepted stock descriptions were made before any precise genetic information was available to test for the possible presence of different races, not to mention genetic distance coefficients.

Cod stocks have well documented histories. Fluctuations in the numbers of cod in each region are influenced by variation in the survival rates of the successive annual broods that supply the recruits to the fisheries. The times and places where cod spawn are generally known. Stocks have characteristic growth curves. Age at first maturation varies from stock to stock. A stock may share a common infestation of parasites. Meristic characters generally show agreement within regions, but may also vary between year classes. Market statistics tend to support the stock description. The bulk of the non-genetic evidence indicates regional stocks perpetuating themselves in comparative isolation.

The picture of geographic ranges and migrations of cod stocks thus built up is supported by the data resulting from the recapture of tagged adults, but some isolated examples of long distance migrations challenge the concept of genetic isolation of regional stocks. One tagged cod made a transatlantic journey (Gulland & Williamson, 1962). Relays of marked cod have covered every stage from Maryland to Novaya Zemlya (Graham, 1956; Tåning, 1934). The drift of pelagic cod larvae is more difficult to trace than adult migrations. A cautious interpretation is that most adult cod conform to the behaviour and morphology of the particular stock to which they recruit. Conceptual confusion about the racial status of the stocks would be much reduced if more genotypic data were used to supplement the earlier descriptions (Marr, 1957; Jamieson, 1974).

To obtain genotypes soluble proteins from tissues are separated by electrophoresis, the advantages and limitations of which are described by Ayala and Valentine elsewhere in this volume. Despite their limitations, electrophoretic methods enable relatively rapid analysis of many genes and hence can define variation in sub-populations within species.

In attempting the racial analysis of Atlantic cod, the whole species is seen as a collective gene pool, each putative race of cod as an assortment of gene frequencies, each individual cod as an unique genotype. An objective statement of the genetic structure of the species would demand the investigation of a large random sample of gene loci over a grid of sub-population samples over all seasons. Such an ideal analysis would not presuppose the equating of races with stocks or with spawning grounds. If tests for genetic balance within sub-samples show too many homozygotes, and tests for heterogeneity between samples show significant departures from expectation, the species appears as more than one genetic race. In practice the availability of field observations encourages subjectivity and arbitrary sub-sampling effects. The available observations on cod genotypes show available population samples screened for a few gene loci with a bias towards loci known to show variable allele frequencies. The effects of distorted sampling may be reduced by placing practical minimal scale limits on unit races. Temporary effects of biased sampling, bottlenecks or local selection may be recognised by attempting to sample the same stock on different occasions.

To date two informative loci, HbI and Tf, have been detected in Atlantic cod. As it happens the frequencies of alleles at the cod HbI locus separate some cod races in the eastern half of the species' Atlantic range and the allele frequencies at the Tf locus separate cod races in the western half of the range. As both those gene loci are polymorphic in all populations of cod sampled where they occur over the

north Atlantic, it is probable that selection conserves the
allele frequencies associated with different areas. The
continued study of cod genotypes applied to racial analyses
may yet show how genetic polymorphisms respond to natural
fluctuations in fish populations. A sustained and concentrated
cod typing effort may even contribute to evolutionary knowledge.
It is possible that those polymorphic loci, which show the
greatest regional variation in allele frequencies, code those
proteins that are most buffeted by evolutionary forces presently moulding the genetic compositions of races within species.

The present and future attempts to identify cod races by
molecular genetic methods may be guided by the findings of
recent evolutionary studies using molecular methods (Ayala,
1975; Avise, 1976; Dobzhansky, 1976). Allele frequencies
at homologous loci are used to measure genetic distances
between taxa. Samples from known populations are more realistic than random samples over whole species, because sampling
problems apply to evolutionists as to fisheries biologists.
Both kinds of investigations strive to measure diversity in
individuals and populations. The initial stages of species
formation perplex the fisheries workers as much as they interest
the evolutionists. A local population may not be expected
to be entirely typical of its own species, because species and
their component races are dynamic. At least some loci may
be expected to vary by chance or by selection.

Evolutionary genetic research is concerned with all degrees
of taxonomic divergence and attempts to sample whole genomes
to get the best measures of genetic distance averaged over loci.
Fisheries biology is less concerned with the fundamental evolutionary forces, and more concerned with knowing which gene loci
give the most sensitive tests for racial differences. Should
the extensive molecular genetic data for _Drosophila_ species
(Ayala _et al_., 1974) prove to be an appropriate model for
genetic variability in local populations in general, fisheries
biology may hope to benefit by the widespread application of
a limited number of selected gene loci which assist in the
recognition of races.

Summary

The historic sub-divisions of the Atlantic cod, _Gadus
morhua_ L., were made by practical reference to regional
fisheries in the absence of any direct genetic evidence about
races. In modern times molecular genetic information has
been applied in the quest to know more about the elemental

units of stock in the cod and in other natural resources.

The present paper describes the separation of 13 co-dominant alleles at the transferrin locus in cod. The use of rivanol precipitation which excludes other soluble plasma proteins before starch gel electrophoresis gave this improved resolution of transferrins. Some notation amendments were necessary.

The extended series of cod transferrin alleles was used in a survey of cod populations sampled on 18 fishing grounds in the north-west Atlantic. The results suggest that at least six races of cod were present in the material surveyed. It is important that fisheries investigators and conservationists should know where adjacent fishing banks support different races of cod.

Acknowledgements

Mr. Lloyd Woolner, of the Lowestoft Laboratory, made the computer program for obtaining coefficients of genetic distance. Dr. Paul J. Odense, Halifax NS provided cod blood sera from Halifax and Grande Riviere. Dr. Tom Cross and Dr. Ron H. Payne, St. Johns, Newfoundland provided the St. Pièrre samples.

REFERENCES

Avise, J.C. 1976. Genetic differentiation during speciation. In, <u>Molecular Evolution</u>. Ayala, F.J. (ed). Sinauer Associates, Inc., Sunderland Massachusetts.

Ayala, F.J. 1975. Genetic differentiation during the speciation process. Evol. Biol. 8, 1-78.

Ayala, F.J., Tracey, M.L., Barr, L.G., MacDonald, J.F., Perez-Salas, S. 1974. Genetic variation in natural populations of five Drosophila species and the hypothesis of the selective neutrality of protein polymorphisms. Genetics 77, 343-384.

Boettcher, E.W., Kistler, P. and Nitachmann, H.S. 1958. Method of isolating the β1 metal-combining globulin from human blood plasma. Nature 181, 490.

Dando, R.P. 1974. Distribution of multiple glucose-phosphate isomerases in teleostean fishes. Comp. Biochem. Physiol. 47B, 663-679.

Dobzhansky, T. 1976. Organismic and molecular aspects of species formation. In, <u>Molecular Evolution</u>. Ayala, F.J. (ed). Sinauer Associates, Inc., Sunderland Massachusetts.

Frydenberg, O., Møller, D., Naevdal, G. and Sick, K. 1965. Haemoglobin polymorphism in Norwegian cod populations, Hereditas 53, 257-271.

Graham, M. 1956. Science and the British Fisheries. In, <u>Sea Fisheries. Their Investigation in the United Kingdom</u>. Graham, M. (ed). Edward Arnold Ltd., London.

Gulland, J.A. and Williamson, G.R. 1962. Transatlantic journey of a tagged cod. Nature 195, 921.

Heincke, F. 1898. Naturgeschichte des Herings, Teil 1, abh. dt. SeefischVer., Band 2, CXXXVI, 128 pp. text, 223 pp. tables.

Hershberger, W.K. 1970. Some physiochemical properties of transferrins in brook trout. Trans. Am. Fish Soc. 99, 207-218.

Jamieson, A. 1967a. Genetic diversity in cod. ICES document CM 1967/F:20

Jamieson, A. 1967b. New genotypes in cod at Greenland. Nature 215, 661-662.

Jamieson, A. 1970. Cod transferrins and genetic isolates. In, Eleventh European Conference on Animal Blood Groups and Biochemical Polymorphisms. Junk, W. (ed). Warsaw 1968.

Jamieson, A. 1974. Genetic tags for marine fish stocks. In, Sea Fisheries Research. Jones, F.R.H. (ed). Elek Science, London.

Jamieson, A. 1975. Protein types of Atlantic cod stocks on the north American banks. In, Isozymes IV. Genetics and Evolution. Markert, C.L. (ed). Academic Press.

Jamieson, A. and Jónsson, J. 1971. The Greenland component of spawning cod at Iceland. (ICES Special Meeting on the Biochemical and Serological Identification of Fish Stocks, Dublin, 1969.) Rapp. P.-v Reun. Cons. perm. int. Explor. Mer, 161, 65-72.

Jamieson, A. and Otterlind, G. 1971. The use of cod blood protein polymorphism in the Belt Sea, the Sound and the Baltic Sea. (ICES Special Meeting on the Biochemical and Serological Identification of Fish Stocks, Dublin, 1969.) Rapp. P.-v Reun. Cons. perm. int. Explor. Mer, 161, 55-59.

Mangaly, K.G. 1974. Electrophoretic and biochemical studies on European hake, Merluccius merluccius L. Ph.D. Thesis, University of East Anglia, U.K.

Marr, J.C. 1957. The problem of defining and recognising sub-populations of fishes. Spec. Scient. Rep. U.S. Fish Wildl. Serv. 208, 1-6.

Møller, D. 1977a. Polymorphism of serum transferrin in cod. FiskDir. Skr. (Ser. Havunders.) 14, 51-60.

Møller, D. 1966b. Genetic differences between cod groups in the Lofoten area. Nature 212, 824.

Møller, D. 1968. Genetic diversity in spawning cod along the Norwegian coast. Hereditas 60, 1-31.

Møller, D. and Naevdal, G. 1966. Transferrin polymorphism in fishes. In, Polymorphismes Biochemiques des Animaux. Tenth European Conference on Animal Blood Groups and

Biochemical Polymorphisms. Institut National de la Recherche Agronimique, Paris.

Nei, M. 1971. Interspecific gene differences and evolutionary time estimated from electrophoretic data on protein identity. Am. Nat. 105, 385-398.

Nei, M. 1972. Genetic distance between populations. Am. Nat. 106, 283-292.

Nyman, O.L. 1965. Inter and intraspecific variations of protein in fishes. K. VetenskSamh. Upps. Arsb. 9, 1-18.

Odense, P.H., Leung, T.C. and Allen, T.M. 1966. An electrophoretic study of tissue proteins and enzymes of four Canadian cod populations. ICES. C.M. 1966. G:14, 1 (mimeo.)

Odense, P.J., Leung, T.C., Allen, T.M. and Parker, E. 1969. Multiple forms of lactate dehydrogenase in the cod, Gadus morhua L. Biochem. Genet. 3, 317-334.

Pichot, P. 1971. Etude électrophorétique des protéines du serum et du cristallin des merlus de la côte nord-ouest Africaine. (ICES Special Meeting on the Biochemical and Serological Identification of Fish Stocks, Dublin, 1969). Rapp P. -v Reun. Cons int. Explor. Mer, 161, 80-85.

Punnett, R.C. 1904. Merism and Sex in Spinax niger. Biometrika 3, 313-362.

Schmidt, J. 1919. Racial studies in fish. III Diallel crosses with trout Salmo trutta L. J. Genet. 9, 61-67.

Schmidt, J. 1930. Racial investigations. X The Atlantic cod Gadus callarias L. and local races of the same. C. L. C. Trav. Lab. Carlsberg 18, (6) 71 pp. plus plates.

Sick, K. 1961. Haemoglobin polymorphism in fishes. Nature 182, 89-896.

Sick, K. 1965a. Haemoglobin polymorphism of cod in the Baltic and the Danish Belt Sea. Hereditas 54, 19-48.

Sick, K. 1965b. Haemoglobin polymorphism of cod in the North Sea and the North Atlantic Ocean. Hereditas 54, 49-69.

Tåning, Å.V. 1934. Survey of long distance migrations of cod in the north-western Atlantic according to marking experiments. Rapp. P.-v Reun. Cons. perm. int. Explor. Mer, 89, 5-11.

Tåning, Å.V. 1937. Some features of the migration of cod. J. Cons. perm. int. Explor. Mer, 12, 1-37.

Templeman, W. 1962. Divisions of cod stocks in the north west Atlantic, Redbook int. Commn N.W. Atlant. Fish 3, 79-123.

Templeman, W. 1976. Biological and oceanographic background of Flemish Cap as an area for research on the reasons for year-class success and failure in cod and redfish. Res. Bull. int. Commn N.W. Atlant. Fish 12, 91-117.

Templeman, W. and Fleming A.M. 1963. Distribution of Lernaeocera branchialis (L.) on cod as an indicator of cod movements in the Newfoundland area. Spec. Publs. int. Commn N.W. Atlant. Fish 4, 318-322.

Templeman, W., Squires, H.J. and Fleming A.M. 1957. Nematodes in the fillets of cod and other fishes in Newfoundland and neighbouring areas. J. Fish. Res. Bd. Can. 14, 831-897.

Tills. D., Mourant, A.E. and Jamieson, A. 1971. Red-cell enzyme variants of Icelandic and North Sea cod (Gadus morhua L.) ICES Special Meeting on the Biochemical and Serological Identification of Fish Stocks, Dublin, 1969. Rapp. P.-v. Reun. Cons. int. Explor. Mer, 161, 73-74.

Wilkins, N.P. 1971. Haemoglobin polymorphism in cod, whiting and pollack in Scottish Waters. Rapp. P.-v. Reun. Cons. perm. int. Explor. Mer, 161, 60-63.

CONTRIBUTORS TO THE N.A.T.O. ADVANCED RESEARCH INSTITUTE

B. Åkesson, Göteborgs Universitet, Zoologiska Institutionen, Fack S-400 33, Göteborg, Sweden.

F. J. Ayala, University of California, Department of Genetics, Davis, California 95616, U.S.A.

G. Bacci, Istituto di Zoologia, Via Academia Albertina, 17, Università di Torino, Torino, Italy.

B. Battaglia, Istituto di Biologia Animale, Via Loredan 10, Università di Padova, Padova, Italy.

J. A. Beardmore, Department of Genetics, University College of Swansea, Singleton Park, Swansea SA2 8PP, U.K.

H-P. Bulnheim, Biologische Anstalt Helgoland, 2 Hamburg 50, Palmaille 9, Federal Republic of Germany.

Catherine A. Campbell[*], Department of Biology, University of California, Santa Barbara, California 93460, U.S.A.

F. B. Christiansen, Institute of Ecology and Genetics, University of Aarhus, Ny Munkegade, Building 550, DK-8000 Aarhus, Denmark.

D. Colombera, Istituto di Biologia Animale, Via Loredan 10, Università di Padova, Italy.

D. J. Crisp, University College of North Wales, N.E.R.C. Unit, Marine Science Laboratories, Menai Bridge, Gwynedd LL59 5EH, Bangor, U.K.

R. Doyle, Dalhousie University, Department of Biology, Halifax, Nova Scotia, Canada.

CONTRIBUTORS

T. Fenchel, Institute of Ecology and Genetics, University of
   Aarhus, Ny Munkegade, Building 550, DK-8000 Aarhus,
   Denmark.

J. F. Grassle, Woods Hole Oceanographic Institution, Woods
   Hole, Massachusetts 02543, U.S.A.

Marianne Holm, Institute of Marine Research, Directorate of
   Fisheries, Box 1870/72, N-5011, Bergen, Norway.

A. Jamieson, Ministry of Agriculture Fisheries and Food,
   Fisheries Laboratory Lowestoft, Suffolk NR33 OHT, U.K.

R. K. Koehn, Department of Ecology and Evolution, State
   University of New York, Stony Brook, N.Y., U.S.A.

Mirjama Krajnović-Ozretić, Institute 'Ruder Boskovic', Centre
   for Marine Research, Rovinj, Yugoslavia.

J. Levinton, Department of Ecology and Evolution, State
   University of New York, Stony Brook, N.Y., U.S.A.

R. Milkman, Department of Zoology, The University of Iowa,
   Iowa City, Iowa 52242, U.S.A.

R. Nobili, Istituto di Zoologia, Via A. Volta, 4, Università
   di Pisa, Pisa, Italy.

E. Rodinó, Istituto di Biologia Animale, Via Loredan, 10,
   Università di Padova, Padova, Italy.

A. Sabbadin, Istituto di Biologia Animale, Via Loredan, 10,
   Università di Padova, Padova, Italy.

R. Scheltema, Woods Hole Oceanographic Institition, Woods Hole,
   Massachusetts 02543, U.S.A.

T. J. M. Schopf, Department of Geophysical Sciences, The
   University of Chicago, 5734 S. Ellis Avenue, Chicago
   Illinois 60637, U.S.A.

R. Seed, Department of Zoology, University College of North
   Wales, Bangor, Gwynedd LL59 5EH, U.K.

J. F. Siebenaller, University of California, San Diego,
   Scripps Institution of Oceanography, La Jolla, California
   92093, U.S.A.

D. O. F. Skibinski, Department of Genetics, University College
   of Swansea, Singleton Park, Swansea SA2 8PP, U.K.

CONTRIBUTORS

M. Solignac, Laboratoire de Biologie et Génétique Évolutives
   C.N.R.S., 91190 Gif-sur-Yvette, France.

J. P. Thorpe, Department of Genetics, University College of
   Swansea, Singleton Park, Swansea SA2 8PP, U.K.

J. W. Valentine, University of California, Department of
   Geology, Davis, California 95616, U.S.A.

R. D. Ward, Department of Genetics, University College of
   Swansea, Singleton Park, Swansea SA2 8PP, U.K.

N. P. Wilkins, Zoology Department, University College,
   Galway, Ireland.

Brigitte Volkmann, Istituto di Biologia del Mare, C.N.R.,
   Riva 7 Martiri, 1364/A, 30122 Venezia, Italy.

* current address

AUTHOR INDEX

Contributors to the present volume are given in capital letters.

Abele, L. G. 330, 365
Agranoff, B. W. 670
AHMAD, M. 469-486
Ahmad, S. 672
AKESSON, B. 563, 564, 573-590, 632
Akroyd, P. 411
Aldridge, W. N. 667
Allard, R. W. 351, 437
Allen, J. A. 328
Ambler, J. W. 103
Ambler, R. P. 5
Anders, F. 530, 543
Anderson, W. W. 333, 638
Ankorion, J. 682
Arnaud, P. M. 330
Arnback, A. C. L. 506
Ar-Rushdi, A. H. 487, 488, 492
Arurkar, S. K. 670
Ashton, P. S. 357
Auclair, W. 496, 197
Aulstad, D. 682, 683, 694
Avise, J. C. 82, 111, 163, 408, 426, 427, 436, 437, 469, 479, 483, 724
AYALA, F. J. 23-51, 55, 56, 84, 96, 123, 137, 163, 284, 323-345, 350, 352, 353-356, 408, 414, 429, 435, 436, 437, 723, 724
Ayles, G. E. 683

Babbel, G. R. 437
BACCI, G. 54, 543, 549-571, 573
Baker, C. M. A. 96, 406
Baker, R. J. 267
Balagot, B. P. 230, 249
Baldwin, J. 668
Balkau, B. J. 190
Ballantine, W. J. 127, 136

Baltzer, F. 496, 550
Band, H. T. 86
Bar, Z. 95
Barrett, A. J. 213, 220
Barrington, E. J. W. 506
Barsotti, G. 460, 462, 479
Baslow, M. H. 668
Bateman, A. J. 647, 650
BATTAGLIA, B. 53-70, 137, 248, 250, 351, 564, 585, 617-636
Bayne, B. L. 6, 223, 224, 230, 231, 234, 247, 248, 250
BEARDMORE, J. A. 72, 74, 123-140, 152, 284, 425-445, 469-486
Beaumont, A. R. 266, 268
Becker, W. A. 684
Beermann, S. 489
Beermann, W. 495
Behme, 601
Benke, G. M. 668, 669
Berger, E. M. 264, 317
Berggren, W. A. 310
Bernstein, S. C. 18, 74
Berrill, N. J. 196, 506, 511
Bertin, L. 389
BISOL, P. M. 53-70, 351, 633
Blair, W. F. 124
Boaden, P. J. S. 267
Bobin, G. 438
Bock, W. J. 289
Bocquet, C. 487, 529, 637, 647
Bodmer, W. F. 280
Boesch, D. F. 352
Boettcher, E. W. 701
Bogyo, T. P. 684
Borden, D. 612
Bortesi, O. 557, 559
Bousfield, E. L. 275, 276
Boyer, J. F. 406
Bozic, B. 585

Bracchi, 594
Braun, H.   494
Bretsky, P. W.   25, 32, 39, 96
Brewer, G. J.   56
Bridges, C. B.   558, 560
Brinton, E.   327, 354
Brodie, B. B.   669
Brown, A. H. D.   475
Brown, J.   289
Brownson, R. H.   670
Brunetti, R.   203
Bruun, A. F.   391
Bryant, E. H.   95
Bullivant, J. S.   381
BULNHEIM, H-P.   529-548
Bushnell, V. C.   381

Calaprice, J. R.   682
Callan, H. G.   530
CAMPBELL, C. A.   39, 157-170, 284, 334, 353, 356
Campbell, J. I.   463
Carmody, G. R.   436
Carson, H. L.   25, 356
Case, S. M.   436
Catamo, A.   74
Caten, C. E.   267
Cavalli-Sforza, L. L.   166, 280, 427
Certain, P.   530
Chamberlin, J. L.   313
Chambers, R.   494, 495
Chappnis, J.   454
Charniaux-Cotton, H.   529
Cheetham, A. H.   316
Cherbonnier, G.   354
Chevassus, E.   682
Chiarelli, B.   391
CHRISTIANSEN, F. B.   124, 171-194, 229, 244, 289, 290, 293
Clarck, 292
Clark, A. R.   54
Clausen, J.   267
Clegg, M. T.   437
Cleret, J. J.   647
Cobbs, G.   18
Coe, W. R.   550
COLOMBERA, D.   197, 487-525

COMPARANI, A.   389-424
Cooper, D. W.   172
Coppage, D. L.   667, 668
Cornfield, A.   166
Cowen, R. C.   33
Coyne, J. A.   16, 17, 18, 111
Crick, F. H. C.   4
CRISP, D. J.   257-273
Crow, J. F.   3, 57, 278
Crumpacker, D. W.   267
Cuenot, L.   498, 502, 504
Curds, C. R.   591, 592, 604

Dall, W. H.   157
D'Ancona, U.   391
Dando, R. P.   72, 700
Darwin, C.   3
Dauterman, W. C.   669
Davis, G. A.   670
Davis, G. E.   8
Deelder, C. L.   389, 391
Delage, Y.   496, 497, 499
Delf, E. M.   136
DE LIGNY, W.   391, 403, 667-678
Delobel, N.   499
Dettbarn, W. D.   672
Diamond, J. M.   353
Dingle, J. T.   213, 220
DINI, F.   591-616
Dobzhansky, T.   23, 123, 330, 630, 662, 724
Donaldson, L. R.   681
Doncaster, L.   496
Donovan, E.   482
Douglas, R. G.   324, 325
DOYLE, R. W.   43, 97, 275-287
Dragesco, J.   591
Drilhon, A.   392, 419
Durazzi, J. T.   372, 382
Dutton, A. R.   264

Edwards, A. W. F.   166
Edwards, Y. H.   81
Ege, V.   390
Ehrman, L.   631, 638
Ekman, S.   96, 392
Eldridge, N.   314
Elson, P. F.   327

AUTHOR INDEX

Emanuelsson, H. 585
Embody, G. L. 680
Emery, K. O. 381
Endler, J. A. 229, 357
Eyring, H. 111

Fairbridge, R. W. 366
Falconer, D. S. 6
Faure-Fremiet, E. 591
FAVA, G. 53-70, 618
Feldman, M. 190
Felton, A. A. 16, 17, 18, 74
Fenaux, R. 503, 505
FENCHEL, T. M. 124, 135, 289-301
Fine, J. M. 392, 419
Fischer, A. G. 324, 327, 365
FISHER, J. B. 365-386
Fisher, L. R. 327, 354
Fisher, R. A. 8
Fleming, A. M. 715
Fleming, R. H. 324
Floderus, M. 509
Flowerden, M. W. 266
Frank, P. W. 327, 355
Fratello, B. 493
FRIER, J.-O. 289-301
Fripp, Y. J. 267
Frydenberg, O. 86, 88, 172, 174, 175, 176, 177, 180, 182, 187, 212, 229, 700
Fujio, Y. 86
Fundiller, D. L. 95, 230, 249
Futch, D. G. 656

Gakstatter, J. H. 667
Gall, G. A. E. 682, 683, 694
GALLEGUILLOS, R. A. 71-93, 408, 409
Gardiner, M. S. 497
Garstang, W. 503
Gause, G. F. 351
Genermont, J. 595, 604
German, J. 496
Gibson, J. R. 667, 668
Giesbrecht, W. 489
Giesel, J. T. 326
Gilbert, C. H. 100

Gilbert, W. 4
Gilles, R. 234
Gillespie, J. H. 74, 86, 95
Ginsburger-Vogel, T. 540-542, 544
Gjedrem, T. 683, 693
Glesener, R. 358
Godlewski, J. E. 496
Gojdics, M. 591
Goldschmidt, R. 558
Golovinskaya, K. A. 680
Gooch, J. L. 43, 95, 96, 150, 229, 264, 266, 307, 317, 350, 351, 352, 353, 356, 406, 436
Gordon, N. 679
Gorman, G. C. 84, 88, 436
Goswami, S. C. 487, 490, 491, 492, 494, 495.
Goswami, U. 487, 490, 491, 492, 494, 495
Gottlieb, L. D. 437, 608
Gould, S. J. 314, 436
Graham, M. 723
Grant, V. 290
Grassi, B. 391, 392, 403
GRASSLE, J. F. 25, 55, 96, 307, 332, 347-364
GRASSLE, J. P. 347-364
Gray, J. 496
Graziani, G. 196, 197, 198
Greenwood, P. H. 409
Gregg, T. G. 306
Gruffydd, L. D. 268
Grunstein, M. 4
Guillon, G. 668
Gulland, J. A. 723
Gupta, K. C. 503

Hadfield, M. G. 305
Haecker, V. 492, 494
Haigh, J. 177
Haldane, J. B. S. 95
Haneline, P. G. 436
Hanson, T. 315
Harding, J. P. 489, 490, 498
Hargis, W. J. 248
Harper, J. L. 267
Harris, H. 81, 87, 425

Hartmann, M. 549, 573
Harvey, E. E. 496
Hayford, C. O. 680
Heberer, G. 488, 490, 495
Hecht, M. K. 314
Heckmann, K. 593, 595, 601, 603, 607, 608, 611
Hedgecock, D. 333
Hedgpeth, J. W. 381
Hedrick, P. W. 146, 332, 355, 427
Heffner, B. 497, 498
Heincke, F. 700
Heller, J. 144, 264, 265
Hendler, G. 353
Hershberger, W. K. 704
Hesey, W. M. 267
Hessler, R. R. 43, 103, 114, 325
Hill, B. A. 509
Hill, W. G. 475
Hillman, R. E. 311
Hincks, T. 438
Hindle, E. 497
Hinegardner, R. 498
Hjorth, J. P. 172, 175, 176, 182, 212
Hochachka, P. W. 95, 111, 112
Hogan, J. W. 668, 670, 671, 672
Hogue, M. J. 499
Holland, H. T. 667
Hollister, C. D. 310
HOLM, M. 679-698
Holmes, R. S. 672
Holmquist, C. 490
Hopkinson, D. A. 81
Horvitz, D. G. 203, 406
Hoskin, G. S. 672
Hubble, S. P. 330
Hubbs, C. L. 97
Hubby, J. L. 74, 414, 425, 435, 469
Hunt, W. G. 436, 483
Hunter, R. L. 669
Hutchinson, G. E. 294, 351
Hylleberg, J. 291, 292
Hyman, L. H. 503

Ihle, J. E. W. 506
Isaacson, J. S. 4
Issel, R. 61
Iwamoto, T. 97

Jackson, J. B. C. 353
Jameson, S. C. 353
JAMIESON, A. 180, 699-729
Jayakar, S. D. 95
Johnson, A. G. 86, 103, 110, 334, 408, 409
Johnson, F. H. 111
Johnson, F. M. 11, 213, 231
Johnson, G. B. 18, 87, 95, 124, 213
Johnson, M. S. 95, 96, 137, 180, 229, 436
Johnson, M. W. 324
Johnson, W. E. 356, 426, 436, 437
Jones, J. W. 391
Jones, M. B. 637
Jonsson, J. 700
Jordan, H. E. 499
Juchault, P. 543
Jumars, P. A. 43, 325, 355

Kafescioglu, I. A. 324
Kanis, E. 683, 693
Karlin, S. 188, 190
Karlson, R. H. 305
Kasas, O. M. 506
Kaufman, W. W. 47, 332, 350, 414
Kilpatrick, C. W. 436
Kim, Y. J. 84, 436
Kimura, M. 3, 4, 278, 414, 415
Kincaid, H. L. 682
Kincaid, T. 157
King, M. C. 163, 437
Kinne, O. 230, 291, 295, 530
Kirpichnikov, V. S. 680
Knight-Jones, E. W. 258
Knowles, C. O. 670, 671, 672
Koehler, R. 16, 17, 18
KOEHN, R. K. 6, 95, 110, 142, 211-227, 229, 230, 231, 249, 266, 284, 392,

AUTHOR INDEX

408, 409, 471
Kofoed, L. H. 249, 292, 293
Kohn, A. J. 330
Kojima, K. 74, 86, 95, 475, 672
KOLDING, S. 289-301
Kornhauser, S. J. 490, 495
Kornfield, I. L. 408, 409
Kossinna, E. 366
Kossman, H. 682
Kosswig, C. 531
KRAJNOVIĆ-OZRETIĆ, M. 667-678
Krimbas, C. B. 670
Krimmel, O. 490
Kruger, P. 492

La Greca, M. 557, 563, 573
Laing, C. 353, 436
Land, K. C. 285
Langley, C. H. 95
LASSEN, H. H. 190, 212, 229-254, 292
Latter, B. D. H. 427
Laven, H. 632
Lawton, J. H. 365
LAZZARETTO-COLOMBERA, I. 57, 487-525
Lea, R. N. 100, 110
Lebour, M. V. 136
Le Calvez, J. 530
Legrand, J-J, 543
Legrand-Hamelin, E. 543
Lejeuz, R. 295
Leser, C. E. V. 285
Leslie, P. H. 278
Leutert, R. 557
Levene, H. 15
Levins, R. 24, 25, 95, 289, 307, 326, 330, 332, 414
LEVINTON, J. S. 55, 95, 96, 151, 190, 211, 229-254
Lewis, J. R. 136, 454, 456, 469, 470, 471
Lewis, R. C. 681
Lewontin, R. C. 71, 74, 135, 307, 414, 425, 475
Li, C. C. 203, 283, 406

Li, W. 427-428, 429, 430, 435
Lieder, U. 682
Lindroth, A. 682
Littlejohn, M. J. 124
Lockwood, A. P. M. 282
Lohmann, H. 506
Lorenz, D. M. 25, 32, 39, 96
Lough, R. G. 313
Loverre, A. 74
Low, P. S. 111, 112, 115
Loya, Y. 352
Lubet, P. 447, 454
Ludke, J. L. 669
Ludtke, A. H. 668
Luecken, W. W. 601
LUPORINI, P. 591-616

MacArthur, R. H. 289, 306, 326, 330, 350, 353
MacIntyre, R. J. 16, 18
Magagnini, G. 594
Maickel, R. P. 669
Makino, S. 497, 498, 499, 500
Malogolowkin, C. 631
Mangaly, K. G. 704
Manwell, C. 96, 406
Markert, C. L. 395, 669
Marr, J. C. 723
Marshall, D. R. 351
Marshall, N. B. 103
Martin, G. 543
Martin, J. P. 436
Massoulie, J. 668
Mathers, N. F. 268
Matschek, H. 490, 494
Matsui, K. 496
Matsumura, F. 672, 673
Matthews, E. 668
Mauchline, J. 327, 354
Maxam, A. 4
May, R. M. 289, 330, 331
Maynard Smith, J. 177, 190
Mayr, E. 260, 479
Mazzarelli, G. 391
McClendon, J. F. 493
McDermid, E. M. 334
McDonald, J. F. 123, 332
McGregor, J. 190

Meise, W.   482
Meluzzi, C.   460, 462, 479
Menard, H. W.   366
Menasveta, D.   681
Mendosa, C. E.   669
Merritt, R. B.   95, 96, 212, 213
Metcalf, R. A.   671, 672
Mettler, L. E.   306
Mileikovsky, S. A.   327
MILKMAN, R.   3-22, 195, 196, 212, 224, 249
Millar, R. H.   195, 511
Millenbach, C.   681
Miller, D. J.   100, 110
Milliman, J.   381
Minganti, A.   509, 510
Minouchi, O.   510
Mitton, J. B.   95, 211, 212, 224, 230, 231, 408
Miyake, A.   607
Moav, R.   680, 681, 682
Møller, B.   295
Møller, D.   406, 690, 700, 701, 705
Moore, H. E.   136
Moore, R. C.   502
Morgan, T. H.   496
MORRIS, S. R.   123-140, 152
Morton, N. E.   57, 166
Moser, H. G.   97, 100
Motoyama, N.   669
MOYNIHAN, E.   141-155
Moyse, J.   136
Muller, H. J.   57
Murdock, E. A.   454, 471
Murphy, D. L.   230
Murphy, L. S.   45, 96, 97, 333, 352, 353, 354,
Murphy, S. D.   668
Muus, B.   292
Myers, P.   436

NAEVDAL, G.   679-698, 705
Naylor, E.   637
Nei, M.   61, 82-84, 86, 88, 89, 110, 160, 165, 168, 427-428, 429, 430, 432, 435, 438,
705, 715
Nelson, K.   333
Nelson-Smith, A.   136
Neumann, D.   632
Nevo, E.   95, 96, 123
Newell, G. E.   135
Newell, N. D.   325
Newell, R. C.   249
Nicol, P. I.   212, 229
Nielsen, J. T.   88
Nigrelli, R. F.   668
Niiyama, H.   497, 498, 499, 500
Nishikawa, S.   497
Niswander, J. D.   105
NOBILI, R.   591-616
Noland, L. E.   591
Norman, J. R.   87
Nyberg, D.   603
Nyman, O. L.   700

O'Brien, R. D.   667, 669, 670, 672
O'Carra, P.   112
Odense, P. H.   700
O'Gower, A. K.   212, 229
Ohnesorge, F. K.   668
Ohno, S.   391
Ohta, T.   4, 5, 17, 71, 414, 415
Oka, H.   202
Okubo, A.   263
Olson, E. C.   409
Olson, R. P.   681
O'REGAN, D.   141-155
Orian, A. J. E.   530
Orlando, E.   495
Ornstein, L.   669
Otterlind, G.   700
Otto, K.   213

Pantelouris, E. M.   391, 403, 419
Parenti, U.   563
Parry, G.   283
Paskausky, D. F.   230
Patton, J. L.   436
Pearcy, W. G.   103
Peck, S. B.   436

AUTHOR INDEX

Pesch, G. 266
Petra, P. H. 16
Pfannenstiel, H. D. 564
Pianka, E. R. 290, 326
Pichot, P. 704
Pielou, E. C. 127, 278
Pinkster, S. 539
Pinney, E. 497
Poisson, R. 530
Potter, J. L. 669
Potts, W. T. M. 283
Powell, A. 11, 231
Powell, J. R. 80, 84, 142, 332, 356, 408
Power, H. W. 595, 603
Prasad, R. 56, 395, 411
Prenant, M. 438
Preston, F. W. 382
Punnett, R. C. 700
Purdom, C. E. 72

Radwin, G. E. 313
Rasmussen, D. I. 212
Rastelli, M. 143
Raup, D. M. 372
Refstie, T. 683, 693
Regal, P. J. 357
Remane, A. 247
Remington, C. L. 479
Richardson, M. E. 355
Richardson, R. H. 355
Richter-Dyn, N. 188
Ricklefs, R. E. 460
Riley, J. D. 89
Robertson, J. 223
Roche, L. 357
RODINÓ, E. 389-424
Rogers, J. S. 427
Rohlf, F. J. 105, 684
Rokop, F. J. 96, 354
Romer, A. L. 409
Ross, H. H. 460
Roychoudhury, A. K. 84, 428, 429, 430, 435
Rubinoff, R. 84
Ruckert, J. 490
Rusch, M. E. 495
RYLAND, J. S. 425-445
Ryman, N. 682

Ryther, J. H. 325

Sabatini, M. A. 493
SABBADIN, A. 195-209
Sacchi, C. 143
Sakai, K. 672, 673
Sala, M. 508
Salser, W. 4
Sampsell, B. 16, 17, 18
Sanders, H. L. 43, 96, 114, 306, 324, 325, 328, 352, 353, 354
Sandifer, P. A. 305, 313
Sanger, F. 4
Santangelo, G. 594
Sanzo, L. 391
Savostyanova, G. G. 681
Schaffer, W. M. 327
Schaperclaus, W. 680
Schedl, P. 4
Scheltema, A. H. 311
SCHELTEMA, R. S. 257, 303-322, 327, 328, 353, 357
Scheurer, P. J. 503
Schlieper, C. 247
Schmidt, J. 180, 188, 389-391, 409, 415, 700
Schoener, A. 327
Schoener, T. W. 124, 293
SCHOPF, T. J. M. 43, 45, 95, 96, 229, 264, 333, 352, 353, 365-386, 406
Schram, T. A. 357
Schroder, D. 365
SEED, R. 230, 447-468, 469, 470, 479, 482
Segerstrale, S. G. 298
Selander, R. K. 4, 42, 47, 80, 103, 212, 332, 333, 350, 353, 408, 414, 426, 436, 437, 483
Sella, G. 557
Sene, F. M. 356
Sepkoski, J. J. 372
Sexton, E. W. 54
Shaw, C. R. 56, 395, 411
Shaw, J. 283
Shuto, T. 314, 315
Sibuet, M. 354

Sick, K. 88, 180, 391, 679, 700
SIEBENALLER, J. F. 95–122, 408
Siegel, R. W. 607, 608, 611
Simberloff, D. 381, 382
SIMONSEN, V. 86, 171–194, 212
Singh, H. 503
Singh, R. S. 16, 18, 74, 111
SKIBINSKI, D. O. F. 469–486
Slatkin, M. 190
Slobodkin, L. B. 306
SMITH, C. A. F. 365–386
Smith, K. 180
Smith, K. L. 103
Smith, M. F. 436
Smith, M. H. 426, 436, 437, 479, 483
Smith, R. I. 276
Smith, S. M. 366
Smithies, O. 669
Smouse, P. E. 355
Snyder, T. P. 150, 264, 436
Sohl, N. F. 316
Sokal, R. R. 105, 684
SOLIGNAC, M. 637–664
Somero, G. N. 47, 55, 95, 96, 100, 101, 103, 110, 111, 112, 115, 334, 415
Sonneborn, T. M. 592, 600, 601, 611
Sorgenfrei, T. 308, 309, 311
Soulé, M. 47, 55, 84, 96, 100, 101, 103, 123, 137, 334, 353, 415
Southward, A. J. 257, 263
Southwood, T. R. E. 351, 358
Sova, C. R. 667
Spassky, B. 630
Spielmann, H. 112
Spiess, E. B. 650
Spight, T. M. 166
Stalker, H. D. 645
Stanley, S. M. 306, 463, 464
Stasek, C. R. 463
Stebbins, G. L. 267
Steele, D. H. 276, 531

Steele, V. J. 276, 531
Stegman, K. 682
Stehli, F. G. 324, 365, 366, 372, 381, 382
Stella, E. 494
Stern, K. 357
Stevens, C. D. 284
Stevens, N. M. 496
Stock, J. H. 539
Stommel, H. 263
Struhsaker, J. W. 308, 317
Suchanek, T. H. 230
Sutcliffe, D. W. 283
Svardsson, G. 679
Svedrup, H. U. 324, 366

Taning, A. V. 708, 723
Taylor, K. M. 509
Templeman, W. 715, 722
Tenca, M. 509
Tennent, H. D. 496, 497, 499
Tesch, F. W. 389–390
Thacker, G. T. 89
Theeder, H. 248
Thoday, J. M. 351, 483
Thompson, D. 72
Thompson, E. A. 166
Thompson, G. 177
Thompson, R. J. 230
Thörig, G. E. W. 18, 213
THORPE, J. P. 425–445, 469
Thorson, G. 257, 258, 267, 305, 313, 315, 327
Throckmorton, L. H. 74, 435, 469
Tills, D. 700
Tilman, D. 358
Tobari, Y. N. 74
Tokioka, T. 506
Tracey, M. L. 406
Traut, W. 530
Tripathi, R. K. 670
Trippa, G. 74
Truffau, M. 604
Tsakas, S. 670
Tsytsugina, V. G. 490
Tucker, D. W. 391
Turano, 212, 230
Turekian, K. K. 328

AUTHOR INDEX

Turner, B. J.   433
Turner, M. E.   284
Turner, R. D.   350
TURNER, R. J.   699-729

Ubisch, M. von.   497, 498
Utter, F. M.   86, 408, 409

VALENTINE, J. W.   23-51, 55, 84, 96, 123, 137, 141, 153, 158, 284, 323-345, 350, 352, 353, 354, 356, 414, 723
Vance, R. R.   328
Van Name, W. G.   503, 506
Van Valen, L.   123
VAROTTO, V.   61, 617-636
Vavra, J.   533
Venier, G.   498
Vermeij, G. J.   136
Vigue, C. L.   213
Vijayan, V. K.   670
Vitturi, R.   510, 511, 512, 514
VOLKMANN, B.   60, 61, 585, 617-636
Volkmann-Rocco, B. see Volkmann, B.
Von Sengbusch, R.   682

Wallace, B.   57
Walne, P. R.   268
WARD, R. D.   71-93, 408, 409
Warren, C. E.   8
Watanabe, H.   202
Watt, W. B.   16
Webster, T. P.   436
Weiss, C. M.   667, 668
Wells, J. W.   372, 382
Whitt, G. S.   411, 672
Whittaker, R. H.   352
Wichterman, R.   595
Widdows, J.   224, 234
WILKINS, N. P.   55, 135, 141-155, 268, 700
Wilkinson, G. N.   113
Williams, A. K.   667
Williams, G. B.   258
Williams, G. C.   110, 212, 229, 258, 267, 355, 358, 392, 393, 403, 419
Williamson, G. R.   723
Wilson, A. C.   5, 163, 437
Wilson, E. E.   496, 497, 499, 500, 501
Wilson, E. O.   289, 306, 326, 350, 353
Wohlfarth, W. G.   681, 682
Wood, L.   248
Woodring, W. P.   310
Woodruff, D. S.   436
Workman, P. L.   105
Wright, S.   3, 259, 283, 474
Wright, T. R. F.   16, 18

Yang, S. Y.   84, 88, 436
Yasumoto, T.   503
Yonge, C. M.   463

Zech, R.   669, 672, 673
Žikić, R.   671, 672
Zimmerman, E. G.   436
Zouros, E.   80, 87
Zuckerkandl, E.   80
Zürcher, K.   669, 672, 673

ORGANISM INDEX

Abudefduf
  troschelii, 334
Acantocyclops
  robustus, 495
  vernalis, 495
Acartia
  centura, 491
  negligens, 491
  plumosa, 491
  spinicauda, 491
Acartiella
  gravelyi, 491
  keralensis, 491
Alcyonidium, 433-434, 437-440
  albidum, 438, 439
  gelatinosum, 439
  hirsutum, 439
  polyoum, 439
  variegatum, 439
Amphiprion
  clarki, 334
Anguilla
  anguilla, 389-420
  rostrata, 390-391, 392, 403, 419
Anisogammarus
  annandalei, 530
Anolis, 436
  lividus 436
  marmoratus, 436
Anomalocera
  patersonii, 490
Anoplarchus
  purpurescens, 137
Antedon
  rosacea, 501, 502
Anthocidaris
  crassispina, 497
Aphelasterias
  nipponica, 498, 499
Aplidium
  aereolatum, 508
  densum, 508
  proliferum, 508
Arbacia
  lixula, 496
  punctulata, 496
  pustulosa, 496
Armadillidium
  vulgare, 543
Ascidia
  callosa, 509
  conchilega, 509
  malaca, 509
  mentula, 509
  paratropa, 509
  virginea, 509
Ascidiella
  aspersa, 509, 511
  patula, 509
  scabra, 509
Asterias
  amurensis, 498, 499
  forbesi, 45, 333, 499
  glacialis, 499
  vulgaris, 44, 333, 499
Asterina
  gibbosa, 499, 502
  pectiniformis, 499
Astropecten
  bispinosus, 500
Avena
  barbata, 351
  fatua, 351

Balanus
  amphitrite, 266
  balanoides, 267
Bathygobius
  ramosus, 334
Biston, 332
Boltenia, 506
  villosa, 510
Bonellia
  viridis, 549, 550, 552, 556, 557, 560
Botrylloides
  leachi, 510
Botryllus, 513
  primigenus, 202
  schlosseri, 195-207
Brachidontes

citrinus, 464
Brissuls
　unicolor, 497
Bugula, 257
　stolonifera, 264
Bursa, 316
　granularis, 308-310

Calanus
　finmarchicus, 490
　gracilis, 490
　helgolandicus, 490
　robustior, 490
Callianassa
　californiensis, 333
Calliostoma
　zizyphinum, 151
Cancer
　gracilis, 333
　magister, 333
Canthocamptus
　staphilinus, 492
　trispinosus, 492
Capitella, 347-359
Centropages
　furcatus, 490
　typicus, 490
Cepaea, 332
Cerion, 436
Chlamys
　opercularis, 266, 268
Ciclhasoma
　cyanoguttatum, 409
Cidaris
　cidaris, 496
Ciona, 506, 511
　intestinalis, 509, 512
　rulei, 509
Clavelina, 511
　lepadiformis, 508, 511, 512
　nana, 508
Clunio marinus, 632
Clypeaster
　rosaceus, 497
Conger
　conger, 408-409, 414-419
Copilia
　denticornis, 495
Coregonus
　lavaretus, 679
Corella, 511
　parallelogramma, 509, 512
　willmeriana, 509
Corvus
　corvus, 482
Coryphaenoides
　acrolepis, 103, 114
Crangon
　franciscana, 333
　negricata, 333
Crassostrea
　gigas, 150, 151, 152, 268
　virginica, 311, 333
Crepidula
　fornicata, 249
　plana, 550, 560
Cribella
　sanguinolenta, 499
Cucumaria
　cucumis, 500
　planci, 500
Culex
　pipiens, 632
Cyclops, 493, 498
　affinis, 494
　albidus, 494
　bicolor, 494
　bicuspidatus, 494
　diaphanus, 494
　fuscus, 494
　gracilis, 494
　insignis, 494
　leuckarti, 494
　modestus, 494
　phaleratus, 494
　prasinus, 494
　serrulatus, 494
　signatus, 494
　strenuus, 494
　vernalis, 494
　viridus, 494-495
Cymatium
　parthenopeum, 305
Cyprinodon, 433
Cyprinus
　carpio, 680
Cystodites
　dellechiaiei, 508

Dascyllus
  reticulatus, 334
Dendrodoa, 511
  grossularia, 510
Diaptomus
  castor, 490
  coeruleus, 490
  denticornis, 490
  gracilis, 490
  laciniatus, 490
  salinus, 490
Diazona, 506
  violacea, 509
Didemnum
  candidum, 508
Diplopteraster
  multipes, 43, 46
Diplosoma
  listerianum, 508
Dipodomys, 436, 437
Distaplia
  rosea, 508
Distomus, 511
  variolosus, 510
Doliolum
  denticolatum, 507
Dosinia
  exoleta, 150, 151
Drosophila, 355, 356, 426, 435, 436, 483, 638, 647, 650
  ananassae, 656
  equinoxialis, 27
  heteroneura, 356
  melanogaster, 11, 18, 231, 558, 560
  pallidosa, 656
  paulistorum, 630–631, 632, 633
  persimilis, 662
  pseudoobscura, 27, 662
  setosimentum, 356
  silvestris, 356
  willistoni, 163, 429, 435, 630
Dunaliella
  salina, 592

Echinaster
  sepositus, 499
Echinocardium
  cordatum, 497
  mediterraneum, 497, 498
Echinus
  acutus, 496
  esculentus, 496
  miliaris, 496
Ecteinascidia
  turbinata, 509
Ectocyclops
  strenzkey, 495
Electra
  pilosa, 437–440
Elminius, 263
Elpidia
  glacialis, 500
Emerita
  analoga, 333
Escherichia
  coli, 5
Etmopterus
  spinax, 700
Eucheta
  marina, 490
Eucyclops
  serrulatus, 495
Eudistoma
  magnum, 508
Euphausia, 45, 55
  distinguenda, 45, 47–48, 334, 354
  mucronata, 45, 47–48, 334, 354
  superba, 45, 47–48, 334, 354
Euplotes, 591, 601, 602, 604, 612
  balticus, 592, 604, 607
  crassus, 592–613
  cristatus, 592–606
  minuta, 592–613
  mutabilis, 592, 604, 607
  Vannus, 592, 593, 595, 600–613

Frieleia
  halli, 43, 46, 334, 354, 356
Fritillaria, 503, 516
  pellucida, 505

Fucus, 267
Fundulus
  heteroclitus, 408, 672

Gadus
  melanogrammus, 679
  morhua, 679, 699-725
Gammarus, 282, 291, 298, 529-530, 544
  chevreuxi, 54, 530
  duebeni, 295, 297, 530-544
  insensibilis, 56-63
  lawrencianus, 275-285
  locusta, 295, 530, 531, 533-534, 539
  oceanicus, 295
  pulex, 530
  salinus, 295, 539
  zaddachi, 295, 531, 533-534
Geomys, 436
Gibbonsia
  metzi, 334
Gibbula
  cineraria, 150, 151
  umbilicalis, 150, 151
Gillichthys
  mirabilis, 334
Glycymeris
  glycymeris, 150, 151
Granulifusus, 314, 316

Hacelia
  attenuata, 499
Halichores, 334
Halocynthia
  papillosa, 510
Hemigrapsus
  oregonesis, 333
Henricia
  nipponica, 498, 499
Hersilia
  apodiformis, 495
Hesperoleucus
  symmetricus, 437
Heterocope
  saliens, 490
  weismanni, 490
Hippoglossus
  stenolepis, 112
Hipponoe
  esculenta, 497
Holothuria
  polii, 500
  tubulosa, 500, 502
Homarus
  americanus, 333
Homo, 437
Hydrobia, 135, 291, 297-298
  neglecta, 291-293
  ulvae, 291-293
  ventrosa, 291-293
Hymenopappus
  artemisiaefolius, 437
  scabiosaeus, 437

Jaera
  albifrons, 637-662

Kareius
  bicoloratus, 86

Labidocera
  acuta, 491
  kroeyeri, 491
  pavo, 490
  pectinata, 490
Laemargus
  maricatus, 493
Laophonte
  setosa, 492
Lavinia
  exilicauda, 437
Lepomis, 436
Leptosynapta
  gallienei, 500
Lernaea
  cyprinacea, 493
Lernaeocera
  branchialis, 715
Leuresthes
  tenuis, 334
Lichomolgus
  forficula, 495
Ligia
  oceanica, 543
Limanda
  limanda, 71-89

ORGANISM INDEX

Limnocalanus
  johanseni, 490
  macrurus, 490
Limulus
  polyphemus, 42, 333
Liothyrella
  notorcadensis, 42-45, 333
Littorina, 284
  littoralis, 142-153, 264, 265
  littorea, 124-138, 142-153, 264, 265
  mariae, 143, 264, 265
  neglecta, 144, 265
  neritoides, 264, 265
  nigrolineata, 124-138, 144, 265
  obtusata, 143, 264
  patula, 144, 265
  picta, 308, 317
  rudis, 124-138, 144, 264, 265, 267
  saxatilis, 142-153, 264, 436
Lupinus
  subcarnosus, 437
  texensis, 437
Lymantria
  dispar, 558, 565
Lyssoclinum
  wegelei, 508
Lytechinus, 4

Macoma, 55
Macrocyclops
  fuscus, 495
Marthasterias
  glacialis, 499
Megacyclops
  viridis, 495
Membranipora
  membranacea, 437-440
Menidia, 436
Mercenaria
  mercenaria, 266
Microcosmus, 506
  polymorphus, 510
  sabatieri, 510
  sulcatus, 510

Mirounga
  leonina, 334
Modiolus
  modiolus, 453, 463-465
Moira
  atropos, 497
Molgula
  manhattensis, 510
Morchellium
  argus, 508
Mugil
  cephalus, 334
Murex
  inornatus, 309, 310
Mus, 436
  musculus, 483
Mya
  arenaria, 151
Myticola
  intestinalis, 495
Mytilus, 6
  californianus, 230
  edulis, 18, 149-150, 151, 152, 211-225, 229-251, 266, 284, 447-466, 469-484
  galloprovincialis, 18, 266, 447-466, 469-484
Myxoderma
  sacculatum, 43, 46

Nassarius, 311, 316
  obsoletus, 266, 305, 307, 436
Nearchaster
  aciculosus, 43, 46
Nitocra
  hibernica, 492
Nucella
  lapillus, 149, 151

Octosporea
  effaminans, 533-539
Oikopleura, 503
  albicans, 505
  dioida, 505
  fusiformis, 505
  longicauda, 505
Oithona

rigida, 495
Oncorhynchus
  tshawytscha, 681
Ophidiaster
  ophidiaster, 499
Ophiocomina
  nigra, 500
Ophioderma
  longicauda, 500
Ophiomixa
  pentagona, 500, 502
Ophiomusium, 355
  lymani, 43, 46, 97, 334, 356
Ophiothrix
  fragilis, 500
Ophryotrocha, 632
  costlowi, 575–588
  hartmanni, 565
  labronica, 563–569, 573–588
  macrovifera, 573–574
  notoglandulata, 573–574
  puerilis, 549–569, 573, 579, 585–588
  robusta, 573, 574
Orchestia, 529–530, 543–544
  cavimana, 540–541
  gammarellus, 540–544
  montagui, 540–542, 544
Orthagoriscola
  muricata, 493
Ostrea
  edulis, 150, 151

Pachycheles
  rudis, 333
Pagurus
  granosimanus, 333
Pan, 437
Pandarus
  sinuatus, 493
Paracentrotus
  intermedius, 497
  lividus, 496, 497, 502
  purpuratus, 497
Paramecium
  aurelia, 592, 611
Patella
  aspersa, 151, 152

coerulea, 550, 557
vulgata, 148, 149, 151, 152
Pegea
  confederata, 506, 507
Pelobates, 123
Perognathus, 437
Peromyscus, 436
Perophora
  listeri, 509
Petrolisthes
  cinctipes, 333
Phallusia
  fumigata, 509
  mammillata, 509
Phestilla
  sibogae, 305
Phoronopsis
  viridis, 42–44, 333
Pinnotheres
  pisum, 454, 455
Platichthys, 72
  flesus, 71–89
  stellatus, 86
Pleuromamma
  abdominalis, 490
  gracilis, 490
Pleuronectes, 71, 72, 89
  platessa, 71–89, 409
Polyandrocarpa, 510
Polycarpa, 511
  gracilis, 510
  pomaria, 510
Polycitor
  adriaticus, 508
Polyclinum
  aurantium, 508
Polysincraton, 511
  lacazei, 508
Porcellidium
  fimbriatum, 54
Porcellio
  dilatatus, 543
Pourtalesia
  jeffreysi, 496, 503, 516
Psammechinus
  microtuberculatus, 496
Pseudomonas
  aeruginosa, 5
Pteraster

jordani, 43, 46
Pyrosoma
  atlanticum, 506, 507
Pyura, 506
  dura, 510
  haustor, 510
  microcosmus, 506, 510
  squamulosa, 510

Rattus, 436
Rhizocephela, 257
Rhopalea
  neapolitana, 509
Rithropanopeus
  harrisii, 351

Salmo
  gairdneri, 694
  salar, 694
  trutta, 700
Salpa
  fusiformis, 507
  maxima, 507
Salvelinus
  fontinalis, 681
Sapphirina
  ovatolanceolata, 495
Sassia
  enode, 309
  remensa, 308
Saxicava
  arctica, 267
Schizomavella
  auriculata, 437-440
  hastata, 437-440
  linearis, 437-440
Schizoporella
  errata, 264
  unicornis, 264
Scoterpes
  copei, 436
Sebastes, 110
  alutus, 334
  caurinus, 334
  elongatus, 334
Sebastolobus
  alascanus, 97-116
  altivelis, 97-116
Sigmodon, 436

Sphaerechinus
  granularis, 497
  parma, 497
  pulcherrimus, 497
Sphaeroma, 291, 297-298
  hookeri, 293-296
  rugicauda, 293-296
  serratum, 54
Spinax
  niger, 700
Spirorbis, 257
  spirorbis, 258
Stephanomeria, 608
Stichopus
  regalis, 500
Stolonica, 511
  socialis, 510
Strongylocentrus, 4
Styela, 511
  clava, 510
  coriacea, 510
  gibsii, 510
  partita, 510
  plicata, 510, 514
Sydnium
  fuscum, 508
  turbinata, 508

Tapes
  decussatus, 149, 151
Temora
  turbinata, 490
Terranova
  decipiens, 715
Tetrahymena, 612
Thais
  emarginata, 309, 310
  haemastoma, 309, 310
  lamellosa, 157-168, 284
Thalia
  democratica, 507
Thelohania
  heredieteria, 533-539
Thomomys, 436
Thunnus
  thynnus, 112
Tigriopus, 54, 56, 62, 64, 66, 351
  brevicornis, 58-63, 492

californicus, 492
fulvus, 58-66, 585
japonicus, 492
Tisbe, 56, 57, 61, 62, 351,
    489, 498, 617
  aragoi, 492
  battagliai, 492
  biminiensis, 57-63, 488,
    492
  bulbisetosa, 493
  clodiensis, 58-63, 492, 585,
    617-634.
  dobzhanskii, 492, 631
  furcata, 492
  gracilis, 492
  holothuriae, 56-66, 488,
    489, 492
  marmorata, 492
  persimilis, 492
  pontina, 492
  pori, 492
  reluctans, 489, 492
  reticulata, 54, 248, 492,
    564
Tortanus
  barbatus, 491
  forcipatus, 491
  gracilis, 491
Toxopneustes
  variegatus, 497, 498
Trematomus
  bernacchii, 334
  borchgrevinki, 334
  hansoni, 334
Tridacna
  maxima, 32-47, 334, 353,
    355, 356
Tridedemnum
  tenerum, 508
Tritonium
  enode, 308, 309, 310

Upogebia
  pugettensis, 333

Venus
  verrucosa, 150, 151

Zoarces

viviparus, 171-191
Zostera, 203, 204, 206

SUBJECT INDEX

Accelerating differentiation, 190
Adaptation, 5, 141-142
Adaptive strategies, 24, 66, 326-340
    see also genetic variation
Alcohol dehydrogenase, 11, 17
Allozymes,
  detection of variants, 18, 25-30, 74
  heat sensitivity of, 11, 17, 142-153.
Aminopeptidase, 57
    see also leucine aminopeptidase
Amphipod, 54, 56, 61, 275, 291, 529-544
Ascidian, 195-207
  chromosomes in, 487, 506-511, 514-516
Asteroid, 43, 45, 46, 353-354, 358

Barnacle, 258, 266
Bivalves, 369-383
Bluegill, 670, 672
Brachiopod, 42-45, 353-354
Breeding system,
  of Euplotes, 592-613
Bretsky-Lorenz hypothesis, 25, 32, 39, 96
Bryozoa, 258, 259, 264, 316, 358, 368-383, 437-440
Bursidae, 310, 312

Carp, 680
Catfish, 670, 672
Character displacement, 124, 135, 289-298
Ciliata, 591
Clines,
  in Littorina, 135-136
  in Mytilus, 211, 212, 229, 230, 249-250
  in Thais, 165

  in Zoarces, 176, 179, 180, 188, 189
  character displacement and, 290
Clone, 195, 592, 601
Cod, 679, 699-725
Copepod, 54-67, 248, 617, 715
  chromosomes in, 487-498, 514-516
Copper tolerance,
  in grasses, 262
Coral, 369-383
Crab, 257,' 351
Cymatiacea, 308-309, 312
Cymatiidae, 310, 312
Cytochrome c, 5

Dab, 71-89
Decapod, 353
DNA,
  non-translated, 5
  and chromosome evolution, 498, 502, 511, 514-515

Echinodermata, 259
  chromosomes in, 487, 498-503, 514-516
Eel, 389-420
Eelpout, 171-191
Electrophoresis, 25-30, 74
  taxonomy and, 425-441, 469-470
Electrophoretic surveys,
  Anguilla, 396-402, 405
  Asteroids, 46
  Botryllus, 205
  Conger, 416-418
  Euphausia, 48
  Frieleia, 46
  Gadus, 709-714
  Gammarus, 59
  Limanda, 75-79
  Liothyrella, 44
  Littorina, 126, 129, 145, 147
  Mytilus, 216, 217, 239-241,

473, 474, 477, 480, 481
Ophiomusium, 46
Phoronopsis, 44
Platichthys, 75-79
Pleuronectes, 75-79
Sebastolobus, 101-109
Thais, 161
Tigriopus, 59
Tisbe, 59
Tridacna, 32-39
Environmental variation,
  classification of, 39-41,
  262
Esterase, 172-191, 264, 266,
  667-673
Euplotidae, 591
Evolution, 303, 317
  chromosomal, 487-516
  Darwinian, 3
  of competing species, 290-
  300
  of mating types, 595
  of Mytilus, 446-465, 469-
  484
  of Pleuronectidae, 87-89
Exposure, 127, 136

Faunal provinces,
  species diversity and, 365-
  383
Fish farming, 688, 693
Fitness, 7-9, 12-15, 137,
  278-280
Flounder, 71-89
Foraminifera, 369-383
Founder effect,
  in Thais, 166
  in Tigriopus, 62

Gammaridae, 529-540, 544
Gastropod, 157, 259, 305-317,
  354, 550
  fossils of, 308-317
Generation path, 259-260, 262
Gene flow, 259, 307, 313, 414-
  415
Gene frequency,
  standard deviation of, 177-
  179, 242-243

Gene regulation, 561-563
Genetic differentiation,
  dispersal ability and, 150,
  212, 356,357
  in Anguilla, 389-420
  in Botryllus, 201-207
  in Gadus, 699-725
  in Mytilus, 211-225, 229-
  251, 469-484
  in Thais, 157-171
  in Tisbe, 248
  in Zoarces, 176-191
Genetic distance, 427-441
  definition of, 83, 705
  evolutionary time and, 84,
  88
  in Gadus, 715, 721
  in Pleuronectidae, 85, 86,
  88-89
  in Sebastes, 110
  in Sebastolobus, 110
  standard error of, 83
Genetic drift, 3-4, 12, 62,
  165-167, 168, 176, 306
Genetic identity, 163, 427-
  441
  definition of, 83, 160, 705
  in Pleuronectidae, 86
  in Sebastes, 110
  in Sebastolobus, 110
  in Thais, 163-164
Genetic load, 57, 64
Genetic similarity, 427-441
  definition of, 428-429
  confidence limits of, 429
  in Alcyonidium, 439-440
  in Schizomavella, 438
  see also genetic
  identity
Genetic variation, see also
  polymorphism, hetero-
  zygosity
  deep sea species and, 96-
  116, 352, 354
  degree of exposure and,
  135, 136
  dispersal ability and, 150
  environmental heterogeneity
  and, 15, 25, 47, 54-67, 81-

SUBJECT INDEX

  82, 87, 95-96, 137, 153, 260, 262, 332, 340, 355-356, 414
  environmental stability and, 32-43
  enzyme function and, 74, 80-81, 86, 87, 336, 338
  enzyme structure and, 80-81, 87
  fecundity and, 267
  gene flow and, 306-307
  grain size and, 336, 350
  measures of, 30-32
  niche similarities and, 82, 87
  niche width and, 123, 135, 138, 152
  population size and, 86, 103, 353, 415
  quantitative, in fish, 679-695
  reproductive strategy and, 264-267
  species coexistence and, 123-138
  species diversity and, 335-339
  trophic resources and, 40-49, 84, 86, 137, 337, 352, 354, 418
Goldfish, 668
Grain, 47, 330, 332, 336, 350, 414-415

Haemoglobin, 4, 17, 172-191, 391, 700, 723
Hake, 704
Halibut, 112
Haplosporidian, 542
Heritability, 278, 680, 681, 695, 700
Heterozygosity, 30-31, 42, 44, 46, 48, 61, 84, 103, 333-334, 356, 408-409
  see also genetic variation, polymorphism
  allozyme heat resistance and, 150
Heterozygote,
  deficiency, 62, 135-136, 268-269, 406, 478, 479, 701, 708, 723
  selection against, 188
Histone IV, 4
Hitch-hiking, 177
Housefly, 670, 672
Hybrid breakdown,
  in Ophryotrocha, 579-581
Hybrid zones, 482
Hybridization,
  in Euplotes, 604-607
  in Jaera, 637-662.
  in Mytilus, 460, 479-484
  in Ophryotrocha, 575-582, 585
  in Pleuronectidae, 88-89
  in Tisbe, 617-634
Hydrostatic pressure,
  influence on proteins, 95, 110, 111-114, 116

Inbreeding depression, 57, 196
Introgression, 482-484
Isolating mechanisms, 124
  ethological, 581-585, 586, 613, 628, 638-662
  reproductive, 483, 617-634
Isopod, 257, 637

K-selection, 326-329, 339
Karyotype,
  in eel, 391
  in marine invertebrates, 487-516, 530
Killifish, 668
K_m,
  definition of, 112
Krill, 45-48, 327, 354

Lactate dehydrogenase, 88, 112-114, 116, 137, 266
Lagoons, 197, 201, 207, 585, 630
Larvacea,
  chromosomes in, 503-505
Larvae,
  dispersal of, 150, 212, 249-251, 257-269, 347

evolution and, 303-318
   species longevity, 303-317
Leslie matrix, 278-280
Leucine aminopeptidase, 211-225, 230-251
   assay for, 214-215
   function of, 213, 223-224
Leucyl alanine peptidase, 264
Limpet, 550
Linkage, 197, 200-201
Linkage disequilibrium, 190, 475

Macrouridae, 103, 114
Malate dehydrogenase, 27, 29, 197-207, 391
Mating type, 592-608, 612
Melanism, 262
Microsporidians, 533-540, 543
Migration, 166, 176
Minimum length spanning network, 166-167
Mixed-function oxidases, 669
Morphological differentiation, 157, 306, 317
Morphological variation, 157, 180, 264
Mouse, 669
Muricacea, 310
Mussel, 6, 18, 149-150, 151, 152, 211-225, 229-251, 266, 284, 447-466, 469-484
Mutation, 3, 16, 306
Myogen, 266
Mytilidae, 447

Nematode, 715
Niche,
   and coexisting species, 289
   overlap, 127-128, 130-133, 135
      measurement of, 127,
   similarities of, 82
   size, 328-331
      sexual dimorphism and, 293
   width, 123, 135
Oat, 351

Olive fruit fly, 670
Ophiuran, 46, 97
Ophiuroid, 353-354
Organophosphate pesticides, 667-673
Osmotic stress, 275-285
   see also salinity
Oysters, 268

Phoronid, 42-44
Phosphoglucomutase, 27-28, 142-153, 172-191
Phosphoglucose isomerase, 142-153, 264
Phylogeny, 82-84, 87-89, 604-608
Plaice, 71-89
Pleiotropism, 650
Pleuronectidae, 72
Polychaete, 347, 354, 358, 549
Polymorphism, 3-18, 32-49, 54, 196-207, 264, 621, 647, 650
   see also genetic variation, heterozygosity
Prosobranch, 291
Protogyny, 196
Pupfish, 433

r - selection, 326-329, 339, 351
Rabbit, 669
Rat, 670, 672
Rattail, 103

Sagebrush, 608
Salinity,
   habitat selection and, 291
   Mytilus and, 211-225, 229-251
   natural selection and, 275-285
   sex determination and, 543-564
   Tisbe and, 248
Salmonids, 680, 695, 700, 704
Sardine, 672
Science,

definition of, 23
Scientific hypotheses, 23-24
Sea urchin, 4
Selection, 3, 6-19, 306-307
  directional, 680-681
  disruptive, 483
  in Mytilus, 212-213, 222-225, 229, 250-251
  in Zoarces, 174, 188
Sex determination,
  in Drosophila, 558
  in Gammarus, 530-540, 543-544
  in Lymantria, 558
  in Ophryotrocha, 549-569
  in Orchestia, 540-544
Sex ratio,
  in Gammarus, 530-539
  in Ophryotrocha, 565, 576, 586
  in Orchestia, 540
  in Tisbe, 620, 623
Shark, 700
Sickle-cell anaemia, 262
Slipper shell, 249
Speciation, 313, 317, 483
  Euplotes and, 612, 613
  Flatfish and, 88
  Jaera and, 654
  Mytilus and, 458-466, 469-470, 479-484
  Ophryotrocha and, 563-568, 585-586
  Tisbe and, 631
Species diversity,
  community stability and, 330-332
  gradients in, 323-325, 352, 365-383
Spirorbidae, 258
Squid, 672
Super-species,
  Jaera and 637-662
  Tisbe and, 617-637
Talitridae, 529, 540-544
Tench, 668
Tetrazolium oxidase, 197-207
Thaididae, 310

Thaliacea,
  chromosomes in, 506-507
Transferrin, 392, 700-725
  typing of, 700-704
Trophic resource stability,
     see genetic variation
Trout, see salmonids
Tuna, 112

Veliger larvae, 305-317

Wahlund effect, 266, 406
White crappie, 672
White fish, 679
Whiting, 679
Winkles, 124-138, 142-153, 264-265
Wobble hypothesis, 4